Thermal Adaptation

Thermal Adaptation

A Theoretical and Empirical Synthesis

Michael J. Angilletta Jr.

OXFORD
UNIVERSITY PRESS

Great Clarendon Street, Oxford OX2 6DP

Oxford University Press is a department of the University of Oxford.
It furthers the University's objective of excellence in research, scholarship,
and education by publishing worldwide in

Oxford New York

Auckland Cape Town Dar es Salaam Hong Kong Karachi
Kuala Lumpur Madrid Melbourne Mexico City Nairobi
New Delhi Shanghai Taipei Toronto

With offices in

Argentina Austria Brazil Chile Czech Republic France Greece
Guatemala Hungary Italy Japan Poland Portugal Singapore
South Korea Switzerland Thailand Turkey Ukraine Vietnam

Published in the United States
by Oxford University Press Inc., New York

First Published 2009

British Library Cataloguing in Publication Data
Data available

Library of Congress Cataloging in Publication Data
Data available

Typeset by Newgen Imaging Systems (P) Ltd., Chennai, India
Printed in Great Britain
on acid-free paper by
CPI Antony Rowe, Chippenham, Wiltshire

ISBN 978–0–19–857087–5 978–0–19–857088–2

10 9 8 7 6 5 4 3 2 1

To Amy, Taylor, and Kyle,
for your exceeding tolerance of my wild ideas.

Preface

The first book that I read as a graduate student was *Foraging Theory* (Stephens, D. W., and J. R. Krebs. 1986, Foraging Theory. Princeton, Princeton University Press). At the time, I was planning to study the evolution of foraging behavior in snakes, and thus wanted to acquire a conceptual background for this work. As fate would have it, my subsequent research did not focus on foraging behavior (or snakes), but the book made a lasting impression on me anyway. The authors laid out a comprehensive and insightful theory, moving from simple models to complex ones, while drawing on relevant empirical data. The book culminated in a rigorous appraisal of the field's accomplishments, driven by the match between theoretical predictions and empirical observations. Had I pursued my interest in foraging biology, the book would surely have become my academic bible. Instead, I soon developed a greater interest in thermal biology, only to discover that no analogous resource existed for this discipline.

I don't want to give you the impression that little was known about thermal biology or that no one had attempted a conceptual synthesis. Neither impression would be accurate. In many respects, thermal biology has progressed tremendously, as one can conclude from reading an early synthesis of this topic (see Cossins, A. R., and K. B. Bowler. 1987, Temperature Biology of Animals. New York, Chapman and Hall). In other respects, though, thermal biology continues to lag behind other disciplines. Let me explain. When it comes to documenting patterns, few disciplines have made more progress than thermal biology. But if we judge progress by our understanding of the processes that generate these patterns, then we have much progress to make. While proximate explanations abound, ultimate explanations remain scarce. This problem stems from the paucity of quantitative models of thermal adaptation. Thermal biologists have described patterns and documented effects to the point that empirical knowledge has far outstripped theoretical understanding. Most papers in thermal biology focus on empirical descriptions or rely on verbal arguments about adaptation. In comparison, consider the discipline of foraging biology that I referred to above; herein lies a trove of quantitative models and experimental tests about evolutionary hypotheses. The basic ideas about the evolution of foraging behavior were formulated in the 1960s and have been gradually developed to the point that the evolutionary impacts of many factors—from constraints on digestion to risks of predation—have been explored in great detail. Few evolutionary processes in thermal biology have been modeled to the same degree, and no major synthesis of evolutionary thermal biology has emerged. My aim is not to cast blame (for I would certainly share in this blame). Rather, I want to see thermal biology keep pace with other biological disciplines. And to do so, we must develop a quantitative theory of thermal adaptation.

Upon closer inspection, a wealth of evolutionary models lie waiting for thermal biologists to adopt as their own. Beginning with Levins (Levins, R. 1968, Evolution in Changing Environments: Some Theoretical Explorations. Princeton, Princeton University Press), biologists have been modeling the evolution of quantitative traits in variable environments for more than four decades. With the growth of this theory, an increasing emphasis has been placed on genetic, developmental, and ecological issues. Many of the models can be applied to understand aspects of thermal adaptation. Although some of these models are mathematically intense, I have distilled from them the essence of what they tell us about adaptation to temperature. For thermal biologists, this book contains a synthesis of theory that should guide research in the coming decade. For evolutionary biologists, the book contains an evaluation of current models of evolution in variable environments.

Throughout the book, you may notice a taxonomic bias. This bias stems from two factors. First, intentional bias resulted from my conceptual focus. The book revolves around the concept of a thermal performance curve, which describes the relationship between body temperature and organismal performance. Most of the book draws on studies of ectotherms because we can measure their performance curves more readily than we can measure those of endotherms. Students of endotherms will find the most relevant theoretical and empirical content in Chapter 4, but other chapters contain relevant theory. Second, some bias inevitably results from the lack of suitable data for certain taxa. In each chapter, I focused on the data most relevant to the theory. Consequently, certain chapters are rich with examples involving animals, plants, or microbes, but no chapter contains an even coverage of all of these groups. I hope this taxonomic bias reflects my effort to evaluate the theory more than my ignorance of the literature (although the latter seems inevitable given the exponential growth of knowledge).

Above all, I tried to produce the kind of synthesis that I had longed for when I was a graduate student. If I succeeded in this effort, my success was due partly to the biologists who mentored me in my formative years: Justin Congdon, Art Dunham, Henry John-Alder, and Peter Petraitis. My views were also shaped by discussions or collaborations with George Bakken, Lauren Buckley, Ray Huey, Rob James, Joel Kingsolver, Bill Mitchell, Carlos Navas, Amanda Niehaus, Peter Niewiarowski, Warren Porter, Leslie Rissler, Mike Sears, Frank Seebacher, and Robbie Wilson. Many colleagues were kind enough to comment on sections of the book or provide information about their research: David Atkinson, George Bakken, Lauren Buckley, Brandon Cooper, Lisa Crozier, Marcin Czarnołeski, Wilfried Gabriel, George Gilchrist, Lumír Gvoždík, Gabor Herczeg, Luke Hoekstra, Ray Huey, Jose Martín, Bill Mitchell, Mike Sears, Ben Williams, and Robbie Wilson. Their suggestions and insights contributed greatly to the final product. Other colleagues provided original images or permission to adapt figures: Martin Ackermann, Steve Beaupre, Gabriel Blouin-Demers, Lauren Buckley, Andy Clarke, Lisa Crozier, Michael Dillon, Wilfried Gabriel, Ted Garland, George Gilchrist, Bruce Grant, Brian Helmuth, Ray Huey, Jacob Johansson, Joel Kingsolver, Mark Kirkpatrick, Michael Kopp, Sue Mitchell, Carlos Navas, Mark New, Camille Parmesan, Steve Piper, Mike Sears, Tim Shreeve, Gian-Reto Walther, and Kazunori Yamahira. Finally, the preparation of this book was greatly accelerated by the help of Angela Borchelt and Phaedra Seraphimidi, who obtained copyright permissions for the figures adapted from other sources.

An undertaking of this magnitude requires the support and encouragement of many people. I am greatly indebted to my colleagues at Indiana State University, who picked up the slack for me during my sabbatical. The chair of my department, Charles Amlaner, kept my teaching load in check during the remainder of my writing. The Lilly Foundation and the Office of the Provost at Indiana State University provided financial support. I owe an equal debt to the students who kept my lab productive despite my reallocation of energy from research to writing: Dee Asbury, Brandon Cooper, Joe Ehrenberger, Matt Schuler, Somayeh Semati, Melissa Storm, and Ben Williams. Most of all, I thank my family—Amy, Taylor, and Kyle. They kindly endured my crazy work schedule and my occasional stress responses. I would not have completed this book without their patience and understanding.

Michael J. Angilletta Jr.
May 24, 2008
Terre Haute, Indiana, USA

Contents

CHAPTER 1

Evolutionary Thermal Biology

1.1 The challenge of evolutionary thermal biology

Of the hundreds of variables that can shape the phenotype of an organism, temperature has undoubtedly captured more than its share of attention.[1] Biologists have linked temperature to everything from temporal patterns of growth, survival, and reproduction (Angilletta et al. 2002a; Huey and Stevenson 1979; Savage et al. 2004) to spatial patterns of body size, population density, and species diversity (Angilletta et al. 2004a; Brown et al. 2004; Wiens et al. 2006). Unlike many other variables that concern biologists, temperature is not just a property of life; it is a property of matter. Nothing escapes its control. Still, the same change in temperature will not affect all organisms equally. Moreover, temperature does not affect the same organism equally at all stages of the life cycle. These phenomena raise serious questions for biologists. To what extent does this phenotypic variation reflect adaptation to temperature? And what processes facilitate or constrain this thermal adaptation? To answer these questions, we must first identify the specific phenotypes that we seek to explain.

Strategies for coping with thermal heterogeneity define a continuum of two dimensions (Fig. 1.1). The first dimension describes the degree to which an organism's performance depends on its temperature (*thermosensitivity* or *thermal sensitivity*), ranging from organisms whose performance depends strongly on temperature (*thermal specialists*) to organisms that perform well over a broad range of temperatures (*thermal generalists*). The second dimension describes the degree to which an organism regulates its temperature (*thermoregulation*), ranging from organisms that maintain a nearly constant body temperature (*perfect thermoregulators*) to organisms that conform to their environmental temperature (*perfect thermoconformers*). All organisms fall somewhere within the continuum bounded by these axes. For example, most mammals and birds are thermal specialists that thermoregulate precisely (Cossins and Bowler 1987). Certain fish, such as those that live in the intertidal zone, are thermal generalists that can undergo dramatic fluctuations in temperature during a single day (Somero et al. 1996). The position of an organism along this continuum can change over time. In fact, an individual can be a specialist during specific seasons but a generalist over the course of a year. Thus, strategies for coping with thermal heterogeneity include irreversible and reversible forms of phenotypic plasticity. These strategies determine the life histories of organisms by limiting their acquisition and allocation of resources. The relationships among thermosensitivity, thermoregulation, and the life history govern the course of adaptation in variable environments.

[1] Thermal biologists often make sweeping statements of this nature, but what evidence do we have to back up this view? I became curious while working on this book, so I performed a simple analysis using the ISI Web of Science. Twelve other well-studied variables were chosen as key words, representing both abiotic and biotic factors: competitors (or competition), energy, food, light, nitrogen, oxygen, parasites (or parasitism), pH, predators (or predation), pressure, salinity, and water. To limit the number of extraneous references, I used the term "biology" in conjunction with the environmental variable (e.g., "temperature and biology"); I have no reason to think this approach would bias the outcome. Since 1945, more than 3568 papers have used the terms "temperature" and "biology." Temperature ranks third on the list, behind water (4369) and light (3723) but well ahead of food (3029), pressure (1920), parasites (1917), oxygen (1769), predators (1621), and other variables (<1500 each).

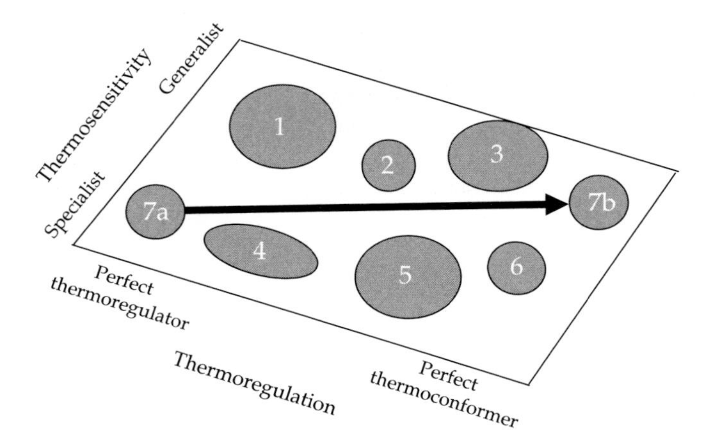

Figure 1.1 Strategies of coping with thermal heterogeneity include different combinations of thermosensitivity and thermoregulation. The shaded regions depict the strategic sets of hypothetical species; intraspecific variation determines the areas of these sets. A strategic set can change seasonally or ontogenetically, as exemplified by the arrow connecting two sets for the same species (7a and 7b).

Although a vast and venerable literature documents the scope of these strategies, we still do not understand *why* certain species exhibit certain phenotypes. Why do some organisms function over a wide range of temperature when other organisms do not? Why do some organisms adjust their thermosensitivity in response to extreme or fluctuating temperatures when other organisms do not? I think most biologists would agree that the answers to these questions involve adaptation. Still, we would be hard pressed to provide rigorous answers at this time. How can we know so much about the thermal biology of organisms, yet know so little about its evolution?

I believe progress in evolutionary thermal biology has lagged behind progress in other evolutionary disciplines because of the paucity and simplicity of our theoretical models.[2] In fact, few investigations have been directly motivated by quantitative models. And even fewer have been motivated by the desire to distinguish between two or more competing models. In the literature on thermal biology, empirical descriptions of patterns probably outnumber quantitative models of evolution by more than 100 to one. This situation partly resulted from historical factors. The study of thermal biology has its roots in ecological physiology (Bennett 1987a), which has always been a relatively empirical discipline.[3] Only recently have ecological physiologists seriously shifted their attention from questions concerning *what*, *when*, and *how* to ponder the question of *why*. This conceptual transition catalyzed the origin of evolutionary physiology, which has flourished in recent decades (Burggren and Bemis 1990; Feder et al. 2000; Garland and Carter 1994; Magnum and Hochachaka 1998). Despite this shift in focus, empirical observations have continued to accumulate far more rapidly than have theoretical ideas. Today, the limited number of quantitative models prevents effective application of the scientific method to evolutionary thermal biology (see Hilborn and Mangel 1997).

In this book, we shall draw on models from the general discipline of evolutionary biology to gain rigorous insights about thermal adaptation. First, we must construct a framework for understanding

[2] Although you may view this statement as a harsh criticism, you should consider the role of quantitative models in a comparable discipline to appreciate my point. For example, biologists who study how organisms adapt to variation in the quantity and quality of food have made tremendous progress over the past few decades, largely because of the development and application of theoretical models (reviewed by Kamil et al. 1987; Stephens et al. 2007; Stephens and Krebs 1986).

[3] As an aside, I believe the controversy surrounding the metabolic theory of ecology stems from its unique attempt to explain a wealth of physiological data with a simple mathematical model (see Brown et al. 2004). Despite the need for caution when invoking such unifying theories, opponents of the metabolic theory seem oblivious to the fact that no alternative model currently offers greater predictive power. Perhaps the strong reaction to this theory reflects the mindset of ecological physiologists as much as it reflects limitations of the models.

thermal adaptation that integrates current theoretical paradigms and modern empirical approaches. Then, we can use this framework to explore the structure of current theory and the level of empirical support; this exploration will begin with models that consider mechanisms operating at the organismal level and build up to models that account for the interactions among organisms. Along the way, we shall draw on biochemical and genetic evidence to validate key assumptions. By synthesizing models of behavioral, physiological, and life-historical evolution, we should advance toward a general theory of thermal adaptation. By the end of the book, we will have addressed many questions concerning thermal adaptation:

1. How do organisms experience thermal heterogeneity in their environment? Do different organisms perceive the same thermal environment differently? If so, how do we define the operative environment experienced by a particular organism? (Chapter 2)
2. Why do some organisms function over a wide range of temperatures, while others function only within a relatively narrow range? (Chapter 3)
3. Why do some organisms regulate their body temperature within narrow limits, while others conform to environmental temperatures? (Chapter 4)
4. Why do some organisms possess a capacity to reversibly or irreversibly alter their sensitivity to temperature, while other organisms do not? (Chapter 5)
5. Despite the diversity of behavioral and physiological responses to temperature, why do general relationships exist between environmental temperature and the life history? (Chapter 6)
6. Can we answer the preceding questions independently, or must we consider the evolutionary interactions among thermosensitivity, thermoregulation, and the life history? (Chapter 7)
7. Can we understand thermal adaptation without considering the interactions among individuals of one or more species? (Chapter 8)
8. Can a theory of thermal adaptation help us to predict the biological consequences of anthropogenic climate change? (Chapter 9)

Any effort to address this many questions in a single book can succeed only with some focus. This focus could take one of several forms—taxonomic, methodological, or conceptual. I have chosen a conceptual focus: the thermal reaction norm. All of the models presented in this book describe the evolution of some type of thermal reaction norm. Therefore, we must start by defining this fundamental concept and recognizing its strengths and limitations.

1.2 Thermal reaction norms

Thermal responses at all levels of biological organization are best described as phenotypic plasticity. The study of phenotypic plasticity has a long and complex history (Sarkar 2004; Schlichting and Pigliucci 1998), but most research in this area has been guided by the concept of a *reaction norm*. A reaction norm describes the relationship between a continuous environmental variable and a continuous phenotypic variable (see reviews by Gotthard and Nylin 1995; Schlichting and Pigliucci 1998; Stearns 1989; West-Eberhard 2003). A thermal reaction norm represents a special case in which the environmental variable is temperature. In practice, any aspect of behavior, physiology, morphology, or life history can be viewed in the context of a thermal reaction norm (DeWitt and Scheiner 2004; Stearns 1989).

Although biologists define plasticity in numerous ways (Debat and David 2001), I define plasticity as the derivative of a reaction norm (Fig. 1.2). For a linear reaction norm, plasticity simply equals the slope of the function. For a nonlinear reaction norm, the concept of plasticity becomes muddled as the slope of the function depends on the temperature of interest. A trait unaffected by temperature has no plasticity, but this trait still has a reaction norm; the slope of this reaction norm equals zero and the intercept equals the mean value of the trait. Given this definition of plasticity, we can compare plasticities among traits or taxa. Moreover, we can quantify the evolution of plasticity for specific traits.

The trait of interest affects the way we conceptualize and quantify a thermal reaction norm. Some traits, such as growth rate, change repeatedly in response to temperature (*labile traits*). Other traits, such as the number of body segments, become fixed during development (*fixed traits*). For a labile trait,

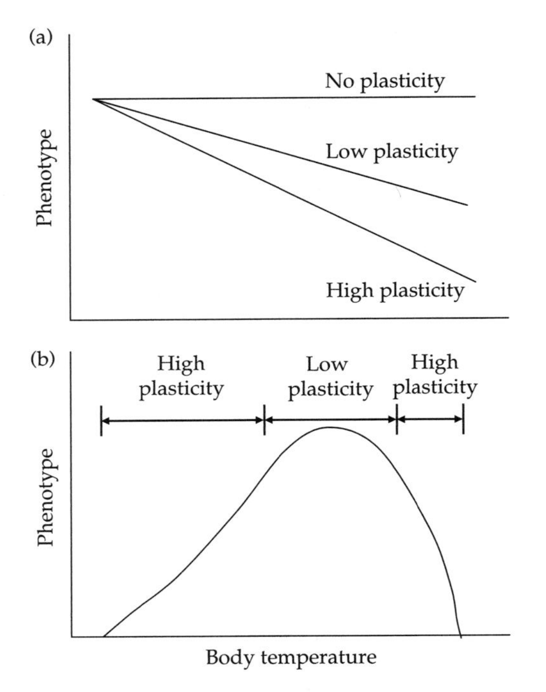

Figure 1.2 The derivative of a reaction norm quantifies the thermal plasticity. For linear reaction norms (a), plasticity remains constant over a broad range of temperatures, but can vary among genotypes. For nonlinear reaction norms (b), the plasticity expressed by a single genotype depends on the range of temperatures.

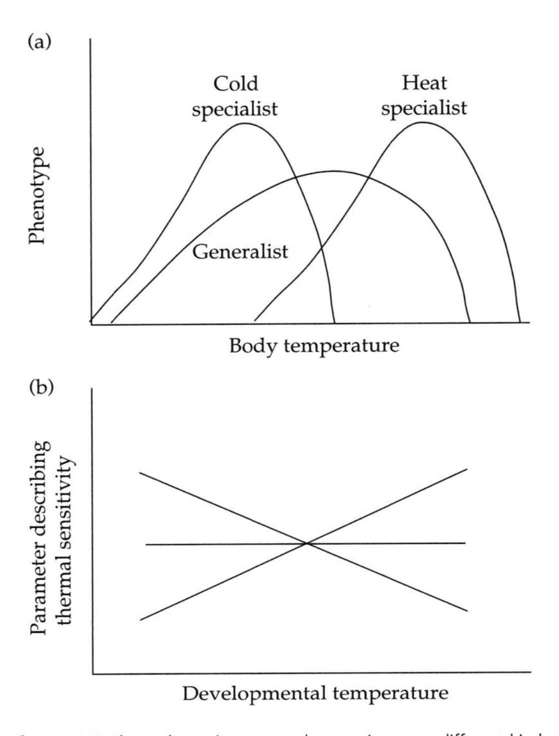

Figure 1.3 Thermal reaction norms characterize many different kinds of traits. (a) For labile traits, instantaneous (or short-term) fluctuations in body temperature affect the phenotype. (b) Thermal reaction norms can change via acclimation; such changes are described as plasticity in the parameters that define the reaction norm shown in (a).

we can measure the change in the trait of an individual exposed to a range of temperatures (Fig. 1.3a). Such studies are usually designed to avoid confounding the temperature and time of measurement (e.g., randomizing the order of measurements solves this problem). Reaction norms for labile traits often depend on the developmental environment of the organism, such as its developmental temperature. In other words, thermal plasticity itself can undergo plasticity! To quantify this "plasticity of plasticity," we can examine the relationship between the developmental temperature and the thermal sensitivity after development (Fig. 1.3b). For a fixed trait, a reaction norm must be estimated from measures of relatives raised at different temperatures. Using a group of distantly related individuals would yield a generalized reaction norm (Sarkar and Fuller 2003), which includes phenotypic variation among genotypes. Using clones or siblings would yield a more accurate reaction norm. Reaction norms for fixed traits usually describe the influence of the mean developmental temperature. If thermal fluctuations are important, we can construct a multivariate reaction norm that describes the phenotype as a function of the mean and the variance of temperature (Fig. 1.4). This diversity of applications underscores the conceptual value of the reaction norm.

How can we know whether thermal reaction norms were produced by adaptation? To acquire such knowledge, we must construct predictive models of evolution and apply these models to the study of thermal plasticity. In other words, we must develop a theory of thermal adaptation. With this goal in mind, the remainder of this chapter focuses on the role of theory in evolutionary thermal biology.

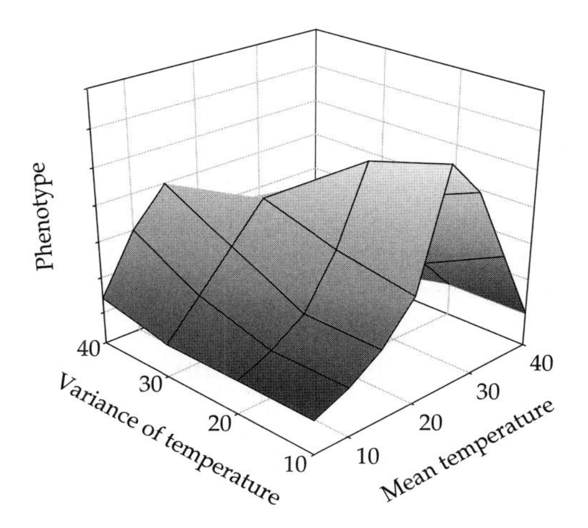

Figure 1.4 A multivariate reaction norm can describe the influence of the mean and the variance of developmental temperature on a fixed trait. In this hypothetical example, a univariate reaction norm would give an incomplete view of thermal plasticity because of the interaction between the mean and the variance of temperature.

1.3 The role of theory

In his book *Evolution in Changing Environments*, Richard Levins (1968) opened with a visionary discussion of the role of theory in biology. He predicted that theoretical biology, an emerging discipline at the time, would soon disappear. I interpret his argument as follows. Science has always been a process of building and evaluating models of how nature works. But some biologists had lost sight of this reality, which eventually fueled a movement to bring theoretical direction to empirical research. Levins reasoned that, as more biologists accepted the need for theory, the distinction between theoretical biology and empirical biology would dissolve. In Levins' own words, "once it [theoretical biology] has made its points part of the thinking of biology in general, it will loose [sic] its identity as a distinct field" (p. 4). Sadly, Levins' admirable vision did not come to pass. Even today, a fairly wide rift separates biologists who conduct theoretical research from those who conduct empirical research. I shall voice an extreme view (hopefully, one that Levins would espouse): scientific understanding advances most rapidly when researchers combine theory and empiricism. At the very least, we need a reciprocal exchange of ideas between theorists and empiricists; theorists should model processes in light of previous observations and empiricists should collect data that address current models. My motivation for writing this book was to bridge the gap between theorists and empiricists who study thermal adaptation. To achieve this goal, I must convince you that progress in evolutionary thermal biology hinges on this connection. As a first step, I should elaborate my view on the interplay between models and data.

A theory consists of a group of models describing related phenomena (e.g., foraging theory). Each model describes a set of mechanisms proposed to explain the phenomena. These models inspire the hypotheses and experiments that capture the attention, time, and energy of biologists. Yet, different forms of modeling exist and not all of these forms can offer the same degree of insight. Indeed, models differ greatly in the clarity of their assumptions and the precision of their predictions.

Most models can be characterized as verbal, graphical, or mathematical. A verbal model is a textual description of the modeler's ideas. Generally, the assumptions of a verbal model are implicit. Furthermore, its predictions would be qualitative and could be inaccurate if the modeler has not processed his assumptions carefully. Because of this imprecision, we cannot easily evaluate and refine verbal models. A graphical model depicts a set of relationships among variables plotted in Euclidean space. Although the assumptions of a graphical model are also implicit, one can infer them by analyzing the graph. The predictions of the model can be qualitative or quantitative depending on whether the modeler has specified numerical values of the variables. Finally, a mathematical model embodies the assumptions of the modeler in an algebraic form. The relationships among variables and the units of measurement are defined explicitly. As such, mathematical models embody clear assumptions and deliver precise predictions.

In this book, I focus primarily on mathematical models rather than verbal or graphical models. This focus does not imply that mathematical models provide the best representations of truth; in fact, they do not (see Lakoff and Núñez 2000). All models, including mathematical ones, represent caricatures of nature more than illustrations of truth (Hilborn and Mangel 1997). But the explicit assumptions of mathematical models enable other biologists to (1) verify the validity of those assumptions and (2) generate competing models by altering the assumptions. Furthermore, mathematical models generate more comprehensive and more precise predictions, which facilitate distinctions among competing models that make qualitatively identical predictions under certain conditions.

We can evaluate a model in one of two ways. First, we can validate the critical assumptions of the model. A critical assumption determines the model's qualitative behavior. Relaxing some assumptions will cause only quantitative changes in behavior, but a critical assumption underlies some unique prediction. If this assumption proves false, the prediction would also be false. Second, we can compare independent observations with the predictions of the model. By independent observations, I mean data that were not used to formulate or parameterize the model. Throughout the book, I adopt both approaches to evaluate models.

Rather than evaluate one model at a time, we gain deeper insights by directly comparing the performances of two or more models (Chamberlin 1890). In the traditional statistical paradigm, the prediction of a single model is pitted against a null hypothesis, which generally states that no pattern or effect should be observed. Presumably, rejection of the null hypothesis implies support for the model (also known as affirming the consequent). Unfortunately, many other models could have made a prediction similar to that of the model in question. Thus, competition between an evolutionary model and a null model provides only weak support for a theoretical idea (Anderson et al. 2000; Oreskes et al. 1994). Competition between two or more evolutionary models always yields a more fruitful outcome. For example, competition between mutually exclusive models eliminates some ideas from the set of possibilities. In this way, each experiment brings us closer to identifying the best available model (Platt 1964). When a set of models describes mutually inclusive mechanisms, we can still infer the relative importance of each mechanism (Quinn and Dunham 1983). Today, we can choose from several statistical paradigms, such as information theory (Burnham and Anderson 2002) or Bayesian theory (Ellison 2004), to directly compare the performance of two or more models (reviewed by Johnson and Omland 2004). These paradigms enable us to formally rank a set of models by their ability to describe the data. But even an informal comparison of models is better than no comparison at all. Therefore, let us examine the kinds of theoretical models available for the study of thermal adaptation.

1.4 Theoretical approaches to evolutionary thermal biology

Three distinct modeling approaches will appear throughout this book: optimality models, quantitative genetic models, and allelic models (Berrigan and Scheiner 2004; Scheiner 1993). Each of these approaches possesses unique strengths, yet each also imposes some limitations. In this section, I outline the basic features of these three approaches. While doing so, I highlight the way that each approach informs us about thermal adaptation.

1.4.1 Optimality models

An optimality model describes the phenotype that maximizes fitness in a given environment (Box 1.1). This goal is accomplished by combining two mathematical functions: (1) a *phenotypic function* that defines the set of possible phenotypes and (2) a *fitness function* that relates these phenotypes to fitness in given environments (Maynard Smith 1978; Mitchell and Valone 1990). Combining the phenotypic and fitness functions enables one to map the fitness landscape—a graphical depiction of the relationship between potential phenotypes and their fitness. A fitness landscape tells us the direction and strength of natural selection given the current phenotype. Under certain conditions (described below), natural selection would ultimately yield the phenotype that maximizes fitness (i.e., the optimal phenotype).

The heart of an optimality model is a set of assumptions about the tradeoffs that constrain the phenotype (Mitchell and Valone 1990), sometimes referred to as *functional constraints* (Arnold 1992). Biologists often consider one of two classes of tradeoffs: *allocation tradeoffs* or *acquisition tradeoffs*. An allocation tradeoff occurs when an increment in resources allocated to one function results in a decrement in resources allocated to other functions (Heino

Box 1.1 Types of optimality models

Optimality models are divided into three classes: static, dynamic, and frequency dependent (Maynard Smith 1978). Below, these classes are described in order of increasing complexity.

Static optimization

Static models optimize one or more variables in an environment that does not change over time. Do not equate static with deterministic; the optimal strategy could depend on probabilities (e.g., the daily probability of surviving), but these probabilities would remain constant. Static optimization requires the use of differential calculus to determine the phenotypic value that maximizes fitness (or some proxy for fitness).

Dynamic optimization

Dynamic models optimize one or more variables in an environment (internal or external) that changes over time. In a dynamic model, the organism makes sequential decisions. These decisions cause changes in variables that describe the state of the environment or the organism. For example, a dynamic optimization of reproduction involves rates of energy intake and expenditure that increase with body size (a phenotypic state variable). Dynamic optimization usually requires the application of optimal control theory, such as dynamic programming.

Frequency-dependent optimization

When the fitness of a strategy depends on the strategies adopted by other individuals (i.e., frequency-dependent selection), one must take a game theoretic approach to optimization (Maynard Smith 1982; Nowak and Sigmund 2004). In game theory, the optimal strategy cannot be invaded by any alternative strategy—hence it is called an *evolutionarily stable strategy*. As we shall see in Chapter 8, the evolutionarily stable strategy often deviates from the optimal strategy under frequency-independent selection.

and Kaitala 1999; Kozłowski 1992). An acquisition tradeoff occurs when an increase in the duration or intensity of foraging results in an increase in the probability of being injured or killed (Lima and Dill 1990). Support for a model indirectly supports the importance of the specified tradeoffs.

When modeling thermal reaction norms, we must consider additional tradeoffs. Models of optimal reaction norms define the degree of plasticity that maximizes fitness in a variable environment. Because all optimality models link environments, phenotypes, and fitness, any model can predict the optimal reaction norm when two conditions hold: (1) the phenotype expressed in one environment does not constrain the phenotypes expressed in other environments and (2) plasticity imposes no decrease in survivorship or loss of energy. We would merely predict the optimal phenotype in each of several environments and connect the points to define a reaction norm. But this approach ignores potentially important constraints on the evolution of thermal reaction norms. For example, a *specialist–generalist tradeoff* occurs when a phenotype that improves fitness at one temperature also reduces fitness at another temperature (Levins 1968; Palaima 2007). Additionally, various costs of plasticity can affect the optimal reaction norm (reviewed by DeWitt et al. 1998; Schlichting and Pigliucci 1998). Hence, the optimal reaction norm in a variable environment often deviates from the optimal phenotype in a constant environment.

Optimality models comprise two kinds of assumptions: hard-core and strategic assumptions (Mitchell and Valone 1990). The hard-core assumptions are the foundation of the optimization research program and thus are the same for all models:

1. Mutation generates a small number of new phenotypes each generation.
2. Different phenotypes yield different fitnesses.
3. Populations of phenotypes typically reside near a stable equilibrium.
4. The relative fitnesses of phenotypes near the stable equilibrium are not influenced by the way these phenotypes are transmitted genetically.

The strategic assumptions depend on the specific problem under investigation and thus differ among models. Strategic assumptions are related to either

(1) the constraints on the phenotype or (2) the way the phenotype determines fitness. As an example of the former, many optimality models of thermal adaptation assume that organisms cannot enhance function at high temperatures without sacrificing function at low temperatures (see Chapter 3). Other optimality models assume that organisms possess a fixed quantity of energy at any time (see Chapter 6). Either of these assumptions would constrain the set of possible phenotypes. As an example of the latter, an optimality model could describe fitness as the product of survivorship and fecundity. Alternatively, one might assume survivorship remains constant and fecundity increases with body size; hence, size at maturity would be considered an adequate proxy for fitness. Strategic assumptions usually determine the qualitative behavior of the model.

The failure of an optimality model to predict the observed phenotypes indicates a violation of at least one strategic or hard-core assumption. First, the hypothesized tradeoffs might be incorrect or incomplete. Second, the mapping between the phenotype and fitness might be incorrect, or fitness might be defined incorrectly. Finally, we might question the hard-core assumptions, which are necessary for selection to maintain the optimal phenotype. Genetic drift impedes adaptation in small populations (Holt 1990), and mutation drives phenotypes away from the optimum when natural selection is weak (Gilchrist and Kingsolver 2001). Persistent gene flow along an environmental cline swamps a population with deleterious genes, despite strong selection in local environments (Lenormand 2002). Moreover, natural selection produces few evolutionary changes when phenotypic variation lacks a genetic basis (Arnold 1992). In the end, we might have to model these additional factors that constrain thermal adaptation. Let us now consider some approaches that emphasize the role of genetic constraints during adaptation.

1.4.2 Quantitative genetic models

Quantitative genetic models enable us to relax some of the hard-core assumptions of optimality models and examine the evolutionary dynamics of traits. In quantitative genetic models, traits are determined by many genes, each of which has a small quantitative effect on the phenotype (Gillespie 2004; Rice 2004). This theory partitions phenotypic variance into genetic and environmental components. In his seminal book, Ronald Fisher (1930) showed that the mean fitness of a population increases in proportion to the genetic component of phenotypic variation. In particular, he distinguished between the additive and nonadditive effects of alleles. Only the additive portion of genetic variation contributes to the evolution of fitness; the nonadditive portion stems from interactions among genes (dominance and epistasis), which are altered by recombination. Thus, the phenotypic variance (V_P) must be partitioned as

$$V_P = V_G + V_E, \tag{1.1}$$

where V_G equals the variance attributable to genetic effects and V_E equals the variance attributable to environmental effects. If we assume only additive genetic effects exist, the heritability of the phenotype (h^2) equals the proportion of variance caused by genetic effects:

$$h^2 = \frac{V_G}{V_P} = \frac{V_G}{V_G + V_E}. \tag{1.2}$$

Thus, the value of a phenotype depends not only on genes but also on environmental factors, such as temperature. When the effect of the environment on the phenotype depends on the genotype, a genotype-by-environment interaction exists. In such cases, a better model for partitioning the phenotypic variance would be

$$V_P = V_G + V_E + (V_{G\times E}) + V_{Error}, \tag{1.3}$$

where $V_{G\times E}$ equals the variance caused by the interaction between genes and the environment—i.e., the genetic variance in phenotypic plasticity. Statistical methods can partition the components of variance and estimate the magnitude of the genotype-by-environment interaction (Lynch and Walsh 1998). With this framework alone, researchers have quantified genetic variation in thermal reaction norms (e.g., Scheiner and Lyman 1991), but predicting their evolution requires (1) a more precise means of estimating genetic variances and (2) a significant advance in modeling selection. Both advances were made during the development of multivariate selection theory (Via and Lande 1985).

Multivariate selection theory enables us to predict the evolution of thermal reaction norms under genetic constraints (reviewed by Arnold 1994). At the root of this theory lies a model of the simultaneous evolution of two or more traits (Lande 1979; Lande and Arnold 1983). This model describes change in the phenotype as a function of selective pressure (estimated by selection gradients) and genetic variation (estimated by genetic variances and covariances):

$$\Delta\overline{\boldsymbol{Z}} = \boldsymbol{\beta}\boldsymbol{G}, \quad (1.4)$$

where $\Delta\overline{\boldsymbol{Z}}$ equals a vector of changes in phenotypic values, $\boldsymbol{\beta}$ equals a vector of selection gradients, and $\boldsymbol{G}$ equals a matrix of genetic variances and covariances. For example, the evolution of three traits would be represented by

$$\begin{bmatrix} \Delta\overline{Z}_1 \\ \Delta\overline{Z}_2 \\ \Delta\overline{Z}_3 \end{bmatrix} = \begin{bmatrix} \beta_1 \\ \beta_2 \\ \beta_3 \end{bmatrix} \begin{bmatrix} G_{11} & G_{12} & G_{13} \\ G_{21} & G_{22} & G_{23} \\ G_{31} & G_{32} & G_{33} \end{bmatrix}, \quad (1.5)$$

where $\Delta\overline{Z}_i$ equals the change in trait i, β_i equals the selection gradient for trait i, G_{ii} equals the genetic variance of trait i, and G_{ij} equals the genetic covariance between traits i and j. A selection gradient equals the partial derivative of relative fitness with respect to the phenotype; in other words, a selection gradient measures the change in fitness that results from a very small change in the phenotype. In the special case of directional selection, the selection gradient equals a constant. In the case of stabilizing or disruptive selection, however, the selection gradient varies with the phenotypic value. The genetic variance in eqn 1.5 equals the additive genetic variance described by Fisher (V_G of eqn 1.2); thus, the rate of adaptation will be proportional to the genetic variance. Many factors can limit the genetic variance of a trait (Table 1.1). The genetic covariance between two traits determines how the selection of one trait will influence the evolution of another trait. For example, when the genetic covariance is negative ($G_{ij} < 0$), selecting a larger phenotypic value of trait j will decrease the phenotypic value of trait i. Genetic covariances result from several mechanisms, including linkage disequilibrium, pleiotropy, and developmental modularity (Lynch and Walsh 1998; West-Eberhard 2003).

Table 1.1 Factors that reduce the genetic variance of a specific trait

1. Persistently small population leads to loss of variation by genetic drift
2. Mutational effects are skewed by mechanistic/physiological constraints
3. Small target of mutation lowers the rate of mutation
4. Mutation rates are unusually low
5. New allelic associations are not frequently formed by recombination
6. The phenotype is canalized during development despite mutations
7. The trait is genetically correlated with another trait under selection
8. The same genotype has the highest fitness in all environments
9. Continuous directional selection leads to fixation of favored alleles

Modified with permission from Blows and Hoffman (2005).

To use multivariate selection theory, we must represent a thermal reaction norm as a set of traits with genetic variances and covariances. Several approaches exist for translating a reaction norm into a set of traits: (1) the character-based approach, (2) the parameter-based approach, (3) the characteristic value-based approach, and (4) the function-based approach. Although these approaches should produce similar results (de Jong 1995), they involve different mechanics.

In the character-based approach, one treats the same trait measured at multiple temperatures as equivalent to multiple traits measured at one temperature (Falconer 1952; Via 1987; Via and Lande 1985). Each element of the G matrix becomes the additive genetic variance of the phenotype expressed at a temperature (G_{ii}) or the additive genetic covariance between the phenotypes expressed at different temperatures (G_{ij}). The vector of selection gradients determines the strengths of selection acting on a single trait at different temperatures.

In the parameter-based approach, one describes the reaction norm with a mathematical function, and treats parameters of the function as equivalent to traits (Scheiner and Lyman 1991). Although any function can be used, mainly linear and polynomial functions have been explored (David et al. 1997; Gibert et al. 1998). For a linear function, the slope represents the plasticity of the phenotype and the intercept represents the mean of the phenotype. As in the trait-based approach, a genetic variance–covariance matrix is used in conjunction with selection gradients to model the evolution of the reaction norm. Selection gradients for the parameter-based approach apply to parameters of the function.

In the characteristic value-based approach, one describes the reaction norm with a mathematical function and then derives characteristic values from this function to serve as traits (Gibert et al. 1998). A subtle but important distinction exists between a parameter and a characteristic value. Parameters, such as a slope and intercept, are estimated by fitting a model to data. Although some parameters have obvious biological meanings, other parameters do not. We can replace biologically meaningless parameters with characteristic values of the function, such as the phenotype at a particular temperature or the mean phenotype over a range of temperatures. The evolution of these characteristic values could be predicted in the same way that one predicts the evolution of the parameters.

In the function-based approach, one treats the reaction norm as a trait with an infinite number of dimensions (Gomulkiewicz and Kirkpatrick 1992). In other words, the thermal reaction norm describes the phenotype at an infinite number of temperatures. The infinite-dimensional approach relies on nonparametric statistics, which have the advantage of relaxing some restrictive assumptions of the other approaches. An infinite-dimensional analysis yields principal components that describe the genetic variance and covariance of the reaction norm. These estimates of genetic variance and covariance can be used to model the response to selection (Kingsolver et al. 2001a).

Regardless of which approach we choose, applying a quantitative genetic model poses unique challenges and requires new assumptions. First, we must know the selection gradients corresponding to the traits that represent the reaction norm. We can obtain these gradients from the fitness landscape of an optimality model (see Section 1.4.1) or estimate them empirically (see Section 1.5.1). Second, we must know the genetic variances and covariances associated with these traits. We can estimate these parameters through breeding experiments or artificial selection (Box 1.2). Regardless of how we obtain genetic variances and covariances, most quantitative genetic models assume these parameters remain constant during evolution. In reality epistasis, selection, recombination, and migration change the genetic variances and covariances of real traits (Archer et al. 2003; Hansen 2006; Pfrender and Lynch 2000; Roff and Mousseau 1999; reviewed by Steppan et al. 2002). Other assumptions arise from the statistical methods used to estimate genetic variances and covariances (Table 1.3). Despite these assumptions, quantitative genetic models provide a valuable complement to optimality models (Roff 1994), as we shall see in later chapters.

1.4.3 Allelic models

As the name implies, an allelic model describes the genes that shape the phenotype. Thus, allelic models pry open the black box of genes created by quantitative genetic models. The classic analytical models of one or two loci (reviewed by Gillespie 2004) have given way to more complex models explored through computer simulations of genetic algorithms (reviewed by Mitchell and Taylor 1999; Schoenauer and Michalewicz 1997).

The details vary greatly from model to model, but the allelic models described in this book all share some general properties. These models focus on a population of diploid organisms, with multiple alleles per locus. Each allele has an additive effect on the genetic value of a trait. For our purposes, the trait will represent some parameter that describes a thermal reaction norm, as in the parameter-based approach of quantitative genetics (see Section 1.4.2). The phenotypic value of this parameter results from the sum of allelic effects and an environmental effect. Because the mean environmental effect equals zero, the mean phenotypic value of the trait equals the mean genetic value. The phenotypic variance equals the sum of the additive genetic variance and the environmental variance (see eqn 1.1).

Initially, a population of genotypes is formed by randomly assigning a set of alleles to each individual. Then, within each generation, individuals survive and reproduce according to their phenotype. Upon reproduction, some probability exists that the genetic value of an allele will change by mutation; most mutations cause very small changes

Box 1.2 Estimating parameters of quantitative genetic models

Quantitative genetic models of thermal adaptation require estimates of genetic variances and covariances. Estimating these parameters for a reaction norm requires a fairly complex experimental design (Scheiner 2002). For labile traits, one can quantify a thermal reaction norm through repeated measures of individuals. For fixed traits, however, one must estimate the reaction norm using clones or siblings, because the phenotype of an individual cannot be measured at multiple temperatures. Quantitative genetic parameters estimated from siblings should be viewed with caution because microenvironmental variation can inflate estimates of genetic variance (Windig et al. 2004). Genetic variances and covariances can be quantified through breeding experiments or artificial selection (Falconer 1989; Lynch and Walsh 1998).

Breeding experiments

Relationships among relatives can be used to estimate genetic variances and covariances. Three kinds of relatives are commonly used: (1) parents and offspring, (2) full siblings, and (3) half-siblings. In each case, males (sires) and females (dams) are randomly mated to produce a set of offspring. These offspring are reared in controlled environments to reduce environmental effects on the phenotype. In the context of thermal adaptation, the trait could be a phenotype measured at a single temperature or a parameter that describes a thermal reaction norm.

The additive genetic variance of each trait can be determined by linear regression or analysis of variance (ANOVA). When a trait is measured in both parents and their offspring, a linear regression of the offspring phenotype on the mean parental phenotype reveals the additive genetic variance. Specifically, the proportion of variance in the offspring phenotype explained by the parental phenotype (r^2) approximates the heritability of the trait (V_G/V_P; see eqn 1.2). Alternatively, the phenotypic variance among full or half-sibling families can be used to estimate the additive genetic variance. For a full sibling design, each male is mated randomly to a single female. A nested ANOVA estimates the proportion of phenotypic variance explained by familial identity (sire or dam). Assuming the maternal environment and genetic dominance do not influence the phenotype, the additive genetic variance equals twice the variance observed among families. When maternal and dominance effects are suspected, a half-sibling design provides a more accurate estimate of additive genetic variance. Each male is mated randomly to several females. A more accurate estimate of the additive genetic variance results because each mother experiences a unique environmental condition and provides a unique genetic complement. A nested ANOVA estimates the proportion of phenotypic variance explained by paternal identity; the additive genetic variance equals four times the variance observed among half-sibling families. By raising individuals from the same family at different temperatures, one can use an ANOVA to estimate the genetic variance of plasticity (Table 1.2).

Artificial selection

The response to artificial selection can also be used to estimate genetic variances and covariances. One starts with a large population of organisms that expresses phenotypic variation in the trait of interest. This population is randomly subdivided to create replicate lines; some lines will experience selection and other lines serve as controls. For each selection line, the researcher selects individuals that have either the highest or the lowest values of the trait.

For example, imagine we wished to know the genetic variance of heat tolerance. In some lines, we would select individuals with a heat tolerance greater than the median (i.e., the top 50%). For other lines, we would select

Table 1.2 Interpreting the results of an analysis of variance applied to a breeding experiment conducted at multiple temperatures

Factor	Description
Genotype or familial identity (V_G)	Genetic variation in the mean phenotype expressed across temperatures
Temperature (V_E)	Mean thermal plasticity of the phenotype
Genotype × temperature ($V_{G\times E}$)	Genetic variation in the thermal plasticity of the phenotype
Error (V_e)	Variation among individuals within genotype at a temperature
Total (V_P)	Phenotypic variance

Modified with permission from Via (1994).

continues

Box 1.2 Continued

individuals with a heat tolerance lower than the median (i.e., the bottom 50%). For control lines, we would select individuals randomly with respect to heat tolerance. In all lines, we would randomly mate selected individuals to produce the next generation. The remaining individuals would be eliminated from each line. With each successive generation, the mean heat tolerance of the selected lines should diverge from the mean heat tolerance of the control lines.

Using eqn 1.4, we could estimate the genetic variance of heat tolerance (G_{ij} or V_G) from the response to artificial selection. This equation tells us the response to selection depends on the selection gradients and the genetic variances and covariances of traits. If we assume no trait other than heat tolerance-experienced selection, the response to selection becomes simple:

$$\Delta\bar{z}_1 = \beta_1 \cdot G_{11}. \tag{1.6}$$

The selection gradient (β_1) would depend on the intensity of selection that we imposed (s) and the phenotypic variance (V_P):

$$\beta_1 = \frac{s}{V_P}, \tag{1.7}$$

where s equals the difference between the mean phenotype of the selected population and that of the original population. Substituting eqn 1.7 into eqn 1.6 and solving for the genetic variance yields:

$$G_{11} = \frac{\Delta\bar{z}_1 V_P}{s}. \tag{1.8}$$

Given eqn 1.8, we can easily estimate the genetic variance from the response to our selective regime and the phenotypic variance. Figure 1.5 illustrates the estimation of $\Delta\bar{z}_1/s$ for fruit flies that were artificially selected for heat tolerance (Gilchrist and Huey 1999).

Furthermore, artificial selection tells us about the genetic covariance between two traits. For example, imagine we wanted to know whether heat tolerance and cold tolerance covaried genetically. After selecting for greater or lesser heat tolerance, we could use the change in cold tolerance to estimate the genetic covariances (G_{ij} and G_{ji}):

$$G_{21} = G_{12} = \frac{\Delta\bar{z}_2 V_P}{s}, \tag{1.9}$$

where V_P and s apply to the selected trait (heat tolerance) and $\Delta\bar{z}_2$ applies to the correlated trait (cold tolerance).

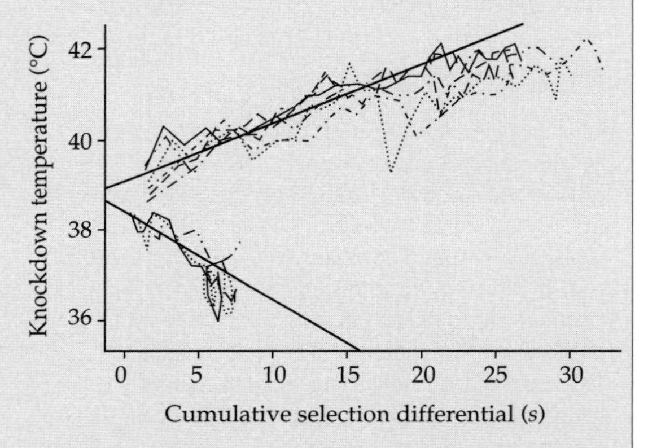

Figure 1.5 Response to selection for heat tolerance (knockdown temperature) in *Drosophila melanogaster.* Replicate lines were selected for greater or lesser heat tolerance. The slope of each relationship equals $\Delta\bar{z}_i/s$, which is proportional to the genetic variance. Adapted with permission from Macmillan Publishers Ltd: *Heredity* (Gilchrist and Huey 1999), © 1999.

Table 1.3 Assumptions of quantitative genetic theory

Assumption	Description
Linearity	Genetic and environmental effects can be described by a linear statistical model
Additivity	No interactions between genes (epistasis) or between genes and environments (environmental effects on the reaction norm)
Normality	Genetic and environmental effects are normally distributed
Constant G matrix	Genetic variances and covariances do not change

Modified with permission from Pigliucci and Schlichting (1997).

in the genetic value but some can have large positive or negative effects (mutational effects are usually drawn from a Gaussian distribution with a mean of zero). A finite population and independent assortment can lead to genetic drift within these simulations. Over time, the mean phenotype of the population evolves as a consequence of this selection, mutation, and drift.

Allelic models confer a major advantage over quantitative genetic models: the genetic variances and covariances can change. Because genes are modeled explicitly, mutation, recombination, drift, and selection change the frequencies of alleles over generations. Ultimately, the trait can evolve to an equilibrial value, such as the evolutionary optimum. Thus, an allelic model provides information about the dynamics and outcome of adaptation. On the downside, allelic models require one to specify not only the fitness landscape but also genomic structure, mutation rate, population size, and mating behavior (Hertz and Kobler 2000). As we shall see, the dynamics of thermal adaptation depend on these genetic and demographic parameters.

1.4.4 The complementarity of theory

All three of the modeling approaches discussed here possess strengths and weaknesses. Optimality models describe fitness landscapes given functional constraints on the phenotype (the phenotypic function) and a hypothetical relationship between the phenotype and fitness (the fitness function). Yet, optimality models assume genetic constraints do not prevent the optimal phenotype from evolving. Furthermore, optimization does not describe the evolutionary dynamics that precede an equilibrial phenotype. In contrast, quantitative genetic and allelic models explicitly account for genetic constraints on phenotypes and describe the dynamics that lead to an evolutionary equilibrium. In these models, however, a fitness landscape must either be assumed or be quantified empirically (Via et al. 1995). Furthermore, evolutionary dynamics depend largely on the assumptions about genetics. In quantitative genetic models, genetic variances and covariances are assumed to be constant even though they are known to vary (Steppan et al. 2002). In allelic models, one must specify the number of loci and alleles along with rates of mutation, recombination, and migration, even though these parameters are usually unknown. Thus, no single approach lacks drawbacks.

In this book, we shall use all three theoretical approaches to study thermal adaptation. The simplest way to draw on their collective strength is to apply these approaches sequentially. Generally, we shall start by applying an optimality approach to a particular phenomenon. When one optimality model fails, another optimality model may succeed because of its more realistic assumptions. When an optimality approach fails altogether, we shall explore the potential explanations offered by quantitative genetic and allelic models. For example, a lack of additive genetic variance might explain why the optimal phenotype has not evolved. Alternatively, migration between populations in different thermal environments might prevent populations from attaining local optima. This progression from simple to complex approaches will provide greater insights than would a focus on any single approach.

1.5 Empirical tools of the evolutionary thermal biologist

Models help us to understand the diversity of phenotypes observed in nature, but simply formulating a model that describes a previous observation will not advance knowledge. The model must predict unobserved phenomena and new observations must be made to confirm this ability. Testing the kinds of models outlined in Section 1.4 requires us to determine the match between predictions and observations. Throughout this book, we shall rely on three distinct approaches to evaluate models: (1) quantifying selection in natural environments, (2) quantifying evolution during selection experiments, and (3) comparing organisms from different environments. These approaches focus on different temporal scales of evolution, ranging from one generation to hundreds of generations. They also provide different strengths of evidence for adaptation and different levels of generality. Let us briefly review each approach to better understand its strengths and limitations.

1.5.1 Quantifying selection

Measures of selection in natural environments can be used to evaluate optimality models and parameterize genetic models. Recall that an optimality model defines the fitness landscape—the relationship between the phenotype and fitness (see Section 1.4.1). This landscape can be described mathematically by parameters called selection gradients. Quantitative genetic and allelic models require knowledge of selection gradients to determine the dynamics of adaptation (see Sections 1.4.2 and 1.4.3). Thus, measures of selection gradients in natural environments can validate critical assumptions of these models. Recently, statistical methods for quantifying selection have undergone rapid development (Box 1.3), enabling one to estimate selection gradients in natural environments.

Sometimes, limited variation in thermal reaction norms prevents one from defining the adaptive landscape in sufficient detail. In such cases, artificial selection can increase the frequency of rare genotypes (Bennett 2003; Conner 2003), thereby expanding the scope of an analysis. Artificially selecting thermal reaction norms requires a different approach than the one used to select a trait expressed at a single temperature. For labile traits, the reaction norm can be measured by exposing each individual to a range of temperatures. For fixed traits, one must estimate the reaction norm by raising clones or siblings over a range of temperatures. In the latter case, the investigator selects individuals from clonal or familial lines that exhibit the desired reaction norm (Scheiner 2002).

In addition to amplifying the frequency of rare genotypes, one can further increase the diversity of phenotypes by engineering ones that do not occur naturally. Both genetic and phenotypic engineering can accomplish this goal. Genetic engineering involves the insertion or deletion of genes to produce changes in the phenotype (see, for example, Tatar 1999). A variety of methods have been developed for genetic engineering in model organisms (reviewed by Tatar 2000). Phenotypic engineering involves either physical or physiological manipulation of the phenotype. In the simplest case, a novel phenotype might be engineered through physical manipulation (Sinervo 1990). More complex approaches involve exogenous delivery of hormones or neurotransmitters (see, for example, Ketterson and Nolan 1992). Both genetic and phenotypic engineering extend the range of phenotypes for analyses of selection (reviewed by Travis and Reznick 1998).

1.5.2 Experimental evolution

Despite the value of short-term measures of selection, observations over longer periods can reveal the dynamics of evolution and possibly even the equilibrial phenotypes. The predictions of evolutionary models have been tested experimentally in artificial mesocosms and natural environments. In the former case, one maintains replicated populations in controlled thermal environments. In the latter case, one establishes replicated populations in a novel natural environment; for mobile organisms, each population must be physically enclosed to prevent migration. In both cases, one compares the evolutionary responses in experimental populations with those in control populations (Box 1.4).

Although both kinds of experiments have the same goal, each has its own advantages and disadvantages. Laboratory experiments directly link environmental temperatures to the evolution of phenotypes (Bennett and Lenski 1999). With a careful experimental design, any difference between the phenotypes of experimental and control lines can be attributed to thermal adaptation (see Box 1.4). However, practical considerations demand organisms with short generations, high fecundity, and tolerance to artificial conditions (e.g., *Drosophila melanogaster* or *Escherichia coli*). Field experiments can reveal novel patterns of thermal adaptation because natural selection depends on the interaction of temperature and other environmental factors. The use of natural environments eliminates the need to care for organisms, but creates difficulties when monitoring populations and interpreting results (see Box 1.4). Although these two kinds of experiment clearly complement one another, most experimental studies of thermal adaptation have occurred in the laboratory.

Box 1.3 Measuring selection gradients

The direction and magnitude of natural selection can be quantified by regression (reviewed by Brodie et al. 1995). A regression model relates the phenotypic value of a trait (z) to the relative fitness of an individual (w; fitness of the individual divided by the mean fitness of the population):

$$w = \alpha + \beta z + \varepsilon, \qquad (1.10)$$

where α equals the intercept, β equals the slope of the relationship between the phenotype and fitness, and ε equals the residual variation (i.e., the variation in fitness not associated with the trait); by assumption, ε is normally distributed with a mean of zero. The selection gradient (β) approximates the strength of directional selection (Fig. 1.6a). The selection gradient can be standardized by the mean or the variance of the trait to facilitate comparisons among traits or taxa (Hereford et al. 2004). Stabilizing or disruptive selection is quantified by adding a second-order term (z^2) to the model; a negative β for the second-order term indicates stabilizing selection whereas a positive second-order term indicates disruptive selection (Fig. 1.6b and c).

Because most traits are correlated with other traits, researchers often use multiple regression to estimate selection gradients. A multiple regression model describes the relationship between two or more traits and the relative fitness:

$$w = \alpha + \beta_1 z_1 + \beta_2 z_2 + \cdots + \beta_n z_n + \varepsilon, \qquad (1.11)$$

where z_n and β_n equal the phenotypic value and the selection gradient of trait n, respectively. The selection gradient calculated by multiple regression describes the effect of trait n on relative fitness, independent of all other traits (Lande and Arnold 1983). In the context of thermal adaptation, the traits would represent phenotypes measured at different temperatures or parameters of a thermal reaction norm (see Section 1.4.2).

In practice, fitness is notoriously difficult to define, much less measure. Ideally, one would measure the lifetime reproductive success of a sexual organism or the population growth rate of a clonal organism, but many other proxies for fitness have been used to quantify selection (Endler 1986; Kingsolver et al. 2001b). When a proxy for fitness is measured during part of the life cycle, selection gradients describe an episode of selection rather than natural selection, per se (for an insightful discussion of these concepts, see Endler 1986).

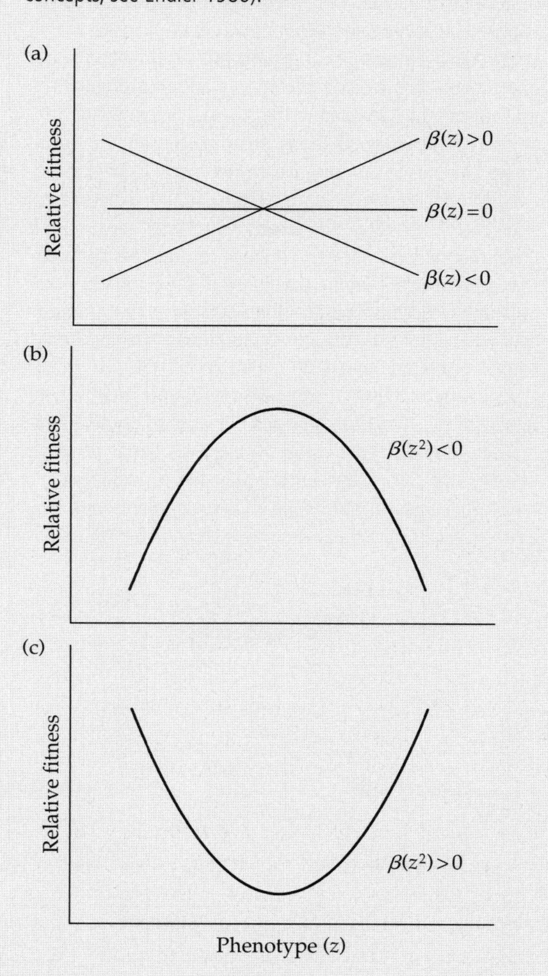

Figure 1.6 Hypothetical fitness landscapes for (a) directional, (b) stabilizing, and (c) disruptive selection. The sign of the selection gradient (β) differs among these functions.

1.5.3 Comparative analysis

Although direct measures of selection are ideal, such observations are impractical or even impossible to accumulate for most species. Given this limitation, the generality of current models must be assessed through comparative analyses. The comparative method has always been the mainstay of biologists (Somero 2000), but we can now choose from a

Box 1.4 Experimental designs for detecting adaptation

Experiments in the laboratory or field enable one to document thermal adaptation, but a careless design can lead to an ambiguous or erroneous result (Rose et al. 1996). An adequate experimental design would enable us to attribute the evolutionary response to natural selection imposed by environmental temperature, as opposed to other selective pressures or genetic drift. In laboratory experiments, one would start with an ancestral population that was already adapted to the artificial environment. In field experiments, populations would be chosen to avoid recent changes in the selective environment. From the ancestral population, one must establish both experimental and control lines (Fig. 1.7). Experimental lines experience a novel thermal environment, whereas control lines continue to experience the thermal conditions of the ancestral population. Because genetic drift within lines can produce spurious patterns or constrain adaptation, each line should consist of a large number of mating pairs (≥50).

As in any experiment, adequate replication underlies the quality of the statistical inference. Replicating observations in a selection experiment poses a major challenge because the effective unit of replication equals the population (or line) rather than the individual. Thus, experiments with only a few lines per thermal environment should be interpreted with caution. That said, most selection experiments presented in this book involved fewer than five lines per temperature. This unfortunate reality stems partly from a tradeoff imposed by logistical constraints; when the availability of space, time, or money limits the sample size, researchers face a tradeoff between having enough individuals within lines to avoid genetic drift and having enough lines within treatments to achieve statistical power.

Inferring thermal adaptation from field experiments requires some caution. In an artificial mesocosm, environmental temperature can be manipulated while all other factors are controlled or randomized among lines. Consequently, one can infer thermal adaptation by comparing experimental lines from a novel thermal treatment with control lines from the ancestral thermal treatment. In a natural environment, many factors covary with environmental temperature. Thus, we might falsely conclude that thermal adaptation occurred if other environmental factors imposed the predicted selective pressure. A more robust inference requires replicated populations in multiple thermal environments, which increases the likelihood that other factors will vary randomly with respect to environmental temperature.

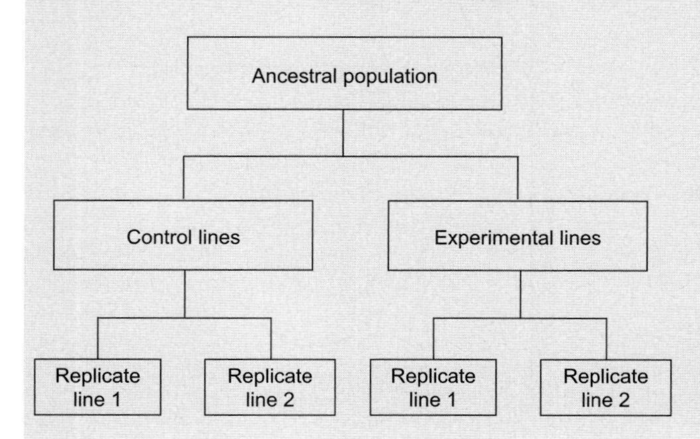

Figure 1.7 Replication of experimental evolution requires multiple lines, each founded from an ancestral population. Individuals are randomly sampled from the ancestral population to establish each line.

bewildering array of statistical procedures for combining phylogenetic, environmental, and phenotypic data to test evolutionary hypotheses. Understanding these procedures will be helpful because much of the theory presented in this book will be tested by the comparative method. Two applications of the comparative method serve our purpose: (1) studies of covariation between traits to test hypotheses about tradeoffs and (2) studies of covariation between environmental temperature and organismal phenotypes to test hypotheses about thermal adaptation.

The simplest comparison involves the study of organisms in extreme environments. Biologists

Box 1.5 Calculating phylogenetically independent contrasts

Imagine a trait, X, that has evolved within a lineage represented by four extant species (A–D). The values of X in these species are interdependent because all four species shared some portion of the evolutionary history (Fig. 1.8). Yet each species also experienced a period of evolution that was independent of the other species. For example, species A and B diverged from a common ancestor (E), and any divergence in X that occurred after these species formed was clearly independent of evolution that occurred elsewhere within the lineage. Therefore, the contrast between the values of X in species A and B (i.e., $X_A - X_B$) is called a phylogenetically independent contrast (Felsenstein 1985). For any set of n data points one can compute $n - 1$ independent contrasts. To compute these contrasts, we need a cladogram depicting only the relationships among extant taxa, branch lengths for the cladogram, and data for two or more variables (Garland et al. 1992).

Contrasts are computed for each pair of taxa, starting at the tips and moving down the tree. The values of X for ancestral taxa, such as E in Fig. 1.8, are estimated as the weighted mean of X for its descendants. For example, the value of X for taxon E (X_E) would be calculated as:

$$X_E = \frac{(1/v_A)X_A + (1/v_B)X_B}{1/v_A + 1/v_B}, \quad (1.12)$$

where v_A and v_B equal the branch lengths leading from the ancestor to taxa A and B, respectively. Because pairs of taxa that diverged more recently will have smaller contrasts, one should standardize contrasts before using them in statistical analyses (Garland et al. 1992). Standardization is quite simple because the expected mean of the contrasts equals zero; therefore, all one has to do is divide the contrasts by their standard deviation.

We can use contrasts in any statistical analysis, but the following restrictions hold: (1) all standard assumptions of parametric statistics that apply to the raw data also apply to independent contrasts and (2) statistical models should not include an intercept because the intercept of a relationship between two sets of contrasts theoretically equals zero (Garland et al. 1992).

The validity of independent contrasts rests on two major assumptions (Garland et al. 2005; Martins and Hansen 1996). First, evolution of a trait must have followed a process that resembles Brownian motion; in other words, the expected change in the trait's value equals zero and the variance of this change scales in proportion to the duration of evolution. This assumption holds for directional selection as long as the optimal phenotype fluctuates over time. Second, the sequence and timings of divergence among taxa are known. Although these assumptions are surely violated in all cases, comparative analyses of raw data require equally restrictive assumptions (Garland et al. 1999).

When conducting comparative analyses of thermal adaptation, researchers often calculate independent contrasts for environmental temperature. Arguably, the use of independent contrasts for environmental temperature is inappropriate because environmental variables do not necessarily possess the same properties as phenotypes (Martins 2000). Nevertheless, Garland and colleagues (1992) noted that contrasts could be computed for a continuous environmental variable if environmental states were passed from ancestor to descendant. This assumption seems reasonable considering that geographic structure among populations ensures that individuals in a relatively cold environment produce offspring that experience a relatively cold environment.

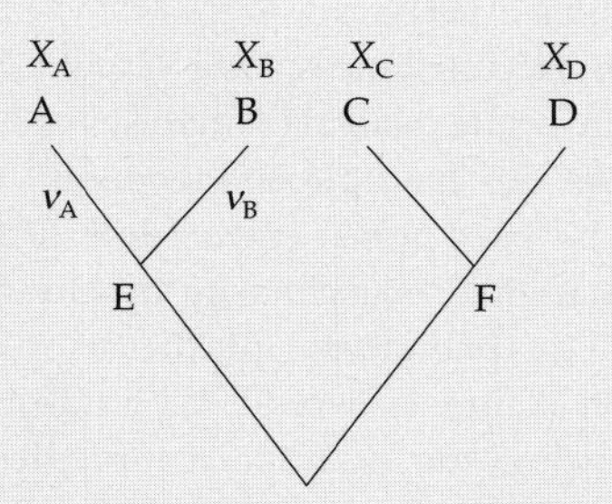

Figure 1.8 This hypothetical cladogram depicts the relationships among four extant taxa, A–D, that differ in some trait X. Ancestral taxa are labeled as E and F. Branch lengths for taxa A and B are labeled as v_A and v_B, respectively.

automatically assume that organisms living in extreme thermal environments, such as Antarctica, have adaptations for doing so (Magnum and Hochachaka 1998). This assumption seems reasonable given most organisms cannot tolerate such thermal extremes. By comparing the phenotypes of these organisms with those of organisms in less extreme environments, we learn how organisms deal with thermal stress. But how extreme must a thermal environment be for us to assume that an organism's thermal physiology reflects adaptation? Because we usually cannot answer this question, comparisons between two taxa from disparate environments provide only weak evidence of thermal adaptation (Garland and Adolph 1994).

A more powerful way to study adaptation is to compare phenotypes of species across a range of environments in a phylogenetic context (Gittleman and Luh 1994; Harvey and Purvis 1991; Huey 1987; Martins 2000; Miles and Dunham 1993). Three methods of phylogenetic comparative analysis have been most commonly used (Garland et al. 2005): (1) independent contrasts (Felsenstein 1985), (2) generalized least-squares models (Martins and Hansen 1997), and (3) Monte Carlo simulations (Martins and Garland 1991). Comparative studies of thermal adaptation have used the method of independent contrasts. This method transforms a set of raw data into a set of new data that meet the assumption of statistical independence. These new data consist of contrasts, which estimate the divergence in a trait between two taxa since their most recent common ancestor (Box 1.5). In principle, independent contrasts can be calculated for either phenotypic or environmental variables (Felsenstein 2002; Garland et al. 1992). Therefore, we can use independent contrasts to relate thermal change to evolutionary change within a lineage.

In certain circumstances, we should rely on analyses of raw data instead of analyses of independent contrasts. For example, independent contrasts should not be used when phenotypic differences among taxa are caused primarily by environmental factors (i.e., phenotypic plasticity). Likewise, independent contrasts would be undesirable if we were testing the quantitative predictions of competing models, particularly when the models predict nonlinear relationships between environmental temperature and phenotypic values. The reason is simple: independent contrasts do not retain information about the actual environmental or phenotypic values—only the difference between the values for sister taxa (see Box 1.5). The extensive use of independent contrasts stems from the overabundance of qualitative predictions in evolutionary biology (i.e., trait X should correlate positively with trait Y). Finally, an analysis of independent contrasts would be an overly conservative test of adaptation if stabilizing selection has maintained the observed phenotypes (Martins 2000). For all of these reasons, we shall interpret analyses of raw data in conjunction with analyses of independent contrasts when the two disagree.

1.6 Conclusions

As an opening to this book, I have tried to lay a foundation on which we can rest the concepts, models, and experiments presented in the coming chapters. The array of phenomena studied by thermal biologists—thermal sensitivity, thermoregulation, and thermal acclimation—naturally lead to the construct of a thermal reaction norm. This construct opens many doors because biologists have spent decades developing tools for the study of reaction norms. By estimating and comparing thermal reaction norms for labile or fixed traits, we can discover general patterns of interest. But to understand the evolution of these patterns, we must do more than accumulate data. We must gather and interpret our data in light of models. Evolutionary biology offers the theoretical and empirical tools needed to dissect these patterns and identify their causes. The evolution of thermal reaction norms can be explored through optimality models, quantitative genetic models, and allelic models. As we shall see, these models offer a wealth of predictions about thermal adaptation. The accuracy and generality of these models can be determined from direct measures of natural selection, experimental evolution in controlled environments, and comparative analyses of phenotypic variation.

Before we begin to use these tools, we must familiarize ourselves with the thermal heterogeneity of natural environments. This variation in temperature over space and time creates the impetus for evolutionary change. For this reason, we shall deal with the topic of thermal heterogeneity in the next chapter.

CHAPTER 2

Thermal Heterogeneity

The temperature of an organism—a quantitative measure of the kinetic energy of its molecules—determines the capacity for heat to flow between the organism and its environment (see Haynie 2001). If the organism possesses the same temperature as its environment, the two are in thermal equilibrium (i.e., no net flow of heat occurs between them). But if the organism and the environment differ in temperature, the warmer entity will lose heat while the cooler one gains it. This flow of heat occurs through a variety of physical processes whose relative importance varies among environments (Gates 1980; Mitchell 1976; Porter and Gates 1969). Aquatic organisms mainly exchange heat with their environment via conduction and convection. The high thermal conductivity and heat capacity of water cause the body temperatures of these organisms to closely follow the temperatures of their aqueous surroundings (Spotila et al. 1992). For terrestrial organisms, conduction, convection, radiation, and evaporation contribute significantly to the body temperature (Fig. 2.1). Below the surface of the ground, the impacts of radiation, convection, and evaporation weaken (as they do in water). These processes cause a net exchange of heat between an organism and its environment as long as the two differ in temperature.

In the real world, organisms rarely enter thermal equilibria with their environments. The motion of the earth combined with radiation from the sun drives a continuous redistribution of heat throughout the planet. Thus, living things must deal with thermal change on a variety of temporal scales. First, environmental temperatures cycle daily because of the periodic exposure to solar radiation; this predictable source of thermal heterogeneity results from the rotation of the earth around its axis. Second, environmental temperatures change rapidly and unpredictably with atmospheric conditions, such as wind speed and cloud cover. Third, environmental temperatures change seasonally because of the tilt of the earth as it orbits the sun. Finally, environmental temperatures change both cyclically and directionally among years. For example, changes in atmospheric and oceanic circulation drive fairly regular phenomena of warming and cooling in the Pacific Ocean, referred to as El Niño and La Niña. In geological history, extensive periods of warming and cooling were triggered by changes in the earth's orientation and orbit. In recent history, humans have perturbed the atmosphere so much that earth has entered a period of warming comparable to those caused by astronomical events. Depending on its location and lifespan, an organism experiences the combined stress of some or all of this thermal change.

In this chapter, we shall consider the patterns of thermal heterogeneity experienced by organisms. As we shall see, these patterns can be quantified with mathematical, physical, or statistical models. In each case, one seeks to define the body temperature of an organism based on variables that influence heat exchange. First, we shall derive a mathematical relationship between environmental conditions and the body temperature of an organism. Then, we shall use this relationship to explore how spatial and temporal variations in environmental conditions create heterogeneity of body temperatures on a global scale. Finally, we shall see how physical and statistical models have enabled researchers to quantify thermal heterogeneity on local scales. The patterns of environmental variation described in this chapter will help us to define the selective pressures on physiology, behavior, and life history in subsequent chapters.

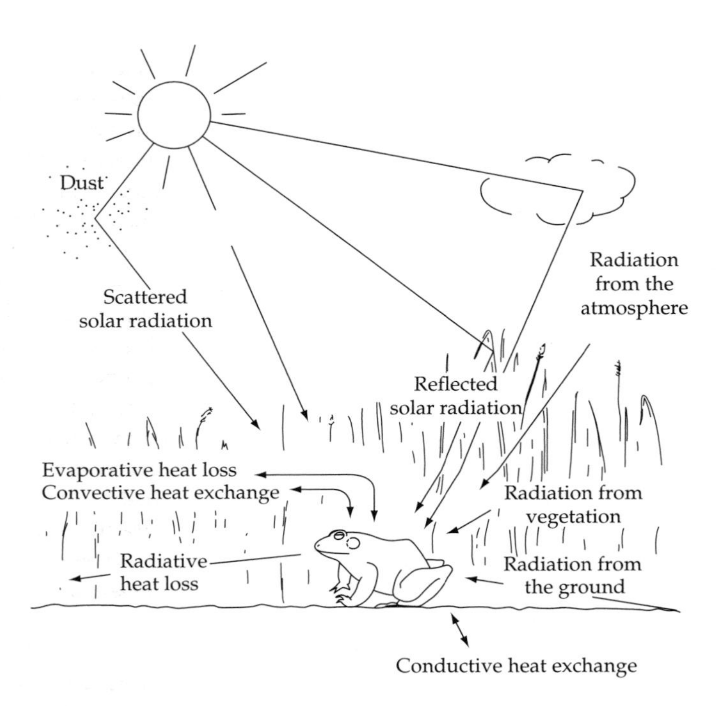

Figure 2.1 Routes of heat exchange between an organism and its environment include radiation, conduction, convection, and evaporation. An organism absorbs several kinds of radiation, including direct radiation from the sun, scattered radiation from the atmosphere, and reflected and thermal radiation from surfaces. Adapted from Tracy (1976) with permission from the Ecological Society of America.

2.1 Operative environmental temperature

When an organism's environment changes, so does its body temperature. Consider a terrestrial organism consisting of a body core and a body surface. This individual gains heat from its metabolic reactions and the radiation absorbed by the body surface. Incoming radiation includes short-wave radiation from the sun (solar radiation) and long-wave radiation from all other surroundings (thermal radiation). The individual loses heat through the emission of radiation from the body surface and the evaporation of water from the respiratory organs and body surface. Heat is also exchanged with the ground and air by conduction and convection, respectively. The transfer of heat by these processes occurs (1) between the core and surface of the organism and (2) between the surface of the organism and its environment (Fig. 2.2). When environmental conditions change, these fluxes of heat change. The organism will heat or cool until it reaches a steady-state temperature. At this steady state, the organism continues to exchange heat with its environment, but gains and losses balance one another.

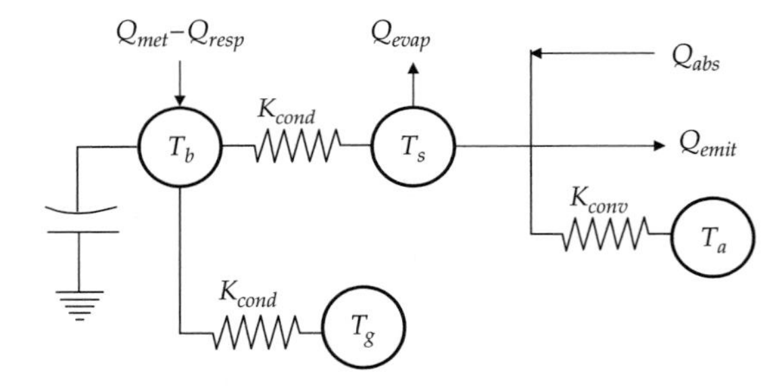

Figure 2.2 A thermal circuit diagram depicts the routes of heat transfer between a terrestrial organism and its environment. Nodes representing the body core, body surface, air, and ground are labeled according to their temperatures (T_b, T_s, T_a, and T_g, respectively). A resistor symbol denotes heat flux by conduction or convection, whereas an arrow denotes any other form of heat flux. Rates of conduction and convection depend on the coefficients of conduction (K_{cond}) and convection (K_{conv}), respectively. The capacitance symbol represents the heat capacity of the animal. Adapted from Bakken (1976) with permission from Elsevier. © 1976 Elsevier Ltd.

To understand how certain environmental conditions influence the body temperature of an organism, we shall apply a concept known as the *operative environmental temperature* (or operative temperature). The operative temperature (T_e) equals the steady-state temperature of an organism in a particular microclimate in the absence of metabolic heating and evaporative cooling (Bakken 1992). As such, this temperature characterizes the thermal environment as perceived by the organism, independent of any physiological thermoregulation.

To define the operative temperature in mathematical terms, Bakken and Gates (1975) modeled the routes of heat flux between an organism and its environment. Gains and losses at the surface of the organism were described as follows:

$$Q_{cond} + Q_{abs} = Q_{evap} + Q_{emit} + Q_{conv}, \quad (2.1)$$

where Q_{cond}, Q_{abs}, Q_{evap}, Q_{emit}, and Q_{conv} equal rates of heat flux caused by conduction, absorbed radiation, evaporation, emitted radiation, and convection, respectively. The term for conduction (Q_{cond}) refers to heat flux between the core and surface of the body; conduction between the body surface and the environment was assumed negligible.[4] Similar fluxes determine the change in the organism's core temperature over time:

$$C\frac{dT_b}{dt} = Q_{met} - Q_{resp} - Q_{cond}, \quad (2.2)$$

where C equals the heat capacity of the organism, T_b equals the core body temperature, t equals time, and Q_{resp} equals the rate of evaporative heat loss during respiration. These rates of heat flux represent averages for the whole organism, which avoids an explicit consideration of spatial heterogeneity or surface area. In reality, these fluxes vary among regions of the organism, but this simplified model will suffice for our purposes (for a more complex model, see Bakken 1981).

At a steady state, the influx and efflux of heat balance one another such that the core and surface temperatures of the organism remain constant over time. To find the steady state, we must find the core temperature that satisfies the conditions set by eqns 2.1 and 2.2, where $dT_b/dt = 0$. Solving for the steady-state temperature becomes much easier when the terms in these equations are linear with respect to temperature. For example, conduction and convection are linear functions of differences between temperatures:

$$Q_{cond} = K_{cond}(T_b - T_s) \quad (2.3)$$

and

$$Q_{conv} = K_{conv}(T_s - T_a), \quad (2.4)$$

where K_{cond} and K_{conv} equal coefficients of conduction and convection ($W\ K^{-1}$), and T_s and T_a equal temperatures of the body surface and air, respectively. In contrast, emitted radiation scales nonlinearly with the temperature of the organism's surface:

$$Q_{emit} = A\sigma\varepsilon T_s^4, \quad (2.5)$$

where A equals the surface area of the organism, σ equals the Boltzmann constant, and ε equals the emissivity of the organism. To simplify calculations of heat transfer, Bakken (1976) approximated this nonlinear term with a linear one:[5]

$$Q_{emit} \cong A\sigma\varepsilon T_a^4 + 4A\sigma\varepsilon T_a^3(T_s - T_a). \quad (2.6)$$

With this approximation, eqn. 2.1 becomes:

$$K_{cond}(T_b - T_s) + Q_{abs} \cong Q_{evap} + (A\sigma\varepsilon T_a^4 + 4A\sigma\varepsilon T_a^3(T_s - T_a)) + K_{conv}(T_s - T_a). \quad (2.7)$$

The approximation of emitted radiation causes little error, except when the body temperature of the organism deviates substantially from the air temperature (Tracy et al. 1984).

Once we have linearized eqn 2.1, we can easily solve for the steady-state temperature of the organism. To do so, we first substitute eqn 2.3 into

[4] One can reasonably ignore conduction between the organism and its environment when little surface area of the organism contacts the ground or when little difference exists between the temperatures of the ground and the body surface. Bakken (1976) presented a more general model of operative temperature that includes conduction between the organism's core and the ground, but subsequent applications of the concept of operative temperature follow the special case derived here (see, for example, Bakken 1985; Bakken 1992; Dzialowski 2005).

[5] Bakken and Gates (1975) used a different linear function to approximate radiation, resulting in a slightly different mathematical definition of operative temperature. Here, I combined the linear approximation of Bakken (1976) with the model of heat exchange proposed by Bakken and Gates (1975). This approach leads to the definition of operative temperature that appears in subsequent publications (G.S. Bakken, personal communication).

eqn 2.2 and rearrange to define T_s in terms of T_b:

$$T_s = T_b - \frac{Q_{met} - Q_{resp} - C\frac{dT_b}{dt}}{K_{cond}}. \quad (2.8)$$

Then, we obtain the steady-state temperature of the organism by substituting eqn 2.8 into eqn 2.7, setting dT_b/dt equal to zero, and solving for T_b:

$$T_b \cong T_a + \frac{Q_{met} - Q_{resp}}{K_{cond}} + \frac{Q_{met} + Q_{abs} - Q_{resp} - Q_{evap} - A\sigma\varepsilon T_a^4}{K_{conv} + 4A\sigma\varepsilon T_a^3}. \quad (2.9)$$

If we assume the organism gains no heat from metabolism and loses no heat by evaporation (or that these fluxes cancel one another), we arrive at the operative temperature:

$$T_e \cong T_a + \frac{Q_{abs} - A\sigma\varepsilon T_a^4}{K_{conv} + 4A\sigma\varepsilon T_a^3}. \quad (2.10)$$

Thus, the operative temperature reflects an increment in body temperature caused by solar radiation, discounted for the heat lost by convection. In a burrow, where an organism receives no solar radiation, the operative temperature equals the ambient temperature (Chappell and Whitman 1990).

For certain organisms, the operative temperature should account for the heat lost by evaporation. For example, amphibians possess a wet epidermis and many species cannot physiologically control their evaporative water loss (Duellman and Trueb 1986; Spotila et al. 1992). The relative humidity of the environment determines the rate of evaporation, and hence influences the body temperature of these organisms (Gates 1980). A biologically relevant measure of operative temperature for such organisms, referred to as the *wet operative temperature*, accounts for evaporative heat loss (for details, see Campbell 1977).

Operative temperatures depend on interactions among absorbed radiation (Q_{abs}), air temperature (T_a), and wind speed (which affects K_{conv}). We can visualize these interactions in the form of a climate–space diagram (Kingsolver 1983; Porter and Gates 1969; Scott et al. 1982; Spotila et al. 1972). These diagrams depict the combinations of environmental conditions for which an organism will experience a certain steady-state temperature. Obviously, these combinations depend on the physical properties of the organism that determine heat exchange (e.g., color, shape, and size). Figure 2.3 shows the combinations of absorbed radiation, air temperature, and wind speed for which a desert iguana (*Dipsosaurus dorsalis*) will reach 30°C. The impact of absorbed radiation and air temperature on the steady-state temperature depends on the wind speed, as depicted by the crossing of the lines. Such plots highlight the interactions among environmental factors that influence the operative temperature. To the extent that these factors vary predictably over space and time, they create patterns of thermal heterogeneity.

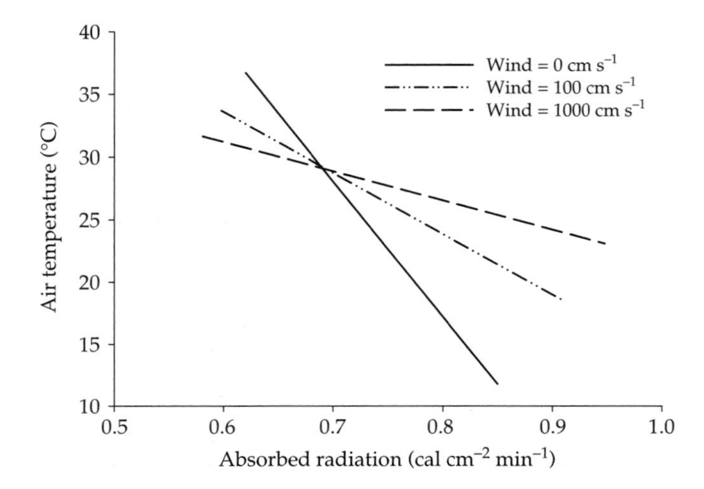

Figure 2.3 The same body temperature occurs under many combinations of radiation load, air temperature, and wind speed. Each line shows the conditions that yield a predicted body temperature of 30°C for a desert iguana, with a mass of 67 g and a surface area of 180 cm^2. Adapted from Porter and Gates (1969) with permission from the Ecological Society of America.

2.2 Global patterns of operative temperature

Global variation in solar radiation and environmental temperature gives rise to latitudinal and altitudinal clines in operative temperatures. The intensity of solar radiation determines the amount of radiation absorbed by a particular organism. The air or water temperature determines the heat exchanged via convection. As solar radiation and environmental temperatures vary dielly and seasonally, so do geographic clines in operative temperatures.

2.2.1 Latitudinal clines

Ultimately, all latitudinal variation in operative temperatures stems from changes in solar radiation. The quantity of solar radiation received by an organism and its environment depends on the time of day, the time of year, and the location on earth (reviewed by Gates 1980; McCullough and Porter 1971). These sources of variation stem from the tilt of the earth's axis and its elliptical orbit around the sun. During June, the northern hemisphere tilts toward the sun, and thus receives more solar radiation that the southern hemisphere. The opposite phenomenon occurs during December, leading to the opposing seasons observed in these regions of the planet. At any location, the rotation of the earth on its axis creates variation in solar radiation on a daily basis; radiation peaks daily at solar noon, when the sun's rays become perpendicular to a particular surface of the earth. These simple principles combine to determine the radiation received at a particular place and time, ${}_hS_0$ (Gates 1980):

$$ {}_hS_0 = \bar{S}_0 \left(\frac{\bar{d}}{d}\right)^2 (\sin\phi \cdot \sin\delta + \cos\phi \cdot \cos\delta \cdot \cos h), \quad (2.11) $$

where $\bar{S}_0$ equals the solar constant (1360 W m^{-2}), d and $\bar{d}$ equal the actual and mean distances respectively between the earth and the sun, ϕ equals the latitude of the surface, δ equals the declination of the sun, and h equals the hour angle of the sun (the angle by which the earth must rotate to bring the point directly perpendicular to the sun's rays).

Given the presence of an atmosphere, only a fraction of solar radiation reaches the surface of the earth; the atmosphere absorbs, reflects, or scatters the remaining radiation. To model the reduction in solar radiation caused by the atmosphere, we must add a coefficient (τ^m) that defines the transmission of radiation:

$$ {}_hS_0 = \bar{S}_0 \left(\frac{\bar{d}}{d}\right)^2 \times (\sin\phi \cdot \sin\delta + \cos\phi \cdot \cos\delta \cdot \cos h)\tau^m, \quad (2.12) $$

where $0 \leq \tau \leq 1$. The exponent m equals the length of the path by which radiation travels through the atmosphere divided by the shortest possible path to sea level.

By integrating eqn 2.2 over time, one can calculate daily or annual radiation load for any latitude. The hour angle changes over the course of a day, and the sun's declination changes over the course of a year. Gates (1980) integrated over the set of hour angles to obtain the daily radiation load; plotting this radiation as a function of latitude and time of year reveals certain patterns (Fig. 2.4). For a realistic range of atmospheric transmissions, the greatest incident radiation occurs at 30°N during the solstice in June and 30°S during the solstice in December (Fig. 2.5a). During the equinoxes in March and September, however, the greatest incident radiation occurs at the equator (Fig. 2.5b). Over the course of a year, the equator receives 2.4 times the radiation that the poles receive (Gates 1980). This variation in solar radiation contributes directly to latitudinal clines in the operative temperatures of organisms (see eqn 2.10).

Solar radiation also affects operative temperatures indirectly by causing variation in the temperatures of air, water, and soil. Global patterns of environmental temperature are most easily inferred from air temperatures, which have been recorded for decades at thousands of locations (Box 2.1). In the northern hemisphere, air temperature decreases strongly with increasing latitude (Fig. 2.7). This latitudinal gradient becomes most pronounced during winter (Addo-Bediako et al. 2000; Bradshaw and Holzapfel 2006; Ghalambor et al. 2006), a phenomenon resulting from the contrasting seasonal patterns of solar radiation at different latitudes (see Fig. 2.4). A global model of daily temperatures constructed by Piper and Stewart (1996) nicely

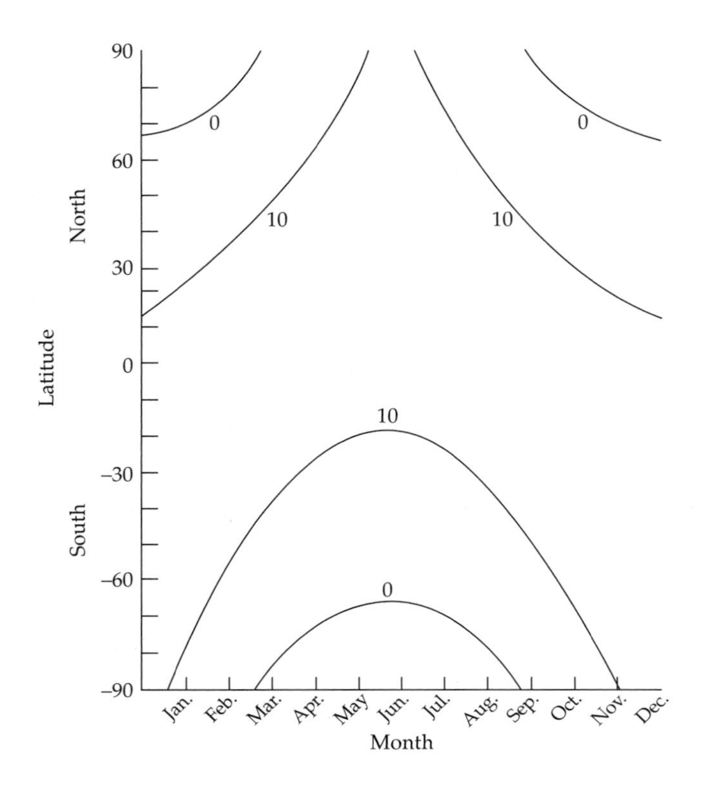

Figure 2.4 The daily intensity of incident radiation varies predictably with latitude and the time of the year. The values plotted here were based on a transmittance of 0.6. Adapted from Gates (1980) with permission from Dover Press.

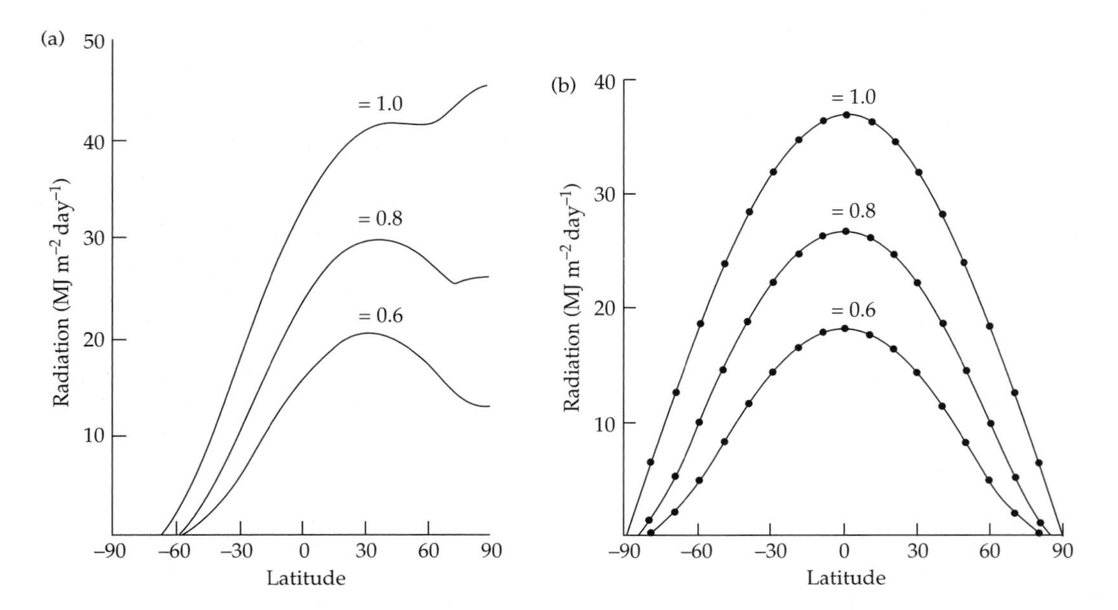

Figure 2.5 The transmission of solar radiation affects the latitudinal clines of solar radiation during the solstice (a) and the equinox (b). Positive values of latitude are for the northern hemisphere. Adapted from Gates (1980) with permission from Dover Press.

Box 2.1 Mapping environmental temperatures on a global scale

Since the production of the first standardized thermometers in the 18th century, the measurement and analysis of temperatures have blossomed into a global enterprise. Today, the World Meteorological Organization, the World Data Center for Meteorology, and the Intergovernmental Oceanographic Commission collectively archive temperatures recorded by roughly 10 000 terrestrial weather stations and several hundred coastal buoys. Additionally, thermal records for the ocean surface have been estimated from satellite images (Kerr and Ostrovsky 2003). To facilitate comparative research, these organizations set guidelines for methodology and accuracy; for example, the World Meteorological Organization defines air temperature as the temperature of a thermometer exposed to air at a height of 1.2–2.0 m and sheltered from direct solar radiation. The efforts of these organizations have generated two beneficial outcomes for thermal biologists. First, we now have excellent models of spatial and temporal variations in air and water temperatures to guide our research. Second, we can use normal temperatures—mean temperatures for a consecutive period of 30 years—to identify anomalous conditions and directional trends. Current estimates of global warming depend on normal temperatures for 1961–1990 (see Chapter 9).

The thermal data collected by weather stations enable one to map environmental temperatures on a global scale. To construct such maps, one must convert the extremely patchy distribution of thermal records into a regularly spaced grid. Climatologists accomplish this feat by fitting regression models to the available data or interpolating temperatures between weather stations (see, for example, Zhao et al. 2005). The global patterns described in this chapter stem from the work of Piper and Stewart (1996) and New and colleagues (1999), who used methods of interpolation to map mean temperatures and mean daily ranges of temperatures for all continents but Antarctica. Below, I briefly contrast the procedures used by these investigators.

Piper and Stewart (1996) used time series of air temperatures from more than 7500 weather stations to create maps of daily mean temperatures and daily ranges of temperature at a spatial resolution of 1° (latitude by longitude). To increase the size of their sample while ensuring consistency among stations, they recalculated mean temperatures as the average of the minimal and maximal temperatures. To simplify their model, they adjusted all temperatures to sea level prior to interpolation. After interpolation, they readjusted the temperature at each location based on the altitude. Their method of interpolation involved distance weighting. For each grid point, the temperature was estimated from temperatures recorded at 5–10 neighboring weather stations; the influence of each record on the predicted temperature was weighted by the proximity of the weather station to the grid point.

In contrast to Piper and Stewart, New and his colleagues used a two-step process to generate maps of monthly mean temperatures and monthly ranges of temperatures at a resolution of half a degree latitude/longitude. First, they modeled the global distribution of normal temperatures for 1961–1990 (New et al. 1999). Because normal temperatures are more accessible than time series, the researchers could include data from more than 12 000 locations. Each temperature was weighted by the variance and the number of years contributing to the normal temperature (maximum = 30 years). Unlike Piper and Stewart, New et al. interpolated over three dimensions, eliminating the need to adjust temperatures for the altitude. To avoid errors caused by the low density of weather stations in some regions, they divided the land surfaces into 22 overlapping tiles (Fig. 2.6) and fitted a separate spline function to the data for each tile. Once New et al. had obtained a grid of normal temperatures, they used records of thermal anomalies (deviations from the normal temperature) to create monthly thermal maps for the period of 1901–1996 (New et al. 2000). Thermal anomalies were interpolated by distance weighting to form a grid for each month. By combining the grid of thermal anomalies with the grid of normal temperatures, New et al. generated a global map of temperatures for each period of time.

Several sources of error influenced the quality of global thermal maps. First, not all measurements of temperature were made at comparable times of day (Piper and Stewart 1996). Deriving daily mean temperatures from minimal and maximal temperatures reduced some bias caused by differences in timing (New et al. 1999). Second, the sheer magnitude of data involved in these analyses virtually guaranteed the existence of typographical errors. Many of these errors were eliminated by careful review of the data prior to analysis. Errors were assumed to exist when monthly mean temperatures failed to follow a seasonal cycle or a mean temperature fell outside the range bounded by the minimal and maximal temperatures. Finally, errors in interpolation might have resulted from a paucity of temperatures at certain latitudes and altitudes.

continues

Box 2.1 Continued

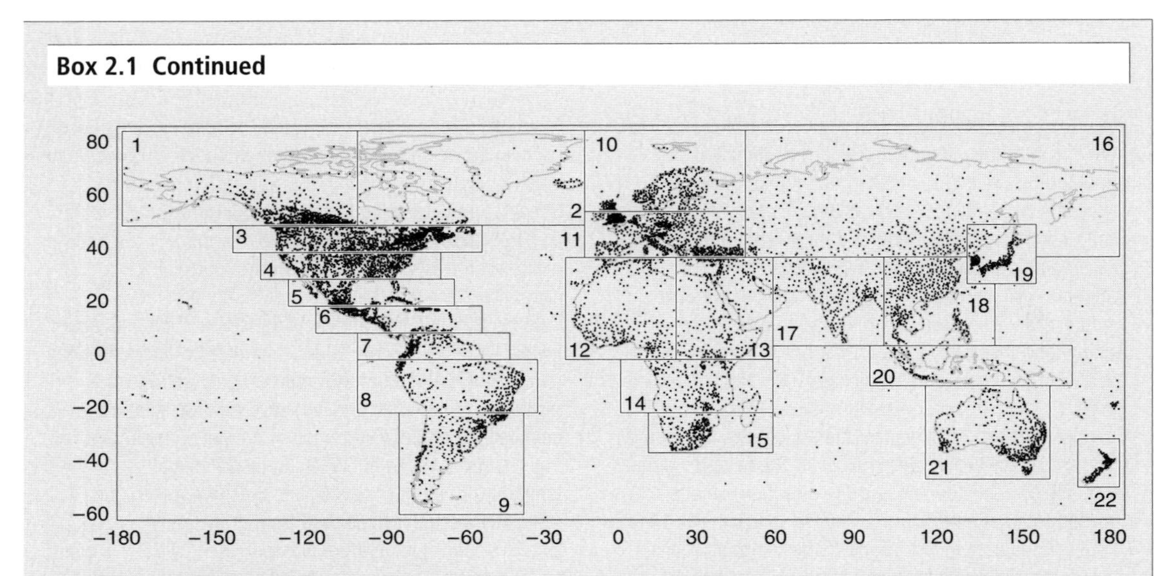

Figure 2.6 Locations map of air temperatures. To increase the accuracy of interpolation, the continents were divided into 22 tiles based on the density of weather stations. Reproduced from New et al. (1999) with permission from the American Meteorological Society.

New et al.'s method of fitting of splines to separate tiles helped to minimize the influence of data-rich regions on data-poor regions. Cross-validation of global thermal maps indicated these potential sources of error were not so problematic as to obscure major patterns of thermal heterogeneity (New et al. 2000).

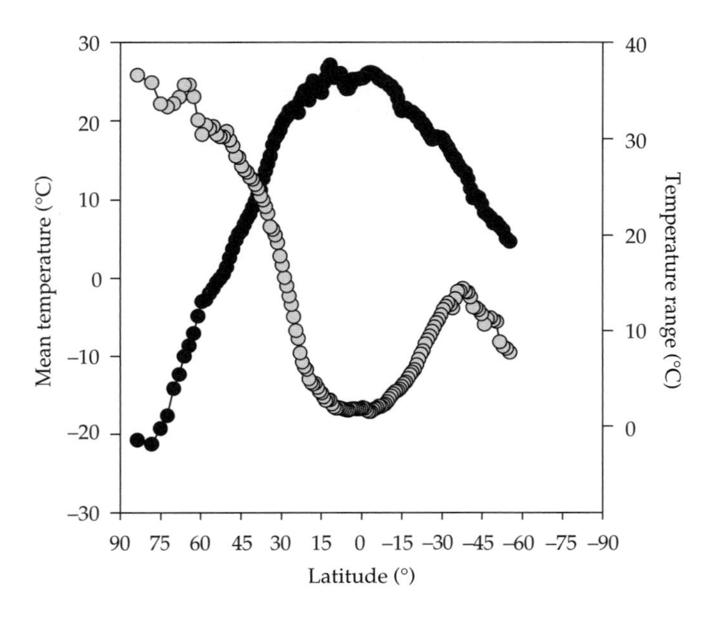

Figure 2.7 Air temperatures of terrestrial environments depend strongly on latitude. Data are the mean air temperatures (black circles) and annual ranges of air temperatures (gray circles) in the New World. Positive values of latitude are for the northern hemisphere. The maximal mean temperature and the minimal thermal range occurs in the tropics. Adapted from Clarke and Gaston (2006) with permission from the Royal Society.

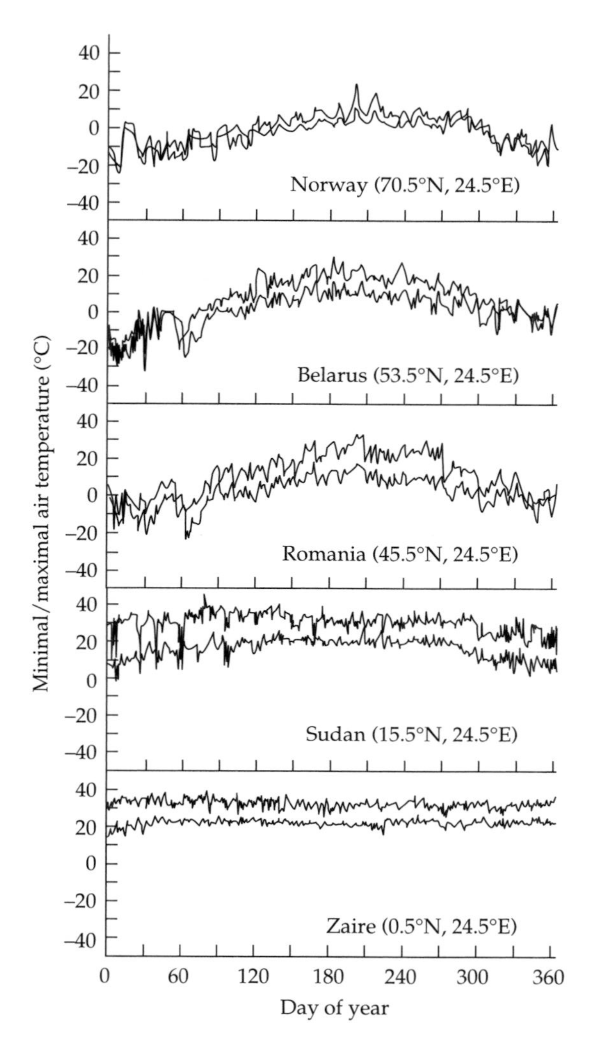

Figure 2.8 Seasonal profiles of daily minimal and maximal temperatures at five latitudes, ranging from the equator to the Arctic. The longitude was held constant at 24.5°E. Modified from Piper and Stewart (1996) by permission of the American Geophysical Union.

illustrates this point (Fig. 2.8). At the equatorial latitude of Zaire (0.5°N), temperatures remain fairly constant throughout the year. Moving from Zaire toward Belarus (53.5°N), the seasonal variation in daily temperatures becomes greater. As one moves even farther northward (70.5°N), the seasonal range of temperatures decreases slightly because summer temperatures never approach those seen at lower latitudes. The data from these localities reflect a very general latitudinal cline in the seasonality of thermal environments (Vázquez and Stevens 2004). In contrast, the southern hemisphere has a weaker latitudinal cline in temperature (Fig. 2.7), owing to the buffering of the small land masses in this hemisphere by vast oceans (Addo-Bediako et al. 2000; Ghalambor et al. 2006).

Aquatic environments exhibit far less variation in temperature because of water's tremendous capacity to store heat. This phenomenon holds for freshwater and marine environments, but is especially exaggerated in large bodies of water such as oceans. For example, the annual range of surface temperatures of the Pacific Ocean never reaches one-half of the annual range observed at the same latitude on land (compare Figs 2.7 and 2.9). This thermal buffering keeps coastal climates warmer than inland ones, particularly in winter (New et al. 2002). Interestingly, while seasonal variation in temperature peaks at high latitudes in terrestrial environments, the peak occurs at intermediate latitudes in the oceans (see Fig. 2.9 and Parmesan et al. 2005). Nevertheless, mean water temperatures of marine and freshwater environments decrease from the tropics to the poles, in a manner that resembles latitudinal clines in air temperature (Clarke and Gaston 2006; Vannote and Sweeney 1980).

2.2.2 Altitudinal clines

At any latitude, a shift in altitude results in predictable changes in solar radiation and air temperature. Both of these changes relate to the relatively low air pressure at a high altitude (air pressure reflects the mass of air above a point in the atmosphere). Recall that the atmosphere absorbs, reflects, or scatters radiation. If all else is equal, a shorter path through the atmosphere results in a greater transmission of radiation. Specifically, the exponent of the transmission coefficient (m of eqn 2.12) decreases linearly with decreasing air pressure (Gates 1980). Therefore, surfaces at a higher altitude receive a greater intensity of solar radiation. Additionally, the low air pressure at a high altitude reduces the air temperature through an adiabatic process (Monteith and Unsworth 1990). As the pressure of rising air decreases it expands and cools. For dry air, the magnitude of adiabatic cooling equals about 10°C per kilometer. Moist air cools less as it rises because of

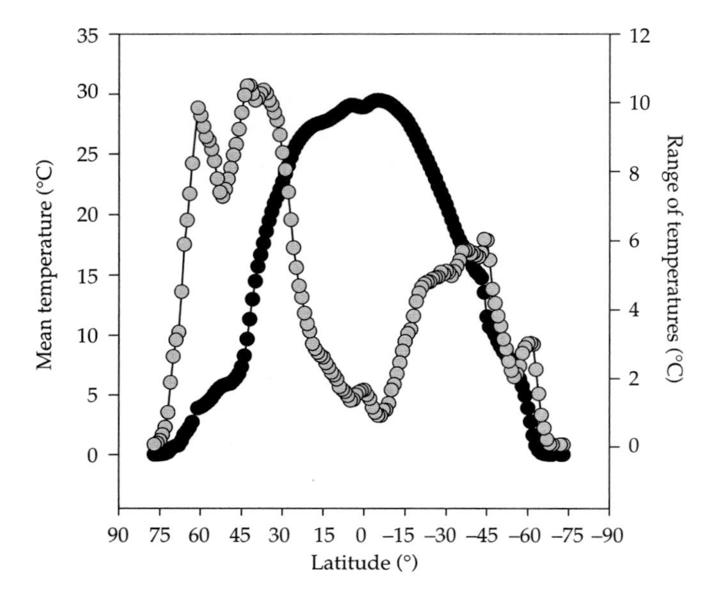

Figure 2.9 Surface temperatures of marine environments depend strongly on latitude. Data are the mean surface temperatures (black circles) and annual ranges of surface temperatures (gray circles) in the Pacific Ocean (170° W). Positive values of latitude are for the northern hemisphere. Latitudinal patterns of ocean temperatures match those of terrestrial air temperatures, except the thermal range of ocean temperatures peaks in each temperate zone and drops sharply toward each polar region. Adapted from Clarke and Gaston (2006) with permission from the Royal Society.

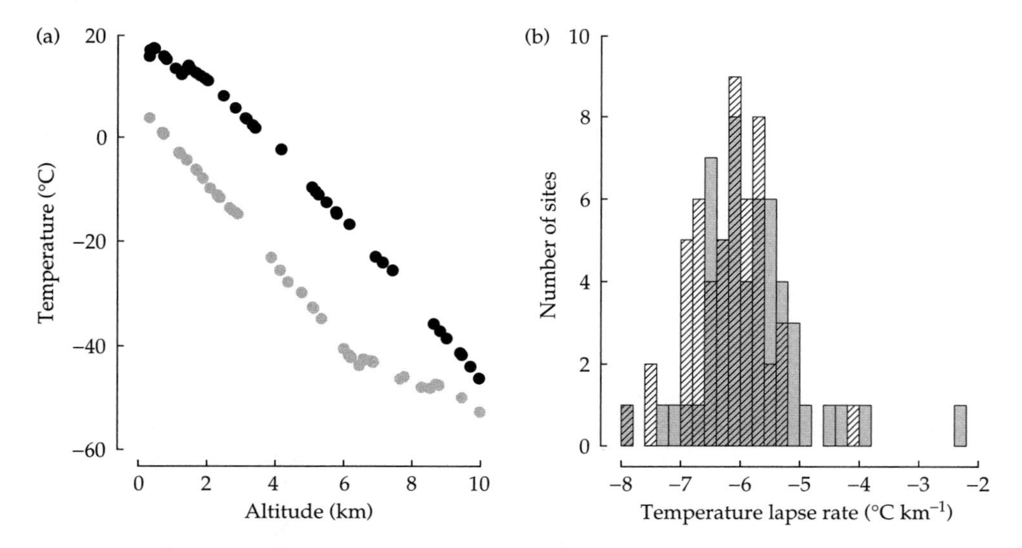

Figure 2.10 Air temperatures decrease with increasing altitude because of adiabatic cooling. (a) Air temperatures at different altitudes measured by a weather balloon released in Germany. Black and gray circles denote measurements taken during summer and winter, respectively. The slope of each line describes the rate of adiabatic cooling. (b) Most rates of adiabatic cooling throughout the globe range from –5°C to –7°C per kilometer of altitude. Hashed and gray bars denote measurements taken during summer and winter, respectively. Adapted from Dillon et al. (2006) by permission from Oxford University Press.

the heat generated by condensation. We express the consequence of this cooling in the form of an environmental lapse rate, which equals the change in the temperature of still air with an increase in altitude. The environmental lapse rate varies among localities, but usually falls in the range of 5–7°C per kilometer (Fig. 2.10). Because of adiabatic cooling, an organism at a high altitude in the tropics can experience an air temperature similar to that experienced by an organism at a low altitude in the temperate zone (Ghalambor et al. 2006). Note, the greater solar radiation and the lower air temperature at higher altitude have opposing effects on the operative temperature of an organism (see eqn 2.10). In general, the effect of air temperature outweighs that of solar radiation, leading to lower operative temperatures at higher altitudes (Adolph 1990; Bashey and Dunham 1997; Beaupre 1995; Diaz 1997; Grant and Dunham 1990; Hertz 1992; Van Damme et al. 1989).

2.3 Quantifying local variation in operative temperatures

Because operative temperature depends on absorbed radiation and environmental temperature, environments at higher latitudes or higher altitudes would offer lower operative temperatures if all other factors were equal. Yet all else is never equal! Regional differences in topographical and meteorological conditions disrupt the latitudinal and altitudinal clines in operative temperature caused by solar radiation and environmental temperature. For example, a uniform warming of air by 1°C along a latitudinal cline was predicted to increase body temperatures of intertidal mussels by anywhere from 0.1°C to 0.9°C, depending on the timing and duration of emersion at each site (Gilman et al. 2006). Similar irregularities in geographic clines result from variations in wind speed, cloud cover, and vegetation density. These factors influence conduction, convection, or radiation to create heterogeneity of operative temperatures on local scales. To really understand the selective pressures imposed by thermal environments, we must model the local variation in operative temperature.

Quantifying local variation in operative temperature poses a major challenge. As we can see from eqn 2.10, the operative temperature depends on interactions among micrometeorological conditions of the environment and physical properties of the organism (see also Gates 1980). Micrometeorological conditions vary greatly over very small scales (see, for example, Bartlett and Gates 1967); therefore, we must have some way of directly or indirectly monitoring these conditions at the numerous microenvironments available to an organism. Furthermore, color, shape, and size vary considerably among organisms. Consequently, models of operative temperature must be tailored for each species, or even for each phenotype of a species. Because of these difficulties, we know less about the variation in operative temperatures within local environments.

What we do know about local heterogeneity comes from the use of mathematical, physical, and statistical models. These models differ in their attention to mechanism and their ease of application. At one extreme, a mathematical model provides a detailed description of heat transfer, but requires careful measurements of many parameters at each location. At the other extreme, a statistical model creates a black box relating environmental characteristics and operative temperatures. Although these models lack an explicit mechanistic link between variables, they offer the potential to predict operative temperatures over much broader scales. At the interface of these two approaches lies the popular approach of measuring operative temperatures with a physical model of the organism. In this section, we shall examine the advantages and disadvantages of each approach.

2.3.1 Mathematical models

The earliest attempts to characterize operative thermal environments relied on mathematical models of heat transfer. As we saw in Section 2.1, we can use the principles of biophysics to derive a mathematical function relating operative temperature to micrometeorological variables and organismal parameters. Researchers have developed similar models for a wide range of organisms, including mollusks (Denny and Harley 2006; Helmuth 1998), insects (Henwood 1975; Kingsolver 1983), amphibians (Tracy 1976), and reptiles (Porter et al. 1973; Seebacher and Grigg 1997). Mathematical models have been used to characterize operative temperatures on both regional and local scales.

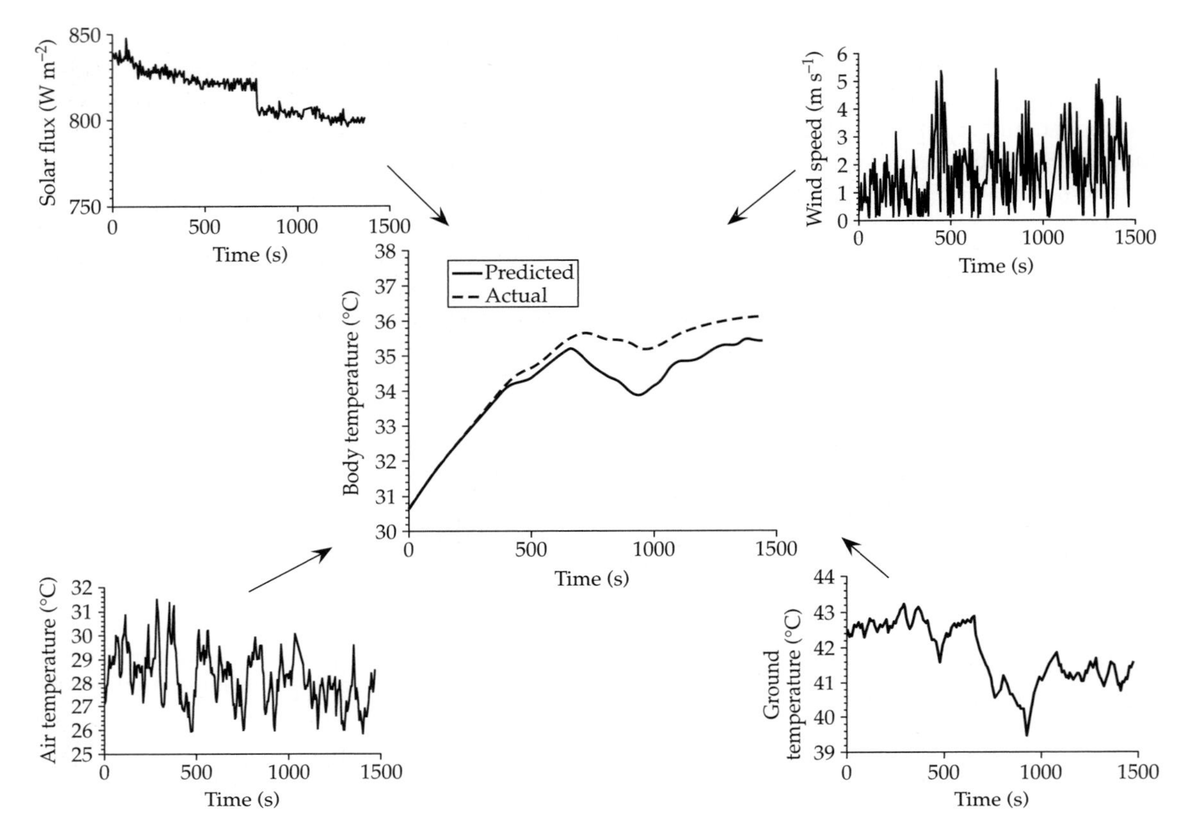

Figure 2.11 Measurements of radiation load, air temperature, ground temperature, and wind speed were used to compute the temperature of a mussel on a rocky shoreline. On average, the predicted temperature fell within 1°C of the actual temperature of a mussel. Reproduced from Helmuth (1998) with permission from the Ecological Society of America.

For example, Porter and colleagues computed geographic variation in operative temperatures of various ectotherms on multiple continents (e.g., Grant and Porter 1992; Kearney and Porter 2004). At the other extreme, Bartlett and Gates (1967) calculated the range of operative temperatures on the surface of a single tree. Mathematical models not only predict operative temperatures, they also predict how evaporative cooling and metabolic heating might cause the body temperature to deviate from the operative temperature (Casey 1992; Helmuth 1998; Tracy 1976).

Although mathematical models have provided tremendous insight regarding the factors that influence operative temperature, they are impractical for mapping thermal environments at a sufficient resolution to understand selective pressures on behavior and physiology (O'Connor and Spotila 1992). To compute operative temperature from a mathematical model, one must know the solar radiation, ground reflectivity, air temperature, ground temperature, and wind speed (see Fig. 2.11). The overwhelming task of obtaining such data for a large number of locations precludes a fine-scale mapping. In most cases, researchers have computed operative temperatures for a limited number of microclimates, such as full sun and full shade (Christian and Bedford 1995; Seebacher and Grigg 1997; Van Damme et al. 1989). As we shall see in later chapters, this coarse-scale mapping of the environment obscures much of the heterogeneity that drives thermal adaptation. For this reason, many researchers have adopted a different approach to study thermal heterogeneity.

Figure 2.12 Physical models of operative temperatures are constructed to mimic the radiative, conductive, and convective properties of an organism. Examples include a copper electroform of a lizard (a), a mussel shell filled with silicone (b), a copper tube designed to mimic a snake (c) and a moist agar model shaped like a toad (d). Because the agar model incorporates evaporative water loss, this type of model records wet operative temperatures. (b) Courtesy of Brain Helmuth; (c) courtesy of Steven Beaupre; (d) courtesy of Carlos Navas.

2.3.2 Physical models

Physical models provide a convenient alternative to mathematical models for mapping thermal heterogeneity on very small scales (Bakken and Gates 1975). Physical models mimic the radiative, convective, and conductive properties of living organisms in the absence of physiological function. As such, the temperature of the model at its steady state equals the operative temperature (Bakken 1976, 1992). In theory, a physical model should have no heat capacity and would reflect a fairly instantaneous measure of operative temperature. In practice, good physical models of operative temperatures respond rather quickly to changes in heat flux relative to the responses of real organisms. To date, researchers have constructed physical models from a wide range of materials, including metallic electroforms, rubber hoses, and dried exoskeletons (Fig. 2.12). For organisms that lack physiological control over evaporative cooling, models with a wet surface have been used to estimate steady-state body temperatures more accurately (Bartelt and Peterson 2005; Navas and Araujo 2000).

When properly constructed, a physical model can predict the body temperature of an organism within 1°C of its actual value (reviewed by Dzialowski 2005). The most accurate models consist of hollow metallic electroforms painted to match the reflectivity of the organism (Fig. 2.12a). Metallic objects of similar dimension but slightly different shape yield larger errors (Walsberg and Wolf 1996). In the rare event that solar radiation and convection contribute little to the heat flux of an organism, simple physical models constructed from commercial temperature loggers might adequately estimate operative temperatures (Vitt and Sartorius 1999). Nevertheless, one should always determine the accuracy of a physical model under a range of conditions that reflect those of the environment (Bakken 1976; Dzialowski 2005).

Because most types of physical models require little time and money to construct, researchers can use these models to record operative temperatures on a much finer scale than they have done with mathematical models (Bakken 1989). For example, one can estimate the frequency of operative temperatures by placing models at randomly determined points (e.g., Grant and Dunham 1990) or at intervals along transects (e.g., Grant 1990). Alternatively, one can place models in specific microenvironments and use the area of each microenvironment to infer the frequency of operative temperatures (e.g., Grant and Dunham 1988). Collectively, these sampling designs have shown that operative temperatures depend on the type of substrate (Adolph 1990), the slope of the ground (Helmuth and Hofmann 2001; Sartorius et al. 2002), the amount of shade (Christian and Weavers 1996; Hertz 1992), and the posture and orientation of the organism (Bakken 1989; Christian and Bedford 1996; Grant and Dunham 1988). For a given location, posture, and orientation, operative temperatures rise throughout the morning and reach a maximum near the middle of the day (Bashey and Dunham 1997; Dorcas and Peterson 1998; Hertz 1992; Sartorius et al. 2002; Schäuble and Grigg 1998). Diel cycles of operative temperatures are more pronounced in terrestrial microclimates that receive solar radiation than they are in shaded or aquatic microclimates (Brown and Weatherhead 2000; Christian and Weavers 1996). Despite these insights, the majority of researchers have sampled environments at very low spatial and temporal resolutions. In a review of published studies, Dzialowski (2005) noted a typical sampling design has included fewer than five physical models deployed for no more than 5 days! Thus, while physical models offer practical advantages over mathematical models, very few researchers have capitalized on their potential to rigorously map operative thermal environments.

2.3.3 Statistical models

A statistical model offers a powerful compromise between a highly mechanistic mathematical model and a purely phenomenological physical model. As seen in Section 2.1, we can use mathematics to describe an operative temperature in terms of heat fluxes between an organism and its environment. Equation 2.10 describes a precise relationship between many independent variables and the operative temperature. Nevertheless, we would need to know the values of these variables at very high spatial and temporal resolutions to map operative temperatures with this mathematical model. Although we cannot easily determine all of these variables at many points in the environment, we can obtain reasonable proxies for these variables at a high resolution. Given these proxies, we could capture a great deal of information about operative thermal environments from a statistical model (Hertz 1992; Howe et al. 2007; Sears et al. 2004). Ultimately, the power of a statistical model depends on the degree to which the independent variables relate mechanistically to operative temperature (or correlate with variables that do). Commonly used proxies include air temperature, wind speed, slope, aspect, and elevation, but some models also include categorical variables such as the type of substrate or vegetation (Box 2.2).

Statistical models enable researchers to create detailed maps of operative thermal environments. For example, Sears and colleagues (2004) constructed a model that related slope, aspect, elevation, wind speed, and air temperature to the operative temperature of a small lizard. The independent variables were obtained from micrometeorological equipment and a digital elevation map. For some locations, operative temperatures were estimated with physical models (see Section 2.3.2). These point samples were used to parameterize the statistical relationship between the independent variables and the operative temperature. Once the model was parameterized, operative temperatures were predicted for an entire landscape at a series of times (Fig. 2.14). These thermal maps have a resolution of less than a meter and a mean error of less than 2°C! This scope and resolution of thermal mapping greatly exceed those achieved with mathematical

Box 2.2 Modeling operative temperatures with artificial neural networks

Identifying the most accurate statistical model to predict operative temperatures can be challenging because these temperatures depend on the interaction of many environmental variables. Traditional statistical approaches, such as multiple regression analysis, often yield poor approximations of nonlinear relationships (Lek et al. 1996). To solve this problem, two groups of researchers have turned to artificial neural networks (Howe et al. 2007; Sears et al. 2004). An artificial neural network discovers the nonlinear relationship between a set of independent variables and a dependent variable. These models offer great flexibility and power because (1) the independent variables can be either continuous or categorical and (2) the relationships between these variables and the operative temperature can take on many different forms (e.g., linear, sigmoidal, and step functions).

Parameterizing an artificial neural network involves a process that mimics learning by a brain (reviewed by Basheer and Hajmeer 2000; Lek and Guégan 1999). The basic structure of the model consists of three interconnected layers: an input layer, a hidden layer, and an output layer (Fig. 2.13). The input layer consists of independent variables, such as air temperature, wind speed, slope, aspect, and elevation. The output layer consists of the operative temperature. In between, a hidden layer transforms the data from the input layer to the prediction at the output layer. This transformation involves a set of coefficients and thresholds that serve as the parameters of the model. The process of learning begins by randomly choosing the values of these parameters. The model then proceeds to train itself by comparing the predicted and observed temperatures and adjusting the parameters to improve the fit. After many iterations of training, an artificial neural network will converge on an excellent description of the data. The trick is to avoid overtraining the model such that it provides a perfect description of the training data but a poor description of novel data. To avoid overtraining, one must periodically confront the model with independent data during the training process. Initially, training improves the ability of the model to predict both the training data and the independent data. But, eventually, training will decrease the model's ability to predict the independent data even though its ability to predict the training data continues to improve. At this point, training should cease and the model should perform well when applied to new data.

In general, artificial neural networks outperform linear models at predicting operative temperatures. For example, general linear models overpredicted low operative temperatures or underpredicted high operative temperatures (Howe et al. 2007; Sears et al. 2004). A potential drawback of an artificial neural network model is that mechanisms remain obscured by the "black box" of the model. Nevertheless, many ecologists have found these models ideal for predicting phenomena when underlying mechanisms exist but cannot be precisely formulated (Lek and Guégan 2000).

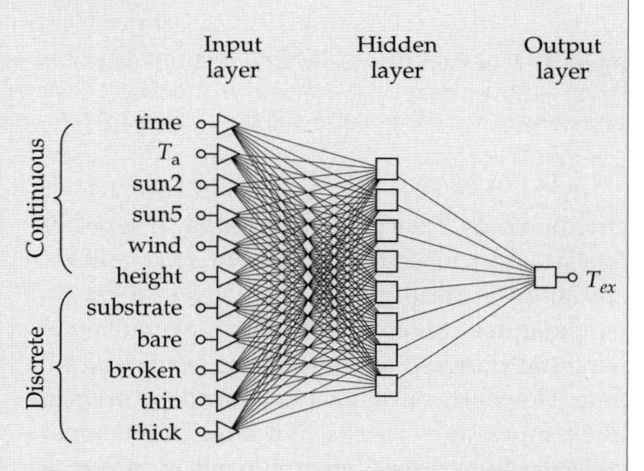

Figure 2.13 The structure of an artificial neural network designed to predict microenvironmental temperatures (T_{ex}) from a combination of continuous and discrete variables. The continuous variables in this model include the time of day (time), air temperature at 1.3 m (T_a), wind speed at 1.5 m (wind), solar radiation over previous 2–5 minutes (sun2 and sun5), and height above the ground (height). Discrete variables include the type of substrate and the degree of ground cover (bare, 0–29%; broken, 30–69%; thin, 70–100% with visible substrate; thick, 70–100% with no visible substrate). Adapted from Bryant and Shreeve (2002) with permission from Blackwell Publishing.

or physical models alone. As we shall see in the chapters that follow, the mapping of fine-scale heterogeneity plays a critical role in understanding the selective pressures on organisms.

2.4 Conclusions

Since its derivation, the concept of an operative temperature has guided efforts to characterize thermal

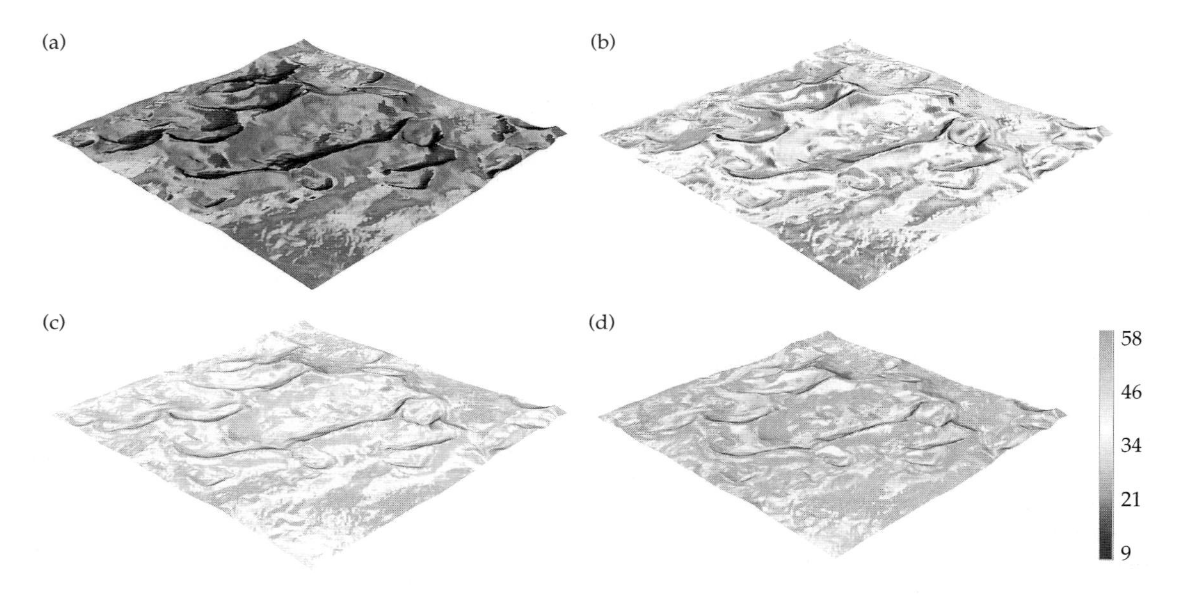

Figure 2.14 Operative temperatures available to small lizards in the Mescalaro sand dunes (New Mexico, USA) change dramatically throughout the morning on a summer's day (M. W. Sears, G. S. Bakken, M. J. Angilletta, and L. A. Fitzgerald, unpublished observations). (a–d) Each map depicts an operative thermal environment over an area of 460 × 320 m at a resolution of 0.5 m. (a) 7 a.m.; (b) 9 a.m.; (c) 11 a.m.; (d) 1 p.m. See plate 1.

environments from the organismal perspective. Spatial and temporal variations in solar radiation and air temperature give rise to changes in operative temperatures. All else being equal, operative temperatures decrease with increasing latitude or altitude. However, substantial heterogeneity of operative temperature exists within local environments, particularly terrestrial environments at higher latitudes. This local heterogeneity results from the physical structure of the habitat, as well as diel, seasonal, and annual changes in meteorological conditions.

Spatial and temporal variations in operative temperatures undoubtedly impose selective pressures on the thermal sensitivities, thermoregulatory strategies, and life histories of organisms. Both the magnitude and predictability of thermal change should determine the evolutionary impact. Diel and seasonal changes can be extreme, but they come with certainty. Stochastic variation among days and years complicates adaptation to predictable cycles. The relative magnitudes of predictable and unpredictable changes vary with latitude. Moreover, the size, mobility, and lifespan of an organism determine its perception of thermal heterogeneity. For example, a fruit fly experiences mostly diel change within its lifetime, but a Galapagos tortoise experiences diel, seasonal, and annual changes. As we shall see in the coming chapters, the course of thermal adaptation depends on the way each organism perceives the thermal heterogeneity of its environment.

CHAPTER 3

Thermal Sensitivity

3.1 Patterns of thermal sensitivity

Anyone familiar with physics and chemistry should be unimpressed by the discovery that life depends on temperature. Temperature constrains the rates of chemical reactions, and biochemical reactions are no exception (Hochochka and Somero 2002). These constraints lead to thermal sensitivities of function at cellular, systemic, and organismal levels (Rome et al. 1992). The thermal sensitivity of biochemical reactions places severe limits on life (Pörtner 2002). Prokaryotic cells function within the narrow range of −5°C to 110°C (Jaenicke 1991, 1993). Eukaryotic cells are even further restricted, tolerating temperatures between −2°C and 60°C (Tansey and Brock 1972). Some organisms can tolerate temperatures outside these ranges, but do not function normally during exposure to such extremes. Yet, each living organism comprises a unique set of reactions. Thus, while the chemistry of life restricts the thermal tolerances of organisms, no two organisms have exactly the same thermal restrictions.

By altering the structures of the reactants or even the nature of the reactions, species have evolved a rich diversity of thermal reaction norms. As an example, consider thermal reaction norms for fitness, which I refer to as *fitness curves*.[6] Barlow (1962) compared the fitness curves of two species of aphids raised in controlled environments (Fig. 3.1a). The fitness of *Myzus persicae* was greatest at 20°C and decreased sharply around this maximum. In contrast, the fitness of *Macrosiphum euphorbiae* was maximized over a wider range of temperatures (20–25°C). Relatively speaking, we could call *M. persicae* a thermal specialist and *M. euphorbiae* a thermal generalist. Similarly, fitness curves varied greatly among fives strains of rotifers (Fig. 3.1b). Differences among fitness curves have been observed for many kinds of organisms. Such observations demand explanation because natural selection should influence the thermal sensitivity of fitness more strongly than it does any phenotype.

Although differences among fitness curves may motivate studies of thermal adaptation, the concept of a fitness curve offers thermal biologists very little advantage for understanding most biological phenomena. First, reaction norms for fitness cannot be measured accurately for many organisms. Fitness is a function of lifetime reproductive success, which obviously cannot be measured more than once per individual. In the examples above, asexual reproduction by aphids and rotifers enabled researchers to estimate the fitness curves of clones. But clones cannot be obtained for the majority of organisms under study. Second, biologists cannot easily measure the fitness of organisms in ecologically relevant environments. In most environments, temperature varies temporally within generations, and we might need to incorporate this variation in the concept of a reaction norm to accurately predict fitness (Fig. 3.2). Moreover, nature contains a plethora of biotic and abiotic factors that one cannot replicate in a laboratory, and these variables can interact with temperature to determine components of fitness (Fig. 3.3). Consequently, we cannot

[6] I avoid the term *tolerance curve* (Levins 1968) because the literature contains conflicting uses of this term. Originally, Levins (1968) defined a tolerance curve as the relationship between an environmental variable and fitness. Since then, this term has also been used to describe the relationship between an environmental variable and survivorship (Gilchrist 1995). The word tolerance can also imply the critical thermal limits to performance (Pörtner 2002; van Berkum 1988).

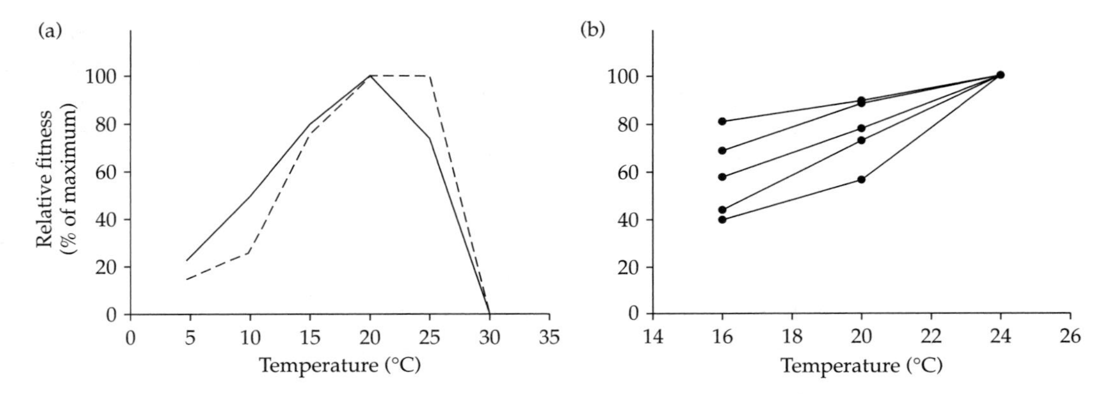

Figure 3.1 Fitness curves vary within and among species. (a) Between two species of aphids, *Macrosiphum euphorbiae* (dashed line) tolerated higher temperatures than did *Myzus persicae* (solid line). Data are from Barlow (1962). (b) The thermal sensitivity of fitness varied tremendously among strains of parthenogenetic rotifers. Data are from Ricci (1991). In both studies, fitness was defined as the Malthusian parameter, *r* (see Box 3.3).

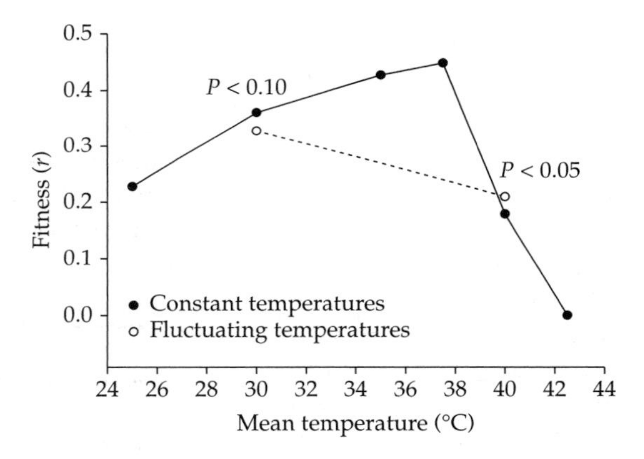

Figure 3.2 Thermal reaction norms for fitness sometimes differ when measured under constant and fluctuating temperatures. In this example, fitnesses measured at two constant temperatures deviated from fitnesses measured in fluctuating environments with the same mean temperatures. Fitness was defined as the Malthusian parameter, *r* (see Box 3.3). Data are from van Huis et al. (1994).

be certain that fitness curves expressed in a laboratory resemble those expressed in nature. Finally, and perhaps most importantly, we generally do not ponder questions about the evolution of fitness, per se. Rather, we wish to understand the evolution of phenotypes that presumably contribute to an organism's fitness.

In this chapter, I focus on thermal reaction norms for organismal performance, which are generally referred to as *thermal performance curves* (Huey and Stevenson 1979). I define performance as any measure of an organism's capacity to function, usually expressed as a rate or probability. Common measures of performance include locomotion, assimilation, growth, development, fecundity, and survivorship[7] (Table 3.1). Less common functions, such as sensory perception (Samietz et al. 2006; Werner 1973) and immunological activity (Hung et al. 1997; Mondal and Rai 2001), can also be considered performances. All of these functions share an important property: they respond rapidly (and usually reversibly) to changes in temperature. We can capture the general characteristics of this response with several parameters (Fig. 3.4): (1) the thermal optimum (T_{opt}); (2) the thermal breadth (or performance breadth); (3) the thermal limits, referred to as the critical thermal minimum (CT_{min}) and the critical thermal maximum (CT_{max}); and the maximal performance (P_{max}).

To study variation in performance curves, we must first be able to quantify them objectively. Several approaches have been used to quantify performance curves. First, a mathematical function can be chosen and its parameters can be estimated by

[7] Other authors have distinguished between curves for survivorship and curves for other performances. For example, Gilchrist (1995) referred to curves for survivorship as tolerance curves. Later, van der Have (2002) referred to the same concept as a viability curve. For simplicity, I chose to treat survivorship as I would any other measure of performance.

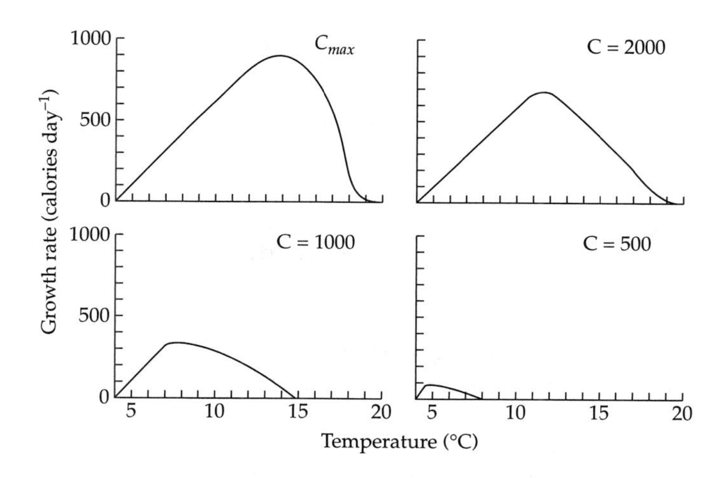

Figure 3.3 The amount of food available to a fish determines its thermal sensitivity of growth rate. Plots are shown for fish consuming 500, 1000, 2000, and maximal (C_{max}) calories per day. Reproduced from Elliot (1982) with permission from Blackwell Publishing.

Table 3.1 Common measures of organismal performance. Examples of each kind of performance reflect relevant measures for various taxa. The sources contain detailed descriptions of the procedures for measuring these performances.

Performance	Examples	Source
Locomotion	Voluntary speed	Gilchrist (1996); Heckrotte (1983)
	Maximal burst speed	Bennett (1980); McConnell and Richards (1955)
	Maximal jumping distance	Navas et al. (1999)
	Maximal sustainable speed	Scribner and Weatherhead (1995); Wilson and Franklin (2000a)
	Maximal distance (endurance)	Angilletta et al. (2002b); Full and Tullis (1990); Ojanguren and Braña (2000)
	Mechanosensory ability	Samietz et al. (2006)
Feeding/assimilation	Probability of capturing prey	Greenwald (1974); Van Damme et al. (1991)
	Ingestion of food/energy	Albentosa et al. (1994); Dutton et al. (1975); Warren and Davis (1967)
	Absorption of water/minerals	Howe and Marshall (2002); Kaufmann and Pough (1982)
	Photosynthesis	Kudo et al. (2000)
	Digestion/passage of food	Avery et al. (1993); Harwood (1979); Nicieza et al. (1994); Waldschmidt et al. (1986)
Growth	Change in length	Boddy (1983); Lebedeva and Gerasimova (1985); Leffler (1972)
	Change in mass/energy	Båmstedt et al. (1999); Brett et al. (1969)
Development	1/duration of developmental stage	Li (2002); Stanwell-Smith and Peck (1998)
	% per day	Lamb and Gerber (1985)
Reproduction	Egg production	Berger et al. (2008); Howe (1962)
	Mating success	Denoel et al. (2005); Wilson (2005)
	Oviposition rate	Carrière and Boivin (1997)
Survival	Probability of survival per unit time	Berrigan (2000); Leffler (1972)
	Time until 50% mortality	Selong et al. (2001)

fitting a statistical model to a set of data (Huey and Stevenson 1979). Choosing among potential models can be difficult, but tools for model selection can help when no a priori reason exists to choose a particular model (Box 3.1). Alternatively, others have adopted a simpler approach based on the concept of a minimum convex polygon (Bauwens et al. 1995; van Berkum 1986). Regardless of the approach, an objective description of the performance curve enables us to determine its key properties.

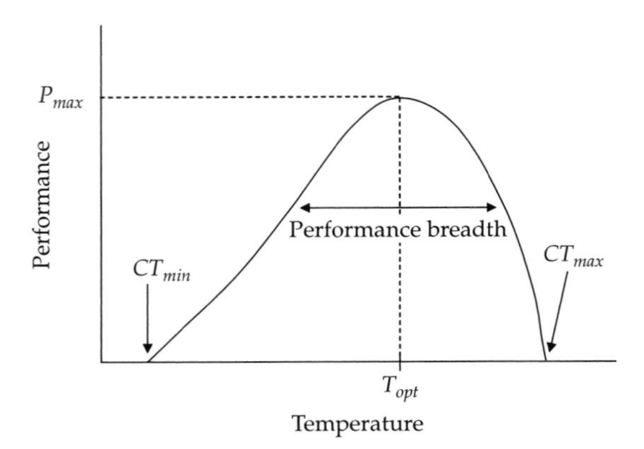

Figure 3.4 A hypothetical performance curve shows the stereotypical optimum at an intermediate temperature. The thermal optimum (T_{opt}), performance breadth, critical thermal limits (CT_{min} and CT_{max}), and maximal performance (P_{max}) are labeled. Adapted from Huey and Stevenson (1979) by permission from Oxford University Press.

Ascertaining the thermal limits of performance can be difficult, particularly when performance cannot easily be measured at extreme temperatures. Often, one must use an indirect means of bounding a performance curve. For example, curves for locomotor performance are generally bounded by some measure of immobility. For lizards, the inability of an individual to right itself has been used to estimate the critical thermal maximum. For fruitflies, the inability to cling to a surface has been used for the same purpose (Berrigan 2000; Huey et al. 1992). Lutterschmidt and Hutchison (1997a) argued that the temperature at which muscles begin to spasm provides the best measure of the critical thermal maximum. Whatever the measure, one should be aware of the assumptions associated with using indirect estimates of thermal limits to define performance curves (Lutterschmidt and Hutchison 1997 b).

Box 3.1 Selecting the best model of a thermal reaction norm

To compare performance curves, we must be able to describe them objectively. This description usually involves fitting a statistical model to empirical data. Several models have been favored by researchers, but how do we know the best model to describe any particular set of data? Instinctively, we might fit several models to the data and choose the model that describes the greatest amount of variation (i.e., the model with the highest r^2). Although this approach seems reasonable at first, further thought raises a serious issue. Literally thousands of models exist to choose from, and some of these models are so complex they would over-fit the data. In other words, a very complex model would not only describe the pattern of thermal sensitivity but also the error in the data (Burnham and Anderson 2002). Consequently, conclusions drawn from this model could be erroneous. What we desire is a model that describes the thermal sensitivity without fitting noise in the data.

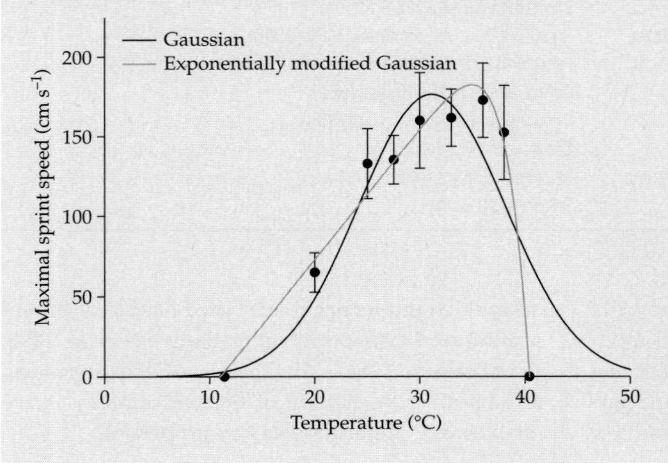

Figure 3.5 The relationship between temperature and sprint speed as described by two statistical models: a Gaussian model and an exponentially modified Gaussian model. A comparison of values of Akaike's information criterion indicated that the Gaussian model provides a better description of the data. Data are the mean sprint speeds of 12 lizards. Adapted from Angilletta (2006) with permission from Elsevier. © 2006 Elsevier Ltd.

continues

Box 3.1 Continued

Information theory provides a more sophisticated approach to describe performance curves (Angilletta 2006). One starts with a set of hypothetical models. These models can be chosen for their general form or derived from some mechanistic understanding (e.g., Schoolfield et al. 1981; Sharpe and DeMichele 1977). Then, one fits each model to the data, obtaining the estimated parameter values and the residual sum of squares. Finally, one calculates the Akaike information criterion (AIC) for each of the models (Burnham and Anderson, 2002):

$$\mathrm{AIC} = -2L + 2K + \frac{2K(K+1)}{N-K-1}, \qquad (3.1)$$

where L equals the maximized log-likelihood value of the model, K equals the number of parameters (including the error term), and N equals the sample size. The maximized log-likelihood value of a model can be computed easily from the model's residual sum of squares (Burnham and Anderson, 2002). The AIC estimates the information lost when using a particular model to describe the data. Therefore, the best the model in the set is the one with the lowest value of AIC.

The AIC differs from a simple measure of fit, such as the r^2, because it also depends on the model's complexity (i.e., the number of parameters). The second and third terms in eqn 3.1 eliminate the bias in fit associated with more complex models. By correcting this bias, we avoid choosing a model that overfits the data. To illustrate this point, consider two models that I used to describe the thermal sensitivity of locomotor performance (Fig. 3.5). Although a simple Gaussian model ($r^2 = 0.70$, $K = 4$, AIC $= 45.2$) describes much less variation than does an exponentially modified Gaussian model ($r^2 = 0.98$, $K = 6$, AIC $= 71.4$), information theory indicates the simple model provides a better description of thermal sensitivity (for details, see Angilletta 2006). This result should satisfy theorists, who often use a Gaussian function to study the evolution of performance curves (see Section 3.3). At the same time, this result should alarm empiricists, who often record performance curves that appear asymmetrical. Because the AIC depends on the sample size, a larger sample might have supported the more complex model. This example underscores the need to collect sufficient data to accurately describe the thermal sensitivity of performance.

Any survey of performance curves within or among species will reveal some variation in the thermal optimum, breadth, and limits of performance. Consider the performances curves shown in Fig. 3.6. For any type of performance, the shape or position of the curve varies among closely related organisms. For example, the lizard *Takydromus wolteri* sprints fastest at 30°C, but its congener *Takydromus sexlineatus* sprints fastest at 34°C (Fig. 3.6a). Likewise, two distinct types of performance need not be equally sensitive to changes in temperature; compared to survivorship, sprint speed proceeds well over a very narrow range of temperatures (Fig. 3.6f). Is such variation in performance curves adaptive? If so, how does this variation influence the ability of organisms to survive and reproduce in their environments?

Many species (or genera) span large portions of the earth. Obviously, species that extend from the tropics to the poles must be capable of tolerating thermal extremes. But how do these species accomplish this feat? Do individuals at the extremes of the range achieve lower fitness than individuals at the center of the range? Or have subpopulations adapted to thermal heterogeneity within the geographic range? Two distinct strategies would enable a species to occur over a broad geographic range. Populations at the extremes of the range could consist of specialists that perform best at their respective environmental temperatures. Alternatively, the global population could consist of generalists capable of functioning over a wide range of temperatures. Which of these strategies, if any, should we expect to evolve? And can this adaptation explain the diversity of performance curves observed among species? To answer these questions, we must consider the processes that define the thermal sensitivity of performance and the evolutionary costs and benefits of altering these processes.

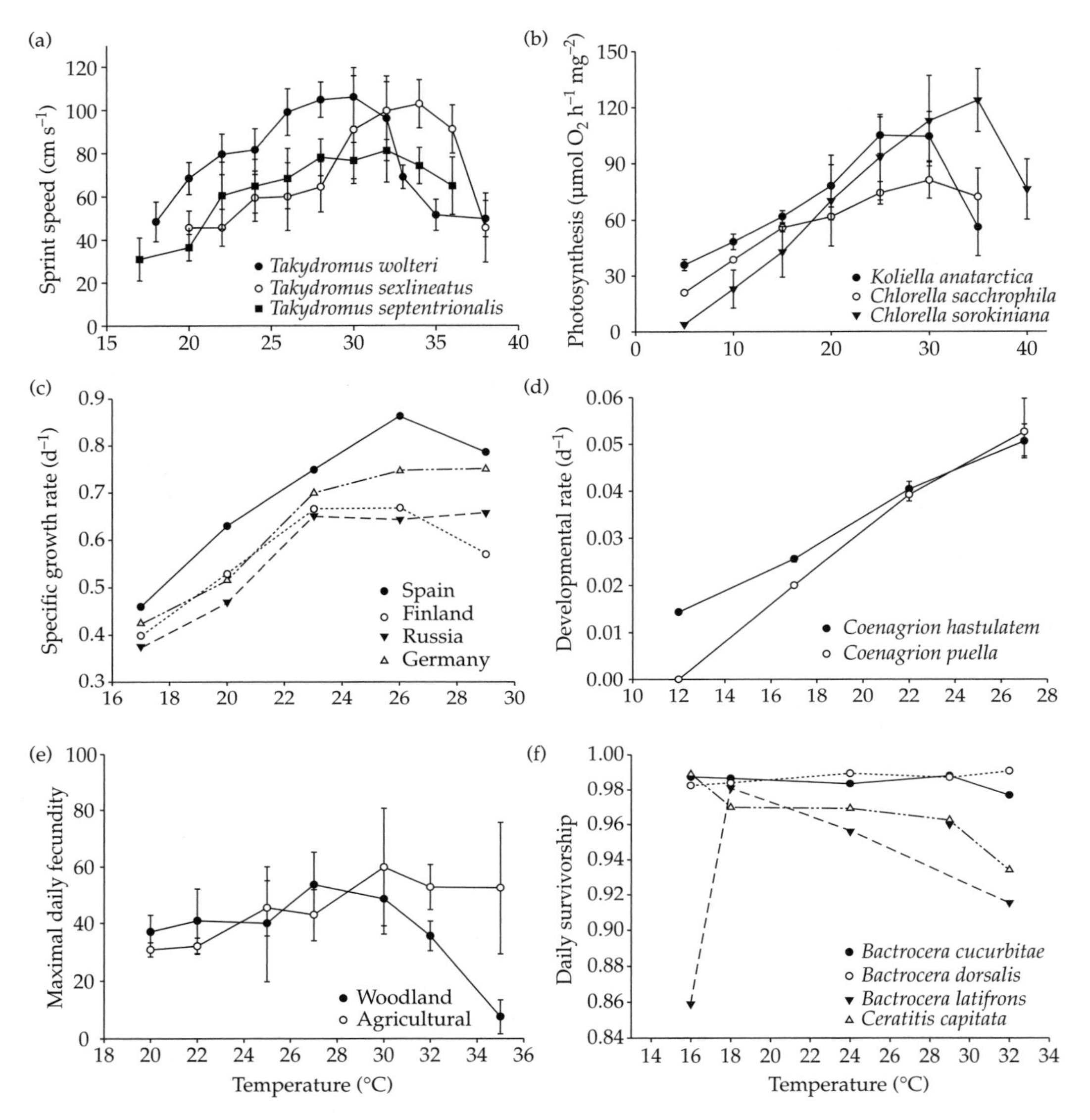

Figure 3.6 Thermal sensitivities of performance vary within and among species. Examples shown here represent common measures of organismal performance: (a) sprinting speeds in three species of lizards (Chen et al. 2003; Ji et al. 1996; Zhang and Ji 2004); (b) gross photosynthesis in three species of algae (Vona et al. 2004); (c) growth rates in four populations of *Daphnia magna* (Mitchell and Lampert 2000); (d) developmental rates (1/duration of larval stage) in two species of damselflies (van Doorslaer and Stoks 2005b); (e) fecundity in two populations of butterflies, *Pararge aegeria* (Karlsson and Van Dyck 2005); and (f) survivorship in four species of fruitflies (Vargas et al. 1996). Where shown, error bars are $\pm$ 2 SE. (e) Adapted with permission from the Royal Society.

3.2 Proximate mechanisms and tradeoffs

What proximate factors determine the shape of a performance curve? The precise answer to this question almost certainly depends on the performance in question. The asymmetrical performance curves for locomotion and growth differ markedly from the more symmetrical performance curves for survivorship (van der Have 2002). The answer also depends on the way in which we measure performance.

Chronic exposure to a temperature usually elicits a different level of performance than acute exposure (Christian et al. 1986; Kingsolver and Woods 1997). Nonetheless, for a particular measure of performance recorded with a particular methodology, we can ask which cellular and systemic factors dictate the thermal sensitivity. Do the same factors determine the thermal optimum and the critical thermal limits? Is the critical thermal minimum necessarily linked to the critical thermal maximum, or can these parameters evolve independently? Likewise, does physiology dictate a genetic correlation between the performance breadth and the maximal performance? As we shall see, modeling the evolution of thermal sensitivity demands tentative answers to these questions (see also Pigliucci 1996). Therefore, before we explore some evolutionary models, let us consider some hypothetical cellular mechanisms for the thermal sensitivity of performance.

3.2.1 Thermal effects on enzymes (and other proteins)

Hochachka and Somero (2002) brilliantly summarized the thermal constraints on the functions of enzymes (see also Fields 2001; Jaenicke 1991; Somero 1995). For any chemical reaction, temperature determines the proportion of reactants that possess the free energy required for reaction; as the temperature increases, more reactants exceed the energy of activation. Enzymes lower the energy required for activation and thus speed the reaction at any given temperature. The function of the enzyme requires both an initial conformation that can bind the substrate(s) and a change in this conformation that creates an ideal environment for the reaction. Because of the diversity of weak interactions that maintain the structures of proteins, either too high or too low a temperature will destroy the necessary conformation to bind substrate. Even a moderate drop in temperature alters the rate of conformational change, leading to inactivation at low temperatures well above that required to denature the enzyme. Thus, extreme temperatures impair enzymatic function and consequently slow biochemical reactions, leading to performance curves for specific enzymes that resemble those for whole organisms (Fig. 3.7).

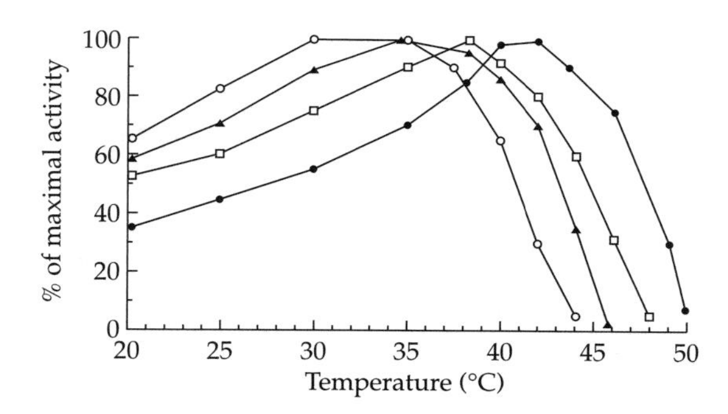

Figure 3.7 Thermal sensitivities of the activity of adenosine triphosphatase (ATPase) resemble sensitivities of organismal performance. Data shown here are for four species of lizards: *Dipsosaurus dorsalis* (solid circles), *Gerrhonotus multicarinatus* (open circles), *Sceloporus undulatus* (triangles), and *Uma notata* (squares). From Licht (1967). Reprinted with permission from AAAS.

The substitution of amino acids can alter the thermal properties of proteins (Fields 2001; Jaenicke 1991). Because the active site of an enzyme requires very specific amino acids to promote covalent reactions, this region generally remains conserved during evolution. Instead, natural selection favors mutations that alter the conformational stability of the enzyme (Hochochka and Somero 2002; Marx et al. 2007). A more flexible structure helps an enzyme to change shape faster during catalysis. However, this instability of conformation comes at a price; a flexible enzyme spends a greater fraction of its time in a conformation that prevents the binding of substrate (Fig. 3.8). Hence, a flexible enzyme could have a lower affinity for its substrate. In general, an enzyme with greater conformational stability functions better at high temperatures, and an enzyme with less conformational stability functions better at low temperatures (Fields 2001; Somero 1995).

Two kinds of enzymatic variants can underlie the diversity of performance curves (Somero 1995). A mutation that alters the structure of an enzyme can produce an orthologous allozyme that enhances performance at some temperatures but reduces performance at others. Duplication of genes can lead to the evolution of paralogous isozymes, which

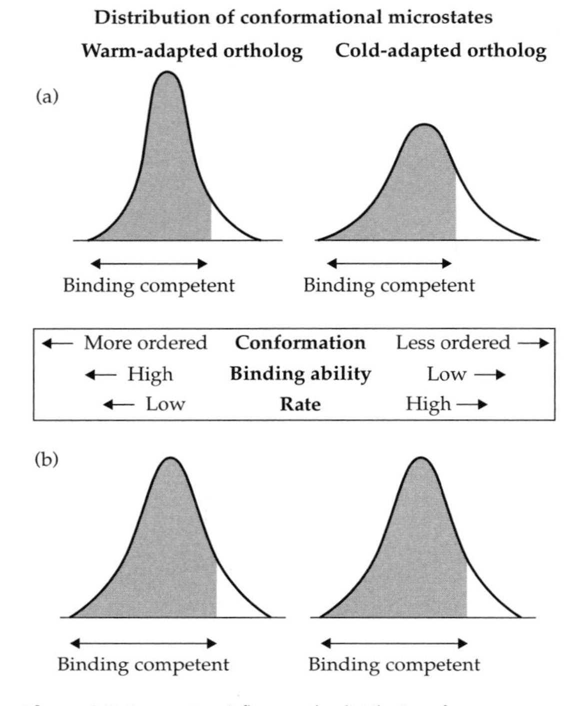

Figure 3.8 Temperature influences the distribution of conformational microstates for orthologs of a protein from warm- and cold-adapted species. (a) At a given temperature, a cold-adapted ortholog possesses a broader distribution of conformational microstates because of its flexible structure. For this ortholog, a larger fraction of molecules exists in conformations that bind ligands poorly (or not at all). However, these less ordered conformations reduce energy barriers for catalytic conformational changes and, therefore, speed enzymatic activity. (b) At their respective environmental temperatures, the intrinsic difference in stability between warm- and cold-adapted orthologs leads to similar distributions of conformational microstates for both orthologs. Adapted from Somero (2003) with permission from Elsevier. © 2003 Elsevier Ltd.

would enable an organism to operate over a wider range of temperatures. First, two identical copies of the gene would exist, and later one or both of these copies could evolve to alter the ancestral function of the enzyme. If one of these isozymes evolved greater or lesser flexibility, the pair would complement each other in a variable environment. However, unless additional resources are used to synthesize isozymes, an organism will still be unable to increase its performance in one range of temperatures without decreasing its performance in another.

Based on correlative evidence, we might conclude that performance curves respond to evolutionary changes in the structures of enzymes (Mitten 1997). For example, orthologous allozymes for lactate dehydrogenase, *LDH-B*^{a} and *LDH-B*^{b}, enable thermal specialization of swimming endurance in the common killifish, *Fundulus heteroclitus* (reviewed by Powers and Schulte 1998). The catalytic efficiency of *LDH-B*^{b} exceeds that of *LDH-B*^{a} at temperatures below 25°C, while the opposite is true at temperatures above 25°C. The function of *LDH-B* influences the level of ATP in erythrocytes, which regulates the affinity of hemoglobin for oxygen. Accordingly, homozygotes for *LDH-B*^{b} sustained faster swimming speeds at 10°C than did homozygotes for *LDH-B*^{a}, but these genotypes swam equally fast at 25°C. At high temperatures, I presume the swimming performance of homozygotes for *LDH-B*^{a} would exceed that of homozygotes for *LDH-B*^{b}, but this comparison has not been made. Consistent with this presumption, *LDH-B*^{a} occurs more frequently in populations that inhabit warm environments than it does in populations that inhabit cold environments. Importantly, neither genotype outperforms the other at all temperatures.

Similarly, variation in the performance curves for growth rate among genotypes of Atlantic salmon (*Salmo salar*) reflects the synthesis of paralogous isozymes (Rungruangsak-Torrissen et al. 1998). A common form of trypsin, TRP-2*100, functions at high temperatures (> 10°C), whereas the variant TRP-2*92 functions at low temperatures (< 9°C). Salmon that synthesized both forms grew fairly well in the range of 4–20°C. A second variant, TRP-1*91, functions over a wide range of intermediate temperatures, such that individuals synthesizing both TRP-2*100 and TRP-1*91 grew faster in the range of 6–17°C than did individuals synthesizing both TRP-2*100 and TRP-2*92. These two patterns of isozyme expression represent thermal specialization and thermal generalization, respectively. Because individuals within populations of salmon differ in the number and concentration of isozymes synthesized under controlled thermal conditions (Rungruangsak-Torrissen et al. 1998), we could hypothesize that performance curves evolved by selection for the synthesis of particular isozymes.

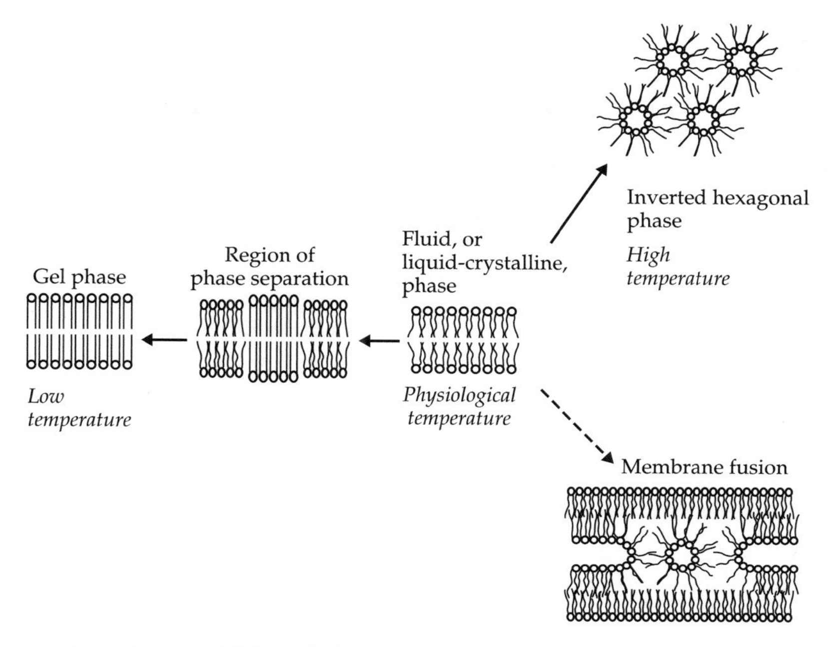

Figure 3.9 The structure of a membrane can shift from a fluid phase at tolerable temperatures (center) to either a gel phase at lower temperatures (left) or an inverted hexagonal phase at higher temperatures (upper right). The last phase ultimately results in fusion of the phospholipid bilayer (lower right). Reprinted, with permission, from the Annual Review of Physiology (Hazel 1995), Volume 57. © 1995 by Annual Reviews (www.annualreviews.org).

Although the evolution of isozymes appears less common than the evolution of allozymes (Fields 2001), both types of biochemical evolution could alter performance curves.

3.2.2 Membrane structure

Temperature also determines the movement and conformation of cellular membranes (Fig. 3.9; Hazel and Williams 1990). On one hand, relatively low temperatures slow the movements of phospholipids, resulting in a gel-like membrane. On the other hand, relatively high temperatures speed these movements too much, disrupting the laminar structure. Passive regulation, active transport, and enzymatic function within membranes depend on a semifluid structure. Therefore, membranes perform the same balancing act between flexibility and stability performed by enzymes. Because of this balancing act, physiological functions involving membranes proceed best within a moderate range of temperatures (reviewed by Hazel 1995).

Although most organisms can maintain the full integrity of their cellular membranes over a breadth of 10–15°C, the optimal temperature depends on the composition of their phospholipids (Hazel 1995; Hazel and Williams 1990). Saturated fatty acids provide less intrinsic fluidity than do unsaturated fatty acids. Consequently, membranes composed of primarily saturated fatty acids function better at high temperatures than membranes composed of primarily unsaturated fatty acids. An organism can produce membranes suited for cold or warm environments by using the appropriate mixture of fatty acids (Fig. 3.10). Still, adaptation to one thermal extreme necessarily leads to maladaptation to the other extreme.

Do thermal effects on membranes limit organismal performance less than or more than thermal effects on enzymes? Although this question seems reasonable, we would have difficulty separating

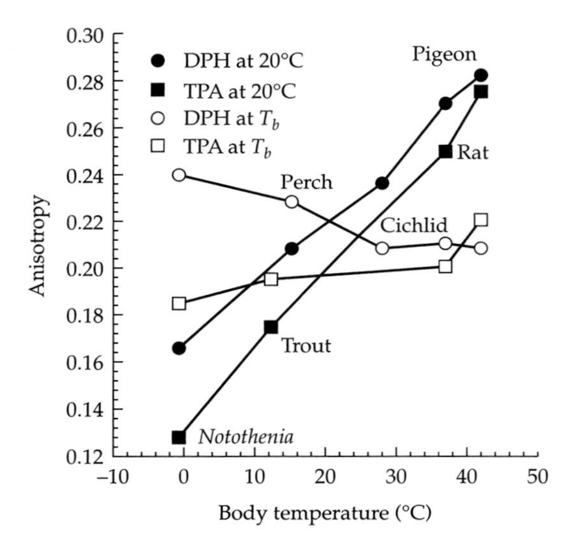

Figure 3.10 Membranes of cold-adapted species (e.g., an Antarctic fish, *Notothenia neglecta*) are far more fluid (i.e., less ordered) than membranes of warm-adapted species (e.g., the rat or the feral pigeon). Species adapted to intermediate temperatures (perch, trout, and cichlid) possess intermediate levels of membrane fluidity. Membrane fluidity was assessed using fluorescent probes: 1,6-diphenyl-1,3,5-hexatriene (DPH) and *trans*-parinaric acid (TPA). A high level of anisotropy reflects a low level of membrane fluidity. Reprinted, with permission, from the Annual Review of Physiology (Hazel 1995), Volume 57. © 1995 by Annual Reviews (www.annualreviews.org).

the functions of membranes and enzymes in real cells. By disturbing the fluidity of a membrane, extreme temperatures likely slow the function of proteins embedded within the phospholipid bilayer (Else and Wu 1999; Wu et al. 2001, 2004). Such disturbances can also affect the function of intracellular enzymes that rely on substrates transported through the membrane. Finally, the permeability of membranes partly determines the cellular milieu in which enzymes must function (Somero 2003). Very likely, the interaction between membranes and enzymes limits the performance of organisms (see, for example, Almansa et al. 2003). With this in mind, we shall now consider a model of thermal tolerance that involves both enzymes and membranes, particularly those associated with aerobic metabolism.

3.2.3 Oxygen limitation

Recently, Pörtner and colleagues (2000) proposed a model of thermal tolerance based on oxygen limitation. This model posits that the thermal limits of performance are set by the temperatures at which aerobic respiration fails to meet energetic needs (Fig. 3.11). At low temperatures, organisms become incapacitated by the inability of mitochondria to generate the ATP required for activity. At high temperatures, ventilation and circulation fall below the level required to supply the mitochondria with sufficient oxygen. This model draws support from the close match between the temperatures that cause death and the temperatures at which organisms shift from aerobic to anaerobic respiration (but see Klok et al. 2004; reviewed by Pörtner 2001).

If this model correctly identifies the cause of critical thermal limits, adaptation to thermal extremes will require changes in mitochondrial, circulatory, and respiratory capacities. A greater density or capacity of mitochondria could lower the critical thermal minimum of an organism (Guderley 2004; Pörtner et al. 2000). Still, such adaptations of metabolic systems would impose a tradeoff. The current view assumes that adaptation to cold environments causes maladaptation to warm environments. This hypothetical tradeoff results from an increase in proton leaking and futile cycling within mitochondria, which increase energy consumption at rest (Pörtner et al. 1998, 2000). The higher resting rate of energy consumption will cause a parallel shift in the critical thermal maximum. This model also predicts a potential tradeoff between the performance breadth and the maximal performance. Although an increase in mitochondrial density would raise metabolic demands at rest, organisms can eliminate this cost by reducing the leakiness of membranes (Pörtner et al. 1998) or the activity of enzymes (Pörtner et al. 2000). Either of these strategies would limit aerobic scope, potentially causing a decrease in the capacity for performance. Thus, the model of Pörtner and colleagues links the thermal limits, performance breadth, and maximal performance in ways that resemble the tradeoffs associated with enzymatic function (see Section 3.2.1).

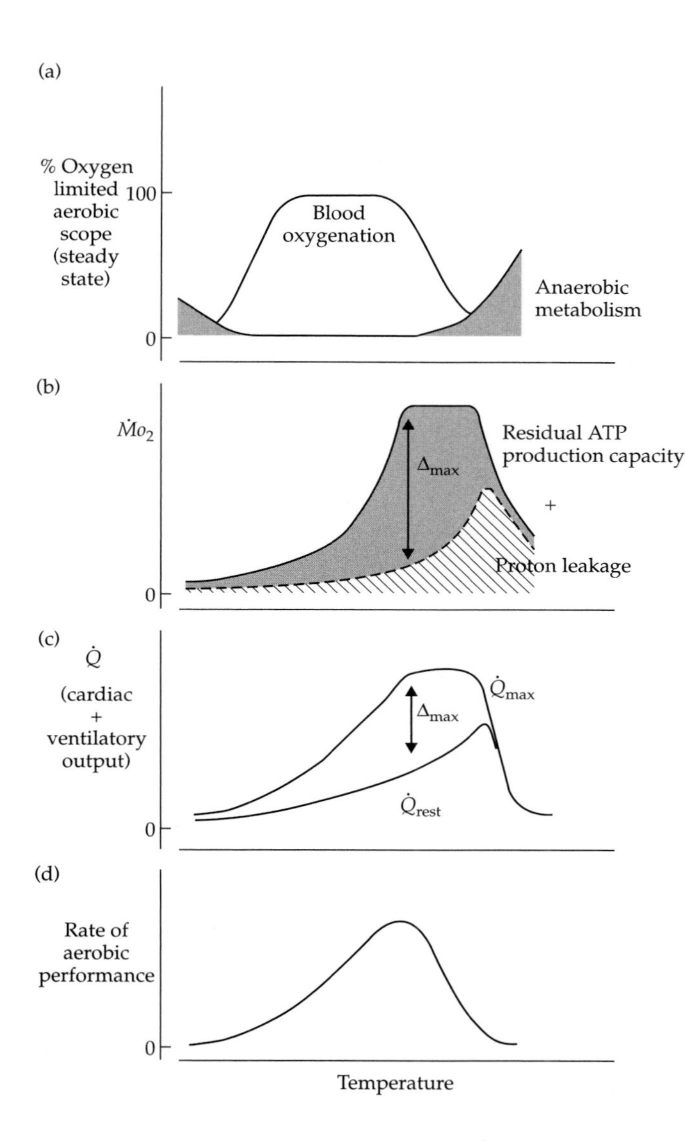

Figure 3.11 A model of thermal tolerance based on limits to aerobic capacity. The mismatch between oxygen demand and supply sets the window of thermal tolerance of metazoans. (a) The onset of anaerobic metabolism sets the critical thermal maximum. (b) At both low and high temperatures, the mitochondria cannot supply sufficient ATP. At low temperatures, the inactivation of mitochondrial enzymes and membranes prevents the production of ATP. At high temperatures, the rising fraction of mitochondrial proton leakage produces a mismatch between the supply and demand of oxygen. (c) Accordingly, the thermal sensitivity of ventilation and circulation, especially the difference between maximal and resting outputs, characterizes limits to oxygen supply and aerobic scope. (d) Thermal sensitivities of mitochondrial, ventilatory, and circulatory functions combine to generate an asymmetric performance curve, with a thermal optimum for performance near the temperature that maximizes aerobic scope. Adapted from Pörtner et al. (2006, *Physiological and Biochemical Zoology*, University of Chicago Press). © 2006 by The University of Chicago.

3.2.4 Conclusions from considering proximate mechanisms

Of the biochemical and physiological processes discussed in this section, which one really limits performance at the organismal level? The answer likely depends on the temporal scale of thermal heterogeneity. Fields (2001) noted that the temperature required to denature a protein can exceed the critical thermal maximum of the organism. Based on this point and other evidence, Pörtner concluded that oxygen limitation offers the most plausible explanation for the thermal tolerances of organisms (Pörtner 2001, 2002). This conclusion may be valid for stable environments where organisms must grow and reproduce during persistent exposure to extreme temperatures. In heterogeneous environments, however, organisms could use anaerobic respiration to endure transient exposures to temperatures outside the window for aerobic scope. Other lines of evidence suggest the function of enzymes as a limiting factor for performance. The induction of heat-shock proteins under thermal stress implies that some loss of enzymatic function could occur within the natural range of temperatures, which presumably exceeds the critical temperature for aerobic scope. van der Have (2002) found a striking match between the predicted thermal limit of enzymatic activity and the observed thermal limit of embryonic survival (Box 3.2). Still, I believe that many factors contribute to the decline in organismal performance at extreme temperatures,

Box 3.2 Proximate basis of thermal sensitivity of development

Thermal performance curves for survivorship during embryonic or larval stages possess a shape that differs from a typical performance curve (van der Have 2002). Instead of being asymmetrical, as in Fig. 3.4, these curves tend to be more symmetrical with very sharp transitions between the performance breadth and the thermal limits (see Fig. 3.6f). According to van der Have, this relationship stems from the reversible inactivation of proteins that drive cellular division. The gradual inactivation of enzymes with increasing or decreasing temperature can result in a thermal threshold for transcription (Fig. 3.12). Undoubtedly, cellular division and differentiation during development requires transcription. If proteins involved in transcription cannot maintain their functional conformation, the embryo or larvae would ultimately perish.

This argument was formalized in a model describing the influence of a single rate-limiting enzyme (or group of enzymes) on development. The upper and lower temperatures at which transcriptional rates should drop were calculated from the Sharpe–Schoolfield model:

$$r(T) = \frac{\rho T P_a}{298.2} e^{\left(\frac{H_A}{R}\left(\frac{1}{298.2} - \frac{1}{T}\right)\right)} \tag{3.2}$$

where r equals the rate of development, T equals temperature in K, R equals the universal gas constant, 298.2 equals the temperature at which no inactivation occurs, H_A equals the enthalpy of activation, P_a equals the probability that the enzyme will be in the active state, and ρ represents a fitted parameter (Schoolfield et al. 1981; Sharpe and DeMichele 1977). Using developmental rates for 14 species, van der Have estimated the temperatures at which activation and inactivation were equally likely (i.e., $P_a = 0.5$). Based on his model of the cell cycle, these temperatures would shut down transcription within cells. A striking match exists between the theoretical limits for development and the empirical limits for survivorship (Fig. 3.13).

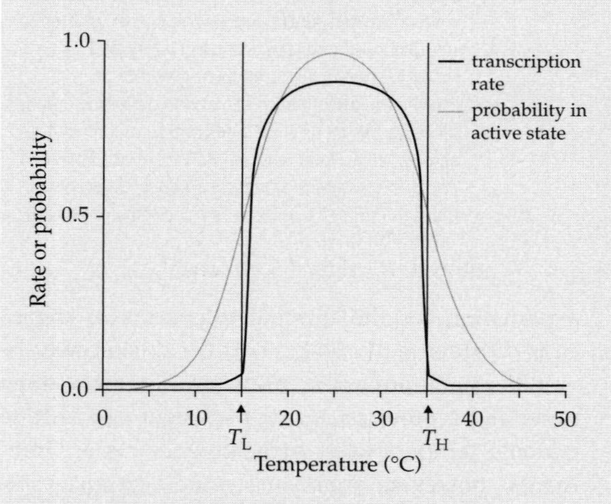

Figure 3.12 Temperature affects the probability that an inducer activates the transcription of an operon. The rate of transcription drops sharply as the temperature approaches the point at which the transducer is equally likely to be active or inactive. Adapted from van der Have (2002) with permission from Blackwell Publishing.

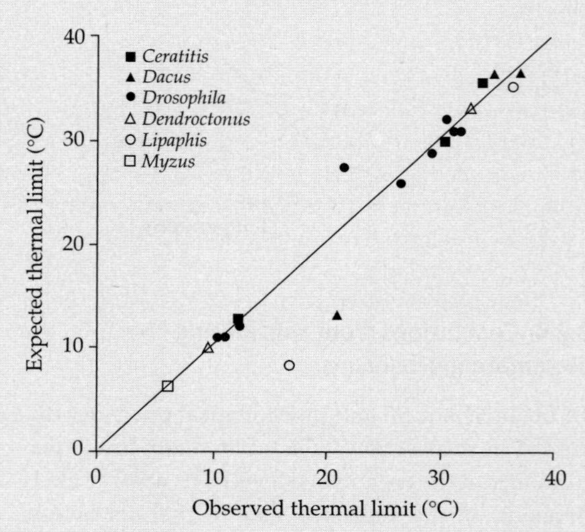

Figure 3.13 An excellent match exists between the observed thermal limits to survivorship and the expected thermal limits of development. Data are for 14 species of insects from six genera. A reference line of $y = x$ is shown. Adapted from van der Have (2002) with permission from Blackwell Publishing.

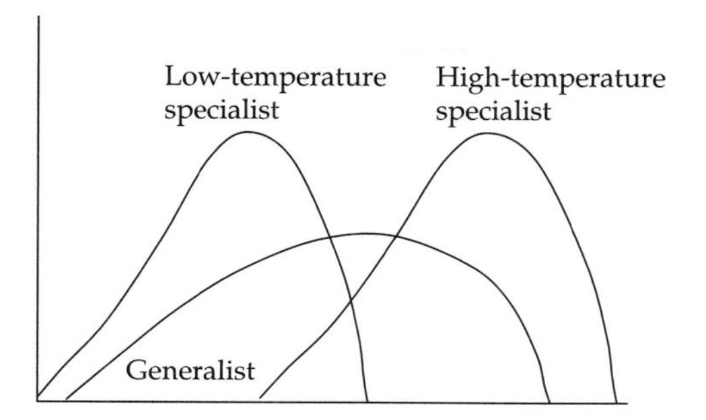

Figure 3.14 Specialist–generalist tradeoffs constrain variation in performance curves. An improvement in performance at low temperature requires a decrement in performance at high temperature, and vice versa. Thus, a specialist for low or high temperature would achieve greater maximal performance than a generalist.

especially given the coadaptation that occurs among different elements of thermal physiology (a topic explored in Chapter 7). Overall, the thermal limits to performance probably depend on a complex interplay between biochemical, cellular, and systemic designs.

Regardless of the primary cause of thermal sensitivity, we can draw at least one conclusion about the potential constraints on performance curves. Most of the cellular mechanisms described above involve a tradeoff between performance at high temperatures and performance at low temperatures. For example, adaptations of enzymes and membranes to low temperatures generally result in maladaptations to high temperatures. Adaptations that enhance the aerobic capacity at low temperatures create an energetic sink at high temperatures. These biochemical constraints should give rise to genetic correlations between parameters of a performance curve. First, a genotype that performs well in cold environments should perform poorly in warm environments. In the context of a performance curve, a shift in the critical thermal minimum should require a parallel shift in the critical thermal maximum. Second, a genotype that functions over a broad range of temperatures (e.g., through the expression of multiple isozymes) should be a jack of all temperatures and a master of none (Huey and Hertz 1984). In other words, an increase in the performance breadth should cause a decrease in the maximal performance. We can classify these predicted correlations as specialist–generalist tradeoffs (see Chapter 1), because an evolutionary increment in performance at one temperature causes an evolutionary decrement in performance at another temperature (Fig. 3.14). Given the cellular mechanisms described above, specialist–generalist tradeoffs should impose ubiquitous constraints on evolution.

3.3 Optimal performance curves

To understand the tremendous variation in performance curves, let us first turn to optimality models. In particular, we shall focus on two distinct models designed to predict the relationship between a performance curve and fitness. Each model predicts the thermal optimum and performance breadth that maximize fitness in a given environment, subject to certain constraints. As we shall see, these models differ in fundamental ways, including the manner in which performance contributes to fitness (Box 3.3). The first model considers the thermal sensitivity of survivorship, or other performances that relate directly to survivorship. The second model considers the thermal sensitivity of fecundity, or other performances that relate directly to fecundity. Nevertheless, both models share many key features, including a strict constraint imposed by specialist–generalist tradeoffs. By examining these models in turn, we shall see how slightly different sets of assumptions lead to very different predictions regarding the optimal performance curve.

3.3.1 Optimal performance curves: survivorship and related performances

Building on the seminal work of Levins (1968), Lynch and Gabriel (1987) modeled the evolution of performance curves for survivorship (i.e., the probability of survival during a unit of time).[8] Their

[8] Lynch and Gabriel envisioned their focus to be the evolution of fitness curves, which they termed tolerance curves. I avoid their interpretation and their vocabulary for good reason. They considered how performance responds to thermal heterogeneity within the life of an individual. Yet fitness is a property of an individual that should be characterized by a single value. Here, I reinterpret their model to understand the evolution of thermal sensitivities of survivorship because this interpretation seems equally valid and more intuitive.

Box 3.3 Performance and fitness

How will performance curves evolve in variable environments? To answer this question, we must understand how the performance of interest relates to fitness.

We might define fitness as the number of offspring produced by an individual during its life. But when estimating the mean fitness of a genotype (or phenotype), we cannot simply record the fecundity of individuals because many individuals will not live to reproduce at all. Therefore, we must also consider the probability that individuals will survive to reproduce.

For practical purposes, we can define fitness as the rate at which a genotype increases within a population. This measure of fitness depends on two demographic variables: age-specific survivorship and age-specific fecundity (Stearns 1992). These variables contribute differently to the fitness of a genotype in stable and growing populations (Kozłowski 1992). In stable populations, fitness equals the net reproductive rate (R_0):

$$R_0 = \int_{x=\alpha}^{\infty} l(x)m(x)dx, \quad (3.3)$$

where $l(x)$ equals survivorship at age x, $m(x)$ equals fecundity at age x, and α is the age at maturity. In a growing population, one must estimate fitness another way because offspring produced early in life contribute more to the increase of a genotype than do offspring produced late in life. For a population with a stable age distribution, fitness equals the Malthusian parameter (r), which we can estimate with Euler's equation:

$$1 = \int_{x=\alpha}^{\infty} e^{-rx} l(x)m(x)\,dx. \quad (3.4)$$

Note that eqn 3.3 defines a special case of eqn 3.4, in which the rate of population growth equals zero (i.e., $r = 0$). Because temperature affects r and R_0 differently (Huey and Berrigan 2001), we need to know the appropriate measure of fitness for each population under investigation. Because many populations remain stable over long periods (Sibly et al. 2007), some modelers feel that R_0 represents a superior measure of fitness (see Kozłowski et al. 2004).

Regardless of the measure we choose, the fitness function reveals an important point about the relationship between performance and fitness. Performance can determine either the survivorship or the fecundity of a genotype at age x. Because $l(x)$ equals the product of survivorship from birth to age x, performances that affect survivorship contribute multiplicatively to the fitness of a genotype. But performances that affect fecundity contribute additively to the fitness of a genotype. As we shall see in this chapter, the distinction between multiplicative and additive contributions to fitness plays a critical role in understanding the evolution of performance curves (see also Gilchrist and Kingsolver 2001).

model makes predictions for an asexual organism with discrete generations, but the main predictions also hold for a sexual organism (Gabriel 1988). This model can also help us to understand the thermal sensitivity of any other performance that directly affects survivorship (e.g., sprint speed).

The model was deigned to determine the optimal performance curve in an environment that varies stochastically over space and time. For simplicity, the body temperature of the organism was assumed to equal its operative environmental temperature. The operative temperature was described by a Gaussian probability distribution. Hence, the organism experienced the mean temperature most often and experienced extremely high and low temperatures rarely. To understand how temporal and spatial sources of variation were combined, imagine a genotype develops, grows, and reproduces in a particular subset of its environment; let us call this subset the microenvironment. This microenvironment has a mean (ϕ_s) and a variance (V_{tw}) of temperature. Assuming microenvironments vary, the mean temperature of each microenvironment, ϕ_s, is drawn from a distribution for the macroenvironment, which has its own mean (ϕ_t) and variance (V_{ϕ^s}) of temperature. Finally, the distribution of temperatures for the macroenvironment, ϕ_t, can change over time. Consequently, two kinds of temporal heterogeneity must be considered:

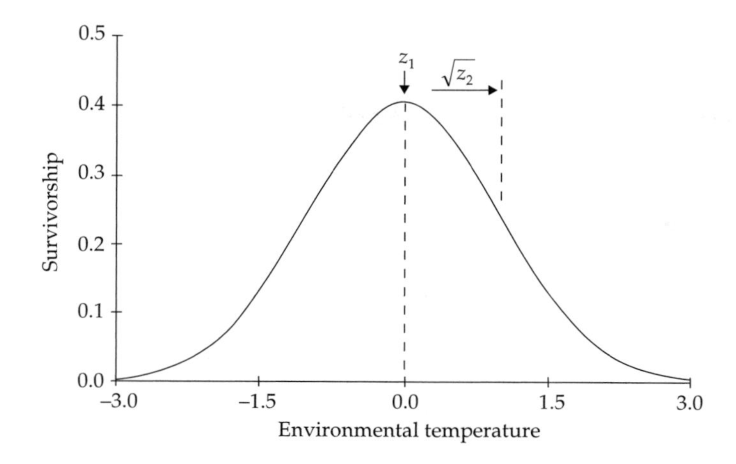

Figure 3.15 Lynch and Gabriel used a Gaussian performance curve to model the evolution of thermal sensitivity. This curve was determined by two parameters: a thermal optimum (z_1) and a performance breadth ($\sqrt{z_2}$). Adapted from Gabriel and Lynch (1992) with permission from Blackwell Publishing.

variance within generations ($V_{\phi^{tb}}$) and variance among generations ($V_{\phi_{tb}}$). As we shall see, these sources of variation do not equally affect the optimal performance curve.

Lynch and Gabriel modeled the performance curve as a Gaussian function (Fig. 3.15):

$$P(z_1, z_2|\phi) = (2\pi z_2)^{-0.5} e^{[-(z_1-\phi)^2/2z_2]}, \quad (3.5)$$

where P equals the performance (i.e., survivorship), z_1 equals the thermal optimum, and z_2 determines the performance breadth. Because this function describes a probability density, the integral (or area under the curve) remains the same for all values of z_1 and z_2. In other words, this function conveniently imposes a specialist–generalist tradeoff because an increase in performance at one temperature necessarily reduces performance at another.

The values of z_1 and z_2 of an individual were determined by genotypic and environmental factors:

$$z_1 = g_1 + e_1 \quad (3.6)$$

and

$$z_2 = g_2 + e_2. \quad (3.7)$$

In these equations, g_1 and g_2 represent genetic determinants of the thermal optimum and performance breadth, respectively, whereas and e_1 and e_2 represent deviations from the genetic values caused by the developmental environment. This developmental noise affects the optimal performance curve very little, but adaptation would proceed more slowly when either e_1 or e_2 differed from zero (Lynch and Gabriel 1987).

The fitness of a genotype, $w(g_1, g_2)$, depends on two things. First, we must consider its fitness in any particular microenvironment (ϕ_s), subject to the temporal variance of temperature ($V_{\phi^{tw}}$). Then, we must consider the probability that a genotype will experience that particular microenvironment, given the spatial variance of temperature (V_{ϕ^s}). Because the product of survivorships at all time intervals defines the survivorship to reproduction, fitness in a particular microenvironment must be calculated as the geometric mean of performance over the duration of a generation (τ):

$$\left[\prod_{i=1}^{\tau} P(g_1, g_2|\phi_s)_i\right]^{1/\tau}. \quad (3.8)$$

The probability of encountering a particular microenvironment was defined as:

$$f(\phi_s|\phi_t) = (2\pi V_{\phi_s})^{-0.5} e^{[-(\phi_s-\phi_t)^2/2V_{\phi_s}]}, \quad (3.9)$$

where $f(\phi_s|\phi_t)$ equals the probability of ϕ_s given ϕ_t. By combining Eqns. 3.8 and 3.9, Lynch and Gabriel defined the expected fitness of a genotype, $w(g_1, g_2|\phi_t)$, given the spatial and temporal

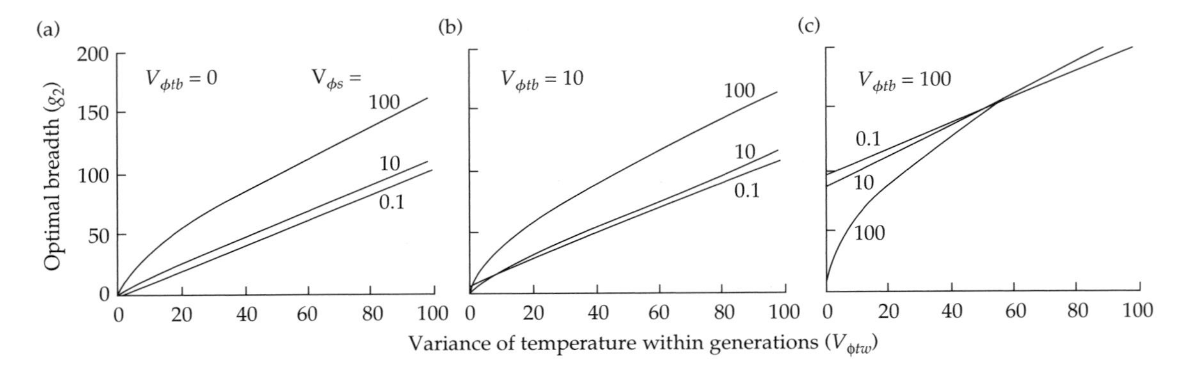

Figure 3.16 Both spatial and temporal variations in environmental temperature affect the optimal performance breadth. (a–c) Each plot shows the effect of temporal variation within generations ($V_{\phi tw}$) for different levels of spatial variation ($V_{\phi s} = 0.1$, 10, or 100); (b The three plots differ in the level of temporal variation among generations ($V_{\phi tb} = 0$, 10, or 100). For these analyses, Lynch and Gabriel assumed the existence of developmental noise in the thermal optimum, but no developmental noise in the thermal breadth. Adapted from Lynch and Gabriel (1987, *The American Naturalist*, University of Chicago Press). © 1987 by The University of Chicago.

heterogeneity of temperature:

$$w(g_1, g_2|\phi_t) = \int_{-\infty}^{+\infty} f(\phi_s|\phi_t) \left[\prod_{i=1}^{\tau} P(g_1, g_2|\phi_s)_i \right]^{1/\tau} d\phi_s, \tag{3.10}$$

In biological terms, fitness equals the probability of surviving until reproduction at age τ. Lynch and Gabriel implicitly assumed that survivorship has no influence on fecundity (i.e., all genotypes produce the same number of offspring). Therefore, the genotype with the greatest probability of surviving until τ could expect the greatest fitness.

Using this fitness function, Lynch and Gabriel calculated the optimal values of g_1 and g_2 for a combination of spatial and temporal heterogeneity in a single generation (Fig. 3.16a). Obviously, the optimal performance curve should be centered on the mean temperature (i.e., $g_1 = \phi_t$). But does spatial and temporal variation affect the optimal breadth of the performance curve? Intuitively, the optimal performance breadth (g_2) increases with increasing thermal heterogeneity. Two other results seem less intuitive. First, developmental noise in the thermal optimum favors a reduction in the performance breadth. This result occurs because developmental noise creates phenotypes with different thermal optima, which reduces the need for a genotype to have an inherently wide performance breadth. Second, spatial and temporal heterogeneity interact to determine the optimal genotype; at high levels of spatial variation, even small levels of temporal variation favor moderate performance breadths (Fig. 3.16a).

Lynch and Gabriel went on to consider the long-term fitness of genotypes with different performances curves. In doing so, they considered the effects of thermal heterogeneity within and among generations on the optimal performance curve. Variation among generations was modeled such that ϕ_t at each generation was drawn from a Gaussian distribution with a variance of V_{tb}. In this case, the fitness of a genotype corresponds to the geometric mean of performance within and among generations (n):

$$w(g_1, g_2) = \lim_{n \to \infty} \left[\prod_{i=1}^{n} w(g_1, g_2|\phi_t) \right]^{1/n} \tag{3.11}$$

By substituting eqn 3.10 into eqn 3.11, Lynch and Gabriel extended their calculation of the optimal performance curve from a single generation to infinitely many generations. Importantly, extreme variation among generations always favors a greater performance breadth than would be favored when the mean temperature remains the same among generations (Fig. 3.16b and c). Especially wide performance breadths are favored when the temperature within generations varies little over space or time. Despite these results, variation within generations has a greater impact on the optimal performance breadth than does variation among generations.

In summary, we can derive three testable predictions from Lynch and Gabriel's model. First, the performance breadth for survivorship should be roughly proportional to the thermal heterogeneity experienced during an organism's life. Second, the widest performance breadths for survivorship should evolve in spatially homogenous environments where temperature varies both within and among generations. Finally, both developmental noise and spatial variation prolong the time required for the optimal performance curve to evolve.

3.3.2 Optimal performance curves: fecundity and related performances

Like Lynch and Gabriel, Gilchrist (1995) modeled the optimal performance curve of an asexual organism with discrete generations. But Gilchrist chose a fitness function that differed fundamentally from the fitness function described in the preceding section. In the model of survivorship, fitness was calculated as the geometric mean of performance over the life of the organism. In Gilchrist's model, fitness was calculated as the sum of performance over the life of an organism. Arguably, performances such as growth and developmental rates contribute to fitness in this additive fashion. If so, we can use Gilchrist's model to understand how certain performance curves evolve by natural selection, particularly when the organism reproduces semelparously or when adult survivorship does not vary with age (see Box 3.3).

Gilchrist envisioned an organism whose body temperature varied dielly and seasonally according to its environment. Diel variation in temperature was stochastic, but seasonal variation was deterministic; specifically, seasonal variation followed a sinusoidal function. Because the organism completed many generations per year, seasonal variation occurred primarily among generations. Using nine combinations of diel and seasonal variations, Gilchrist determined the optimal performance curve for different levels of thermal heterogeneity within and among generations (Fig. 3.17).

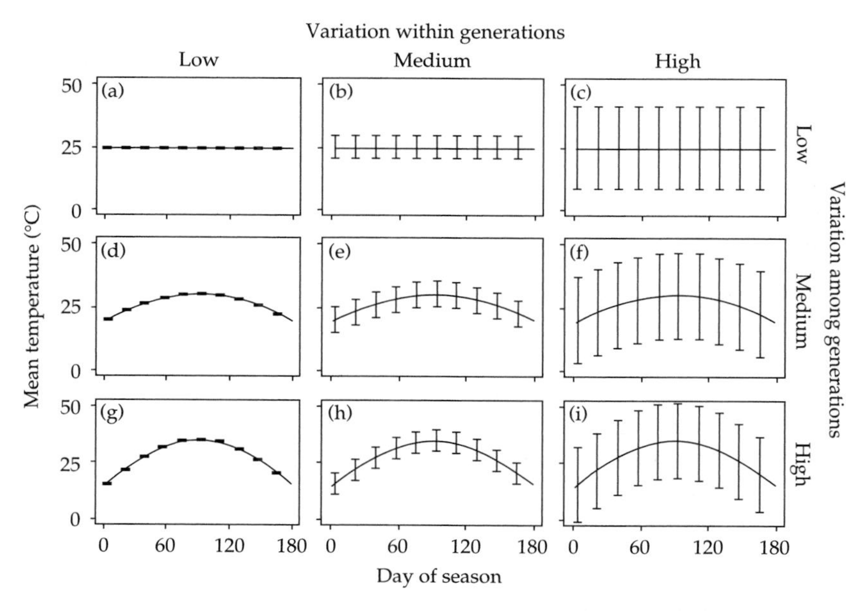

Figure 3.17 Gilchrist modeled the evolution of thermal sensitivity under nine combinations of daily and seasonal variations in temperature. These combinations created different levels of thermal heterogeneity within and among generations. Variation within generations depended on stochastic daily changes (error bars) and the sinusoidal seasonal changes. Variation among generations depended entirely on seasonal changes. Adapted from Gilchrist (1995, *The American Naturalist*, University of Chicago Press). © 1995 by The University of Chicago.

Instead of using a Gaussian function, Gilchrist modeled the performance curve using an asymmetric function (e.g., Fig. 3.4), which was characterized by a performance breadth (B) and critical thermal limits:

$$P(T_b, B, CT_{max}) = 4\left[e^{\rho(T_b - CT_{min})} - e^{[(\rho B) - 1.2\rho(CT_{max} - T_b)]}\right] \quad (3.12)$$

where T_b equals the body temperature, B equals the range of temperatures bounded by the CT_{min} and the CT_{max}, and ρ determines the area under the curve. To impose specialist–generalist tradeoffs, Gilchrist maintained a constant area under the performance curve by setting the value of ρ according to the other variables.

As mentioned above, the fitness of a phenotype equaled the sum of performance throughout life. Because the temperature varied within and among days, the optimal strategy maximizes the sum of performance given the frequency distribution of temperatures in the current generation (f_g). Thus, the fitness for a single generation equals the product of the frequency distribution and the performance curve:

$$w = \sum_{T_b = CT_{min}}^{CT_{max}} f_g(T_b) \cdot P(T_b, B, CT_{max}). \quad (3.13)$$

The long-term fitness of a genotype must be defined as the geometric mean of fitnesses for all generations (n):

$$\overline{W} = \left\{\prod_{g=1}^{n}\left[\sum_{T_b = CT_{min}}^{CT_{max}} f_g(T_b) \cdot P(T_b, B, CT_{max})\right]\right\}^{1/n} \quad (3.14)$$

Using eqn 3.14, Gilchrist determined the optimal performance curve in the nine environments shown in Fig. 3.17.

The majority of environmental conditions favor a thermal specialist with the smallest possible performance breadth (Fig. 3.18). The thermal optimum of this specialist should roughly equal the most frequent body temperature. This result makes perfect sense when the temperature remains relatively constant over time. But, surprisingly, an

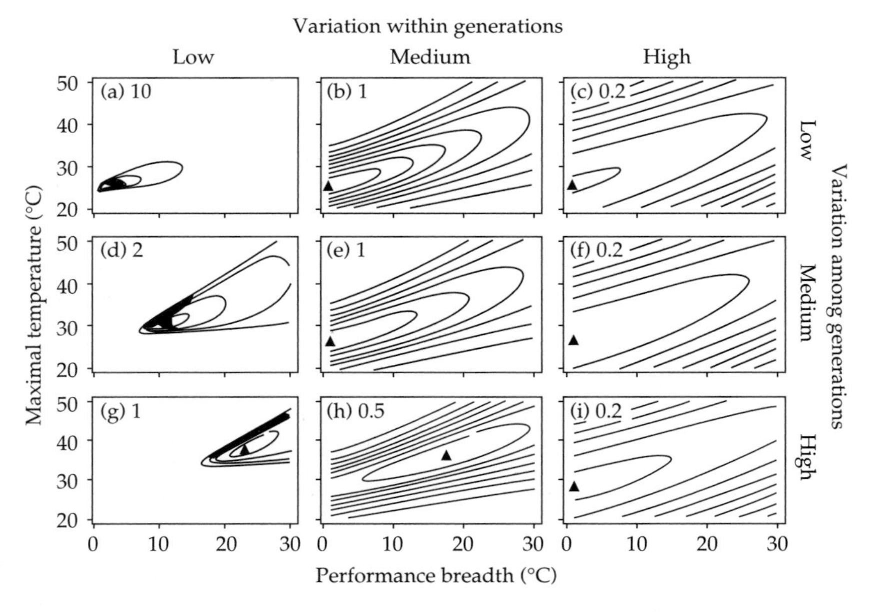

Figure 3.18 Most patterns of thermal heterogeneity explored by Gilchrist (1995) favor the evolution of thermal specialists. These fitness landscapes correspond to patterns of thermal heterogeneity shown in Fig. 3.17. Contour lines delineate the fitness of the various performance curves; the interval of the contour lines resides to the right of each plot's label (e.g., a 10). The triangle on each fitness landscape marks the optimal performance curve. Adapted from Gilchrist (1995, *The American Naturalist*, University of Chicago Press). © 1995 by The University of Chicago.

extreme specialist also achieves the greatest fitness when temperature varies moderately or even greatly within generations. This conclusion holds as long the variation among generations remains low. In fact, the optimal performance breadth tightly follows the ratio of variation within generations to variation among generations. Specialists are more fit than generalists when the variation within generations equals or exceeds the variation among generations (Fig. 3.18). The cause of this result should be obvious; a thermal specialist would become extinct during a generation when temperatures fall outside its performance breadth. Keep in mind, this model does not consider how spatial variation in temperature might preserve such genotypes.

Gilchrist went on to consider how a bimodal distribution of temperatures would affect the optimal performance curve. Arguably, this extension of the model more accurately represents the thermal heterogeneity of most terrestrial environments and certain aquatic environments. The optimal performance curves were surprisingly similar to those of the initial model; extreme specialists possessed the greatest fitness unless variation among generations exceeded variation within generations. Nevertheless, we should expect the thermal optimum for performance to be centered on the modal temperature instead of the mean temperature. Because of the asymmetry of the performance curve, a thermal optimum equal to the upper mode confers greater fitness than a thermal optimum equal to the lower mode. If temperature does not vary, a thermal optimum equal to the lower mode confers the same fitness as a thermal optimum equal to the upper mode. We shall return to these points in Chapter 7.

3.3.3 Contrasting the two models

We have analyzed two models that predict optimal performance curves in variable environments. Armed with these models, we can try to understand the diversity of performance curves described in Section 3.1. Before doing so, let us briefly compare and contrast the ideas captured by these models.

Although both models consider a similar problem, each represents a unique set of assumptions about the nature of thermal heterogeneity, the constraints on the phenotype, and the relationship between the phenotype and fitness. These assumptions define the environmental, phenotypic, and fitness functions of each model (Table 3.2). Not surprisingly, different assumptions lead to different predictions. In particular, the models provide radically different perspectives on the importance of temporal heterogeneity. Lynch and Gabriel's model predicts that the performance breadth will be directly related to the thermal heterogeneity within generations. In contrast, Gilchrist's model predicts that the performance breadth will be more affected by the relative amounts of thermal heterogeneity within and among generations. While these models differ in several ways, one seems most relevant to this result: the way in which performance contributes to fitness. Therefore, we should expect to observe broader thermal breadths of survivorship in organisms subject to greater variation in temperature. At the same time, we should not expect to see this same relationship when comparing thermal breadths of other performances.

Despite this discrepancy, both models predict that the optimal performance curve will enable maximal performance at the most frequent body temperature (or very close to this temperature). Therefore, support for either model would depend on the degree to which the thermal optima for various performances match the mean or modal temperatures. For example, if we compared genotypes from warm and cold environments, we should observe a pattern similar to the one shown in Fig. 3.19a. Alternatively, we might observe only partial adaptation to local environments (Fig. 3.19b), which would weakly support these models.

Finally, we might observe a difference in performance curves between genotypes that provides no support for these models. In particular, both models would be refuted if genotypes from one environment performed as well as or better than genotypes from other environments at all temperatures. I refer to such genotypes as masters of all temperatures. When these masters of all temperatures come from the colder environment (Fig. 3.19c), the pattern of intraspecific variation is called countergradient variation because the genetic effects on performance counter the environmental effects on performance (Conover and Schultz 1995). When these masters

Table 3.2 A comparison of two models of optimal performance curves. Major assumptions underlying the environmental, phenotypic, and fitness functions of each model are listed.

Model component	Lynch and Gabriel's model	Gilchrist's model
Environmental function	• Temperature is normally distributed.	• Temperature is either normally distributed or bimodally distributed.
	• Mean temperature varies stochastically among generations.	• Mean temperature varies systematically among generations.
	• Mean temperature varies spatially.	• Mean temperature does not vary spatially.
	• Body temperature equals environmental temperature.	• Body temperature equals environmental temperature.
	• Temperature does not depend on other environmental variables.	• Temperature does not depend on other environmental variables.
Phenotypic function	• The performance curve is symmetrical.	• The performance curve is skewed left.
	• The integral of the performance curve remains constant.	• The integral of the performance curve remains constant.
	• Both genes and the environment (i.e., developmental noise) affect the performance curve.	• Developmental noise does not occur.
	• The performance curve cannot acclimate to temperature.	• The performance curve cannot acclimate to temperature.
Fitness function	• Fitness for a single generation equals the product of performance over time.	• Fitness for a single generation equals the sum of performance over time.
	• Fitness over many generations equals the geometric mean fitness.	• Fitness over many generations equals the geometric mean fitness.

of all temperatures come from the warmer environment (Fig. 3.19d), the pattern is called cogradient variation. Neither pattern should occur when the performance curves evolve according to specialist–generalist tradeoffs.

3.4 Using models to understand natural patterns

Many species have large geographic ranges, spanning broad latitudinal or altitudinal gradients. As the range of a species expands, populations at the leading edge encounter novel thermal environments. In general, populations at high latitudes or altitudes experience colder environments than do populations at low latitudes or altitudes (Chapter 2). Populations at different latitudes or altitudes also experience different amounts of temporal variation in temperature. Based on the models described in the preceding section, we should expect very different selective pressures over the range of a species. Provided sufficient genetic variation and limited gene flow exist, performance curves would diverge among populations (Grether 2005; Kawecki and Ebert 2004). Therefore, performance curves should vary predictably along latitudinal and altitudinal clines.

To make useful predictions about geographic variation in performance curves, we must first consider how the performance of interest contributes to fitness. Consider the five performances most commonly measured: survivorship, locomotion, development, growth, and reproduction. How could each of these performances affect the fitness of an organism? The effect of survivorship is obvious; survivorship to maturation and survivorship between successive reproductive events factor heavily in calculations of fitness (see Box 3.3). Furthermore, Lynch and Gabriel's model was specifically designed to predict the evolution of performance curves for survivorship (see Section 3.3.1). Using this model, we could make two predictions about the thermal sensitivity of survivorship: (1) colder environments, such as those found at high latitudes and altitudes, should lead to the evolution of lower thermal optima and (2) more variable environments, such as those found

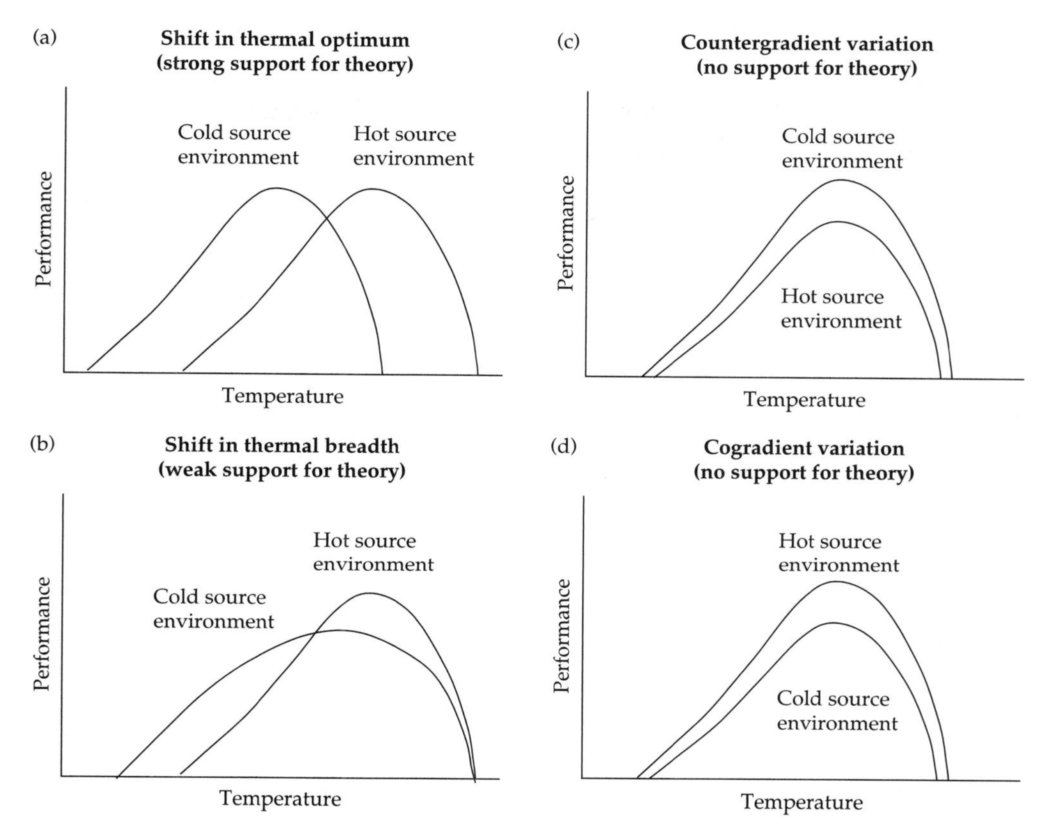

Figure 3.19 One of four patterns could emerge when comparing thermal performance curves between populations. (a) Evidence that thermal optima have diverged between populations strongly supports theory. (b) Evidence that performance curves have diverged and a specialist–generalist tradeoff occurred weakly supports theory. (c and d) Countergradient or cogradient variation does not support theory because thermal optima have not diverged and a specialist-generalist tradeoff has not occurred.

at high latitudes, should lead to the evolution of wider performance breadths.

Now, consider the remaining performances. A greater capacity for locomotion can enhance foraging (Greenwald 1974), dominance (Robson and Miles 2000), and survivorship (Christian and Tracy 1981; Husak 2006; Jayne and Bennett 1990). If locomotor performance contributed significantly to survivorship, we should expect its thermal sensitivity to also vary according to Lynch and Gabriel's model. If locomotor performance contributes to fitness more through fecundity than through survivorship, we should expect its thermal sensitivity to vary according to Gilchrist's model. The same would be true for development, growth, and reproduction, which seem to contribute additively to fitness. The sum of development over time determines the developmental stage at the end of that time; reproduction generally requires developmental transitions among embryonic, larval, and adult stages. Likewise, the sum of growth over these stages determines the body size at maturation, which strongly influences the fecundity of an organism (Roff 2002). Finally, reproductive performance contributes directly to fecundity.

When performance contributes to fitness via its effect on fecundity rather than its effect on survivorship, we should expect slightly different patterns of variation along thermal clines. Although the thermal optimum should still decrease with increasing latitude or altitude, the performance breadth should vary only under certain conditions. Specifically, geographic variation in the performance breadth should depend on the relative magnitude of intra- and intergenerational variation in temperature (see Section 3.3.2). Obviously, these two components of variance depend on the lifespan and habitat of an organism. Short-lived aquatic species, such as *Daphnia magna*, experience relatively stable temperatures within generations because of the high heat capacity of water, but experience more variable temperatures among generations because of seasonal changes in water temperature. In such species, performance breadths for development, growth, and reproduction should increase with increasing latitude. For long-lived species ($\geq$ 1 year), and particularly those in terrestrial habitats, performance breadths should remain relatively narrow over a latitudinal range.

Armed with these specific predictions for real environments, we can see how well optimality models explain intra- and interspecific variation in performance curves. To keep things organized, we shall concentrate on the five performances described above, beginning with survivorship and ending with reproduction.

3.4.1 Survivorship

The mere existence of organisms in extreme thermal environments, where most cannot survive, implies such organisms have adapted to these conditions (Clarke 1991). At one extreme, Antarctic fish swim in water cold enough to kill other fish (Wilson et al. 2001). At the other extreme, certain bacteria dwell in waters that would boil virtually any organism (Brock 1994). As assumed by current models, species adapted to extreme temperatures usually function poorly at moderate temperatures (see, for example, Wilson et al. 2001). Adaptation to thermal extremes seems to have come with a cost. This kind of qualitative support for the theory is reassuring, but the real utility of a theory comes from its ability to make predictions over a broad range of conditions. To determine whether current models provide this power, we must see how well they can predict intra- and interspecific variation in performance curves.

Ideally, we would use direct comparisons of performance curves for survivorship to evaluate the predictions of Lynch and Gabriel's model. Surprisingly, such data are limited (Table 3.3). Although many researchers have documented relationships between temperature and survivorship (see review by Angilletta et al. 2004a), they have generally done so for one or two populations at a time. The paucity of well-designed comparative studies prohibits general inferences about the evolution of these performance curves; however, available studies paint a fuzzy picture. Moore (1940) made the first effort to formally relate the thermal breadth of survivorship to the range of environmental temperatures. His summary of published data suggested aquatic animals possess much narrower breadths for survivorship than do terrestrial animals. More recently, at least one intraspecific analysis of survivorship supported the predicted relationship between thermal heterogeneity and performance breadth; in the beetle *Hyperaspis notata*, the performance breadth of a strain from high latitude was broader than that of a strain from low latitude (Dreyer et al. 1997). In the vast majority of intraspecific comparisons, however, little support can be gleaned for our predictions. Comparisons of survivorship over narrow (Drent 2002; Lankford and Targett 2001) or wide ranges (Ayres and Scriber 1994; Otterlei et al. 1999; Reisen 1995) of temperatures yielded no covariation between thermal heterogeneity and performance breadth. In fact, two latitudinal comparisons of tadpoles from the same species yielded different results. For a pair of populations of *Rana temporaria* (Stahlberg et al. 2001), tadpoles from a cold environment survived better at a low temperature than did tadpoles from a warm environment. For another pair of populations of the same species (Merila et al. 2000), tadpoles from the colder environment survived worse at a low temperature and better at a high temperature! These more recent comparisons within species and genera fly in the face of Moore's early study.

Despite the inability of Lynch and Gabriel's model to explain variation in performance breadth, the model still could explain variation in thermal limits of survivorship (reviewed by Garland and Carter 1994; Vernberg 1962). Many studies report direct measures of the thermal limits or indirect measures of thermal tolerance (Table 3.4). Indices of thermal tolerance include the temperature at which

Table 3.3 Comparative studies of the thermal sensitivity of survivorship. In most cases, individuals were raised from birth in a common garden (CG = Y), but in a few cases individuals either were acclimated to identical conditions for some number of days or were not acclimated (CG = N).

Organism	Cline	CG	Measure(s)	Pattern	Source
Canadian tiger swallowtail (*Papilio canadensis*)	Latitude	Y	Survivorship at 12–30°C	No difference between populations	Ayres and Scriber (1994)
Atlantic croaker (*Micropogonias undulatus*)	Latitude	14 days	Survivorship at 1–5°C	No difference between populations	Lankford and Targett (2001)
Common frog (*Rana temporaria*)	Latitude	Y	Survivorship at 10–20°C for 15 days	Larvae from a cold environment survived better at 10°C than larvae from a warm environment.	Stahlberg et al. (2001
Common frog (*Rana temporaria)*	Latitude	Y	Survivorship of larvae at 14°C and 22°C	Larvae from a cold environment survived better at 22°C and worse at 14°C than larvae from a warm environment.	Merila et al. (2000
Beetle (*Hyperaspis notata)*	Latitude	Y	Survivorship at 15–34°C	B_{80} was broader for the strain from high latitude than it was for the strain from low latitude.	Dreyer et al. (1997)
Baltic clam (*Macoma balthica*)	Latitude	Y	Survivorship at 10–20°C	No difference between populations	Drent (2002)
Mosquito (*Culex tarsalis*)	Latitude	Y	Survivorship at 15–35°C	Survivorship of females from a warm environment was higher at 20°C and 25°C than survivorship of females from a cold environment.	Reisen (1995)
Atlantic cod (*Gadus morhua*)	Latitude	Y	Survivorship at 4–14°C	No thermal or population trends	Otterlei et al. (1999)

Notes: B_{80}, 80% performance breadth.

animals lose motor function (knockdown temperature or critical thermal maximum), the duration of exposure to an extreme temperature before loss of function (knockdown time), and the time required to recover from exposure to an extreme temperature (recovery time). Because these measures are quick and easy to make on large numbers of individuals, many data exist to evaluate the predictions of optimality theory. We might deduce much about survivorship from these indirect measures of thermal tolerance. For example, we could interpret the knockdown temperature as an acute thermal limit to survivorship (Cooper et al. 2008); in other words, survivorship equals zero at the knockdown temperature. Knockdown and recovery times probably correlate with the thermal limit to survivorship under chronic exposure, but should be interpreted with caution (Hoffmann et al. 2003).

Certain geographic patterns of thermal tolerance make sense in light of Lynch and Gabriel's model (Table 3.5). Early studies of reptiles and amphibians focused on variation in the critical thermal maximum, which arguably correlates with the acute thermal limit of survivorship. In some taxa, the CT_{max} tends to be greater for individuals that inhabit

Table 3.4 Common indices of the tolerance of extreme temperatures. Examples were chosen to illustrate applications in diverse taxa. Indices of freeze tolerance (e.g., supercooling point) were deliberately omitted because they do not reflect adaptation to temperature per se (Clarke 1991).

Measure	Description	Example
Lethal temperature	The temperature that causes the death of 50% of the individuals in a sample (abbreviated as LT_{50})	Stillman and Somero (2000)
Lethal time	The time required for 50% of the individuals in a sample to die from exposure to a temperature (also abbreviated as LT_{50}, but should not be confused with the lethal temperature)	Berrigan (2000)
CT_{min} or CT_{max}	Minimal or maximal temperature at which motor function ceases or muscular spasms occur (often used to place an upper or lower bound on a performance curve)	Gaston and Spicer (1998); Miller and Packard (1977); Winne and Keck (2005)
Heat-shock survivorship	Probability of survival following a brief exposure to a high temperature	Sorensen et al. (2005)
Cold-shock survivorship	Probability of survival following a brief exposure to a low temperature	Gilchrist et al. (1997)
Knockdown temperature	The minimal temperature that causes an individual to lose motor function during heating	Huey et al. (1992)
Knockdown time	The time required for an individual to lose motor function during exposure to a high temperature	Huey et al. (1992)
Chill-coma temperature	The maximal temperature that causes an individual to lose motor function during cooling	Gibert and Huey (2001)
Chill-coma recovery	The time required for an individual to recover from loss of motor function following exposure to a low temperature	Castaneda et al. (2005)

warmer environments (Hertz 1979; Hertz et al. 1979; Licht et al. 1966a; Miller and Packard 1977; but see Spellerberg 1972). In other taxa, both the CT_{max} and CT_{min} decrease with increasing latitude (Snyder and Weathers 1975). In both situations, the thermal limits of performance appear to have shifted in the direction of the environmental temperature. Still, other patterns make no sense. For example, several investigators found no divergence in the CT_{max} or lethal limits along latitudinal clines (Gaston and Chown 1999; Klok and Chown 2005; Winne and Keck 2005). Overall, lower thermal limits seem to vary more closely with latitude than do upper thermal limits, which could stem from the steeper latitudinal cline in minimal environmental temperature (Ghalambor et al. 2006). The CT_{min} of dung beetles decreased with increasing latitude, despite no cline in CT_{max} (Gaston and Chown 1999). Furthermore, Casteñada and colleagues (2005) observed a very strong relationship between the minimal environmental temperature and the chill-coma recovery time among four populations of woodlice. Importantly, most of these studies involved a small number of populations or species and included no phylogenetic control for pseudoreplication (see Box 1.5).

More rigorous studies included larger samples or phylogenetic control. Stillman and Somero (2000) focused on latitudinal variation in the upper lethal temperature of crabs in the genus *Petrolisthes*. They found that crabs from warm environments survived acute exposures to higher temperatures than did crabs from cold environments (Fig. 3.20a). The same result was obtained when independent contrasts were used (Fig. 3.20b), indicating that adaptation of the thermal limit of survivorship has occurred within this genus. Despite this strong support for the current theory, the most comprehensive intraspecific study to date provided little support. Addo-Bediako and colleagues (2000) compared lower lethal temperatures and critical thermal maxima among species of insects. They found no evidence that the CT_{max} varied latitudinally. Although lower lethal temperatures decreased at the northern latitudinal extreme, this decrease in the mean resulted from an increase in the variance. Many species from very high latitudes exhibited lower lethal temperatures as high as those of species from low latitudes.

A more comprehensive picture of the evolution of thermal tolerance has emerged from comparisons among populations and species of *Drosophila*

Table 3.5 Comparative studies of thermal tolerance, including lethal thermal limits (LT_{50}, CT_{min}, or CT_{max}), heat resistance (knockdown temperature or time), and cold resistance (chill-coma temperature or recovery time). In most cases, individuals were raised from birth in a common garden (CG = Y), but in a few cases individuals either were acclimated to identical conditions for some number of days or were not acclimated (CG = N).

Organism(s)	Cline	CG	Measure(s)	Pattern	Source
Algae					
Valonia utricularis	Latitude/longitude	Y	lower and upper lethal temperatures (LT_{50}s)	Lower and upper LT_{50}s were positively related to minimal and maximal environmental temperatures, respectively.	Eggert et al. (2003a)
Insects					
Drosophila buzzatii	Altitude	Y	Knockdown time at 37°C; survivorship after exposure to 40.5°C for 1 hour; recovery from 12 hours at 0°C	Knockdown time was negatively related to altitude; survivorships after heat and cold shocks were unrelated to altitude.	Sorensen et al. (2005)
	Latitude	Y	Knockdown time at 40°C; survivorship after exposure to 40.5°C for 1 hour	Knockdown time was negatively related to latitude; in females, survivorship after heat shock was positively related to latitude (no relationship existed for males).	Sarup et al. (2006)
	Altitude	Y	Knockdown time at 40°C	Knockdown time was greater for populations from lower elevation.	Sorensen et al. (2001)
Drosophila melanogaster	Latitude	Y	Loss of male fertility (CT_{max})	Tropical species had a higher CT_{max} than did temperate species.	Rohmer et al. (2004)
	Latitude	Y	Times to recover from exposures between –7°C and 2°C	Temperate populations recovered faster than subtropical populations at low temperatures.	David et al. (2003)
	Latitude	Y	Survival at –2°C for 2 hours	Survival was positively related to latitude.	Hoffmann et al. (2002)
	Latitude	Y	Recovery from 8 hours at 0°C	Recovery time was negatively related to latitude.	Hoffmann et al. (2002)
	Latitude	Y	Knockdown time at 39°C	Knockdown time was negatively related to latitude.	Hoffmann et al. (2002)
	Latitude	Y	Time to recover from exposure to 0°C for 16 hours	Recovery time was negatively related to latitude.	Ayrinhac et al. (2004)
	Latitude	Y	Loss of righting response (CT_{min})	CT_{min} was negatively related to latitude.	Gibert and Huey (2001)
Drosophila serrata	Latitude	Y	Time to recover from exposure to 0°C for 8 hours	Recovery time was negatively related to latitude.	Hallas et al. (2002)
Drosophila spp.	Latitude	Y	Lower and upper lethal temperatures	Cold tolerance was greater for species from higher latitudes, but heat tolerance was unrelated to latitude.	Ohtsu et al. (1998)
Drosophila spp.	Latitude	Y	Lethal time (LT_{50}) at high temperature, resistance to knockdown, and knockdown temperature	Heat tolerances were greater for populations from lower latitudes.	Berrigan (2000)
Drosophila spp.	Latitude	Y	Time to recover from exposure to 0°C for 16 hours	Temperate species recovered faster than tropical species.	Gibert et al. (2001)

Continued

Table 3.5 Continued

Organism(s)	Cline	CG	Measure(s)	Pattern	Source
Drosophila and *Scapto-drosophila* spp.	Latitude	Y	Lower and upper lethal temperatures	Cold tolerance was greater for species from higher latitudes, but heat tolerance was unrelated to latitude; within species, cold tolerance was greater for strains from higher latitudes.	Kimura (2004)
Embryonopsis halticella	Latitude	N	Lethal limits	No differences between populations	Klok and Chown (2005)
Lycaena tityrus	Altitude	Y	Time to recover from exposure to −20°C for 6 min; knockdown time at 47°C	Cold tolerance increased with increasing altitude; heat tolerance was greater in populations from low and middle altitudes than in a population from high altitude.	Karl et al. (2008)
Insecta	Latitude	N	CT_{max} and lower and upper lethal temperatures	Lower lethal temperature was very low at high northern latitudes; upper lethal temperature and CT_{max} were unrelated to latitude.	Addo-Bediako et al. (2000)
Scarabaeidae	Altitude	4 days	Loss of righting response (CT_{min}) and onset of spasms (CT_{max})	CT_{min} was negatively related to latitude; CT_{max} was unrelated to latitude.	Gaston and Chown (1999)
Crustaceans					
Orchestia gammarellus	Latitude	51 days	Lethal temperature and lethal time	Individuals from a warm environment tolerated heat better than individuals from a cold environment.	Gaston and Spicer (1998)
Petrolisthes spp.	Latitude	N	Upper lethal temperature (LT_{50})	LT_{50} was positively related to maximal environmental temperature.	Stillman and Somero (2000)
Porcellio laevis	Latitude/altitude	14 days	Times to recover from exposures between 0°C and 7°C	Recovery time was positively related to minimal environmental temperature.	Castañeda et al. (2005)
Amphibians					
Pseudacris triseriata	Altitude	4 days	Loss of motor function (CT_{max})	CT_{max} was negatively related to altitude.	Miller and Packard (1977)
Reptiles					
Amphibolurus spp.	Latitude/altitude	N	Duration of survival at 46°C	CT_{max} was positively related to maximal field body temperature.	Licht et al. (1966a)
Anolis krugi	Altitude	N	Onset of spasms and loss of righting response (CT_{max})	CT_{max} was negatively related to altitude.	Hertz (1979)
Anolis semilineatus	Altitude	N	Onset of spasms and loss of righting response (CT_{max})	CT_{max} was negatively related to altitude.	Hertz (1979)
Anolis gundlachi	Altitude	N	Onset of spasms and loss of righting response (CT_{max})	CT_{max} was negatively related to altitude.	Hertz et al. (1979)
Nerodia rhombifer	Latitude	Y	Onset of spasms and loss of righting response (CT_{max})	No relationship with latitude	Winne and Keck (2005)
Zootoca vivipara	Altitude	N	Loss of righting response (CT_{min} and CT_{max})	No differences between populations	Gvoždík and Castilla (2001)

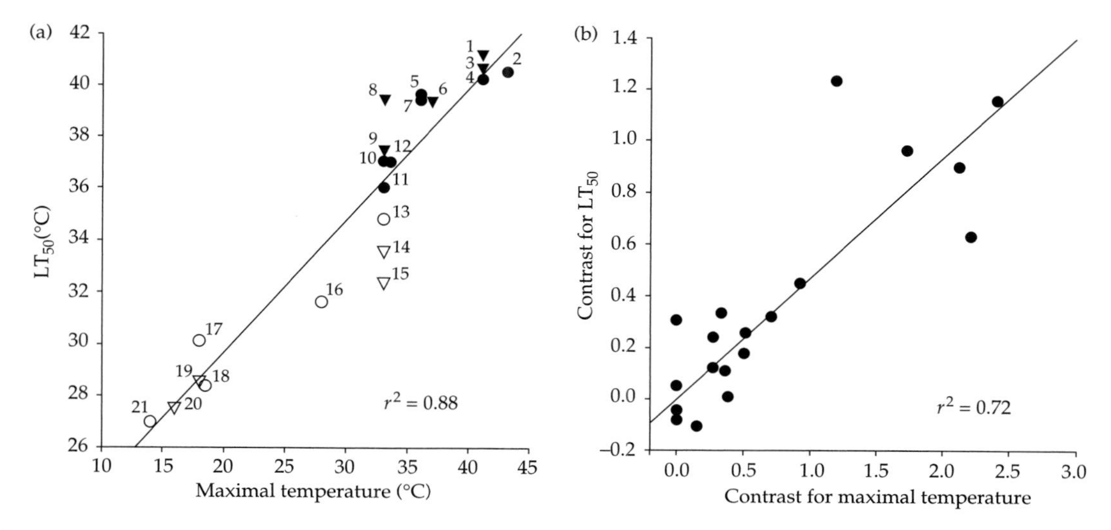

Figure 3.20 Among species of porcelain crabs, the lethal temperature (LT_{50}) was strongly related to the maximal temperature of the microenvironment. This relationship held for analyses of raw data (a) and independent contrasts (b). In the plot of raw data, each point represents the LT_{50} of a species: 1, *Petrolisthes gracilis*; 2, *P. tridentatus*; 3, *P. armatus*, northern Gulf of California; 4, *P. armatus*, Panama; 5, *P. holotrichus*; 6, *P. hirtipes*; 7, *P. platymerus*; 8, *P. crenulatus*; 9, *P. sanfelipensis*; 10, *P. agassizii*; 11, *P. haigae*; 12, *P. galathinus*; 13, *P. granulosus*; 14, *P. cabrilloi*; 15, *P. cinctipes*; 16, *P. laevigatus*; 17, *P. violaceus*; 18, *P. tuberculatus*; 19, *P. manimaculis*; 20, *P. eriomerus*; 21, *P. tuberculosus*. Adapted from Stillman and Somero (2000, *Physiological and Biochemical Zoology*, University of Chicago Press). © 2000 by The University of Chicago.

(reviewed by Hoffmann et al. 2003). Most studies to date accord with our theoretical predictions. For three species, intraspecific comparisons of chill-coma recovery revealed that tolerance of low temperatures has adapted to local climates. Flies of both *D. melanogaster* and *D. serrata* from high latitudes recovered faster than conspecific flies from low latitudes (Ayrinhac et al. 2004; Hallas et al. 2002). *Drosophila buzzatii* exhibited a similar cline in chill-coma recovery over its altitudinal range (Sorensen et al. 2005). Variation among species of *Drosophila* can also be explained by the theory. Intuitively, tropical species of *Drosophila* require far more time to recover from chill coma than do temperate species (Fig. 3.21). A phylogenetic comparative analysis by Gibert and Huey (2001) revealed correlated changes in latitude and the lower thermal limit during the evolution of a major clade; as expected, species that encountered higher latitudes have evolved a lower chill-coma temperature. Collectively, these data provide strong evidence that thermal limits in species of *Drosophila* have adapted to local thermal environments. Nevertheless, the same data also indicate that adaptation to one thermal extreme does not necessarily cause maladaptation to the other. For example, tolerances to heat and cold seem to have evolved in parallel within *D. melanogaster* (Hoffmann et al. 2002), but have evolved independently within *D. buzzatii* (Sorensen et al. 2005). Unfortunately, we do not know whether the apparent increases in the breadth of thermal tolerance have imposed any costs of maximal performance.

In summary, measures of survivorship under chronic exposures to constant temperatures do not support the current theory, whereas measures of thermal limits during acute exposures partially support the theory. Generally, thermal optima do not decrease and performance breadths do not increase with increasing latitude as predicted by Lynch and Gabriel's model. In contrast, measures of lethal temperatures, knockdown times, and recovery times often covary predictably with latitude and altitude. In some instances, latitudinal variation in the lower thermal limit occurs in the absence of latitudinal variation in the upper thermal limit, suggesting greater thermal heterogeneity at high latitudes has led to the evolution of thermal generalists. Perhaps measures of acute thermal tolerance relate more to

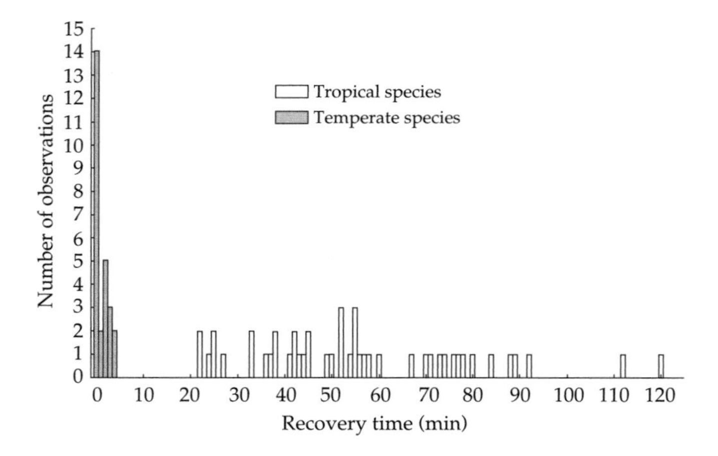

Figure 3.21 Tropical species took far longer than temperate species to recover from an exposure to 0°C for 16 hours. Data are means for 26 temperate and 48 tropical species of *Drosophila*. Adapted from Gibert et al. (2001) with permission from Blackwell Publishing.

phenotypes under selection in natural environments than do measures of chronic thermal tolerance.

3.4.2 Locomotion

The thermal sensitivities of jumping, swimming, walking, and sprinting have been estimated for many species (Table 3.6). In amphibians, the thermal sensitivities of jumping and swimming vary predictably within and among species. Wilson (2001) compared jumping performances among populations of the marsh frog, *Limnodynastes peronii*, along both latitudinal and altitudinal clines in eastern Australia. As expected, frogs from cool environments jumped better at low temperature and worse at higher temperatures than did frogs from warm environments; the performance breadth also appeared to be broader in cooler environments, but measurements were not taken over a broad enough range to compare performance breadths quantitatively. A similar study was conducted by John-Alder and colleagues, who compared jumping performances among species of North American frogs; they observed broader performance breadths in species that bred in colder environments. Extensive comparisons of swimming performance have been made among populations and species of frogs in the Andes of South America (reviewed by Navas 2006). Navas (1996a) showed that four species of frogs from high elevations had lower thermal optima and broader performance breadths than did their congeners from low elevations. In two of the four cases, this expansion of the performance breadth seems to have come at a cost of maximal performance (Fig. 3.22), indicating a specialist–generalist tradeoff. Similar adaptations along altitudinal clines have been observed within and among other species of anurans (Navas 2006).

For a combination of historical and logistical reasons, the most comprehensive comparisons of thermal sensitivities have focused on the maximal sprint speeds of lizards (reviewed by Angilletta et al. 2002a). Briefly, one stimulates an individual to run along a linear track and records the greatest velocity achieved during the run. Usually, one records the fastest of several such runs as the maximal sprint speed. This procedure is repeated at different temperatures to construct a performance curve for each individual. An early study by Hertz and colleagues suggested the thermal sensitivity of sprint speed in an agamid lizard (*Stellio stellio*) had not adapted to thermal variation along an altitudinal cline. Later, more extensive investigations produced evidence of thermal adaptation. Among five species of *Anolis*, the thermal optimum for sprinting was correlated with the median field body temperature, and the 95% performance breadth[9]

[9] Although mathematical definitions of the performance breadth vary among models, empiricists typically estimate this parameter as the range of temperatures that enables performance that equals or exceeds some percentage of the maximal performance. The two most common estimates are the 95% performance breadth and the 80% performance breadth.

Table 3.6 Comparative studies of thermal performance curves. These exemplary studies were chosen for their reasonable samples along thermal clines (at least three populations per cline or multiple clines) and sufficient resolutions of thermal performance curves (at least five temperatures). In most cases, individuals were raised from birth in a common garden (CG = Y), but in some cases individuals either were acclimated to identical conditions for some number of days or were not acclimated (CG = N).

Performance	Organism(s)	Cline	CG	Pattern	Source
locomotion					
Sprinting	Anoles	Latitude/altitude	N	T_{opt} increased with increasing field body temperature; B_{80} increased with increasing variation in body temperature.	van Berkum (1986)
Sprinting	Starred agama	Altitude	N	None	Hertz et al. (1983)
Swimming	Various genera of frogs	Altitude	N	T_{opt} was lower for each congener at a higher altitude.	Navas (1996a)
Jumping	Tree frogs	Latitude	N	B_{80} increased with increasing variation in body temperature.	John-Alder et al. (1988)
Jumping	Striped marsh frog	Latitude/altitude	N	T_{opt} increased with increasing environmental temperature.	Wilson (2001)
Rolling	Isopods	Latitude	N	T_{opt} increased with increasing environmental temperature.	Castañeda et al. (2004)
Sprinting	Lizards	Latitude/altitude	N	T_{opt} and CT_{max} increased with increasing field body temperature; CT_{min} was unrelated to T_{opt} and CT_{max}.	Huey and Kingsolver (1993)
Development					
Embryonic	Common frog	Latitude	Y	None	Laugen et al. (2003b)
Larval	Wood frog	Altitude	Y	Countergradient variation	Berven (1982)
Larval	Green frog	Latitude/altitude	Y	Countergradient variation	Berven and Gill (1983; Berven et al. (1979)
Larval	Pea aphid	Latitude	Y	None	Lamb et al. (1987)
Assimilation/growth					
Wet mass	Marine algae	Latitude/longitude	Y	B_{80} increased with increasing range of environmental temperatures.	Eggert et al. (2003a)
Cell number	Bacteria	Latitude	Y	T_{opt} was related to water temperature.	Hahn and Pockl (2005)
Dry mass	Trees	Latitude	Y	T_{opt} was related to latitude.	Cunningham and Read (2003a)
Photosynthesis	Pondweed	Latitude	Y	none	Pilon and Santamaría (2002)
Photosynthesis	Aquatic macrophytes	Latitude	N	T_{opt} was related to latitude.	Santamaría and van Vierssen (1997)
Cell number	Symbiotic algae	Latitude	Y	T_{opt} was related to latitude.	Sun and Friedmann (2005)
Cell number	Marine algae	Latitude	Y	T_{opt} did not vary, but B_{80} and thermal limits were widest for algae from the most stable environment.	De Boer et al. (2005)

Continued

Table 3.6 Continued

Performance	Organism(s)	Cline	CG	Pattern	Source
Dry mass	Water flea	Latitude/longitude	Y	Cogradient variation	Mitchell and Lampert (2000)
Energy	Atlantic halibut	Latitude	40–50 days	Countergradient variation	Imsland et al. (2000)
Length	Medaka	Latitude	Y	Countergradient variation	Yamahira et al. (2007)
Length/mass	Silverside	Latitude	Y	Countergradient variation	Yamahira and Conover (2002)
Wet mass	Arctic charr	Latitude	Y	none	Larsson et al. (2005)
Reproduction					
number of infective juveniles	parasitic nematodes	worldwide	Y	none	Grewal et al. (1994)

Note: T_{opt}, thermal optimum for performance; B_{80}, 80% performance breadth; CT_{min}, critical thermal minimum; CT_{max}, critical thermal maximum.

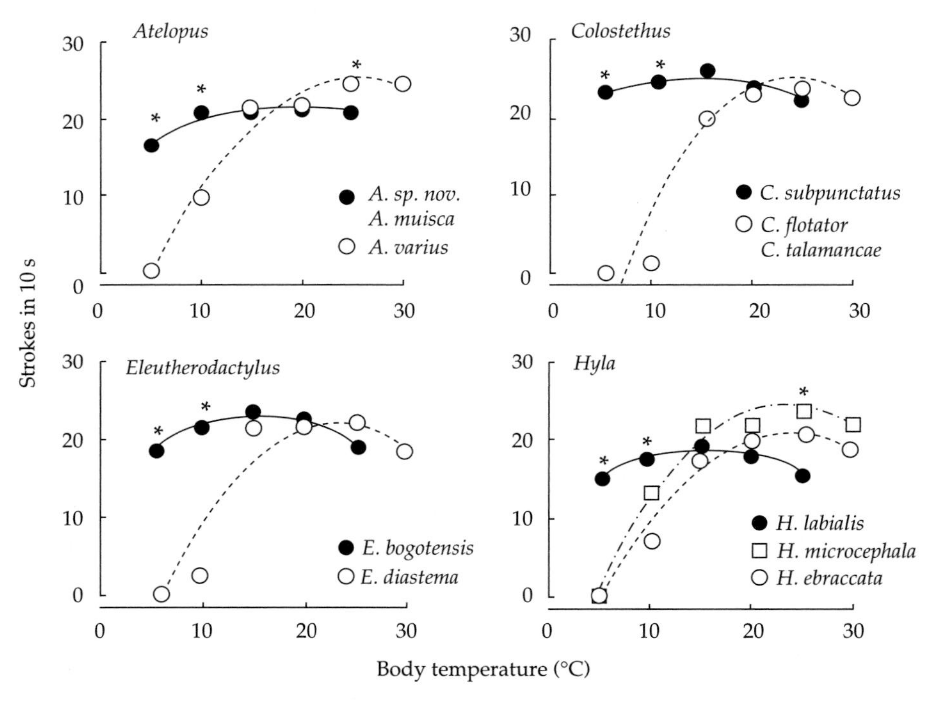

Figure 3.22 The thermal sensitivity of swimming performance differed between congeners in four genera distributed along an altitudinal gradient. Solid and open symbols denote species from high and low altitudes, respectively. Adapted from Navas (1996a, *Physiological and Biochemical Zoology*, University of Chicago Press). © 1996 by The University of Chicago.

was correlated with the range of field body temperatures (van Berkum 1986). An even larger study comprising 18 species of lizards representing four families lent additional support to the current theory (van Berkum 1988). Among these species, the 95% performance breadth and the range of temperatures between the thermal limits ($CT_{max} - CT_{min}$) were correlated with the range of field body temperatures.

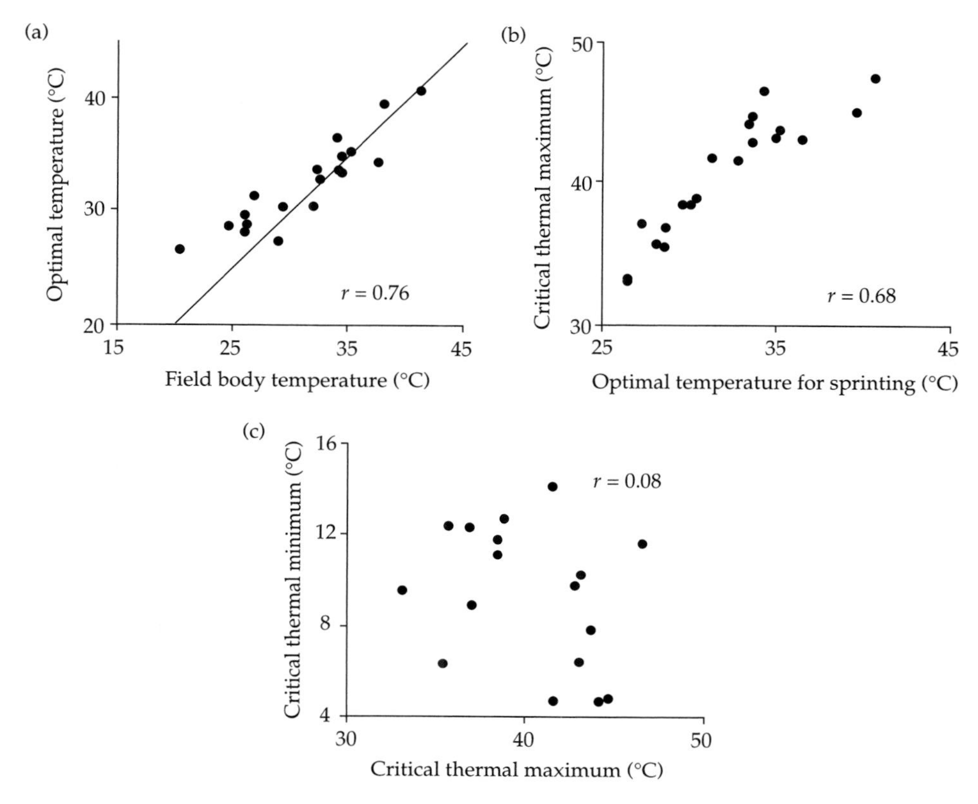

Figure 3.23 Evolutionary correlations were detected between thermal characteristics of 19 species of lizards. (a) The thermal optimum for sprinting evolved in concert with the mean field body temperature; the solid reference line has an intercept equal to zero and a slope equal to 1. (b) The critical thermal maximum evolved in concert with the thermal optimum for sprinting. (c) The critical thermal limits evolved independently. Adapted from Huey and Kingsolver (1993, *The American Naturalist*, University of Chicago Press). © 1993 by The University of Chicago.

Interestingly, the 80% performance breadth was unrelated to variation in field body temperature.

In general, does the thermal optimum evolve to match the mean temperature experienced by lizards? Compiling data from these and other studies, Huey and Kingsolver (1993) conducted the largest comparative analysis of performance curves for sprinting. Moreover, they used a robust phylogenetic hypothesis to compute independent contrasts for these variables. Contrasts of thermal optima correlated strongly with contrasts of field body temperature (Fig. 3.23a), supporting the prediction that the thermal optimum should track the temperature experienced most frequently. If sprint speed were to evolve according to current models, we should also expect corresponding shifts in the CT_{min} and CT_{max}; however, only the CT_{max} evolved in concert with the thermal optimum (Fig. 3.23b and c). Thus, species with higher thermal optima can sprint over a wider range of temperatures than species with lower thermal optima. This result disagrees with the notion that specialist–generalist tradeoffs constrain the evolution of this performance curve. Such a result also sharply contrasts the patterns observed in frogs, whose evolutionary history seems to have been constrained by specialist–generalist tradeoffs (see above).

As with the thermal sensitivity of survivorship, the thermal sensitivity of locomotion sometimes evolves according to the models described in Section 3.3, but does not always do so. For some taxa, important discrepancies exist between the predicted and observed relationships among thermal heterogeneity, thermal optima, and performance breadths. The agreement between theory and observation will get considerably worse as we consider other performances.

3.4.3 Development

From earlier in Section 3.4, recall two predictions about the thermal sensitivity of development. First, the thermal optimum should match the mean (or modal) temperature. Second, the performance breadth should scale according to the relative magnitudes of thermal heterogeneity within and among generations; for short-lived species, the performance breadth should increase with increasing latitude because seasonal variation would occur primarily among generations.

Unfortunately, optimality models do little to help us understand latitudinal variation in the thermal sensitivity of development. Although few comparative studies of development have adequately defined the shapes of performance curves, none of these studies suggests these curves have evolved according to Gilchrist's model (see Table 3.6). A few other studies provide either partial or indirect support for the model. For example, green frogs from a montane population developed faster at low temperatures but slower at high temperatures than did frogs from a lowland population (Berven et al. 1979). Similarly, copepods have diverged along latitudinal clines; genotypes from high latitude developed faster at low temperatures than genotypes from low latitude, and vice versa at high temperatures (Lonsdale and Levinton 1985).

More often, investigators have uncovered either no latitudinal pattern or an unexpected pattern. Clones of pea aphids (*Acyrthosiphon pisum*) from five populations did not possess very different thermal optima or breadths, and neither parameter was related to latitude or temperature (Lamb et al. 1987). In fact, variation among clones within populations was greater than variation among populations. Hoffmann and colleagues (2005) raised genotypes of the mouse-ear cress (*Arabidopsis thaliana*) from 65 populations throughout the world. Their results were exactly the opposite of what we should expect given thermal adaptation; plants from cold environments developed slower at 14°C and faster at 22°C than plants from warm environments.

By comparing performance curves for development along latitudinal and altitudinal transects, investigators have uncovered many examples of countergradient and cogradient variations (Fig. 3.19c and d). Cogradient variation has been documented in common garden studies of insects, including ant lions (Arnett and Gotelli 1999) and fruit flies (Norry et al. 2001). Countergradient variation has been documented in similar studies of reptiles and amphibians. In two species of lizards, embryos from cold environments developed faster under cool and warm cycles of temperature than embryos from warm environments (Oufiero and Angilletta 2006; Qualls and Shine 1998). In frogs (*Rana* spp.), countergradient variation occurs on both macro- and microgeographic scales. Among five populations from latitudes ranging from 55° to 70°N, tadpoles at higher latitudes developed faster at constant temperatures (14–22°C) (Laugen et al. 2003a; Merila et al. 2000). Along altitudinal transects, tadpoles from mountain ponds developed faster over wide ranges of temperatures than did tadpoles from lowland ponds (Berven 1982; Berven and Gill 1983). On a microgeographic scale, tadpoles from shaded ponds developed faster at two constant temperatures than tadpoles from unshaded pond (Skelly 2004). The abundance of evidence for cogradient and countergradient variations challenges our theory, which assumes a jack of all temperatures will be a master of none. If we take these patterns at face value, we must conclude a genotype can be a master of all temperatures.

3.4.4 Growth

Geographic variation in the thermal sensitivity of growth rate has been documented for many species (see Table 3.6). Here, I define growth broadly to include everything from the change in mass (or length) to the assimilation of energy from the environment (photosynthesis). As with developmental rate, we can explain very few observations using insights from Gilchrist's model.

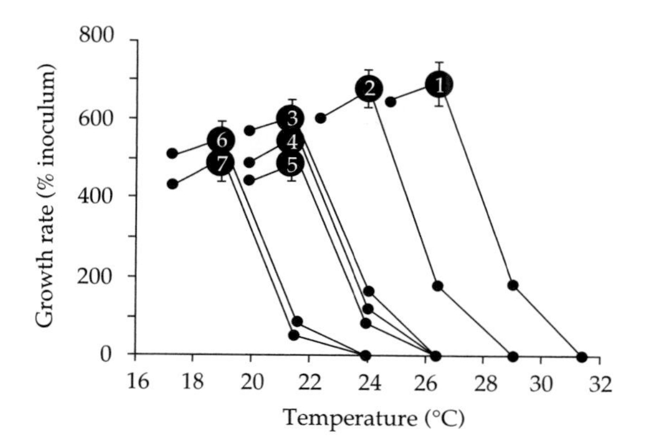

Figure 3.24 The thermal optimum for the growth of symbiotic algae (*Trebouxia irregularis*) isolated from lichens of the genus *Cladina* varied with latitude. The relationship between the temperature of the environment during summer and the thermal optimum was quite strong ($r^2 = 0.74$, $P < 0.005$). Source localities: 1, Florida, USA; 2, North Carolina, USA; 3, Rhode Island, USA; 4, Oaxaca, Mexico; 5, Ontario, Canada; 6 and 7, Alaska, USA. Adapted from Sun and Friedmann (2005) with kind permission from Springer Science and Business Media. © Springer-Verlag 2005.

Studies of growth, in terms of biomass or length, have yielded very weak support for the current theory. The best support comes from a study of symbiotic algae (*Trebouxia irregularis*) isolated from seven populations of lichens (*Cladina* spp.) along a latitudinal transect (Sun and Friedmann 2005). The performance breadths of the algae were remarkably uniform for all populations, but the thermal optimum correlated strongly with latitude (Fig. 3.24). In other words, algae from cold and warm environments were specialized for low and high temperatures, respectively. Several other studies provide mixed support. Australian trees from temperate rainforests exhibited slightly lower thermal optima than did trees from tropical rainforests. Nevertheless, performance curves for temperate and tropical trees overlapped greatly (Cunningham and Read 2003a). Similar patterns of thermal adaptation were observed in fish (Belk et al. 2005; Schultz et al. 1996) and frogs (Berven et al. 1979; Stahlberg et al. 2001), but divergences in thermal optima between populations were minimal and genotypes grew submaximally within their natural range of environmental temperatures.

Many other researchers have documented patterns that cannot be explained by Gilchrist's model. In most cases, genotypes from one environment outperform genotypes from other environments over a wide range of temperatures (i.e., countergradient or cogradient variation). For example, water fleas (*Daphnia magna*) spanning Finland to Sicily exhibited cogradient variation, with no evidence of divergence in the thermal optimum for growth rate among populations (Fig. 3.25c). Countergradient variation has been observed in a wide range of animals, including insects (Van Doorslaer and Stoks 2005a), crustaceans (Lonsdale and Levinton 1985), and fish (Imsland et al. 2000; Jonassen et al. 2000; Yamahira et al. 2007). The best studied case of countergradient variation occurs within the Atlantic silverside, *Menidia menidia* (Conover and Present 1990). Genotypes from high latitudes grow as fast as or faster than genotypes from low latitudes. A similar pattern has also been documented within the more narrowly distributed congener, *M. peninsulae* (Fig. 3.25d). Finally, certain patterns correspond to neither cogradient nor countergradient variation but still cannot be explained by Gilchrist's model. For example, genotypes of herbaceous plants (*Vicia pisiformis* and *V. dumetorum*) from intermediate latitudes exhibited faster growth over a range of temperatures than did genotypes from high and low latitudes (Black-Samuelsson and Andersson 1997). Furthermore, marine algae (*Fibrocaspa japonica*) from very stable waters off the coast of New Zealand exhibited a wider performance breadth than those of conspecifics from more variable waters off the coasts of Germany and Japan (De Boer et al. 2005).

Thermal sensitivities of photosynthesis also seem difficult to explain with Gilchrist's model. Although thermal optima and performance breadths vary greatly among plants (reviewed by Berry and Bjorkman 1980; Davison 1991), rigorous comparative studies have generated some unexpected patterns. Cunningham and Read (2002) compared the thermal sensitivities of net photosynthetic rate between temperate and tropical trees. Consistent with the theory, trees from higher latitudes underwent maximal rates of photosynthesis at lower temperatures than trees from lower latitudes. Because a generation for a tree spans many seasons, thermal heterogeneity among seasons should have

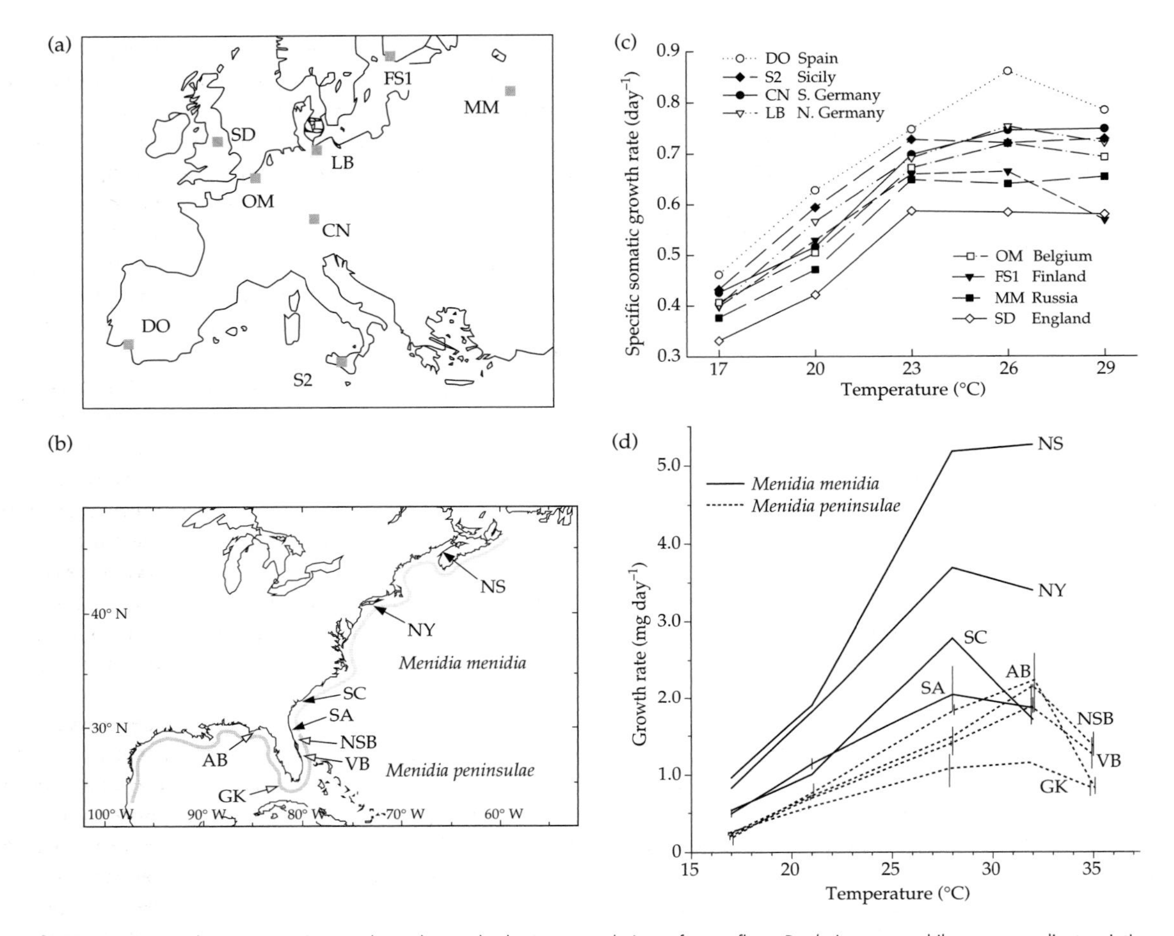

Figure 3.25 Cogradient variation in growth rate has evolved among populations of water fleas, *Daphnia magna*, while countergradient variations in growth rate have evolved among populations of Atlantic silversides, *Menidia menidia* and *Menidia peninsulae*. (a) Sites throughout western Europe where water fleas were collected: Munich, Germany (CN); Doñana, Spain (DO); Storgrundet, Finland (FS1); Lebrade, Germany (LB); Moscow, Russia (MM); Oude Meren, Belgium (OM); Sicily (S2); South Dalton, UK (SD). (b) Sites near the eastern coast of the USA where fish were collected: Nova Scotia (NS); New York (NY); South Carolina (SC); and St. Augustine (SA), Apalachee Bay (AB), New Smyrna Beach (NSB), Vero Beach (VB), and Grassy Key (GK), Florida. (c and d) The thermal sensitivities of growth, measured in terms of mass. Plots for *Daphnia magna* were reproduced from Mitchell and Lampert (2000) with permission from Blackwell Publishing. Plots for *Menidia* spp. were reproduced from Yamahira and Conover (2002) with permission from the Ecological Society of America.

resulted in the evolution of thermal specialists (see Section 3.3.2). Yet trees from more seasonal environments possessed a wider thermal breadth of photosynthesis (Fig. 3.26). Thermal sensitivities of photosynthesis by aquatic macrophytes are even more difficult to explain. Clones of *Potamogeton pectinatus*, collected over latitudes of 42° to 68°, possessed very similar performance curves (Pilon and Santamaria 2002). Worse still, species of aquatic macrophytes collected from high latitudes had higher thermal optima than did species collected from low latitudes—the opposite of what should emerge from thermal adaptation (Santamaria and van Vierssen 1997).

3.4.5 Reproduction

Relatively few researchers have investigated performance curves for reproduction along geographic

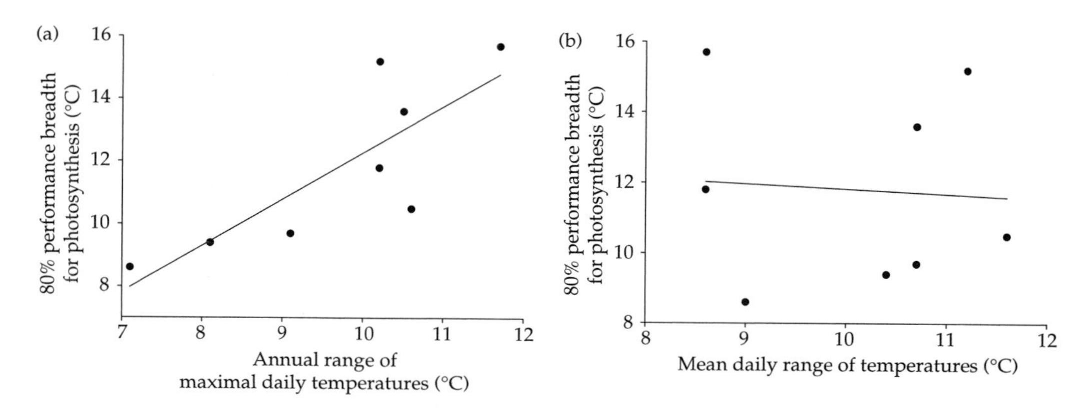

Figure 3.26 Performance breadth for photosynthesis increased with increasing thermal heterogeneity during the year (a), but was unaffected by the magnitude of thermal heterogeneity during the day (b). Data are from Tables 2 and 3 of Cunningham and Read (2002).

clines (see Table 3.6). The most extensive comparison was made by Grewal and colleagues (1994), who compared the performance breadths of 12 species of parasitic nematodes from different regions of the world. Performance breadths differed between species from the same locality but did not differ among species from different localities. Hence, Grewal concluded thermal adaptation had not occurred. In contrast, recent studies of butterflies on a microgeographic scale yielded less surprising results. Butterflies from warm agricultural habitats laid more eggs at high temperatures and fewer eggs at low temperatures than did butterflies from cool wooded habitats (see Fig. 3.6e); this pattern held both within species (Karlsson and Van Dyck 2005) and between species (Karlsson and Wiklund 2005). Lastly, the thermal limits of fertility differed between males of temperate and tropical strains of *D. melanogaster*; the CT_{min} and CT_{max} of tropical males exceeded those of temperate males (David et al. 2005; Rohmer et al. 2004). Although these studies are too few to draw a general conclusion, Gilchrist's model explains some, but not all, of these data.

3.4.6 Why do certain patterns differ from predicted ones?

The simple models described in Section 3.3 make many assumptions, some of which will be invalid for any species. These assumptions fall into four broad categories: (1) assumptions about the state of the environment, (2) assumptions about the nature of phenotypic constraints, (3) assumptions about the relationship between the phenotype and fitness, and (4) assumptions about the presence of genetic variation. First, we might have incorrectly related the thermal conditions of natural environments to the conditions of the models, including the mean, mode, or variance of operative environmental temperature. Second, the models might have incorrectly defined have the phenotypic set, including the range of phenotypes that organisms can express (e.g., thermal optima and performance breadths) and the tradeoffs that constrain the covariation between traits (e.g., specialist–generalist tradeoffs). Third, the models might have incorrectly defined the fitness set, including the timing or nature of how performance contributes to fitness. Finally, the hard-core assumptions of optimality models might have been invalid for some of the organisms considered above; from Chapter 1, recall that several assumptions must be made to infer evolutionary equilibria from optimality models, including the existence of genetic variation in performance curves (see Section 1.4.1). When one or more of these assumptions are violated, we can expect real patterns to differ from predicted patterns. Given the mixed support for the predictions of optimality models (Table 3.7), we should expect that some of the assumptions of the theory do not apply very generally to the species that have been studied thus far. By questioning our assumptions, we can learn how to develop more accurate models of the evolution of thermal

Table 3.7 A summary of comparative studies of thermal performance curves listed in Tables 3.3 and 3.6. The majority of these studies provided evidence of evolutionary divergence between populations, but only a minority of them provided support for the predictions of optimality theory

Performance	Case studies	Evidence of evolution	Support for theory
Survivorship	8	4	2
Locomotion	7	6	6
Growth	11	9	5
Development	4	2	0
Reproduction	1	0	0

sensitivity. Therefore, the remainder of this chapter focuses on the validity of these assumptions and the models that might provide deeper insights.

3.5 Have we mischaracterized thermal clines?

To predict geographic variation in performance curves, we assumed some knowledge of the temperatures experienced by organisms. Although clear latitudinal and altitudinal clines in temperature exist (see Chapter 2), temporal and spatial heterogeneity in real environments likely differ from the conditions assumed in current optimality models. Unlike the normally distributed temperatures of Lynch and Gabriel's model, temporal variation is bimodally distributed. Furthermore, temperatures in natural environments are correlated over time because of seasonal changes in climate. Gilchrist's model accounts for the bimodal distribution and the temporal autocorrelation of temperature, yet this model also fails to explain many empirical observations. Perhaps the bigger problem is that both models assume that organisms conform to the operative temperature of their environment. Yet thermoregulation decouples the mean temperature of an organism from the mean temperature of its environment. In doing so, thermoregulation can inhibit the evolution of performance curves along geographic clines (Huey et al. 2003). In Chapter 4, we shall take a closer look at the evolution of thermoregulation. Lastly, neither optimality model considers the possibility that environmental temperature could change directionally over time, such as the recent warming of many regions throughout the planet. In Chapter 9, we shall return to this topic and examine the consequences of continuous warming for the evolution of performance curves.

Do these complexities really explain why geographic patterns do not conform to the predictions of optimality models? Or is there something fundamentally wrong with another aspect of the theory? To answer these questions, we can consider the results of two kinds of experiments: reciprocal transplants and laboratory selection (Kassen 2002). In a reciprocal transplant experiment, one moves a subset of individuals from one thermal environment to another thermal environment, and vice versa. If each population has adapted to its respective environment, genotypes will achieve greater fitness in their native environment than they will in a foreign environment. Furthermore, the native population should outcompete the transplanted population in each environment (Kawecki and Ebert 2004). Still, transplant experiments provide only weak support for models of thermal adaptation because environments differ in more ways than temperature. Laboratory selection experiments offer more powerful tests of evolutionary models. In a selection experiment, one exposes replicated populations to controlled temperatures and observes the evolutionary responses (Section 1.5.2). Thus, we can easily infer the selective environment and relate the evolutionary responses to our theoretical predictions. Nevertheless, artificial environments of the laboratory lack many of the important elements of natural environments (predators, parasites, competitors, etc.), preventing us from generalizing conclusions from laboratory experiments. Because both reciprocal transplant and laboratory selection experiments have distinct advantages and disadvantages, let us examine examples of each to see whether they support the predictions of optimality models.

3.5.1 Reciprocal transplant experiments

Reciprocal transplant experiments have revealed strong evidence of adaptation to local environments. Schluter (2000) reviewed more than 40 experiments in which the performance or fitness of native and transplanted genotypes were compared in multiple environments. In all but eight experiments, native genotypes outperformed transplanted genotypes in each environment. Some degree of adaptation to the local environment seems the norm rather than the exception. But to what extent has this adaptation been driven by thermal heterogeneity? If thermal adaptation has occurred, we should expect a correlation between the relative performance of a genotype and the thermal conditions of the environment. Such a correlation was observed when populations of the carline thistle (*Carlina vulgaris*) were transplanted among five regions of Europe (Becker et al. 2006). The fitness of plants was greatest in their native region and decreased as the difference in temperature between the native and transplant regions increased. An experiment conducted by Chapin and Chapin (1981) provided weaker support for the current theory, in the sense that only one of two performances followed the expected pattern. These researchers transplanted populations of a sedge (*Carex aquatilis*) among five locations, ranging from an Arctic site (4°C) to a subalpine site (17°C). Only Arctic plants survived in the Arctic, but these same genotypes did not survive in three of the warmer sites. Roughly speaking, growth rates in localities where multiple genotypes survived were consistent with cogradient variation.

Other experiments yielded results consistent with countergradient variation or no variation in performance curves. Cleavitt (2004) reciprocally transplanted fragments of a subalpine moss (*Mnium arizonicum*) and a lowland moss (*M. spinulosum*). In both environments, the subalpine species survived as well as or better than the lowland species. Furthermore, quantum yield was greater for subalpine adults in both environments. By transplanting tadpoles between lowland and mountain ponds, Berven (1982) observed faster growth (but not faster development) of montane genotypes in both environments. Reciprocal transplants of lizards (Sorci et al. 1996) and frogs (Morrison and Hero 2003a) along altitudinal gradients yielded no evidence of adaptation to local environments. Overall, reciprocal transplant experiments confirm the conclusion we drew from a survey of comparative data: simple optimality models cannot account for some of the variation in thermal sensitivity observed along geographic clines.

3.5.2 Laboratory selection experiments

Laboratory experiments enable researchers to directly explore the evolutionary consequences of specific selective environments (see Chapter 1). To date, laboratory selection of short-lived organisms—viruses, bacteria, and fruit flies—have contributed to our understanding of thermal adaptation. In each experiment, a population was maintained at a single constant temperature for many generations to enable adaptation to the artificial conditions of the laboratory. Then, the population was subdivided to form selection lines. Each line was exposed to one of several thermal environments for another series of generations. At the end of this period, the performance curves of these lines were compared. For each experiment, optimality models predict specific outcomes; for example, selection at a constant temperature should produce a thermal specialist that performs maximally at this temperature.

Bennett, Lenski, and their colleagues have conducted some of the most elegant experimental studies of thermal adaptation. Taking advantage of *Escherichia coli*'s short generations, ease of culture, and genetic markers, these investigators designed a selection experiment that meets the rigorous standards required to infer adaptation (see Box 1.4). First, selection lines were established from populations that had been raised under laboratory conditions for 2000 generations; thus, the investigators were unlikely to have confounded adaptation to artificial conditions with adaptation to environmental temperature. Second, sufficient replication was used to generate confidence intervals for the mean response to selection. Third, genetic markers enabled the researchers to eliminate the possibility that contamination or migration caused the observed patterns. Finally, a remarkable feature of *E. coli* makes these studies of adaptation truly unique: *E. coli* survives prolonged exposure to freezing temperatures, enabling Bennett and colleagues to store ancestral populations and directly compare the performances of ancestral and derived lineages.

Bennett et al. (1992) used a population of *E. coli* that had been cultured at 37°C for 2000 generations. The population was raised in flasks of medium and was diluted daily through serial transfers. After an initial period, the population was divided into 24 selection lines, each of which was assigned to a thermal environment. Six lines were shifted up by 5°C (42°C lines), six lines were shifted down by 5°C (32°C lines), and six lines were cycled daily between 32°C and 42°C (32–42°C lines); a fourth set of six lines remained at 37°C to enable the researchers to consider further adaptation to laboratory conditions. These lines were maintained at their respective thermal regime for 2000 generations. After this period of selection, the thermal sensitivities of fitness were described in two ways. First, they estimated absolute fitness in terms of clonal growth. Second, they calculated the fitness of the derived line relative to its ancestor by resuscitating the ancestral genotype and growing it alongside the descendants (Bennett and Lenski 1993, 1996; Bennett et al. 1992).

During the experiment, all lines adapted to their thermal environment. In fact, adaptation of the thermal optimum was nearly perfect! Thermal optima for lines that evolved at 32°C, 37°C, and 42°C equaled 30°C, 37°C, and 42°C, respectively. Moreover, lines in the fluctuating environment evolved a thermal optimum equal to the upper mode (42°C), which corresponds exactly to the prediction of Gilchrist's model. In my mind, these matches between theoretical predictions and empirical observations are the most impressive evidence of thermal adaptation to date. Still, some important predictions of the theory could not be confirmed. Optimality models assume that adaptation to the mean or modal operative temperature will come at the expense of fitness at other temperatures. This phenomenon was seen in some lines but not others. Adaptation to 20°C was accompanied by a decrement in fitness at 42°C, but the converse was not true. In fact, most lines maintained their tolerance ranges despite the expectation that specialists would have evolved at constant temperatures (Fig. 3.27). Perhaps the duration of selection was insufficient for specialists to evolve. In a similar experiment, selection for 20 000 generations at 37°C resulted in increasing specialization over the first 10 000 generations, with significant increases in specialization occurring only after the first 2000 generations (Cooper et al. 2001).

Huey and colleagues (1991) compared the performances of flies that had evolved at 16.5°C and 25°C for over 4 years. They found only weak support for the predictions of Gilchrist's model. Flies that evolved at 16.5°C developed faster at 16.5°C and slower at 25°C than flies that evolved at 25°C. Nevertheless, differences in developmental rates were small and both sets of selection lines developed faster at 25°C than they did at 16.5°C. Hence, any shift in the thermal optimum was likely to be very slight. Possibly, the thermal limits of survivorship responded to selection because flies that evolved at

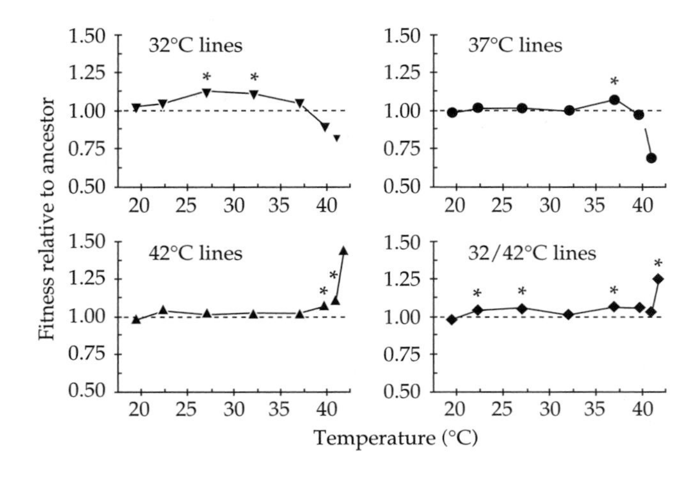

Figure 3.27 Experimental evolution of *Escherichia coli* at 32°C, 37°C, 42°C, or 32–42°C for 2000 generations increased fitness in the selective environment without decreasing fitness in other environments. Plots show the mean fitness of selected lines relative to the mean fitness of the ancestral line over a range of temperatures. Asterisks denote temperatures at which the fitness of the selected lines differed significantly from the fitness of the ancestral line. Adapted from Bennett and Lenski (1993) with permission from Blackwell Publishing.

16.5°C were less likely to survive a brief exposure to 39.5°C than were flies that evolved at 25°C. One year following this study, the lines at 25°C were divided to create an additional set of lines, which were shifted to 29°C. After 4 additional years of evolution, James and Partridge (1995) remeasured developmental times in the three sets of lines. At temperatures ranging from 16.5°C to 29°C, the lines that evolved at 16.5°C developed faster than the lines that evolved at either 25°C or 29°C. Countergradient variation, which has been documented often among natural populations (see Section 3.4), had evolved in the laboratory! The survivorship following heat shock still differed predictably among lines. However, the knockdown temperature, the survivorship following cold shock, and the thermal optimum for sprinting differed very slightly or not at all (Gilchrist et al. 1997).

Recently, viruses have been added to the family of organisms used to study experimental adaptation to temperature. Holder and Bull (2001) permitted two bacteriophages to evolve through serial transfers at a constant temperature. Each phage was introduced into a culture flask with an excess of bacterial cells and was incubated for a period of time. Each infected bacterial cell lysed to yield approximately 100 viral progeny, such that the viral population increased exponentially during incubation. Following incubation, a subsample of the viral population was transferred to a new flask of naïve bacterial cells and the cycle continued. Large populations (10^6) were used to minimize genetic drift during the experiment.

Selective regimes were tailored to each phage. The investigators knew one of the phages (G4) tolerated high temperatures poorly; therefore, they cultured this phage at 37°C for 40 transfers, then at 41.5°C for 50 transfers, and finally at 44°C for 50 transfers. The other phage (ϕX174) was cultured directly at 44°C for 50 transfers. At certain intervals, the fitnesses (or doubling rates) of the phages were measured at 37°C and 44°C to assess whether adaptation had occurred. The fitness of the G4 phage increased slowly and steadily throughout the experiment, but the fitness of the ϕX174 phage increased very sharply within five generations and remained relatively constant thereafter (Fig. 3.28). During adaptation to 44°C, the ϕX174 phage suffered a slight decrease in fitness at 37°C, but the G4 phage suffered no decrease in fitness at this ancestral temperature. Possibly, loss of fitness at lower temperatures occurred but went undocumented. Following up on this possibility, Knies and colleagues (2006) analyzed the pattern of adaptation in the G4 phage using

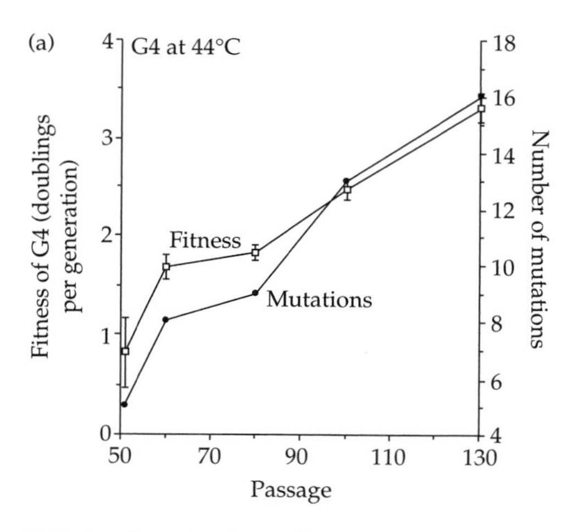

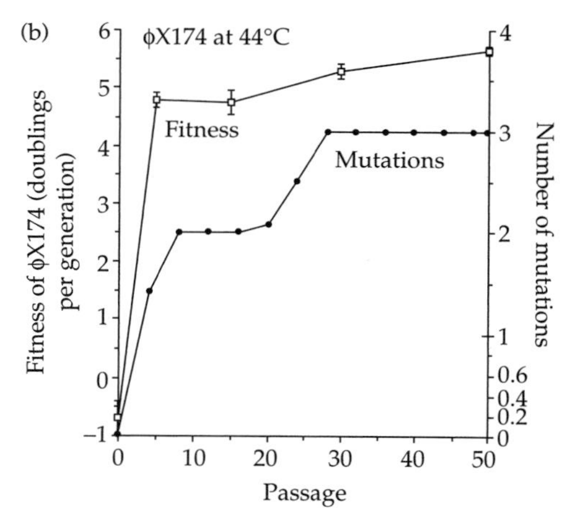

Figure 3.28 Experimental evolution of bacteriophages at 44°C revealed slow and steady adaptation by one phage (a) and rapid adaptation by another (b). Each plot shows the accumulation of mutations (filled circles) in a phage, along with the change in fitness (open circles). Error bars denote standard errors of the mean of four assays. Adapted from Holder and Bull (2001) with permission. All copyright is retained by the Genetics Society of America.

Box 3.4 Template mode of variation: a novel statistical method for describing variation among performance curves

Assuming performance curves have a common form or template, one can describe variation in these curves among genotypes in terms of the shifting or stretching of this template. Kingsolver (2001) described three modes of variation: hotter–colder, faster–slower, and specialist–generalist (Fig. 3.29). The hotter–colder and faster–slower modes describe horizontal and vertical shifts, respectively. The specialist–generalist mode describes a simultaneous horizontal stretch and vertical shift.

Izem and Kingsolver (2005) showed how one can estimate the percentage of variation associated with each of the modes. First, a template model must be fitted to data for multiple genotypes. These investigators used a simple model with three parameters of interest:

$$P_i(T_b) = a_i + \frac{1}{b_i} P\left[\frac{1}{b_i}(T_b - m_i)\right], \tag{3.15}$$

where P_i equals the performance of genotype i, and the parameters a_i, m_i, and b_i set the mean performance, thermal optimum, and performance breadth, respectively, of the template. By examining differences in these parameters among genotypes, one can determine the degree to which each of the three modes contributes to variation in the performance curve.

Importantly, the specialist–generalist mode should not be confused with the specialist–generalist tradeoffs discussed throughout this chapter. The specialist–generalist mode defines a pattern of variation in the performance curve, whereas specialist–generalist tradeoffs define a class of mechanisms that lead to such patterns. Actually, both the faster–slower and the specialist–generalist modes can result from specialist–generalist tradeoffs. Therefore, the concepts of modes and tradeoffs are related but not equal.

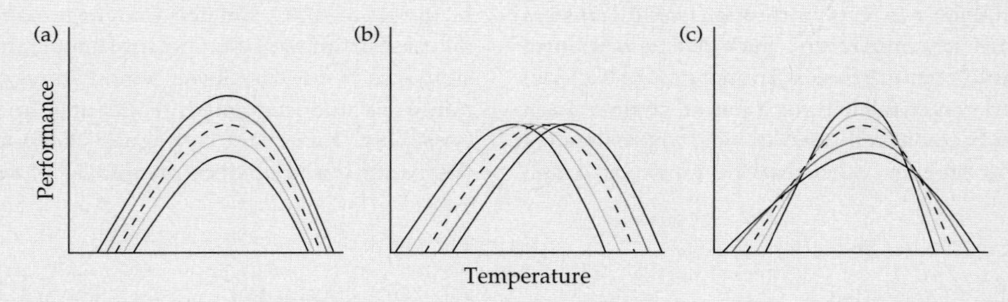

Figure 3.29 Using template mode of variation, one can partition the variation in a performance curve among three modes: (a) faster–slower, (b) hotter–colder, and (c) specialist–generalist. These modes reflect changes in the parameters of a template function. Adapted from Knies et al. (2006) with permission from the Public Library of Science.

a method known as template mode of variation (Box 3.4). They obtained samples of this phage from the ancestral and selected populations and measured the thermal sensitivity of fitness over a much broader range of temperatures (27–44°C). A shift in the thermal optimum accounted for almost 50% of the evolutionary change in the thermal reaction norm. This analysis nicely demonstrated that adaptation to high temperature caused maladaptation to low temperatures, as assumed by optimality models.

3.5.3 Conclusions from reciprocal transplant and laboratory selection experiments

Reciprocal transplant and laboratory selection experiments have corroborated the failure of optimality models to explain the evolution of thermal sensitivity. Many reciprocal transplant experiments revealed evidence of local adaptation, but adaptation to one environment does not always cause maladaptation to other environments. Selection experiments with *E. coli* and *D. melanogaster* demonstrated conclusively that a jack of all temperatures need not be a master of none. Furthermore, the experiment with *D. melanogaster* showed that thermal optima sometimes evolve slowly even in the face of strong selective pressure. Given the designs of these experiments, we cannot attribute the mismatch between theory and observation to poor knowledge of the thermal environment. Instead, we must question other assumptions of optimality models.

3.6 Have phenotypic constraints been correctly identified?

Two constraints define the phenotypic functions of the optimality models described in this chapter: (1) performance curves cannot change by acclimation and (2) performance curves must maintain a constant area during evolution. The first constraint prevents the hypothetical organism from adjusting its thermal sensitivity of performance during development. In Chapter 5, we shall consider the potential for thermal acclimation and examine models that relax this constraint. The second constraint prevents an organism from adapting to one temperature without losing function at another temperature. Because specialist–generalist tradeoffs have been a major focus of this chapter, let us now consider the empirical evidence for this constraint.

3.6.1 A jack of all temperatures can be a master of all

Much evidence stands against the idea that specialist–generalist tradeoffs universally constrain performance curves. Numerous examples can be found among the comparative studies discussed in Section 3.6. Even the earliest comparisons of performance curves indicated that a jack of all temperatures could also be a master of all; for example, Huey and Hertz (1984) found that the rank order of performance among individuals of several species was fairly conserved across a range of temperatures (ca. 20°C). Much experimental work, including both quantitative genetic and artificial selection experiments, now corroborates this comparative evidence. A comparison of wasp strains raised under common environmental conditions revealed little or no correlation between the performance breadth and the maximal performance (Carriere and Boivin 1997), whereas a comparison of aphid families revealed a negative genetic correlation for males but no genetic correlation for females (Gilchrist 1996). Among 29 genotypes of *Daphnia pulicaria*, no relationship existed between the thermal breadth of fitness (r) and the maximal fitness (Palaima and Spitze 2004). Using template mode of variation (Box 3.4), Izem and Kingsolver (2005) discovered that only half of the variation among families of butterflies reflected some form of specialist–generalist tradeoff; the remaining half included variation in the height of the performance curve (13%) and variation that could not be explained by their model (33%). Selection experiments have also provided mixed support for the importance of specialist–generalist tradeoffs. Selection for greater performance at one temperature does not always cause a correlated decrease in performance at other temperatures (Anderson et al. 2005; Bennett and Lenski 1993; Carrière and Boivin 2001). As concluded from the survey of comparative data (Section 3.4), a jack of all temperatures does not have to be a master of none.

3.6.2 The proximate basis of performance determines tradeoffs

Recall that specialist–generalist tradeoffs arise from the thermal adaptation of enzymes, membranes, and organelles (see Section 3.2). Despite the sound biochemical basis for specialist–generalist tradeoffs, species can apparently avoid these tradeoffs during evolution. How do some genotypes become masters of all temperatures? And what other kinds of tradeoffs prevent these genotypes from taking over the world?

When performance curves differ between genotypes, the proximate mechanisms that generate this difference can be quite complex. These mechanisms not only include changes in the structures of enzymes, but can also include changes in concentrations (Somero et al. 1996). Furthermore, enzymes do not function in a vacuum; they interact intimately with their surroundings and these interactions affect the thermal sensitivity of enzymatic function. Additional stability emerges from interactions between enzymes and other molecules, such as ions, substrates, coenzymes, biogenic amines, and ATP-consuming chaperones (Fields 2001; Hochochka and

Somero 2002; Jaenicke 1991; Somero 2003). Greater concentrations of paralogous isozymes or extrinsic stabilizers could expand the thermal breadth of performance. Finally, the number, size, or quality of organelles, cells, tissues, and organs should also influence performance.

But any evolutionary increase in performance comes at a price! Enzymes and their stabilizers cost energy to produce or sequester. Organelles, cells, and tissues cost energy to produce and maintain. The energy expended on such adaptations to temperature must be diverted from other functions. When these kinds of thermal adaptation occur, an organism must acquire additional energy to continue performing multiple functions or reallocate energy among competing functions. For instance, an individual that maintains higher concentrations of enzymes could enhance performance at extreme temperatures without a loss of performance at moderate temperatures, but would have less energy to devote to other cellular processes (an allocation tradeoff). If we assume the same individual forages more to offset this energy demand, we should expect this behavior to impose a greater risk of predation or parasitism (an acquisition tradeoff). Therefore, every modification of a performance curve results in some type of tradeoff (Box 3.5).

To appreciate the importance of these tradeoffs, consider some of the proximate mechanisms for variation in the thermal sensitivity of growth rate. In many cases, populations in colder environments have evolved faster growth at relatively low temperatures (see Table 3.6). Yet the mechanisms that underpin these adaptations can be quite diverse. Thermal adaptation of enzymes decreases the concentrations necessary for anabolic and catabolic processes, leaving more energy available for somatic growth. Growth efficiency also increases when organisms divert energy from cellular maintenance, such as protein turnover or ion transport (Bayne 2004; Hawkins 1991, 1995; Hulbert and Else 2000). Finally, intense rates of foraging and consumption fuel rapid growth.

Accordingly, species have taken multiple routes to the same evolutionary endpoint. In some species, rapidly growing genotypes allocate more energy to growth and less to maintenance (Imsland et al. 2000; Lindgren and Laurila 2005; Oufiero and Angilletta 2006; Robinson and Partridge 2001). In other species, genotypes achieve rapid growth through high rates of energy intake (Jonassen et al. 2000; Nicieza et al. 1994). The evolution of growth rate often involves simultaneous changes in food consumption and growth efficiency (Billerbeck et al. 2000; Jonsson et al. 2001; Lonsdale and Levinton 1989; Stoks et al. 2005; Van Doorslaer and Stoks 2005a). Artificial selection has underscored the interactive role of behavior and physiology in the evolution of growth rate. Aquacultural breeding programs have produced slow- and fast-growing lines of commercially important species. In two species of oysters, rapidly growing genotypes had higher rates of food consumption, lower rates of maintenance metabolism, and lower costs of somatic growth (Bayne 1999, 2000). Tradeoffs arising from an increase in feeding or a reduction in maintenance can affect the fitness of these genotypes in certain environments (Box 3.6). Because a diversity of proximate mechanisms can transform the genotype into the phenotype, we cannot assume all evolutionary changes will give rise to specialist–generalist tradeoffs.

Once we begin to consider the interactions among acquisition, allocation, and specialist–generalist tradeoffs, we have entered the realm of life-history theory. This body of theory describes the way in which strategies of acquisition and allocation evolve. In Chapters 6 and 7, we shall see how life-history theory can help us to understand the evolution of performance curves.

3.7 Do all performances affect fitness?

At first, asking whether performances such as survivorship and reproduction affect fitness sounds silly. Still, optimality models assume a very specific relationship between performance and fitness. To find the optimal performance curve, both Lynch and Gabriel (1987) and Gilchrist (1995) assumed fitness was directly related to performance. In other words, these models propose a form of directional selection characterized by a positive linear selection gradient (see Fig. 1.6a). Does directional selection really approximate the true nature of selection? Perhaps

Box 3.5 Tradeoffs associated with the evolution of thermal performance curves

Recently, my colleagues and I summarized the kinds of tradeoffs that constrain the evolution of performance curves (Angilletta et al. 2003). These tradeoffs are best illustrated by an example: imagine performance curves for growth and reproduction. Suppose a mutation enables a genotype to grow more rapidly at low temperatures. The mutant could grow rapidly because it allocates more energy to growth than the ancestor would; however, the mutant's increase in growth rate would decrease its reproductive rate (Fig. 3.30a). Life historians refer to such relationships as allocation tradeoffs (Zera and Harshman 2001). Alternatively, the mutant could grow more rapidly by acquiring more energy, such that no energy is diverted from reproduction (Fig. 3.30b). In this case, the mutant could enhance growth rate and reproductive rate simultaneously (Reznick et al. 2000). This mutant will likely experience an acquisition tradeoff, in which an increase in the duration or intensity of feeding increases the probability of death by predation (Gotthard 2000; McPeek 2004; Stoks et al. 2005). Finally, the mutant could grow more rapidly by assimilating and using resources more efficiently (Wieser 1994). This form of adaptation would require changes in enzymatic structure that impose a specialist–generalist tradeoff (Fig. 3.30c). Importantly, genotypes can combine these different mechanisms to further enhance performance, but multiple tradeoffs would occur.

When a performance curve evolves, both the derivative and the integral of the function can change (i.e., the shape and the height can evolve). Mechanisms associated with specialist–generalist tradeoffs influence the derivative of the curve without changing its integral (see Fig. 3.30c), whereas mechanisms associated with allocation or acquisition tradeoffs influence the integral of the curve. Because a performance curve results from many interacting mechanisms, we cannot always understand tradeoffs by comparing the shapes of curves among genotypes. Evolutionary thermal biologists must closely examine the proximate mechanisms by which genotypes express different performance curves.

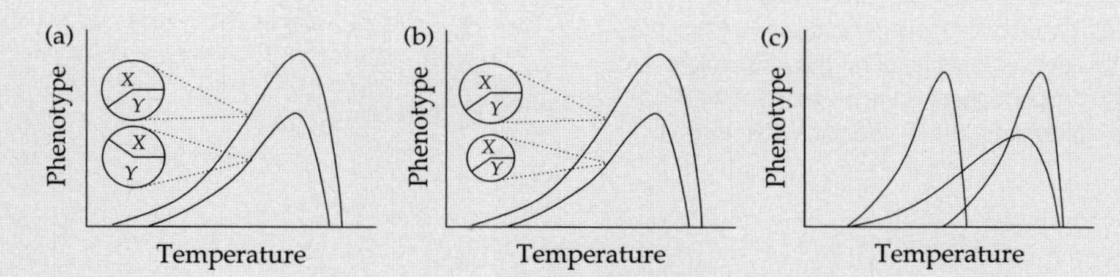

Figure 3.30 Three classes of tradeoffs associated with variation in thermal performance curves. Each plot depicts two or more performance curves; the pies depict the quantity of resources allocated to one function (X) versus another function (Y) at a given temperature. (a) Given a fixed quantity of resources, a difference in the allocation of resources between X and Y (e.g., growth and reproduction) can cause a difference in performance at all temperatures. An evolutionary change in performance resulting from function X (e.g., growth rate) would be negatively correlated with an evolutionary change in a performance resulting from function Y (e.g., fecundity). (b) A difference in the acquisition of resources can cause a difference in performance at all temperatures. In this case, an evolutionary change in the performance resulting from function X could be positively correlated with an evolutionary change in a performance resulting from function Y. (c) Performance at certain temperatures can be enhanced by thermal specialization, as a genotype can perform more efficiently within a narrow range of temperatures or less efficiently within a broader range of temperatures. Such specialist–generalist tradeoffs are commonly imposed by assuming the area under the performance curve remains constant (see Section 3.3). Adapted from Angilletta et al. (2003) with permission from Elsevier. © 2003 Elsevier Ltd.

under certain conditions submaximal performance would lead to a greater expectation of fitness, even if it reduces survivorship or reproduction during the short term. Consider an even more outrageous idea: variation in certain performances might not contribute to fitness at all.

To evaluate these ideas, we must quantify selection gradients. Unfortunately, few people have successfully measured selection gradients for performance (Kingsolver et al. 2001b). The reason is simple: researchers often use performance (e.g., survivorship) as a proxy for fitness. This practice

Box 3.6 Evolutionary tradeoffs: a case study of fish

The investigations of countergradient variation in the growth rate of the Atlantic silverside (*Menidia menidia*) constitute one of the best case studies illustrating the tradeoffs associated with the evolution of performance curves. At all temperatures, fish from Nova Scotia grow faster than fish from South Carolina because they consume more food and convert a greater fraction of their food to body mass (Billerbeck et al. 2000; Present and Conover 1992). Presumably, the costs associated with the evolution of growth rate can be reduced by combing behavioral and physiological strategies. Furthermore, these strategies enable northern fish to grow relatively quickly over a wide range of temperatures.

Still, Conover and colleagues presented compelling evidence of a tradeoff between growth rate and swimming speed (Fig. 3.31). When starved, the rapidly growing fish from Nova Scotia were unable to swim as fast as the slowly growing fish from South Carolina. Feeding further increased the difference in swimming speed between the two groups. As a consequence, fish from Nova Scotia suffered a greater risk of predation in staged encounters with predators than fish from South Carolina (Fig. 3.32). In addition to the obvious tradeoff between feeding and survivorship, the intrinsically slower swimming of northern fish might have resulted from a poor quality of muscle caused by allocation tradeoffs. Interestingly, the mechanisms that underlie variation in the performance curve depend on the phylogenetic scope of analysis. Although variation among populations of *M. menidia* was caused by different strategies of acquisition and allocation, variation between *M. menidia* and its southern congener, *M. peninsulae*, was driven partially by thermal specialization (Yamahira and Conover 2002).

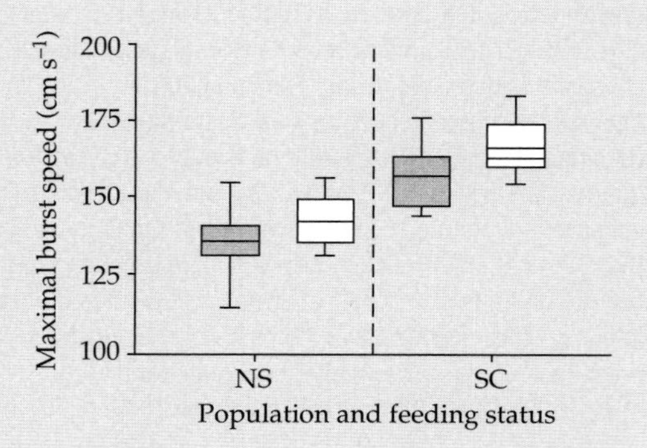

Figure 3.31 Atlantic silversides from Nova Scotia (NS) were unable to swim as fast as fish from South Carolina (SC), regardless of whether they were fed (filled boxes) or fasted (open boxes). Reproduced from Billerbeck et al. (2001) with permission from Blackwell Publishing.

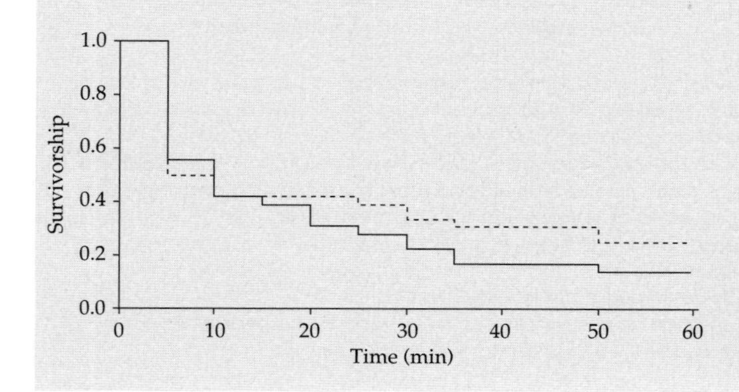

Figure 3.32 Survivorship of Atlantic silversides from South Carolina (dashed line) exceeded that of fish from Nova Scotia (solid line) during staged encounters with predators. Reproduced from Lankford et al. (2001) with permission from Blackwell Publishing.

assumes the very thing we are trying to determine—namely, that fitness depends directly on performance. In reptiles, proxies for fitness only sometimes correlate with locomotor performance (reviewed by Angilletta et al. 2002a). We know virtually nothing about selection gradients for other performances, let alone selection gradients for their thermal sensitivity.

In fact, despite the hundreds of thermal sensitivities reported in the literature, only one group of researchers has ever estimated a selection gradient for a performance curve. Kingsolver and colleagues (2007) quantified performance curves for relative growth rate in caterpillars (*Pieris rapae*). Because these caterpillars grow incredibly fast during late instars, these investigators were able to record growth by the same individuals over a range of temperatures. These acute thermal sensitivities differed from chronic thermal sensitivities (Kingsolver and Woods 1998) and better reflect the temporal scale of thermal variation in nature (Kingsolver 2000). After determining the performance curve for each individual, they exposed individuals to one of two thermal regimes: a cool cycle and a warm cycle (Fig. 3.33a). Fitness was estimated as the product of developmental rate, adult survivorship, and adult mass. Thus, the authors were able to quantitatively evaluate the assumption of directional selection.

To better appreciate the results of this experiment, let us briefly consider how selection should act on each region of the performance curve. Kingsolver and Gomulkiewicz (2003) proved that, under certain conditions, the strength of selection should depend directly on the frequency with which individuals experience a particular temperature. Therefore, selection gradients for the caterpillars should have been greatest at 13°C and 19°C in the cool and warm environments, respectively (Fig. 3.33a). In stark contrast to this expectation, selection gradients were maximal at the same temperature in both environments (Fig. 3.33b and c).

Why did the predicted selection gradients poorly match the observed selection gradients? Honestly, I do not know. But several considerations regarding the effect of performance on fitness seem worthwhile. First, performance could influence both survivorship and fecundity in complex ways. If so, the relationship between performance and fitness might depend on the ontogenetic stage of the organism or environmental variables other than temperature. Second, temporal autocorrelation of body temperatures would affect the optimal performance curve when the time-course of performance matters. Third, demographic stochasticity generally leads to a realized fitness that differs from the expected fitness (Burger and Lynch 1995). Consequently, simulation models of evolution provide valuable complements to optimality models.

3.8 Does genetic variation constrain thermal adaptation?

When using optimality models, we must assume populations in different environments reside at their respective fitness maxima. This primary assumption envelopes two secondary assumptions. First, each population must contain sufficient additive genetic variance to respond to selection. Second, thermal environments must remain fairly consistent over time for each population to reach its evolutionary equilibrium. Either low genetic variance or relatively weak selection could prevent the evolution of optimal genotypes. Additionally, gene flow among populations in different thermal environments could prevent each population from adapting to local conditions. We can explore the consequences of relaxing these core assumptions through genetic models of evolution.

Quantitative genetic and allelic models explicitly consider the genetic constraints on adaptation. In Chapter 1, we discussed several ways one could model the genetics of a performance curve (see Section 1.4.2). One way would be to focus on the genetic variance of performance at each of many temperatures. Another way would be to focus on the genetic variance of parameters that describe the performance curve, such as the thermal optimum, performance breadth, and thermal limits. Finally, we could treat the entire performance curve as an infinite dimensional trait and estimate its genetic variance–covariance function. Each of these approaches uniquely defines the traits under consideration. Because performance curves are continuous functions, researchers have generally adopted the second or third approach.

Regardless of how one chooses to model genetics, the evolution of a performance curve depends directly on genetic variances and covariances (Arnold 1992, 1994). The optimal performance curve will evolve eventually, as long as some genetic variance persists and genetic covariances are small enough to permit independent evolution of the performance curve (Gomulkiewicz and Kirkpatrick

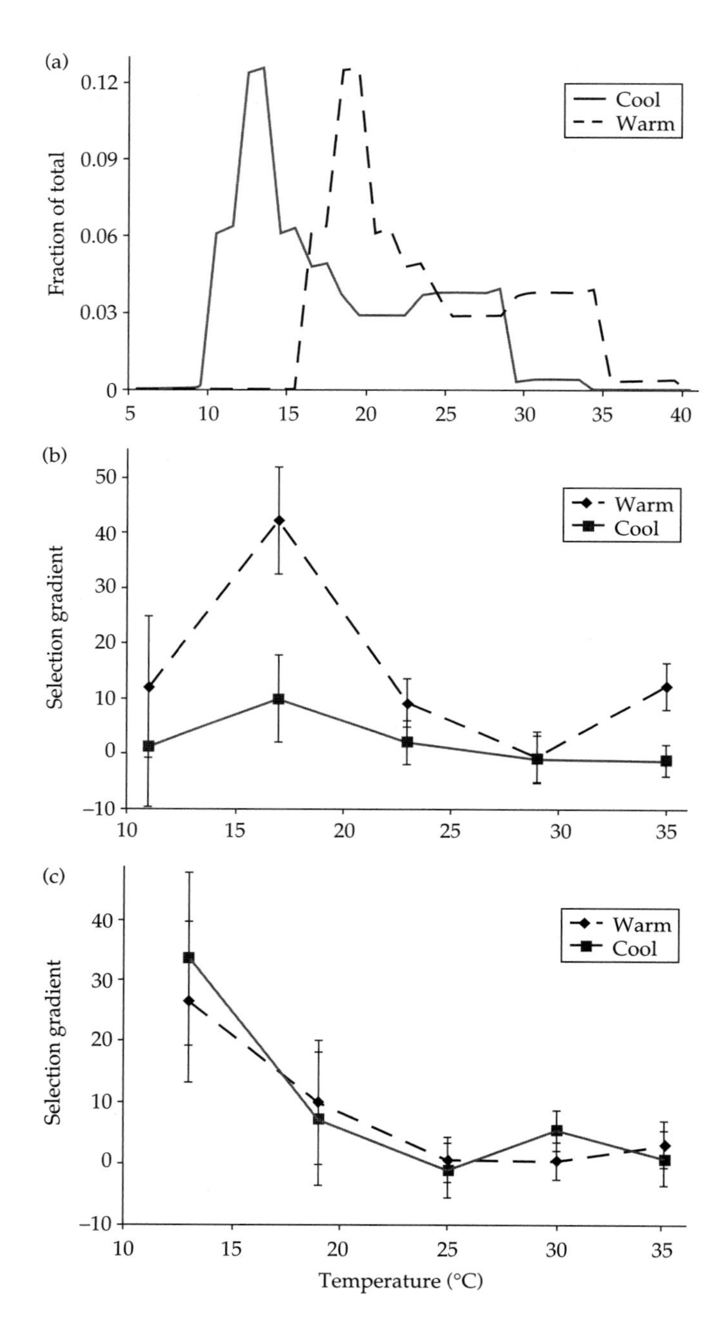

Figure 3.33 Selection gradients for growth rate at different temperatures did not match the predictions of theory. Selection gradients were calculated for caterpillars (*Pieris rapae*) in cool and warm environments (a) during two separate experiments (b and c). To compute selection gradients, fitness was estimated as the product of survivorship, developmental rate and adult mass. Reproduced from Kingsolver (2007, *The American Naturalist*, University of Chicago Press). © 2007 by The University of Chicago.

1992; Scheiner 1993; Via and Lande 1985). Nevertheless, the trajectory of evolution can be circuitous and hundreds of generations might be required to reach an evolutionary equilibrium (Via 1987). Therefore, genetic models provide valuable insights about the dynamics of adaptation (Gomulkiewicz and Kirkpatrick 1992; Via 1994).

In this section, we shall examine both quantitative genetic and allelic models of evolution (see Chapter 1). Quantitative genetic models combine selection gradients with genetic variances and covariances to project evolutionary changes. Allelic models consider the impact of mutation, drift, and selection on the course of evolution. Such models were constructed by extending the optimality models of Lynch and Gabriel (1987) and Gilchrist (1995) to include a genetic basis for the performance curve. To apply either kind of model, one must replace the assumption that populations reside at their evolutionary equilibria with assumptions about genes, mutations, and drift. Given these new assumptions, one can simulate the evolution of a performance curve. Below, we shall draw on such simulations to learn how genetic constraints influence the dynamics of adaptation.

3.8.1 A quantitative genetic model based on multivariate selection theory

In a landmark paper, Kingsolver and colleagues (2001a) showed how one can predict the evolution of a performance curve by simply combining knowledge of selection gradients and genetic constraints. These researchers treated a performance curve as an infinitely dimensional trait (see Section 1.4.2). Rather than focus on a discrete matrix of genetic variances and covariances, they focused on a continuous function:

$$G(T,\theta), \tag{3.16}$$

which describes the genetic variance of performance at any temperature T and the genetic covariance between any pair of temperatures T and θ. Figure 3.34 shows a genetic variance–covariance function estimated from variation in growth rate among full-sibling families of butterflies (*Pieris rapae*). Correspondingly, a function of selection gradients, $\beta(\theta)$, describes the linear selection gradients over an infinite set of temperatures. Figure 3.35 shows three selection gradient functions for larval growth rate in the same species.

The response to selection depends on genetic variances and covariances. The direct effect of selection results from the selection gradient and the genetic variance at T. Indirect effects of selection result from selection gradients at temperatures other than T and genetic covariances between T and θ. These genetic covariances could reflect a combination of several different biological processes (see Box 3.5). By integrating the product of the genetic variance–covariance function and the selection gradient function over all temperatures,

$$\Delta\bar{P}(T) = \int G(T,\theta)\beta(\theta)d\theta, \tag{3.17}$$

Kingsolver and colleagues summed the direct and indirect effects of selection on performance at temperature T (Fig. 3.36).

Many assumptions must be made to implement this quantitative genetic approach. To estimate the genetic variance–covariance function, we must assume that phenotypic variation among families excludes the effects of dominance, epistasis, and shared environments (see Chapter 1). To estimate the selection-gradient function, we must define fitness appropriately and quantify it accurately. Furthermore, both functions must remain constant during the period in which we wish to explore the evolutionary dynamics. If either function changes appreciably, the dynamics of evolution will also change. Indeed, we should expect both functions to change during evolution. The genetic variance–covariance function changes systematically under strong selection and genetic drift because mutation and gene flow might not be sufficient to maintain genetic variation (Steppan et al. 2002). Likewise, the selection gradient function will change when the fitness landscape possesses maxima and minima (i.e., peaks and valleys); in such cases, both linear (directional) and nonlinear (stabilizing) selection gradients must be estimated over a suitable range of performance. In the simple example described by Kingsolver and colleagues, only directional selection was considered. Given the assumptions required, this quantitative genetic approach would be most useful for predicting the evolution of performance curves over a few generations.

3.8.2 A genetic model for survivorship and related performances

Gabriel (1988) extended the optimality model of Lynch and Gabriel (1987) to include genetic constraints on the evolution of the thermal sensitivity of survivorship. The environmental, phenotypic, and

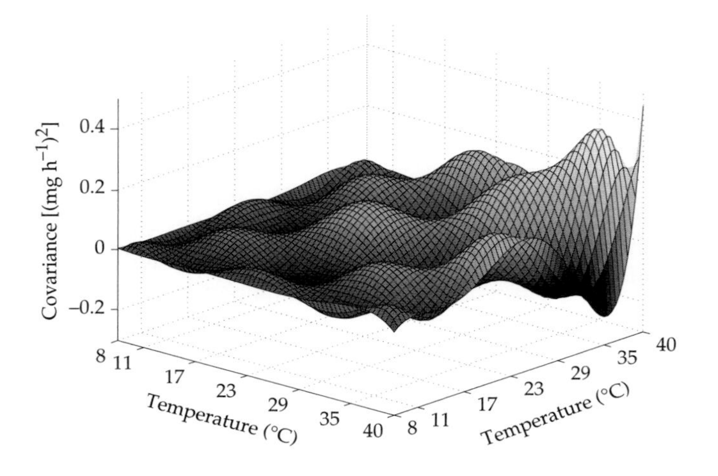

Figure 3.34 A genetic variance–covariance function defines the genetic constraints on the evolution of a performance curve. This function was estimated from variation in the thermal sensitivities of larval growth rate among families of butterflies, *Pieris rapae*. Reproduced from Kingsolver et al. (2004) with permission from Blackwell Publishing.

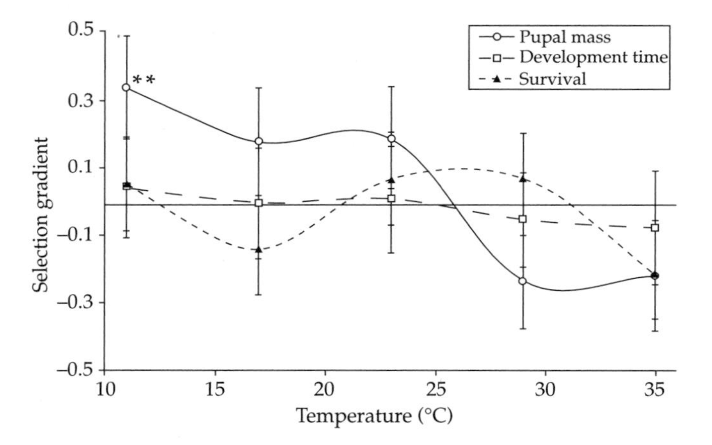

Figure 3.35 A selection-gradient function defines the strength of selection for performance at each temperature. These functions were estimated for the larval growth rate of a butterfly, *Pieris rapae*. To calculate selection gradients, pupal mass, developmental time, and survival were used as three proxies for fitness. The selection gradient at 11°C, estimated from pupal mass, differed significantly from zero (as indicated by asterisks). Adapted from Kingsolver et al. (2001a) with kind permission from Springer Science and Business Media. © 2001 Kluwer Academic Publishers.

fitness functions followed Lynch and Gabriel (1987), with the exception that genetic values of the performance curve were modeled via quantitative genetics. Recall from Section 3.3.1 that the performance curve of Lynch and Gabriel's model was defined by the genotypic values g_1 and g_2. In Gabriel's model, each of these values was defined by the allelic contribution of 20 diploid loci. The environment and the performance curve determined survival to reproduction. Surviving individuals were randomly culled to 250 individuals and these survivors produced the next generation. Reproduction could occur through parthenogenesis or sex, depending on the simulation. Mutation of alleles occurred with a probability of 0.1 per genome.

Using this model, Gabriel asked how the performance curve would respond to a major shift in the mean or variance of operative environmental temperature. For example, if a drastic reduction in the mean temperature occurs, we might expect the thermal optimum (g_1) to evolve to match the new mean. Although our expectation was confirmed through simulations, the population required hundreds of generations to attain the new optimal value of g_1 (Fig. 3.37a). During adaptation, a greater performance breadth (g_2) evolved to enhance performance of the genotype while the thermal optimum approached the mean environmental temperature (Fig. 3.37b). Once the optimal value of g_1 had evolved, the performance breadth returned to its

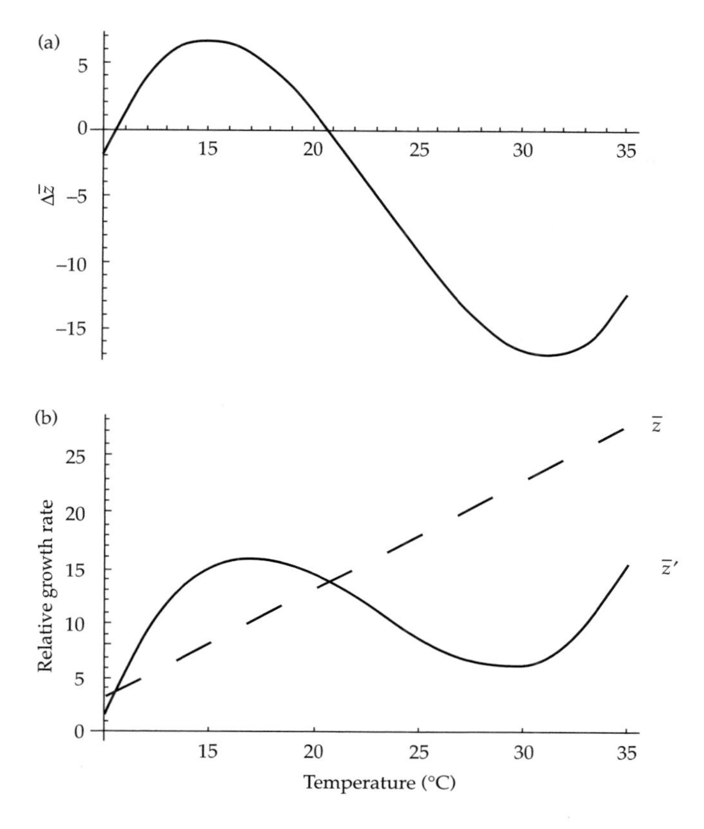

Figure 3.36 Predicted evolutionary response of the performance curve for larval growth rate in a butterfly, *Pieris rapae*. (a) Estimates of the selection-gradient function (Fig. 3.35) and the genetic variance–covariance function (Fig. 3.34) were used to calculate the expected change in the performance curve. (b) The predicted performance curve after selection (solid line) differs markedly from the performance curve before selection (dashed line). Adapted from Kingsolver et al. (2001a) with kind permission from Springer Science and Business Media.

original value. Now consider a drastic reduction in the variance of temperature among generations. In this case, we should expect the value of g_2 to decrease correspondingly. Again, the process of adaptation required hundreds of generations. Not surprisingly, adaptation in both scenarios proceeded more rapidly when sexual reproduction occurred more frequently.

3.8.3 A genetic model for fecundity and related performances

Gilchrist (2000) also extended his optimality model to consider the consequences of mutation, selection, and drift for the evolution of performance curves. Instead of defining distinct loci with allelic contributions to performance, he considered genotypic values for parameters of the performance curve (CT_{max} and B; see Section 3.3.2). The parameters of the performance curve were ultimately determined by genotypic and environmental values, the latter being normally distributed with an expected value of zero. The genotypic values changed by some quantity during mutation. Gilchrist allowed replicate populations to evolve for 20 000 generations under each of the nine environmental states depicted in Fig. 3.17.

Under the combined processes of selection, mutation, and drift, adaptation required many, many generations. In fact, Gilchrist observed a more pronounced lag than we might have expected from Gabriel's simulations (Fig. 3.38). In certain environments, the performance breadth was still suboptimal after 20 000 generations! Adaptation occurred most slowly when temperature varied greatly within generations because specialists achieved only slightly greater fitness than generalists (see Fig. 3.18). Other factors also contributed to the rate

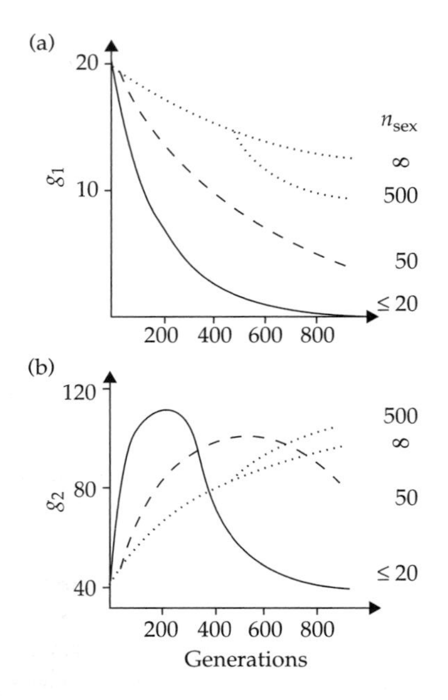

Figure 3.37 Following an abrupt change in the mean environmental temperature, the thermal optimum (g_1) and the performance breadth (g_2) took hundreds of generations to approach their new optima. Various frequencies of sexual reproduction (n_{sex}) were explored, ranging from at least once per 20 generations (≤ 20) to never (∞). Adapted from Gabriel (1988) with kind permission from Springer Science and Business Media. © Springer-Verlag 1988.

of adaptation. In all environments, a genetic correlation between CT_{max} and B prevented populations from directly ascending the fitness landscape. This correlation resulted from two assumptions of the model:

1. the performance curve was an asymmetric function (eqn 3.12) rather than a symmetric function as in Lynch and Gabriel's model (eqn 3.5);
2. the performance curve evolved because of genes for thermal plasticity rather than genes for performance at different temperatures (i.e., the parameter-based approach was used instead of the character-based approach; see Chapter 1).

If these assumptions were relaxed, adaptation would have occurred more rapidly (Gilchrist 2000). The first assumption seems appropriate for most types of performances, but the second assumption has stimulated controversy (Via et al. 1995).

3.8.4 Predictions of quantitative genetic models depend on genetic parameters

Quantitative genetic models tell us that adaptation depends on the genetic variance and the genetic architecture of thermal sensitivity. So what do we know about these things in real populations?

In most populations examined to date, some degree of genetic variance in thermal sensitivity has been found. Genetic variances have been estimated from similarities among relatives and responses to selection (see Chapter 1). Most studies have focused on parameters of the performance curve such as performance breadth or thermal limits (including indices of thermal tolerance, such as knockdown temperature). Gilchrist (1996) used a breeding experiment to estimate genetic variation in the performance breadth of male aphids ($h^2 = 0.19, P = 0.06$), but could not detect genetic variation among female aphids. The same population possessed no genetic variation in the thermal optimum ($h^2 = -0.08$, $P = 0.32$). Similar experiments were used to detect genetic variation in the thermal sensitivity of growth rate in populations of caterpillars (Kingsolver et al. 2004) and fish (Yamahira et al. 2007). Selection experiments have revealed substantial genetic variance of thermal tolerance in populations of *Drosophila* and other insects (Anderson et al. 2005; Bubli et al. 1998; Carriere and Boivin 2001; Folk et al. 2006; Gilchrist and Huey 1999; Loeschcke and Krebs 1996). In contrast, Baer and Travis (2000) observed no response to artificial selection of the critical thermal maximum in a poeciliid fish (*Heterandria formosa*). Given that most experiments have detected some degree of genetic variation, one might be tempted to conclude that genetic constraints on adaptation are relatively unimportant. However, two major caveats must be considered. First, the observed genetic variances and covariances do not permit all kinds of evolutionary responses. For example, variation among families of caterpillars suggests the performance breadth can evolve more readily than the thermal optimum (Izem and Kingsolver 2005). Yet, genetic variation among families of fish suggests the mean performance can evolve more readily than either the thermal optimum or the performance breadth (Yamahira et al. 2007). Second, genetic variation

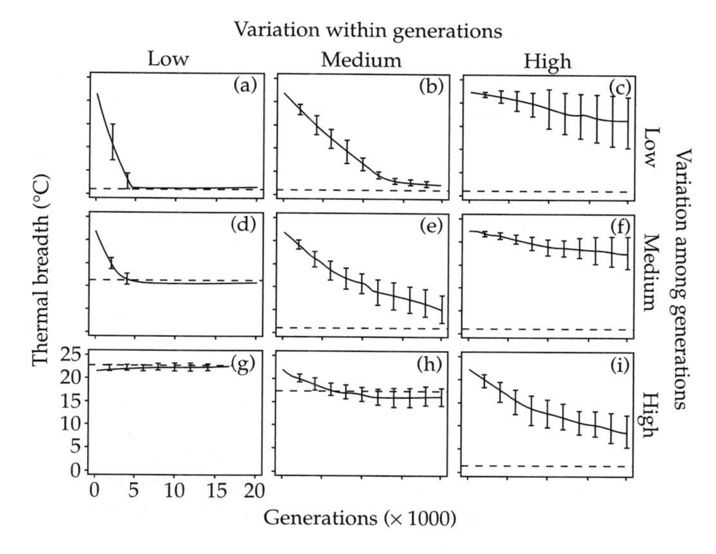

Figure 3.38 Simulated evolution of thermal breadth in the nine environments shown in Fig. 3.17. Error bars are standard deviations of the mean response for five replicate populations. The dashed line marks the optimal breadth. Adapted from Gilchrist (2000) with permission from Elsevier. © 2000 Elsevier Ltd.

in the performance curve depends on the environmental context (Kingsolver et al. 2006). Third, and most importantly, these studies focus on a tiny set of species. Clearly, much more research will be necessary to draw general conclusions about the role of genetic constraints in the evolution of thermal sensitivity.

As natural selection and genetic drift proceed, mutation must replenish the genetic variance required for adaptation. Theory indicates genetic variances and covariances for some traits will change markedly over time (Gilchrist et al. 2000; Jones et al. 2003). But how many mutations must occur to sustain thermal adaptation? How many loci contribute to the shape of a performance curve? Are differences in thermal sensitivity caused by many mutations with small effects or a few mutations with large effects? These questions become critical as we try to understand how genetic variation persists during adaptation. Studies of viruses have begun to shed some light on these issues. Apparently, two or three mutations accounted for the majority of adaptation to the extreme temperature of 44°C (Holder and Bull 2001). Interestingly, the fitness benefit of these mutations depended on the genetic background; a mutation improved fitness most when the fitness of the background genotype was lowest (Bull et al. 2000). Many more studies of this kind will be required to characterize the nature of genetic constraints on thermal adaptation.

3.9 Does gene flow constrain thermal adaptation?

Until now, we have implicitly assumed that natural selection, mutation, and genetic drift occur in a thoroughly mixed gene pool. Even in Lynch and Gabriel's model, in which offspring dispersed throughout a spatially heterogeneous landscape, all individuals experienced the same degree of spatial and temporal heterogeneity. But what if a population was more like a set of spatially and genetically structured subpopulations? How would gene flow between subpopulations influence the evolution of performance curves?

Gene flow can prevent populations from adapting perfectly to their environments (Ronce and Kirkpatrick 2001). In the context of thermal physiology, two distinct phenomena can emerge. Thermal optima for performance might diverge between subpopulations, but a balance between selection and migration will maintain some degree of maladaptation to each environment. Alternatively, similar thermal optima might exist between subpopulations, such that only one of the subpopulations adapts well to its local environment. The latter phenomenon, referred to as migrational meltdown, occurs when migration between populations differs dramatically (Ronce and Kirkpatrick 2001).

How does migration affect the evolution of thermal physiology when a species colonizes a novel environment? To address this question, we can draw on recent models by Holt and colleagues (2003, 2004). These investigators viewed the novel environment as a sink because a population of well-adapted genotypes would become maladapted following dispersal from the current range. For simplicity, they assumed a "black-hole" sink, to which some individuals emigrated each generation but from which no individuals ever returned. Their model described a generic trait whose value contributed to fitness ($\overline{w}$) via survivorship:

$$\overline{w} = P_{max}\exp^{\left[-\frac{(\bar{g}-\theta)^2}{2(\sigma^2+V_P)}\right]}, \quad (3.18)$$

where P_{max} equals the survivorship of a perfectly adapted genotype, $\bar{g}$ equals the mean genotypic value, θ equals the optimal phenotype in the sink, V_P equals the phenotypic variance, and $1/\sigma^2$ determines the strength of selection. For our purposes, we can think of this trait as a symmetrical performance curve with a thermal optimum equal to g and a performance breadth proportional to σ^2. Holt and colleagues explored the process of adaptation with continuous analytical models and individual-based simulations; the latter approach also included stochastic demographic processes resulting in genetic drift and possibly extinction.

The probability of adaptation depends on many factors, including the degree of initial maladaptation, the size of the population, the rate of immigration, the variation in temperature, and the thermal performance breadth. Under any set of conditions, the population would be more likely to adapt when its mean thermal optimum more closely corresponds to the mean operative temperature of the environment. A critical level of maladaptation exists, above which adaptation never occurs because the population shrinks severely during selection. For a given level of maladaptation, the population would be more likely to adapt when more individuals immigrated per generation (Fig. 3.39a). A wide performance breadth or temporal variation in temperature would also promote adaptation. Moderate stochastic variation in temperature promotes adaptation, but extreme stochastic variation inhibits adaptation because certain periods cause severe maladaptation (Fig 3.39b). When temperatures are temporally autocorrelated or cycle periodically, the population adapts more readily because longer stretches of good periods enable the population to rebound. This result has interesting implications for

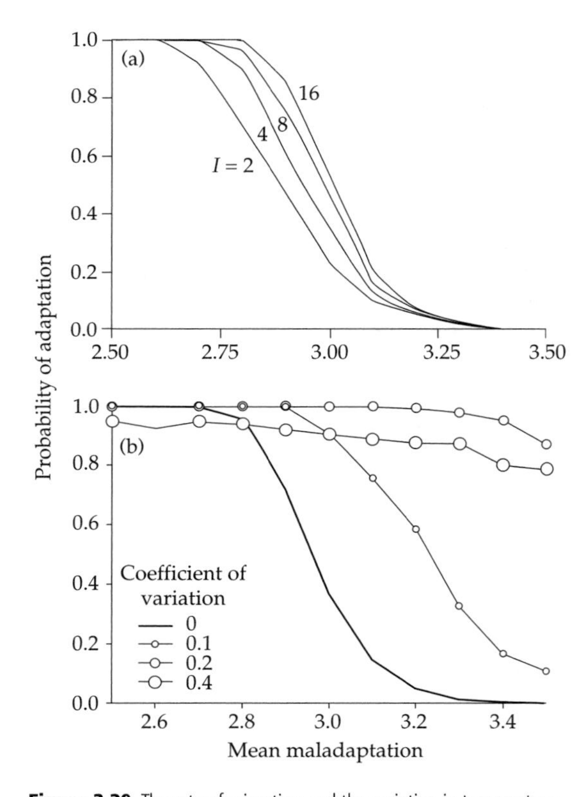

Figure 3.39 The rate of migration and the variation in temperature affect the probability of adaptation to a novel environment. (a) For any degree of maladaptation, the probability of adaptation increases as the number of immigrants per generation (*I*) increases. (b) For any degree of maladaptation, the population would be more likely to adapt when environmental temperature varies moderately over time. (a) Adapted from Holt et al. (2004) with permission from the Royal Society. (b) Adapted from Holt et al (2003, *The American Naturalist*, University of Chicago Press). © 2003 by The University of Chicago.

adaptation during range expansion because temporal variation in temperature differs latitudinally (see Chapter 2). Temporal variation should facilitate adaptation at low latitudes, but inhibit adaptation at high latitudes.

Overall, migration appears to be a double-edged sword. In small populations, immigrants help to prevent extinction during the process of adaptation. In moderate to large populations, however, the influx of maladapted genotypes prevents perfect adaptation. Some degree of maladaptation will always persist when migration occurs. Thus, gene flow initially enhances the probability of adaptation during a range expansion but reduces this probability once adaptation has begun.

3.10 Conclusions

Throughout this chapter, we have seen that the current theory provides a limited understanding of the evolution of thermal sensitivity. Although many mechanisms of thermal adaptation should impose a specialist–generalist tradeoff, optimality models that incorporate this constraint often fail to explain empirical observations. The mismatch between theory and data seems particularly great for the thermal sensitivities of growth and development, which often follow patterns of countergradient or cogradient variation. We cannot solely attribute the failures of the theory to a naïve view of thermal heterogeneity because reciprocal transplant and laboratory selection experiments have also revealed the failings of optimality models. We now have sufficient evidence for the need to consider additional constraints when modeling fitness landscapes, such as acquisition and allocation tradeoffs. We also need to pay more attention to evolutionary processes that limit rates of adaptation, such as mutation, drift, and migration. The slow rates of adaptation predicted by quantitative genetic and allelic models of performance curves might explain why experimental evolution at constant temperatures has not produced thermal specialists. Genetic models promise to guide our understanding of the limits to adaptation if we can obtain the data needed to parameterize these models for specific populations. Nevertheless, both optimality and genetic models of performance curves make critical assumptions that we know to be false. First, these models assume that organisms fail to thermoregulate in the face of severe thermal heterogeneity. Second, they assume the thermal sensitivity of performance cannot acclimate in response to temperatures experienced during ontogeny. A more advanced theory would need to account for these strategies of thermal adaptation. In route to such a theory, we next consider the evolution of thermoregulation and the evolution of thermal acclimation.

CHAPTER 4

Thermoregulation

Although environmental temperatures vary tremendously over space and time, most organisms regulate their body temperature through behavior, physiology, and morphology. By thermoregulation, I do not mean the temperature of an organism simply differs from that of its surroundings. Rather, a thermoregulator maintains a particular mean or variance of body temperature (Fig. 4.1), using neural mechanisms to sense and respond to its environment (Bicego et al. 2007). Although this phenomenon appears most pronounced in endotherms such as birds and mammals, virtually all organisms exhibit some degree of thermoregulation. For example, most organisms use behavior or physiology to avoid lethal extremes, and many ectothermic animals maintain a precise body temperature despite variation in environmental temperature (Beitinger and Fitzpatrick 1979; Brattstrom 1979; Danks 2004; Heinrich 1981; Huey 1982; Lagerspetz and Vainio 2006). Even slime molds will migrate toward a specific environmental temperature when provided with the opportunity (Whitaker and Poff 1980). In fact, the boundaries between poikilotherms and homeotherms have blurred considerably because we now know that many "homeotherms" exhibit diel and seasonal variation in body temperature (Prinzinger et al. 1991).

In this chapter, we shall examine models of optimal thermoregulation and use these models to understand the behavior of organisms in natural and artificial environments. We begin by considering the potential benefits and costs of thermoregulation and then focus on a model that maximizes the net energetic benefits. We then expand our focus to include models that also consider nonenergetic benefits and costs. Finally, we shall contemplate the origin and maintenance of endothermic thermoregulation. With minimal interpretation, models of optimal thermoregulation apply equally well to ectotherms and endotherms.

4.1 Quantifying patterns of thermoregulation

The earliest studies of thermoregulation were focused on the relationship between ambient temperature and body temperature. Using this approach, researchers attempted to distinguish between thermoregulators and thermoconformers (Fig. 4.2). A perfect thermoregulator would possess an invariant body temperature over a range of air or water temperatures. On the other hand, a perfect thermoconformer would possess a body temperature equivalent to the ambient temperature. Imperfect thermoregulators exhibit some intermediate pattern. Using these criteria, we would conclude that a variety of organisms thermoregulate to some degree under natural conditions, including insects (Harrison et al. 1996; Verdu et al. 2004), amphibians (Duellman and Trueb 1986), reptiles (Van Damme et al. 1990), and fish (Baird and Krueger 2003). More importantly, though, different species seem to adopt very different thermoregulatory strategies in similar environments, as evidenced by variation in the slope of this relationship among sympatric species (Fig. 4.3).

Although the relationship between air temperature and body temperature has been the most widely used index of thermoregulation, Heath (1964) cautioned thermal biologists to carefully consider the meaning of this relationship. He placed beer cans (filled with water) in the sun and recorded their temperatures throughout the day. The cans were generally hotter than the surrounding air, despite their obvious lack of a thermoregulatory strategy (Fig. 4.4). Not surprisingly, the physical properties

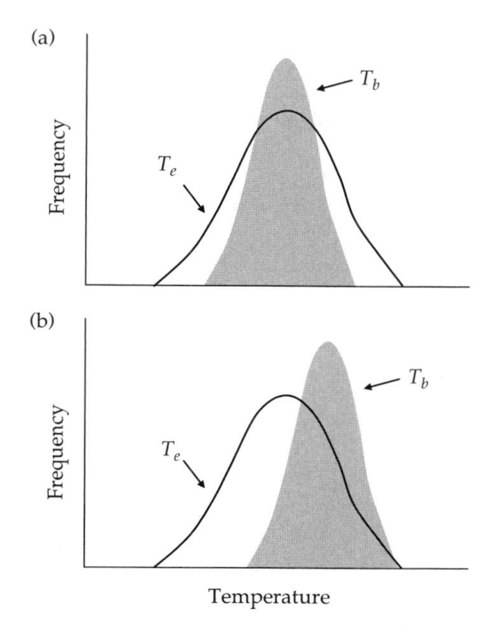

Figure 4.1 (a) Thermoregulation can reduce the variance of body temperature (T_b) relative to variance of operative environmental temperature (T_e). (b) Alternatively, thermoregulation can alter the mean and reduce the variance of body temperature relative to the mean and variance of operative environmental temperature.

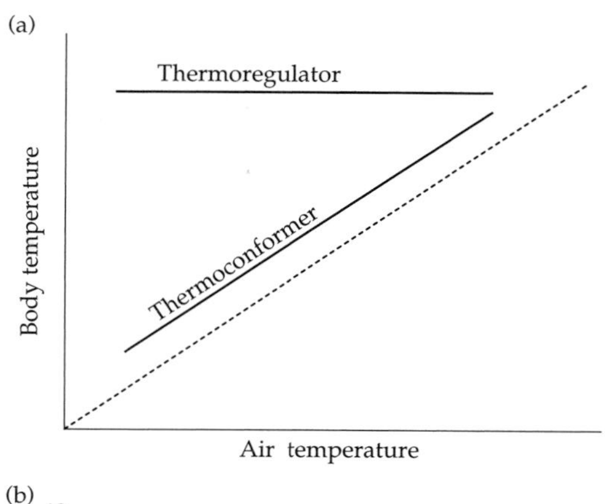

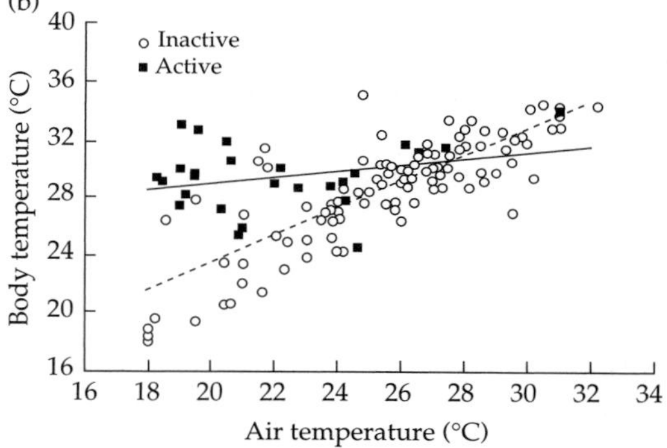

Figure 4.2 (a) The relationship between air temperature and organismal temperature might indicate some form of thermoregulation. (b) For example, the temperatures of active lizards fell within a narrow range, regardless of the air temperature. However, the temperatures of inactive lizards depended strongly on air temperature. Adapted from Bauwens et al. (1999) with permission from Blackwell Publishing.

of a can, or an organism, differ markedly from those of air. Heath's message did not fall on deaf ears. By the 1970s, researchers were combining observations of behaviors and temperatures to deduce whether animals were actually thermoregulating (see, for example, Huey et al. 1977). And by the 1980s, many researchers began to examine thermoregulation more rigorously with the help of operative temperature models (see Chapter 2). In the simplest case, the temperatures of animals were compared with the operative temperatures of their surroundings (Christian et al. 1998; Van Damme et al. 1989). In more sophisticated efforts, researchers attempted to model the frequency distribution of operative temperatures (Grant and Dunham 1988, 1990; Newell and Quinn 2005). If the distribution of body temperatures deviated from the distribution of operative temperatures, researchers concluded that the organism thermoregulated in its natural environment (see Fig. 4.5). Still, we cannot know whether these patterns reflect a regulation of body temperature because other activities might better explain deviations between potential and realized body temperatures. For example, operative temperature could covary with another environmental resource, such as water, food, or refuge. For this reason, we need to eliminate confounding factors that occur in natural environments if we want to know something about the thermal preference of an organism.

The best evidence of thermal preference comes from measuring the body temperatures of organisms in thermal arenas (Box 4.1). These arenas consist of linear gradients or joined compartments that offer an organism the ability to choose its temperature. Such simple, controlled environments offer the advantage that one can easily quantify the frequency

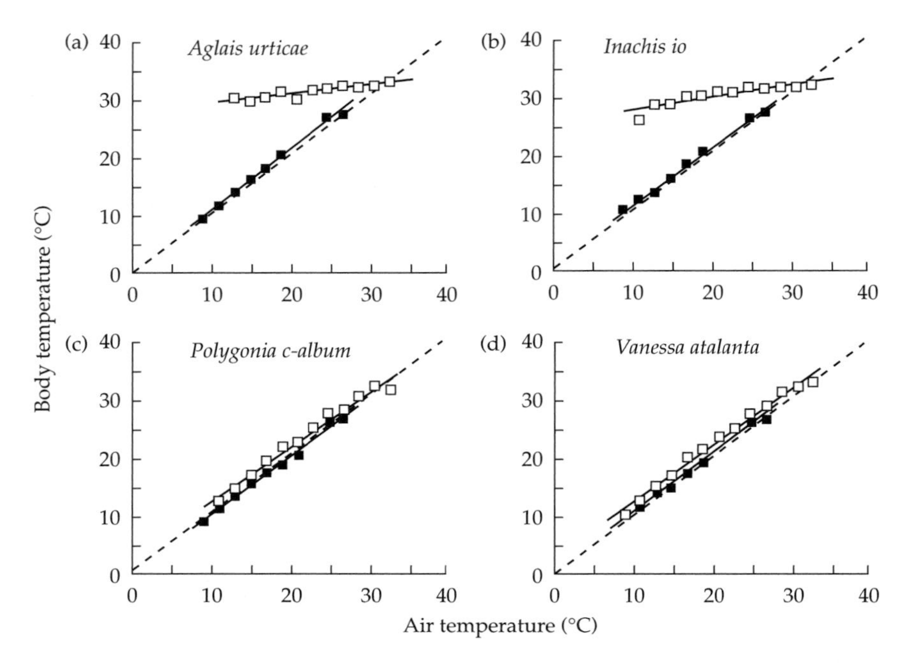

Figure 4.3 Relationships between air temperature and body temperature varied greatly among species of butterflies and depended strongly on environmental conditions. On sunny days (open symbols), *Aglais urticae* and *Inachis io* maintained a narrow range of body temperature despite a broad range of air temperatures, whereas *Polygonia c-album* and *Vanessa atalanta* conformed to air temperatures. On cloudy days (filled symbols), temperatures of all four species conformed to air temperatures. Adapted from Bryant et al. (2000) with kind permission from Springer Science and Business Media. © Springer-Verlag 2000.

of operative temperatures. By placing individuals in an environment where only temperature varies over space, researchers have confirmed distinct and repeatable thermal preferences. Although a variety of factors can influence the preferred body temperature, this measure can also be a remarkably repeatable property of a species. For example, Angilletta and Werner (1998) recorded a preferred body temperature for the gecko *Christinus marmoratus* that differed by less than a degree from the preferred body temperature recorded for the same species about three decades earlier (Licht et al. 1966b). Preferred body temperatures provide valuable information about thermoregulatory strategies in environments where no ecological constraints exist. By combining this information with measures of operative temperatures and body temperatures in natural environments, we can discover whether organisms regulate their temperature in circumstances where many factors can influence behavior.

Eventually, the development of thermal indices by Hertz and colleagues (1993) set a new standard for quantifying thermoregulation in natural environments (Box 4.2). The opportunity for thermoregulation was captured by the index d_e, which equals the difference between the mean operative temperature of the environment and an animal's preferred temperature (or preferred range of temperatures). The accuracy of thermoregulation was captured by the index d_b, which equals the difference between the mean temperature of the animal and its preferred temperature. Variations on this theme have been proposed, but all preserve the idea that d_e adequately estimates the quality of the thermal environment. Although these thermoregulatory indices emerged relatively recently, thermal biologists have already documented considerable variation in the accuracy and effectiveness of thermoregulation in natural environments (see, for example, Christian and Weavers 1996).

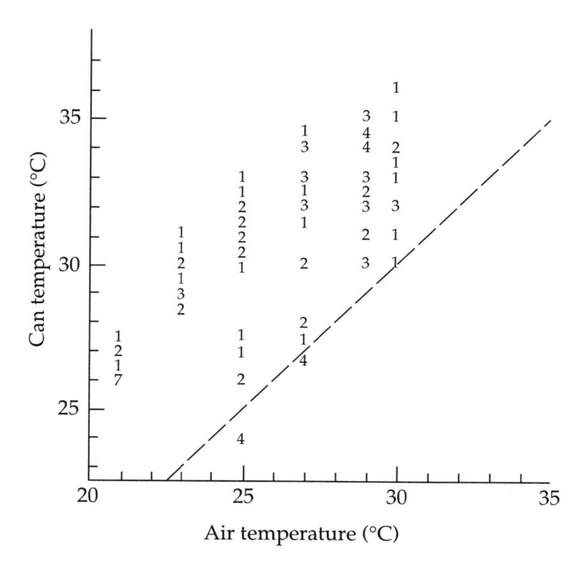

Figure 4.4 A difference in temperature between an organism and its environment does not necessarily imply thermoregulation. For example, the temperatures of beer cans differed from those of the surrounding air by as much as 8°C (numerals indicate the number of records at each temperature). This pattern likely resulted from the passive orientation of the cans to the changing angle of solar radiation; upright cans should have intercepted the most radiation during early and late hours of the day, when the air was relatively cool (Huey et al. 1977). From Heath (1964). Reprinted with permission from AAAS.

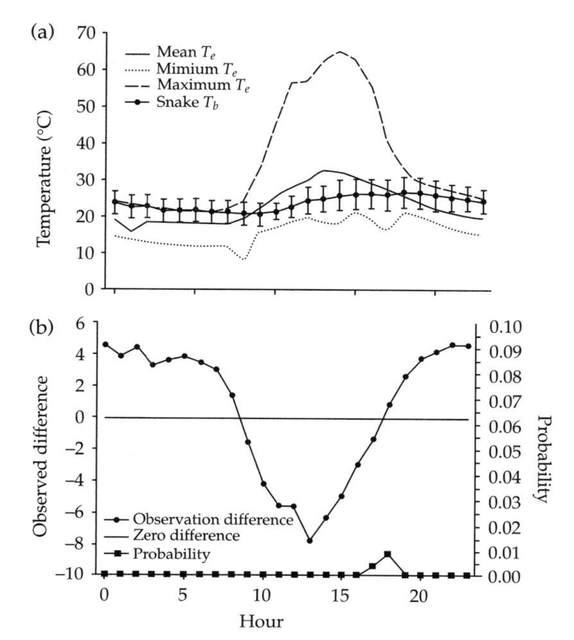

Figure 4.5 During August 1997, timber rattlesnakes in Arkansas maintained a distribution of body temperatures that differed from the distribution of operative temperatures. (a) The mean body temperature exceeded the mean operative temperature in the morning and the evening, but the reverse was true at midday. (b) The observed difference between the mean body temperature and the mean operative temperature (left axis, circular symbols) was statistically significant at almost all hours of the day. This plot also shows the probability of observing an equal or greater difference by chance (right axis, square symbols). Given the number of comparisons, a probability of <0.002 corresponds to a type I error rate of 0.05. Adapted from Wills and Beaupre (2000, *Physiological and Biochemical Zoology*, University of Chicago Press). © 2000 by The University of Chicago.

Accurate thermoregulation reflects the seemingly adaptive integration of behavioral, physiological, and morphological strategies. Yet, these strategies almost certainly involve costs that could offset the benefits of thermoregulation. How can we determine the adaptive value of any degree of thermoregulation? To do so, we could take an optimality approach, as we did to study the evolution of thermal sensitivity (Chapter 3). Before we can develop an optimality model of thermoregulation, we must consider the kinds of benefits and costs associated with thermoregulatory strategies.

4.2 Benefits and costs of thermoregulation

4.2.1 Benefits of thermoregulation

Undoubtedly, thermoregulation evolved as a means of enhancing performance in variable environments. An organism that regulates its body temperature can ensure a maximal (or nearly maximal) level of performance. Because thermoregulation buffers the heterogeneity of the environment, effective thermoregulators could also evolve greater performance through thermal specialization (see Chapter 3). Still, an important question remains. Given the potential to evolve a low thermal optimum for performance, why do many organisms maintain relatively high body temperatures? In other words, why regulate the mean and variance of body temperature when an organism could more easily regulate the variance and conform to the mean environmental temperature?

The high body temperatures of endotherms and many ectotherms probably serve to overcome

Box 4.1 Measuring preferred body temperatures

When placed in artificial thermal gradients, organisms generally maintain body temperatures within a narrow range. Because an organism in such an environment lacks physical and ecological constraints on thermoregulation, the temperature selected presumably reflects its preference, or the *preferred body temperature*. Although many authors consider the mean temperature as the preferred body temperature, a model based on upper and lower set points appears to better fit the distributions of temperatures in these experiments (Barber and Crawford 1977). In this case, one might choose to consider a preferred range, rather than a single preferred temperature (see, for example, van Berkum et al. 1986).

Preferred body temperatures must be measured in artificial conditions that enable an organism to select from a wide range of equally accessible microclimates. These conditions usually require a thermal arena, consisting of a linear (Licht et al. 1966b), circular (Bowker 1984), or patchy (Pulgar et al. 1999) distribution of operative temperatures. To create a uniform gradient, a conductive surface, such as a continuous metal floor, would be heated at one end and cooled at the other end. Aquatic gradients require more sophisticated designs than terrestrial ones (see Wallman and Bennett 2006). Generally, nonthermal resources (e.g., refuge) should occur uniformly within the gradient or should not occur at all; otherwise, movement within the gradient could reflect the compromised use of thermal and nonthermal resources. Controlling the distribution of nonthermal resources can be trickier than one might think. For example, lizards prefer a different temperature when exposed to a combination of visible and infrared radiation instead of infrared radiation alone (Sievert and Hutchison 1988). Likewise, snails prefer specific combinations of temperature and light intensity; in one experiment, snails thermoregulated well in steep gradients of visible light, but did not do so in shallow gradients of visible light (Rossetti and Cabanac 1985). Based on these and similar observations, we might choose to decouple these two forms of radiation within a thermal gradient; however, animals in natural environments generally bask in sunlight consisting of both visible and infrared wavelengths. Therefore, no gradient can simultaneously reflect the natural conditions of thermoregulation and eliminate the confounding effects of nonthermal resources. When in doubt, we could compare the movements of animals in an artificial thermal gradient with their movements in the same enclosure without the thermal gradient; no predictable pattern of movement should be observed when in the absence of the thermal gradient (see, for example, Fig. 4.6). Alternatively, we could use a null model to predict the movements of a thermoconformer and compare this prediction with the observed behavior (Anderson et al. 2007).

Once an organism enters a thermal gradient, it may spend some period habituating to artificial conditions before thermoregulating normally. Exploratory behavior could lead to inaccurate estimates of preferred body temperature unless recordings are postponed for some period. Often, researchers introduce individuals to the experimental conditions at least 1 day prior to measurements. If the period of habituation were uncertain, body temperature could be monitored continuously until fluctuations appear to give way to a fairly stable value; this stable value after habituation has been termed the final preferendum (Reynolds and Casterlin 1979). Once an individual habituates to the experimental conditions, it usually maintains a similar body temperature over the course of several days (Arnold et al. 1995; Dohm et al. 2001; Le Galliard et al. 2003), suggesting the preferred body temperature constitutes a real property of an organism. Furthermore, genetic variation in preferred body temperature has been documented through common garden experiments (Freidenburg and Skelly 2004) and laboratory natural selection (Good 1993).

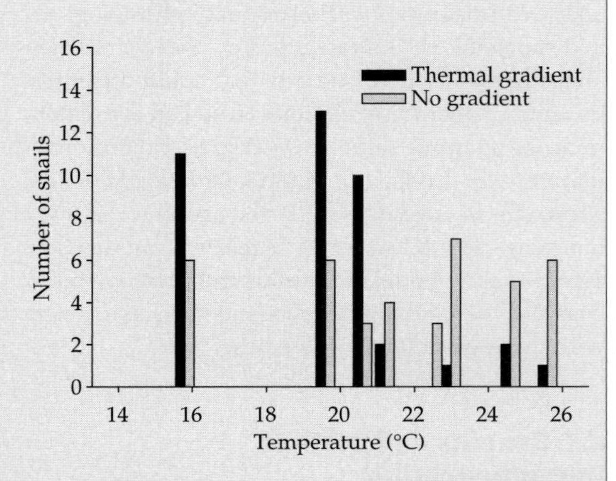

Figure 4.6 Thermoregulation can be inferred from the behaviors of animals in the presence or absence of a thermal gradient. For example, the distribution of snails in a thermal gradient became more uniform when the thermal gradient was turned off. Data from Gerald and Spezzano (2005).

Box 4.2 Indices of thermoregulation

Hertz and colleagues (1993) developed a way to quantify thermoregulatory behavior, which combines information about the preference of body temperatures in an artificial thermal gradient and the availability of operative temperatures in a natural environment. They presented three questions relevant to the study of thermoregulation. First, how precisely does an organism regulate its body temperature when freed from ecological constraints? To answer this question, they used the range of body temperature selected in an artificial thermal gradient, called the *set-point range*. Most researchers consider the central 50% or 80% of preferred body temperatures to define the set-point range. Second, how accurately does an organism maintain its body temperature within the set-point range? To answer this question, they calculated an index known as the *accuracy of thermoregulation* (d_b), which equals the mean absolute deviation of body temperature from the set-point range. The index d_b provides a way to compare the extent to which body temperatures in natural environments correspond to preferred temperatures. Finally, how well do thermoregulatory strategies actually enhance the accuracy of thermoregulation? This question arises because a low d_b could result from one of two causes: (1) an organism regulates it body temperature within its set-point range or (2) operative temperatures generally fall within the set-point range. To establish that accurate thermoregulation (i.e., a low d_b) results from behavioral or physiological regulation, one must have some knowledge of the operative thermal environment. Armed with such knowledge, Hertz and colleagues calculated an index known as the *thermal quality of the environment* (d_e), which equals the mean deviation of operative temperature from the set-point range.

Using the values of d_b and d_e, they calculated a final index known as the *effectiveness of thermoregulation* (E):

$$E = 1 - \frac{\overline{d_b}}{\overline{d_e}} \qquad (4.1)$$

One can infer the relative intensity of thermoregulation from the value of E. Random use of thermal microclimates yields an E equal to zero and perfect regulation of body temperature yields an E equal to 1.

To demonstrate the value of these indices, Hertz and colleagues (1993) compared the thermoregulatory behaviors of several species of Puerto Rican anoles. Figure 4.7 shows a subset of their data. In August, the preferred temperatures of *Anolis cristatellus* occurred rather infrequently in its environment at high elevation ($d_e = 5.0$). Despite this fact, lizards maintained a mean body temperature that was closer to the set-point range than was the mean operative temperature ($d_b = 2.5$), resulting in a moderate effectiveness of thermoregulation ($E = 0.5$). In contrast, *A. gundlachi* experienced an environment in which its preferred temperatures occurred very frequently ($d_e = 0.7$). Consequently, the extremely accurate thermoregulation by this species ($d_b = 0.6$) could have easily resulted from random use of microenvironments, which is reflected in its low effectiveness of thermoregulation ($E = 0.1$). These data illustrate how thermoregulatory indices enable one to distinguish between a thermoregulator and a thermoconformer in a natural environment.

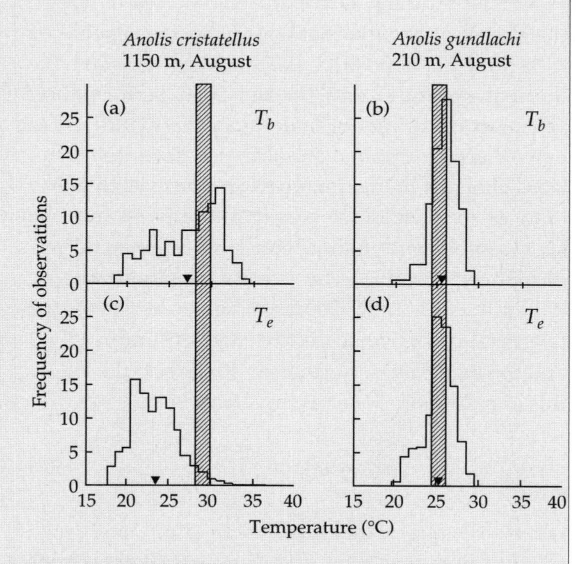

Figure 4.7 Thermoregulatory behavior differed between two species of lizards inhabiting different environments. (a–d) The distributions of body temperatures (T_b) and operative temperatures (T_e) for *Anolis cristatellus* and *A. gundlachi*. The set-point range of each species is delineated by the shaded region. *Anolis cristatellus* thermoregulated both accurately and effectively, whereas *A. gundlachi* appeared to maintain its preferred temperature without thermoregulation. Adapted from Hertz et al. (1993, *The American Naturalist*, University of Chicago Press). © 1993 by The University of Chicago.

thermodynamic constraints on biochemical reactions. As discussed in Chapter 3, organisms in hot environments should evolve relatively stable enzymes. Certain biochemical structures provide functional stability at high temperatures, but they also slow the rate of catalysis, K_{cat} (Hochochka and Somero 2002). Nevertheless, the catalytic rate increases with increasing temperature. When the effects of enzymatic structure and environmental temperature combine, the advantage of a high body temperature more than outweighs the disadvantage of a stable enzymatic structure (Hochochka and Somero 2002; Somero 2004). In other words, *hotter is better* for the performance of enzymes.[10] Scaling up this logic, an organism with a high thermal optimum should outperform an organism with a low thermal optimum.

We can test this hypothesis using comparative or experimental data. Comparative studies of multiple clones or populations within species enable us to search for a positive relationship between the thermal optimum and the maximal performance. Selection experiments that cause the evolution of new thermal optima should also lead to correlated changes in maximal performance. With either kind of evidence, we can quantitatively test the idea that thermodynamics constrain organismal performance. Based on the current metabolic theory (Gillooly et al. 2001, 2002; Savage et al. 2004), the relationship between the thermal optimum (T_{opt}) and the maximal performance (P_{max}) should follow this expression:

$$P_{max} \propto e^{-E/kT_{opt}}, \quad (4.1)$$

where E equals the mean activation energy of rate-limiting reactions and k equals Boltzmann's constant.

How well do thermodynamic constraints on biochemical function translate into organismal performance? To my knowledge, Eppley (1972) provided the first evidence that hotter is better by comparing maximal growth rates of algal species (Fig. 4.8). In the decades that followed, many other comparative studies of performance curves accumulated. A survey of these relevant studies suggests that hotter is usually better (Table 4.1). Moreover, all four studies that compared rates of population growth (a common estimate of fitness) strongly supported the existence of a thermodynamic constraint. Still, the magnitude of the thermodynamic effect often deviates significantly from that predicted by the metabolic theory. For example, Frazier and colleagues (2006) compared rates of population growth for 65 species of insects. The effect of the thermal optimum on the maximal performance was greater than that predicted by the metabolic theory; in other words, hotter appeared even better than expected from thermodynamic considerations.

If hotter is truly better, we should expect a correlated increase in maximal performance when

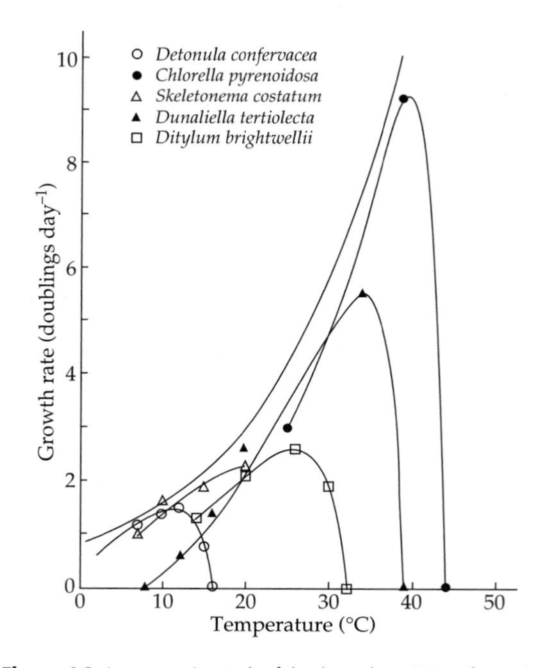

Figure 4.8 A comparative study of the thermal sensitivity of growth rate provided strong evidence of a thermodynamic effect on maximal performance. Data are population growth rates of fives species of algae grown in laboratory cultures. Adapted from Eppley (1972) with permission from the National Marine Fisheries Service.

[10] In the literature, the term "hotter is better" or "warmer is better" has been used in different contexts. For example, Bennett (1987b) used the term "warmer is better" to describe an increase in performance associated with an acute change in body temperature. In contrast, I use this term to describe an increase in maximal performance with an evolutionary change in the thermal optimum.

Table 4.1 A review of comparative studies of thermal performance curves suggests that hotter is better. For each case, the correlation (*r*) between the thermal optimum and the maximal performance is given, along with the number of populations or species used to estimate this correlation (*N*). Because correlations were computed on an Arrhenius scale, a negative correlation supports the hypothesis that hotter is better.

Organisms	Performance	*r*	*N*	Source
Frogs	Locomotion[2]	0.10	5[1]	Navas (1996a,b)
Marsh frogs	Locomotion	−0.96	5	Wilson (2001)
Lizards	Locomotion[2]	−0.42	10[1]	Bauwens et al. (1995)
Lizards	Locomotion[2]	−0.21	7	van Berkum (1986)
Mites	Locomotion[3]	0.39	5	Deere and Chown (2006)
Fungal symbionts of lichen	Growth	0.53	4	Sun and Friedmann (2005)
Algal symbionts of lichen	Growth	−0.82	7	Sun and Friedmann (2005)
Fish	Growth[2]	−0.10	23	Asbury and Angilletta (unpublished)
Arctic charr	Growth[2]	−0.62	11	Larsson et al. (2005)
Trees	Photosynthesis	0.72	8	Gratani and Varone (2004)
Trees	Photosynthesis	−0.18	8	Cunningham and Read (2002)
Pea aphids	Development	−0.66	5	Lamb et al. (1987)
Endoparasites	Parasitization	0.25	12	Grewal et al. (1994)
Wasps	Parasitization	−0.31	26	Carrière and Boivin (1997)
Algae	Fitness	−0.92	4	Eppley (1972)
Actinobacteria	Fitness	−0.85	6	Hahn and Pockl (2005)
Water fleas	Fitness	−0.69	29	Palaima (2002)
Insects	Fitness	−0.50[4]	65	Frazier et al. (2006)

Notes: Fitness was estimated as the rate of population growth.
[1] Some taxa were omitted because thermal optimum was not resolved.
[2] Performance was adjusted for body size.
[3] Performance was averaged among acclimation treatments.
[4] The correlation coefficient was estimated from independent contrasts.

selection leads to a higher thermal optimum. Unfortunately, most selection experiments either did not include enough measures of performance to estimate thermal optima (Huey et al. 1991; James and Partridge 1995) or were too short to produce significant shifts in thermal optima (Gilchrist et al. 1997). In an exceptional study, Santos (2007) allowed replicate populations of *Drosophila subobscura* to evolve at a constant temperature of 13°C, 18°C, or 22°C. He then measured the net fitness of flies from each line at all three temperatures. Not only did the thermal optimum diverge among these lines, but so did the mean fitness. Consistent with the patterns observed in comparative studies, flies that evolved at 22°C had the highest mean fitness, whereas flies that evolved at 13°C had the lowest. In an equally impressive study, Cooper and colleagues (2001) compared the thermal sensitivities of maximal growth rate between ancestral lines and selected lines of *Escherichia coli* after 20 000 generations at 37°C (see Chapter 3 for details). The ancestral lines grew maximally at 40°C. Not surprisingly, the selected lines experienced a decrease in their thermal optimum, such that their new optimum matched the temperature of their environment. Contrary to the hypothesis, the selected lines grew faster at their low thermal optimum than the ancestral lines did at their high thermal optimum. Furthermore, a study of mutational effects refuted a necessary link between the thermal optimum and the maximal fitness; mutants of *E. coli* whose performance curves were shifted toward higher temperatures did not display a higher maximal fitness relative to their ancestors (Mongold et al. 1999). Importantly, these patterns contrast the pattern derived by comparing two natural isolates of *E. coli* (Bronikowski et al. 2001). Hence, support for the hypothesis from comparative studies might be countered by evidence from selection experiments involving the same species.

Despite the mounting evidence that hotter is better, the thermodynamic effect differs considerably among species and performances (see Table 4.1). Furthermore, a comparative study of extreme thermophiles reaffirms that life has its limits; the maximal growth rate of cyanobacteria increased with an increasing thermal optimum up to 60°C, but decreased above this temperature (Miller and Castenholz 2000). This finding implies that extremely high temperatures impose a physical stress that outweighs any positive effect on catalytic rates. Clearly, we need more research to understand variation in the thermodynamic effect and the mechanisms that enable certain lineages to overcome this constraint. But for now, the majority of data suggest that organisms benefit from adaptation to high temperatures combined with precise thermoregulation.

4.2.2 Costs of thermoregulation

Although the benefits of thermoregulation seem fairly straightforward, the costs can be quite complex. We shall consider three kinds of costs: energy expenditure, mortality risk, and missed opportunities. Both energy expenditure and mortality risk can easily be converted into demographic consequences. Energy expended for thermoregulation cannot be used to grow or reproduce. Thus, current and future fecundity would suffer during periods of intense thermoregulation. A greater risk of mortality would directly affect fitness by reducing the likelihood of future reproduction. Missed opportunities seem more difficult to quantify because one needs to know what activities conflict with thermoregulation and how these activities contribute to fitness. If thermoregulation interferes with mating, an organism would suffer a decrement in fecundity. But a conflict between thermoregulation and feeding could reduce the survivorship or fecundity of thermoregulators. Cost–benefit analyses of thermoregulation are complicated by the fact that organisms use many mechanisms to thermoregulate (Chappell and Whitman 1990; Lustick 1983), each of which imposes particular costs. Let us briefly consider behavioral, physiological, and morphological mechanisms and the kinds of costs they can impose.

Behavioral mechanisms of thermoregulation are most readily discernable. Many kinds of animals shuttle between hot and cold patches to thermoregulate. For example, lizards preferentially perch on exposed rocks during the morning and evening but move under bushes during midday (Bauwens et al. 1996). Similarly, fish shuttle within thermally stratified waters (Snucins and Gunn 1995), and Japanese beetles shuttle between sun and shade (Kreuger and Potter 2001). Even plants can track the position of the sun to maximize their absorption of solar radiation (Ehleringer and Forseth 1980). When the body temperature of a species remains constant along a geographic cline, shifts in behavior clearly play an important role. The lizard *Sceloporus occidentalis* perches on trees at low elevation, its congener *S. graciosus* remains on the ground at high elevation, and both species prefer a mixture of microhabitats at mid-elevation (Adolph 1990). When transplanted to a common environment, lizards from all three elevations used the same microhabitats (Asbury and Adolph 2007). In contrast, genetic factors contribute to geographic differences in the thermoregulatory behaviors of some grasshoppers (Samietz et al. 2005). In addition to overt shuttling, small shifts in posture and orientation can facilitate radiative and convective heat exchange between an organism and its environment (Ayers and Shine 1997; Bauwens et al. 1996; Coelho 2001; Heath 1965; Heinrich 1990; Munoz et al. 2005). Behavioral thermoregulation imposes energy expenditure, mortality risk, and missed opportunities; for example, organisms shuttling between microclimates will spend more energy, attract more predators, and startle more prey than will organisms remaining still (Huey 1974, 1982; Huey and Slatkin 1976).

Physiological mechanisms—including circulation, evaporation, and metabolism—offer more subtle means of thermoregulation (Bartholomew 1982; Kammer 1981). The flow of blood (or hemolymph) distributes heat between the core and surface of the body, determining rates of conduction and convection (Grigg and Seebacher 1999; Seebacher and Grigg 2001; Smith 1979). Organisms can quickly lose heat by shunting blood to body parts with high surface area relative to volume, such as the head or limbs (Dzialowski and O'Connor 2004; May 1995).

Greater cardiac output demands more energy, and shunting of blood likely alters the metabolic capacity of peripheral tissues. Alternatively, countercurrent systems enable the flow of blood to the periphery while retaining heat within the core (Holland et al. 1992). Through evaporation, organisms can dump heat at the expense of water. Because water loss affects performance (reviewed by Chown et al. 1999; Lillywhite and Navas 2006; Tsai et al. 1998; Walvoord 2003), organisms tend to reserve evaporation as a strategy of last resort; for instance, waterproof frogs tracked ambient temperature up to 40°C and then used evaporative water loss to prevent further increases (Shoemaker et al. 1987). To aid in evaporative cooling, organisms can seek out regions of low humidity when exposed to extreme heat (Prange and Hamilton 1992). Behaviors, such as panting or tongue lashing, can also enhance the rate of evaporation (Roberts and Harrison 1998). Finally, metabolism generates heat at the expense of chemical energy. This strategy enables endothermy in plants (Patino et al. 2000), insects (Heinrich 1974), fish (Block and Finnerty 1994), birds (Prinzinger et al. 1991), and mammals (Grigg et al. 2004). Bees increase the frequency of their wingbeats as the temperature of the environment decreases (Harrison et al. 1996; Roberts and Harrison 1998). Likewise, birds and mammals undergo shivering thermogenesis when their body temperature drops below a critical level (Scholander et al. 1950a). The costs of physiological thermoregulation depend on the specific mechanism, but include both energy expenditure and mortality risk.

Morphology also serves to regulate body temperature. Changes in color—whether rapid, seasonal, or developmental—enable organisms to adjust their solar absorptivity (Clusella-Trullas et al. 2008; Hazel 2002; Kingsolver and Huey 1998; Lacey and Herr 2005; Walton and Bennett 1993). Adjusting the position of fur or feathers alters heat exchange by convection and radiation (Cossins and Bowler 1987; Scholander et al. 1950b). Even when morphology changes slowly, the current morphology plays an indirect role in thermoregulation by influencing the effectiveness of behavioral and physiological strategies. Size and shape have the greatest impact on the capacity for thermoregulation (Bartholomew 1981; Stevenson 1985). For example, the use of blood flow to regulate heating and cooling requires a substantial size (see Dzialowski and O'Connor 2001). Furthermore, larger insects generate more metabolic heat during flight (Bishop and Armbruster 1999). Very large organisms possess a thermal inertia that dampens fluctuations in body temperature and could even result in homeothermy without effort (Fig. 4.9, see also Seebacher et al. 1999). But bigger is not always better; small butterflies can use postural changes to alter radiative and convective heat fluxes more effectively than large butterflies (Kemp and Krockenberger 2004). Morphologies associated with thermoregulation, such as color and size, require energy to produce and maintain. Dark color results from the deposition of pigments in the skin, fur, or feathers (Cooper and Greenberg 1992; Riley 1997). Organisms must incur some cost of producing pigments for thermoregulation, given the loss of melanin associated with poor diet (Talloen et al. 2004) and life underground (Jeffery 2006; Mejia-Ortiz and Hartnoll 2005; Parzefall 2001). Large size requires energy for growth and maintenance at the expense of other activities. Even when a small size facilitates thermoregulation, organisms can save energy on growth but might suffer loss of performance in foraging,

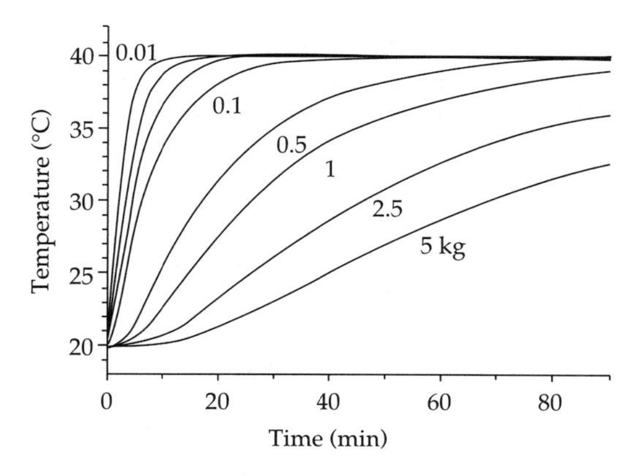

Figure 4.9 The rates of heating and cooling depend on the mass of an organism. This plot shows calculated rates of heating for a range of body masses. Large organisms (≥5 kg) heat and cool so slowly they experience inertial homeothermy. Adapted from Seebacher and Shine (2004, *Physiological and Biochemical Zoology*, University of Chicago Press). © 2004 by The University of Chicago.

defense, or reproduction. Therefore, morphological traits designed for thermoregulation can impose energy expenditure, mortality risk, and missed opportunities.

Importantly, organisms generally thermoregulate by a combination of behavior, physiology, and morphology, switching rapidly between suites of strategies (Box 4.3). In fact, the same pattern of thermoregulation can arise from different mechanisms in closely related organisms. Crepuscular cicadas thermoregulate using metabolic heat, but diurnal cicadas thermoregulate using behavior (Sanborn et al. 2004). This complexity challenges our abilities to understand the evolution of thermoregulation. Although the gross benefits of thermoregulation depend on body temperature alone, the costs of thermoregulation depend on the suite of mechanisms that an organism adopts in a particular environment.

4.3 An optimality model of thermoregulation

The prevailing theory of thermoregulation stems from Huey and Slatkin (1976), who modeled the optimal degree of thermoregulation given energetic costs and benefits. They reasoned the cost of thermoregulation could outweigh the benefits, particularly in environments where thermal resources were scarce. This argument was formalized in an optimality model, which predicts thermoregulatory behavior when the energetic costs and benefits are known. Additional costs, such as predation risk and missed opportunities, were also discussed but were not modeled explicitly (Huey and Slatkin 1976; Huey 1982). Here, I shall derive their original model focusing on energetic costs and benefits.

The model captures the thermoregulatory strategy of an organism with a parameter k, which

Box 4.3 Thermoregulation by butterflies: a case study of integrated mechanisms

Lepidopterans rely on a combination of behavior, physiology, and morphology to thermoregulate. As larvae, individuals develop their color based on the environmental temperature (Nice and Fordyce 2006). Dark coloration, provided by melanin, helps caterpillars in cold environments to raise their temperature through basking. As their body temperature rises, they seek shade within vegetation and also engage in physiological mechanisms of cooling (Fig. 4.10). As adults, butterflies combine wing melanism with postural adjustments to raise their temperature prior to flight. Three forms of basking have been described: dorsal, lateral, and reflectance basking (Casey 1981; Kemp and Krockenberger 2002). Each of these strategies relies on an interplay between behavior and morphology for effectiveness (Kingsolver 1987). During dorsal basking, an individual spreads it wings flat to allow solar radiation to fall directly on the thorax. Most of the heat gain occurs as the individual absorbs solar radiation at the thorax or conducts heat to the thorax from the basal portions of the wings. Therefore, melanism on the basal regions of the dorsal surface will speed heating during dorsal basking. During lateral basking, an individual closes its wings and orients perpendicular to solar radiation. In this position, melanin in the basal regions of the ventral surfaces enhances heating. During reflectance basking, an individual opens its wings at an angle to reflect solar radiation from the dorsal surface onto the thorax. A lack of melanin along distal regions of the wings enhances reflection. Heinrich (1990) disputed the significance of reflective basking, suggesting that heating was instead mediated by dorsal basking, with elevated wings to minimize convection. Regardless, this alternative interpretation still involves the integration of behavior and morphology.

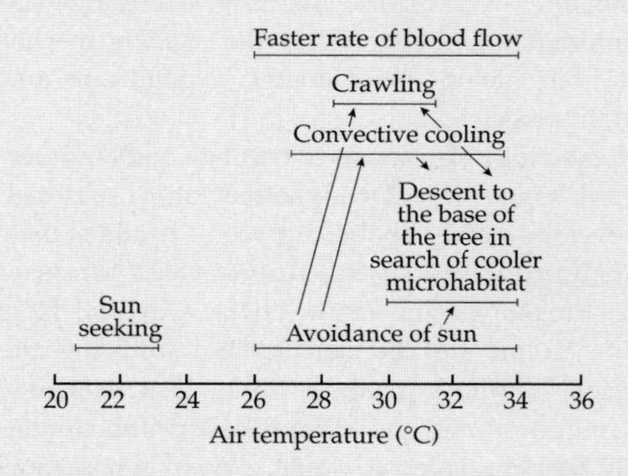

Figure 4.10 An ethogram depicts the suite of mechanisms used for thermoregulation. This ethogram was constructed for larvae of the moth *Antheraea assama*. Arrows depict transitions between thermoregulatory mechanisms. Reproduced from Bardoloi and Hazarika (1994) with permission from Blackwell Publishing.

determines how closely the body temperature (T_b) tracks the environmental temperature. Originally, Huey and Slatkin considered ambient temperature, but we can replace this variable with the operative temperature (T_e). Therefore, k determines thermoregulation according to the following relationship:

$$T_b = T_{opt} + k(T_e - T_{opt}). \quad (4.2)$$

By choosing a value of k between 0 and 1, we can generate strategies that range from perfect thermoregulation to perfect thermoconformation (Fig. 4.11a).

The benefit of thermoregulation (b) depends on the body temperature of the organism (Fig. 4.11b). Because the body temperature depends on the thermoregulatory strategy and operative temperature, we can define the benefit of thermoregulation more explicitly as

$$b(T_b(k, T_e)). \quad (4.3)$$

This benefit represents the energy gained from adopting strategy k in the microenvironment characterized by a certain operative temperature.

Because temperature varies spatially, we must use information about the thermal macroenvironment to calculate the expected benefit of thermoregulation. The expected benefit depends on the probability density function for operative temperature, $f(T_e)$; this function describes the chances of encountering specific operative temperatures. Integrating the benefit of thermoregulation weighted by the probability density function yields the expected benefit of thermoregulation ($\bar{b}$):

$$\bar{b} = \int b(T_b(k, T_e)) \cdot f(T_e) dT_e. \quad (4.4)$$

An organism that maintains a body temperature equal to its thermal optimum (i.e., $k = 0$) would maximize its expected benefit. However, we should not necessarily expect perfect thermoregulation to evolve because this strategy doubtless involves costs. The optimal strategy would maximize the net benefit, which equals the expected benefit minus the expected cost.

Huey and Slatkin modeled the cost of thermoregulation (c) as a function of the difference between the target temperature of the organism and the operative temperature of the microenvironment (Fig. 4.11c):

$$c(T_b - T_e). \quad (4.5)$$

Because the target body temperature depends on the thermoregulatory strategy and the operative temperature, we can define the cost of thermoregulation

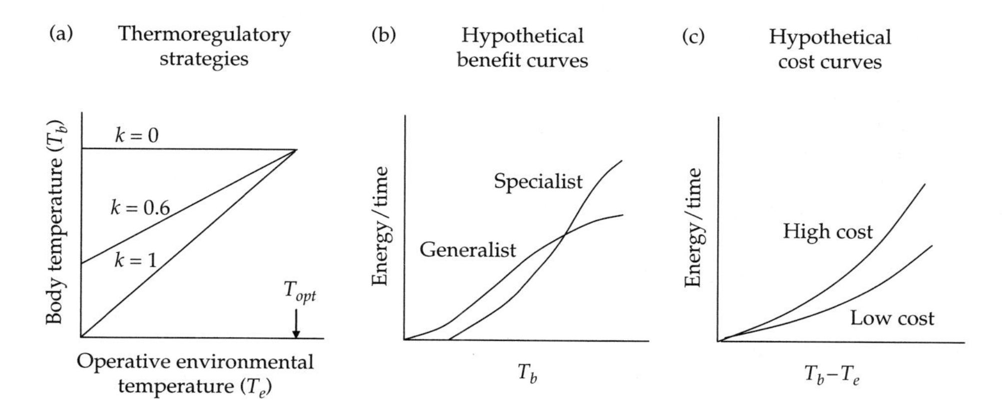

Figure 4.11 Modeling the optimal strategy of thermoregulation requires precise definitions of the potential strategies and their respective costs and benefits. (a) The thermoregulatory strategy was modeled as a linear relationship between the operative environmental temperature (T_e) and the body temperature (T_b). The slope of this relationship (k) varied between 0 and 1, with a slope of 0 defining a perfect thermoregulator and a slope of 1 defining a perfect thermoconformer. (b) The rate of energy gain was assumed to increase as the body temperature approached the thermal optimum; two hypothetical curves are shown. (c) The energetic cost of thermoregulation was assumed to increase as the operative temperature deviated from the target body temperature; two hypothetical curves are shown. Adapted from Huey and Slatkin (1976, *The Quarterly Review of Biology*, University of Chicago Press). © 1976 by The University of Chicago.

more explicitly as:

$$c(T_b(k, T_e) - T_e). \quad (4.6)$$

This cost represents the energy required to reach a microenvironment in which $T_e = T_b(k, T_e)$, given the operative temperature of the current microenvironment. As with the expected benefit, the expected cost of thermoregulation ($\bar{c}$) for a particular macroenvironment must be calculated by integrating the weighted cost over all possible microenvironments:

$$\bar{c} = \int c(T_b(k, T_e) - T_e) \cdot f(T_e) dT_e. \quad (4.7)$$

Thus, the expected cost depends on the thermoregulatory strategy of the organism and the probability densities of operative temperatures. The thermoregulatory strategy determines the target body temperature that an organism will try to attain when faced with a given operative temperature. The probability density function determines the chance of encountering particular operative temperatures. Additionally, the energetic cost of thermoregulation depends on other properties of the organism and its environment. With respect to the organism, locomotor ability, thermal perception, and spatial memory will affect the energetic cost of locating a preferred microclimate. With respect to the environment, the size, connectivity, and autocorrelation of thermal patches will also matter (Fig. 4.12).

Once we have formulated the expected benefit and cost of a thermoregulatory strategy, we can calculate the net benefit (B):

$$B = \bar{b} - \bar{c} = \int \left[b(T_b(k, T_e)) - c(T_b(k, T_e) - T_e) \right] \cdot f(T_e) dT_e. \quad (4.8)$$

All else being equal, the optimal thermoregulatory strategy maximizes the net benefit. By varying certain parameters and holding all others constant, Huey and Slatkin explored the optimal strategy k for various forms of b and c. Take another look at the functions shown in Fig. 4.11. By superimposing those functions for b and c on a single plot, the net benefit of a perfect thermoregulator or a perfect thermoconformer can be depicted graphically (Fig. 4.13). From such plots, we can assess whether perfect thermoregulation, perfect conformation, or

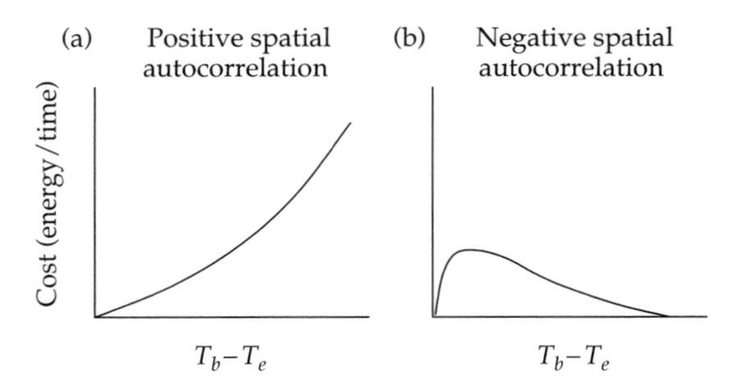

Figure 4.12 The cost curve associated with thermoregulation depends on the spatial structure of the thermal environment. (a) With positive spatial autocorrelation, the current microenvironment of the organism likely resembles the surrounding microenvironments. Consequently, an organism in an undesirable microenvironment (large value of $T_b - T_e$) would need to move a great distance to locate a preferred microclimate. (b) With negative spatial autocorrelation, the current microenvironment of the organism likely differs greatly from the surrounding microenvironments. Consequently, an organism in an undesirable microenvironment would need to move only a short distance to locate a preferred microenvironment.

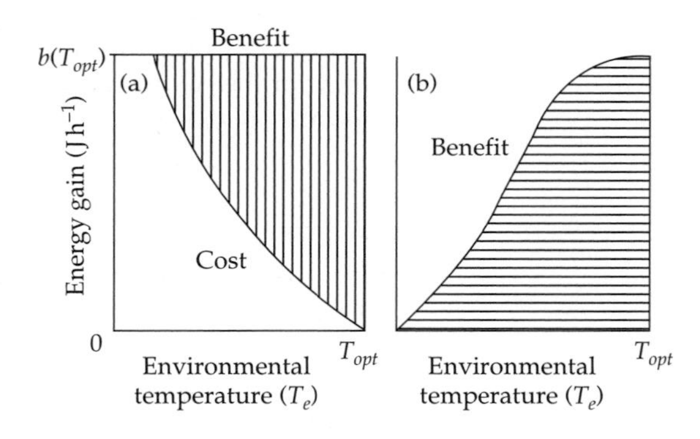

Figure 4.13 The benefit and cost of thermoregulation depend on the strategy adopted by the organism. These hypothetical curves are based on the functions shown in Fig. 4.11. The hatched area depicts the difference between the benefit and the cost. (a) A perfect thermoregulator ($k = 0$) will experience the same benefit, $b(T_{opt})$, in all microenvironments, but will experience a greater cost as the operative temperature of the microenvironment (T_e) falls below the thermal optimum for energy gain (T_{opt}). (b) A perfect thermoconformer ($k = 1$) will experience no cost in any microenvironment, but will experience less benefit as the operative temperature of the microenvironment falls below the thermal optimum for energy gain. Adapted from Huey and Slatkin (1976, *The Quarterly Review of Biology*, University of Chicago Press). © 1976 by The University of Chicago.

some intermediate strategy would maximize the net rate of energy gain.

Using this model, Huey and Slatkin reasoned a decrease in the environmental temperature should cause a decrease in the mean body temperature of an optimal thermoregulator. This prediction follows from their choice of phenotypic function: recall, a simple linear function was used to describe the relationship between environmental temperature and body temperature (eqn 4.2). Consequently, mean body temperature must decrease as mean environmental temperature decreases, unless both warm and cold environments favor perfect thermoregulation. Broad comparative surveys of amphibians and reptiles support this prediction (Andrews 1998; Feder and Lynch 1982). Nevertheless, we should infer the cause of these patterns with caution. If the benefit and cost curves vary among environments, optimal thermoregulation within environments would not necessarily lead to the anticipated pattern. Moreover, we do not know whether individuals in cold environments exhibit low body temperatures because of optimal thermoregulation or because of thermal constraints. For example, consider two environments that differ in their mean operative temperatures: a warmer environment that enables an organism to reach its thermal optimum for performance, and a colder environment that does not (Fig. 4.14a). Although an individual could thermoregulate perfectly ($k = 0$) in either environment, this strategy would confer a greater gross benefit in the warm environment than it would in the cold environment (Fig. 4.14b). Clearly, the gross benefit of thermoregulation depends on the range of body temperatures that an organism can attain in its current environment. The same conclusion holds for the cost of thermoregulation. When the thermal optimum cannot be attained, the cost curve for a perfect thermoregulator would be referenced to the closest attainable body temperature (Fig. 4.14c). Even if both environments did enable an individual to attain its thermal optimum, the cost of thermoregulation would still differ between environments. The environment in which the thermal optimum occurs more frequently should impose a lesser cost of thermoregulation (Fig. 4.14d–f). Huey and Slatkin did not envision this situation because they assumed the cost curve did not depend on the probability distribution of operative temperatures (see eqn 4.5). In general, benefit and cost curves should vary considerably along latitudinal and altitudinal clines. To make useful predictions from the model, we must know the precise forms of these curves.

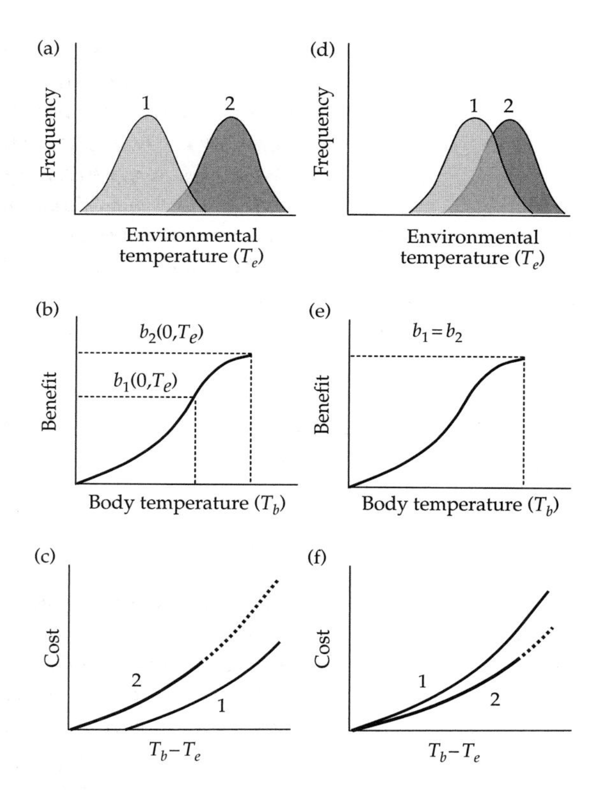

Figure 4.14 The realized benefit and cost of thermoregulation depend on the distribution of operative temperatures (T_e) in the environment. (a) Distributions of operative temperatures are shown for two hypothetical environments. The colder environment (1) does not permit the organism to attain its thermal optimum. (b) Because of constraints on body temperature, the realized benefit of perfect thermoregulation in the warm environment, $b_2(0, T_e)$, exceeds the realized benefit in the cold environment, $b_1(0, T_e)$. (c) The cost of thermoregulation in the cold environment (1) exceeds the cost in the warm environment (2). Note that the cost curve for the cold environment is shifted to reflect the difference in the target body temperature during thermoregulation. (d–f) These plots are the same as (a–c), except now both environments permit the organism to attain its thermal optimum, and thus $b_1(0, T_e) = b_2(0, T_e)$. In (f), the difference between the two cost curves reflects the higher frequency of preferred microclimates in the warm environment (2), relative to the cold environment (1).

Better descriptions of benefits and costs are also needed for experimental tests of the model. For example, Herczeg and colleagues (2006) applied

Huey and Slatkin's model to explain variation in thermoregulation by lizards in three different experimental arenas (Fig. 4.15). In arenas where most of the operative temperatures were either well above or well below the preferred range, lizards thermoregulated very accurately and effectively (mean $d_b \leq 1$ and mean $E > 0.8$). Given the high values of d_e, lizards presumably thermoregulated by shuttling between hot and cold patches. In an arena where all of the operative temperatures were below the preferred range, lizards thermoregulated least accurately and effectively (mean $d_b = 4$ and mean $E = 0.05$); because lizards in the cold arena could not attain a preferred temperature no matter how they behaved, Herczeg and colleagues wisely calculated d_b and E using the closest available temperature to the set-point range. Given these results, we might conclude lizards in the coldest arena chose not to thermoregulate effectively because the cost of thermoregulation was too high. But this interpretation ignores the fact that the cost and benefit of thermoregulation depend on thermal constraints. Assuming the thermal optimum for performance lies within the set-point range, the gross benefit of thermoregulation in the coldest arena was lower than the gross benefits in the other two arenas (see Fig. 4.14b). Furthermore, the costs likely differed among arenas because a lizard in the cold arena could not have attained its thermal optimum (see Fig. 4.14c). Without better estimates of the benefit and cost curves for lizards in these specific arenas, we cannot fully understand the behaviors documented in this experiment.

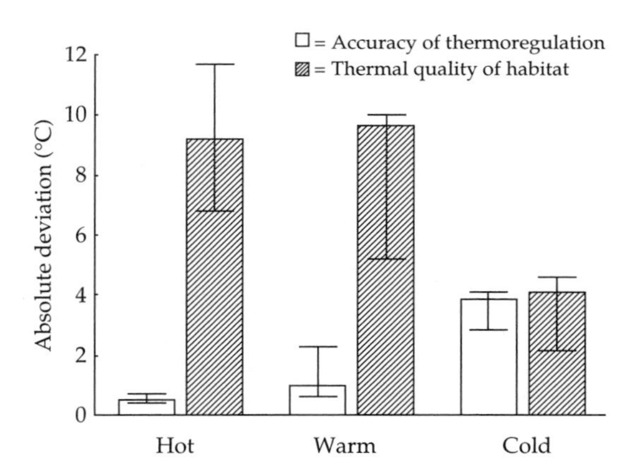

Figure 4.15 Indices of thermoregulation for *Zootoca vivipara* in three different thermal arenas: hot, warm, and cold. In the cold arena, indices were calculated by using the maximal operative temperature instead of the lower boundary of the set-point range. Data are medians ± quartiles of the accuracy of thermoregulation (d_b) and the thermal quality of habitat (d_e). Adapted from Herczeg et al. (2006) with kind permission from Springer Science and Business Media. © Springer-Verlag 2006.

In summary, thermal optima and operative temperatures alone cannot tell us the optimal strategy of thermoregulation because we need additional information to calculate the benefits and costs of thermoregulation. Unfortunately, the paucity of additional information limits our ability to apply this optimality model to the real world. Despite this problem, we shall do our best to use Huey and Slatkin's model to understand the thermoregulatory behaviors of organisms in artificial and natural environments.

4.4 Do organisms thermoregulate more precisely when the benefits are greater?

The optimal level of thermoregulation increases ($k \rightarrow 0$) as the gross benefit of thermoregulation (b) increases. Two properties of a performance curve determine the relative benefits of thermoregulation and thermoconformation: the performance breadth and the maximal performance (see Fig. 3.4). As the performance breadth increases, small deviations of body temperature from the thermal optimum impose a lesser loss of energy gain. For a given cost curve, the net benefit of thermoconformation for a generalist would exceed that for a specialist when the environmental temperature differs from the thermal optimum; consequently, a generalist should thermoregulate less precisely than a specialist. Furthermore, a decrease in the maximal performance (energy gain) decreases the net benefit of thermoregulation at all environmental temperatures, which also tends to favor less precise thermoregulation. Therefore, a change in the shape or height of the performance curve can shift the optimal strategy away from perfect thermoregulation. Is there any empirical evidence for these phenomena?

Despite the hundreds (if not thousands) of performance curves that have been quantified, we currently lack the data to know whether generalists and specialists thermoregulate differently. In cases where we know something about thermoregulatory strategies, we generally do not know the thermal sensitivities of energy gain. In cases where we know the thermal sensitivities, we generally do not know enough about the thermoregulatory strategies. We cannot merely compare the body temperatures of specialists and generalists because they usually inhabit different environments. Consequently, we must account for the different costs and constraints as well as the benefits.

Existing data enable us to explore the effect of maximal performance on thermoregulatory behavior with greater confidence. The rate of energy gain at any temperature depends on the quantity and quality of food, which depend on the interactions among the thermoregulator, its competitors, and its prey. Although exact rates of energy gain are difficult to measure under natural conditions, maximal rates can easily be manipulated in artificial or natural environments. In the simplest way, we could measure the preferred body temperatures of animals that were systematically fed or fasted (Regal 1966). Because starved animals gain no energy from thermoregulation, we should expect fasting to alter the behavior of an animal in a thermal gradient. Numerous investigators have compared the preferred body temperatures of recently fed individuals with those of fasted individuals (Table 4.2). In general, fasted individuals select lower body temperatures than do fed individuals. In one experiment in which several levels of ration were used (Mac 1985), the mean body temperature in a thermal gradient decreased as the ration decreased; this graded response accords with Huey and Slatkin's model. One of two factors could explain these effects of fasting on thermoregulation (Huey 1982). First, fasted individuals gain less energetic benefit from thermoregulation. Alternatively (and perhaps more likely), the optimal temperature for energy gain decreases as the rate of feeding decreases (see Fig. 3.3). In either case, these phenomena suggest organisms respond rapidly to changes in their thermal sensitivity of performance.

The thermal responses to feeding observed in artificial thermal gradients should lead us to expect similar behaviors under natural conditions. Two teams of researchers have supplementally fed animals and recorded their body temperatures in natural environments. Both of these experiments focused on snakes, whose body temperatures were recorded intermittently by radiotelemetry. Because these experiments took place in natural environments, the researchers also had to consider thermal heterogeneity over time; to do so, they compared both body temperatures and indices of thermoregulatory effectiveness (see Box 4.2) before and after feeding. After Brown and Weatherhead (2000) fed water snakes (*Nerodia sipedon*), the snakes did not increase their body temperature or their effectiveness of thermoregulation (E). These same researchers also observed snakes in an artificial thermal gradient, and saw no significant change in body temperature associated with feeding. Both results contrast reports by other researchers who did observe increases in body temperature ranging from 2°C to 8°C after feeding (Lutterschmidt and Reinert 1990; Sievert and Andreadis 1999). Likewise, after Blouin-Demers and Weatherhead (2001) fed rat snakes (*Pantherophis obsoleta*), the snakes did not maintain higher body temperatures relative to operative temperatures. This result contrasts the increase in preferred body temperature after feeding that was observed by the same researchers (see Table 4.1). Nevertheless, snakes in their natural environment did increase their effectiveness of thermoregulation after feeding, suggesting they had worked harder to maintain the same body temperature. Certain snakes shifted habitats after feeding, apparently to facilitate thermoregulation. Of 14 snakes that were fed in the interior of the forest, all but three shifted to the edge of the forest. In contrast, only two snakes that were fed at the edge moved to the interior. The differences between observations in artificial and natural environments could reflect uncontrolled variation in feeding rate or thermoregulatory cost, which cannot be avoided in natural environments.

The quality of food also alters the benefit of thermoregulation. By definition, food of low quality results in slower energy gain than food of high quality. Thus, we should expect animals eating food of low quality to thermoregulate less precisely than animals eating food of high quality. This hypothesis has been tested experimentally in both artificial and natural environments. Nussear and colleagues (1998) fed chuckwallas either a high-fiber diet or a low-fiber diet and measured their body

Table 4.2 Starvation generally lowers the preferred body temperatures of ectotherms.

Species	Effect of fasting on preferred temperature	Source
Insects		
Panstrongylus megistus	Decreased ≈0.1°C per day	Pires et al. (2002)
Triatoma infestans	Decreased by ≈0.25°C per day	Minoli and Lazzari (2003)
Fishes		
Dasyatis sabina	1°C lower	Wallman and Bennett (2006)
Girella laevifrons	Selected higher temperatures more frequently and lower temperatures less frequently	Pulgar et al. (1999)
Rutilus rutilus	No effect on temperature during photophase; temperature during scotophase decreased by ≈0.2°C per day	van Dijk et al. (2002)
Salmo salar	2°C higher	Morgan and Metcalfe (2001)
Salvelinus namaycush	Decreased linearly with decreasing ration	Mac (1985)
Amphibians		
Bufo marinus	No effect	Mullens and Hutchison (1992)
Bufo woodhousii	No effect during most times of day, but 2°C lower during early scotophase (1900–2300 h)	Witters and Sievert (2001)
Desmognathus fuscus	No effect	Moore and Sievert (2001)
Triturus dobrogicus	2–3°C lower	Gvoždík (2003)
Reptiles		
Anolis carolinensis	0.6°C lower in females and 1.3°C lower in males	Brown and Griffin (2005)
Charina bottae	2°C lower	Dorcas et al. (1997)
Morelia spilota	3°C lower	Slip and Shine (1988)
Nerodia sipedon	8°C lower	Lutterschmidt and Reinert (1990)
Nerodia sipedon	2°C lower	Sievert and Andreadis (1999)
Nerodia sipedon	No effect	Brown and Weatherhead (2000)
Pantherophis guttatus	6°C lower	Sievert et al. (2005)
Pantherophis obsoleta	2°C lower	Blouin-Demers and Weatherhead (2001)
Thamnophis sirtalis	4°C lower in one subspecies, but no effect in another subspecies	Lysenko and Gillis (1980)
Trimeresurus stejnegeri	5°C lower when in a gradient with water and refugia; no effect in an empty gradient	Tsai and Tu (2005)

temperatures in a thermal gradient. In contrast to our expectation, the preferred body temperature did not differ between lizards on different diets. In a field experiment, Underwood (1991) recorded the thoracic temperatures of honeybees as they arrived at feeders containing concentrated syrup or dilute syrup. As expected, bees were hotter during flight when they fed on the more concentrated syrup. Finally, Pulgar and colleagues (2003) fed fish either a high-quality diet of bivalves or a low-quality diet of algae and recorded their preferred body temperatures. Fish consuming bivalves selected higher temperatures (16–19°C) than did fish consuming algae (10–13°C). Importantly, fish not subjected to a thermal gradient did not prefer any particular portion of the aquarium, suggesting the fish were actually regulating their temperature during the experiment.

Energy gain relies not just on the availability of food, but also on the ATP required to fuel its capture, ingestion, digestion, and absorption. Under conditions of hypoxia, the aerobic capacity of an organism could fall below that required to sustain metabolic activities. All else being equal, a shortage of oxygen would reduce the benefit of thermoregulation, as does a shortage of food. Based on this argument, an organism should select a lower body temperature in a hypoxic environment to minimize metabolic demands. Experimental studies of preferred body temperature strongly support this prediction. Hypoxia reduces the preferred body temperatures of crustaceans, fish, and reptiles (Petersen et al. 2003; Rausch et al. 2000; Schurmann et al. 1991; Wiggins and Frappell 2000). Furthermore, the effect of hypoxia on thermoregulation depends on the levels of hemoglobin present in tissues; water fleas with abundant hemoglobin exhibited a small reduction in preferred body temperature during hypoxia compared to water fleas with less hemoglobin (Wiggins and Frappell 2000).

Taken together, manipulations of food availability, diet quality, and oxygen concentrations confirmed that ectotherms do respond rapidly to changes in the energetic benefits of thermoregulation. Additional tests of this hypothesis could involve indirect manipulations of energy gain by changing the number of competitors. Although we have only meager experimental evidence of the indirect effects of competition on thermoregulation (see, for example, Diego-Rasilla and Perez-Mellado 2000), these effects should parallel those caused by direct manipulation of food intake (Huey and Slatkin 1976). Detecting the indirect effects of competition will require adequate control for temporal variation in environmental temperatures and directs effects of intraspecific aggression (see Sections 4.6 and 4.7)

4.5 Nonenergetic benefits of thermoregulation

Although energy gain clearly depends on body temperature, thermoregulation offers benefits that cannot be quantified in terms of energy. Just consider the variety of performances that depend on body temperature (see Table 3.1). Nonenergetic performances (e.g., development, locomotion, and immunocompetence) mediate indirect effects of body temperature on survivorship or fecundity. Depending on the environmental context, any or all of these performances could be considered benefits of thermoregulation. For example, male cicadas use endothermy to enhance mate calling at night, when atmospheric conditions promote acoustic communication (Sanborn et al. 2003). When a single temperature maximizes both energetic and nonenergetic benefits (Fig. 4.16), a simple cost–benefit model will suffice. But organisms cannot always maximize

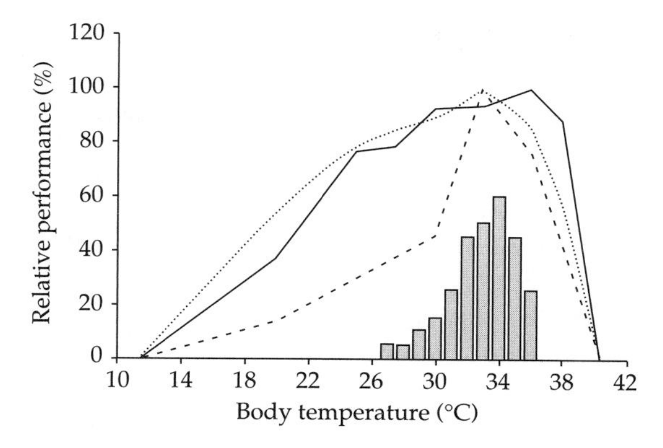

Figure 4.16 Thermal sensitivities of sprint speed (solid line), endurance (dotted line), and metabolizable energy intake (dashed line) in the eastern fence lizard, *Sceloporus undulatus*. The distribution of preferred body temperatures in this species is depicted by the solid bars (scale is relative). Adapted from Angilletta et al. (2002b) with permission from Elsevier. © 2002 Elsevier Ltd.

every performance within a narrow range of temperatures (Huey 1982). How should organisms behave when the thermal optimum or performance breadth differs between two or more performances? To answer this question, we must consider the conflicts and compromises that arise when maximizing fitness. Two examples will shed light on this topic: thermoregulation during infection and thermoregulation during pregnancy.

4.5.1 Thermoregulation during infection

Infection often elicits a change in the thermoregulatory strategy of an animal. In many cases, higher body temperatures are preferred—a phenomenon generally referred to as fever. Both endotherms and ectotherms are capable of sustaining fever, the latter group using behavioral rather than physiological mechanisms (see, for example, Sherman et al. 1991). Substantial evidence supports the view that fever enhances a host's immunocompetence and impairs a pathogen's growth (reviewed by Kluger 1979; Kluger et al. 1998). For example, Ouedraogo and colleagues (2004) noted that locusts (*Locusta migratoria*) infected by a fungus basked more frequently than did uninfected locusts. By manipulating the potential for basking, these researchers quantified a substantial increase in survivorship associated with thermoregulation. The longer an infected locust could thermoregulate, the more likely it was to survive the course of the experiment (21 days). Further studies provided evidence that thermoregulation enhanced survivorship by raising hemocyte levels and impairing fungal growth (Ouedraogo et al. 2003). Given the phylogenetically widespread occurrence of fever (see Kluger 1979), the thermal optimum for immunological performance probably exceeds that for other performances; if so, fever could represent a shift in the optimal thermoregulation caused by a gain in survivorship that more than compensates for the loss of other performances.

Likely, a more complex relationship exists between temperature and host–pathogen interactions. Although high temperatures can enhance immunocompetence during early stages of infection, these same temperatures can speed the growth of the pathogen during later stages (Warwick 1991). This problem would seem most relevant when the heat tolerance of the pathogen exceeds that of the host. In such cases, we might expect infected individuals to select relatively low temperatures to slow the growth of the pathogen. Some combinations of host and pathogen do cause the host to select lower temperatures during infection. For example, snails (*Lymnaea stagnalis*) reduced their preferred body temperature from 25°C to 20°C when infected with either of two species of trematodes, but retained a preference for 25°C when infected with a third species (Zbikowska 2004). Perhaps, these different responses would make sense if we knew the thermal sensitivities of the host and parasites. Still, not all organisms reduce their body temperature during infection when the growth of the pathogen would suffer. Although tent caterpillars inoculated with a nucleopolyhedrovirus died faster at higher temperatures, inoculated and uninoculated individuals preferred relatively high temperatures, which accelerated the death of infected individuals (Frid and Myers 2002). As with fever, a reduction in body temperature during infection probably causes some loss of performance for the host. Therefore, thermoregulation during infection reflects a compromise between maximizing survivorship and maximizing other performances.

4.5.2 Thermoregulation during pregnancy

Another excellent example of a conflict imposed by different thermal optima involves thermoregulation during pregnancy. Both oviparous and viviparous animals may change their thermoregulatory behavior when they carry developing offspring (Table 4.3). Beuchat and Ellner (1987) hypothesized that this behavior results from a mismatch between the thermal optima for mothers and embryos. In this case, optimizing embryonic performance enhances current reproduction whereas optimizing maternal performance enhances both current and future reproduction. Using the thermal sensitivities of maternal and embryonic performances, they modeled the optimal body temperature for pregnant lizards of a viviparous species (*Sceloporus jarrovi*).

Beuchat and Ellner assumed the temperature during gestation was the only factor that influenced the performance of a mother and her embryos. The performance of the mother was modeled as the product of survivorship (p_x) and growth (δ); growth contributed to future reproduction because litter size increased linearly with increasing body size. The performance of embryos was modeled as their probability of survival to birth. Given the litter size at age x, the probability of embryonic survival could be used to calculate the thermal sensitivity of fecundity (m_x). Empirical data from laboratory experiments were used to parameterize the thermal sensitivities of $p_x \cdot \delta$ and m_x in *Sceloporus jarrovi* (Fig. 4.17). The optimal body temperature—that is, the one that maximizes the fitness of the mother—resides somewhere between the body temperature that maximizes current fecundity (m_x) and the body temperature that maximizes future fecundity ($p_x \cdot \delta$). This optimum depends on the expected variance of body temperature because deviations from the mean body temperature reduce fitness asymmetrically (see Fig. 4.17). For *S. jarrovi*, the optimal body temperature was predicted to be 31.4°C under the observed variance in body temperature and 34°C for no variance in body temperature. The former value very nearly matches the mean body temperature of pregnant females (32°C).

Patterns of thermoregulation during infection and pregnancy clearly illustrate the need to consider all

Table 4.3 Viviparous reptiles tend to increase their body temperature when gravid, but the behaviors of oviparous reptiles vary among species. For each study, the effect of gravidity on the mean body temperature is given. Studies published prior to 1980 were reviewed by Shine (1980).

Species	Setting	Effect	Source
Viviparous			
Chalcides ocellatus	Laboratory	Higher	Daut and Andrews (1993)
Charina bottae	Field	Higher	Dorcas and Peterson (1998)
Crotalus viridis	Field	Same	Charland and Gregory (1990)
Crotalus viridis	Field	Higher	Graves and Duvall (1993)
Hoplodactylus maculates	Laboratory	Higher	Rock et al. (2000)
Mabuya multifasciata	Laboratory	Lower	Ji et al. (2007)
Malpolon monspessulanus	Field	Higher	Blázquez (1995)
Nerodia rhombifera	Field	Higher	Tu and Hutchison (1994)
Nerodia sipedon	Field	Higher	Brown and Weatherhead (2000)
Sceloporus bicanthalis	Laboratory	Same	Andrews et al. (1999)
Sceloporus grammicus	Field	Lower	Andrews et al. (1997)
Sceloporus jarrovi	Laboratory	Lower	Mathies and Andrews (1997)
Thamnophis elegans	Laboratory	Higher	Gregory et al. (1999)
Thamnophis elegans	Field	Higher	Charland (1995)
Thamnophis sirtalis	Field	Higher	Charland (1995)
Zootoca vivipara	Laboratory	Lower	Gvoždík and Castilla (2001)
Oviparous			
Antaresia childreni	Laboratory	Higher	Lourdais et al. (2008)
Hemidactylus frenatus	Field	Same	Werner (1990)
Lepidodactylus lugubris	Field	Same	Werner (1990)
Natrix natrix	Field	Lower	Isaac and Gregory (2004)
Podarcis muralis	Field	Lower	Braña (1993)
Podarcis muralis	Laboratory	Same	Braña (1993)
Sceloporus aeneus	Laboratory	Same	Andrews et al. (1999)
Sceloporus gadoviae	Field	Same	Lemos-Espinal et al. (1997)
Sceloporus scalaris	Field	Higher and lower[1]	Smith et al. (1993)
Sceloporus undulatus	Laboratory	Same	Angilletta et al. (2000)
Sceloporus undulatus	Field	Same	Gillis (1991)
Sceloporus virgatus	Field	Lower	Smith and Ballinger (1994)
Uromastyx philbyi	Lab	Higher	Zari (1998)
Urosaurus ornatus	Field	Same	Smith and Ballinger (1995)

[1]The effect of reproductive condition on body temperature differed between two populations.

potential benefits of thermoregulation. Even when multiple performances proceed best at a single body temperature, their thermal sensitivities might differ radically (see Fig. 4.16). By considering both energetic and nonenergetic benefits, Beuchat and Ellner constructed a model that predicted thermoregulation more accurately. Similar models could be developed for thermoregulation during infection or for other scenarios involving thermoregulatory conflicts.

4.6 Do organisms thermoregulate less precisely when the costs are greater?

A major motivation for Huey and Slatkin's model was the energetic cost of thermoregulation that arises from shuttling between preferred microclimates. Both the frequency distribution and the spatial distribution of operative temperatures determine this cost. Thermoregulation would be less costly when preferred microclimates occur more

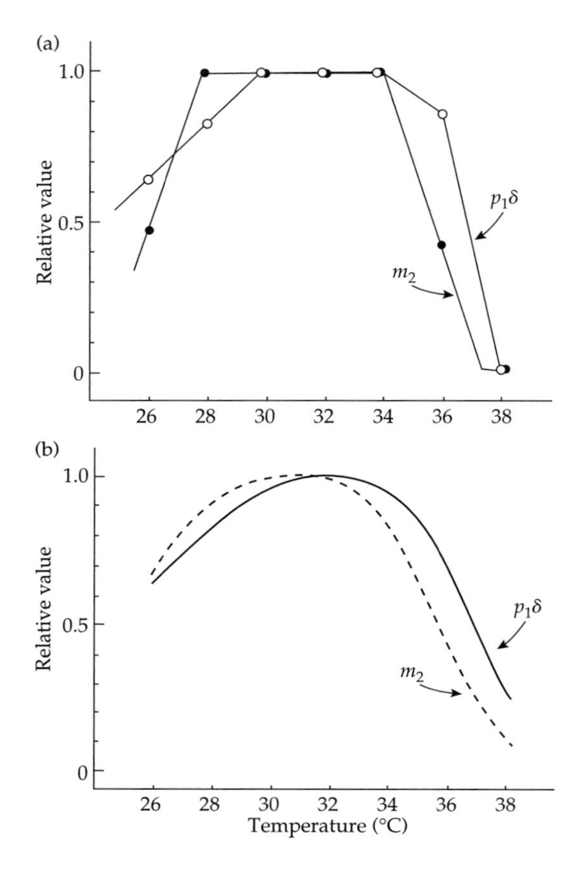

Figure 4.17 (a) Observed relationships between mean body temperature (SD $= 0°$C) and estimates of the current fecundity (m_2) and the future fecundity ($p_x \cdot \delta$) of pregnant lizards. (b) Modeled relationships between mean body temperature (SD $= 1.4°$C) and estimates of current fecundity (m_x) and future fecundity ($p_x \cdot \delta$). Adapted from Beuchat and Ellner (1987) with permission from the Ecological Society of America.

frequently in space. Furthermore, the movement required to locate a preferred microclimate should decrease as the environment becomes patchier (i.e., negatively autocorrelated). Because of diel shifts in the angle of solar radiation, organisms must move among ephemeral patches to maintain their preferred body temperatures throughout the day. These activities are the implicit cause of the energetic cost of thermoregulation in Huey and Slatkin's model.

The optimal level of thermoregulation decreases ($k \rightarrow 1$) as the cost of thermoregulation (c) increases. Several factors could raise the cost of thermoregulation. As the height of the cost curve increases, the net benefit of thermoregulation decreases at all environmental temperatures, which tends to favor less precise thermoregulation. A higher cost curve could result from organismal factors such as inefficient locomotion or inadequate thermoperception, or from environmental factors such as the patchiness of the thermal landscape. Additionally, the optimal level of thermoregulation can decrease as operative temperatures decrease. In colder environments, the difference between the operative temperature and the thermal optimum increases; this effect would increase the cost of thermoregulation in certain environments (see Fig. 4.12a). Importantly, we must assume the spatial autocorrelation of operative temperatures does not change, even though the mean operative temperature does.

The best evidence that energetic costs affect thermoregulatory behaviors comes from an experiment conducted by Withers and Campbell (1985). They constructed an experimental apparatus known as a shuttle box. The box contained an incandescent bulb at each end, which was activated when an animal broke a beam created by a pair of photocells. Once activated, the bulb would emit heat for a specified period but could not be reactivated until the animal moved to the opposite side of the box and activated the other bulb. Thus, an animal had to shuttle between the two sides of the box to maintain a body temperature that exceeded the ambient temperature. The frequency of shuttling required for thermoregulation depended on how long a bulb remained lit once it was activated. By varying this period between 30 and 180 seconds, Withers and Campbell directly manipulated the energetic cost of thermoregulation. When desert iguanas (*Dipsosaurus dorsalis*) were placed in the shuttle box, they behaved exactly as predicted by Huey and Slatkin's model (Fig. 4.18). Iguanas shifted from nearly perfect thermoregulation at 180 seconds of light per activation to nearly perfect thermoconformation at 30 seconds! To this day, Withers and Campbell's ingenious experiment remains the best demonstration of the impact of energetic costs on thermoregulation.

When assessing the role of energetic costs in generating diel, seasonal, and geographic variation

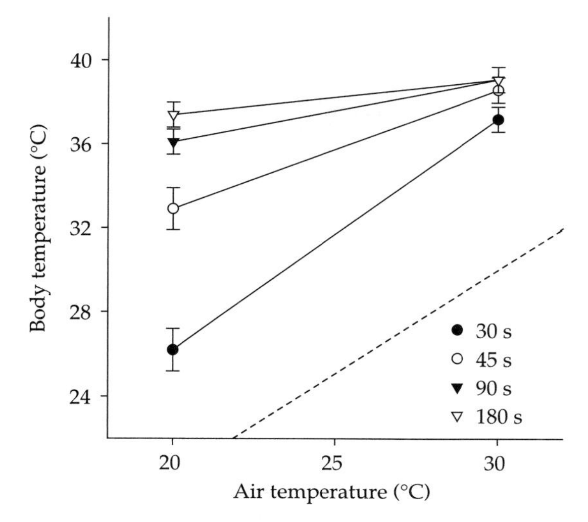

Figure 4.18 A lower energetic cost of thermoregulation led to a greater degree of thermoregulation by desert iguanas. Error bars represent 2 standard errors. Body temperatures of desert iguanas were measured in an arena that required shuttling at certain intervals to maintain a radiative heat source. The 30 second interval demanded the most energetic expenditure and the 180 second interval required the least. The dashed line represents the hypothetical slope of a thermoconformer. When placed in a thermal gradient, where no shuttling was required for thermoregulation, the same iguanas preferred 38.7±0.4°C. Data are from Withers and Campbell (1985).

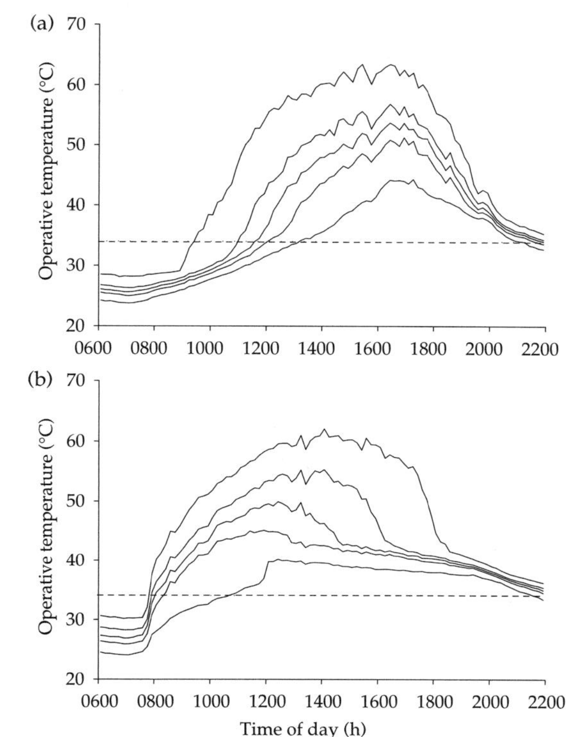

Figure 4.19 The diel patterns of operative temperatures for *Sceloporus merriami* differed dramatically between the (a) east and (b) west sides of a canyon in Big Bend National Park, Texas, USA. In each plot, the solid lines denote minimal, third quartile, median, first quartile, and maximal operative temperatures. The dashed line marks the thermal optimum for energy gain (Beaupre et al. 1993). During the majority of the day, a lizard could not attain this thermal optimum on either side of the canyon. Adapted from Grant (1990) with permission from the Ecological Society of America.

in body temperature, a major problem emerges: researchers rarely record sufficient data to estimate the benefits and costs of thermoregulation. In most studies to date, operative temperatures were sampled haphazardly or selectively (see Chapter 2), precluding an accurate estimate of the thermal quality of the environment. Furthermore, researchers rarely present the frequency distribution of operative temperatures, preventing us from teasing apart energetic costs and thermal constraints. The following example illustrates this problem. Grant (1990) compared the thermoregulation of lizards (*Sceloporus merriami*) on two sides of a canyon in southwest Texas. In the morning (1000–1200 hours), the thermal optimum for energy intake was readily attainable on the east side of the canyon, but was rarely attainable on the west side (Fig. 4.19). As predicted by the theory, lizards on the east side thermoregulated more precisely during this period of the day. However, lizards on both sides of the canyon were unable to attain their thermal optimum during most of the day. Without this detailed knowledge of the relative abundance of operative temperatures over space and time, we could easily draw a false conclusion about thermoregulation.

For lack of better information, we might assume the cost of thermoregulation increases as preferred microclimates become rare. Indeed, thermoregulation in forested and open habitats have been compared, under the assumption that thermoregulation costs more energy in forested habitats because basking sites are less abundant (see, for example, Hertz 1974; Huey 1974). Similarly, body temperatures of a species have been compared along latitudinal and altitudinal clines. The mean body temperature decreases with increasing latitude or altitude in some species, (Hertz 1981, 1983; Kiefer et al. 2005; Pianka 1970; Van Damme et al. 1989), but does not do so in other species (Angilletta 2001; Hertz and Huey 1981). Because these intraspecific comparisons were not coupled with random sampling of operative temperatures, we have no way of really knowing whether colder environments imposed greater energetic costs of thermoregulation.

By considering frequency distributions of operative temperatures, Blouin-Demers and Nadeau (2005) provided the first broad and rigorous evaluation of the impact of energetic costs on thermoregulation in natural environments. They reasoned that the index d_e provided a good measure of the cost of thermoregulation because a large d_e means the environment is (on average) much hotter or colder than the organism prefers (see Box 4.2). This reasoning makes sense if we envision a costly environment as one in which an animal must exert greater effort to find preferred microclimates. If fewer sites lie within the preferred range of temperatures, these sites should be harder to locate (and might also be farther from other important resources, such as food and refuge). Assuming d_e reflects the increase sharply relative cost of thermoregulation, we should expect the accuracy of thermoregulation (d_b) to increase sharply as d_e increases (i.e., $\Delta d_b > \Delta d_e$; Fig. 4.20a). Plotting independent contrasts for 22 species of lizards, Blouin-Demers and Nadeau observed just the opposite (Fig. 4.20b). Moreover, Row and Blouin-Demers (2006) discovered that similar plots of d_b versus d_e for individual snakes (*Lampropeltis triangulum*) also differed from the predicted relationship.

If Withers and Campbell's experiment provided such strong evidence that energetic costs influence thermoregulation, why do thermoregulatory behaviors in natural environments seem to contradict this conclusion? Blouin-Demers and Nadeau hypothesized that individuals cannot afford to thermoconform when operative temperatures differ greatly from their preferred temperature because the loss of benefit associated with thermoconformation exceeds the greater cost of thermoregulation. This idea has some merit but cannot be validated without accurate measures of benefit and cost curves in the relevant environments. Additional possibilities

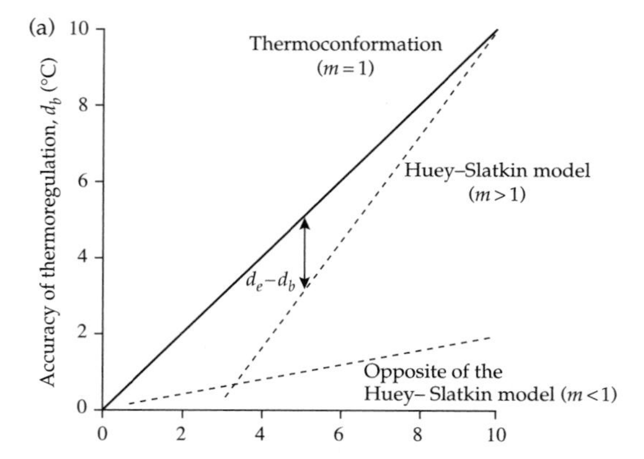

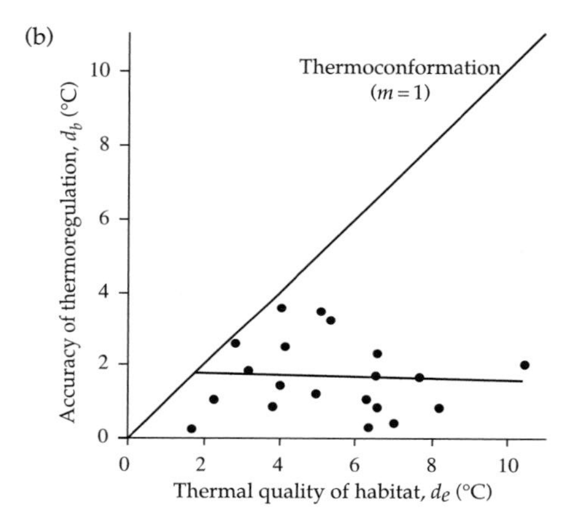

Figure 4.20 (a) Using the thermal quality of the environment (d_e) to estimate of the cost of thermoregulation, Blouin-Demers and Nadeau predicted the slope (m) of the relationship between d_e and the accuracy of thermoregulation (d_b). The bold line depicts the relationship for a perfect thermoconformer ($m = 1$), whereas the dashed lines depict relationships presumably consistent ($m > 1$) and presumably inconsistent ($m < 1$) with Huey and Slatkin's model. (b) The observed relationship between d_e and d_b opposes the predicted relationship. Adapted from Blouin-Demers and Nadeau (2005) with permission from the Ecological Society of America.

become clear in light of the assumptions made by these investigators. First, they assumed the thermal sensitivity of energy gain was the same for all animals. Second, they assumed preferred body temperature equaled the thermal optimum for energy gain. Third, they assumed d_e adequately captured the energetic cost of thermoregulation. If any of these assumptions proves invalid, we should question the inference drawn from the comparative data.

Finally, both the optimality model and the comparative study ignore nonenergetic costs of thermoregulation. Costs of predation risk and missed opportunities also influence the optimal thermoregulation and these costs have not been integrated with the energetic cost.

Even if we attribute a paramount significance to energetic costs, we will have a difficult time constructing a cost curve from measurements of d_e because this index ignores the spatial structure of the environment. After all, d_e tells us only the mean difference between the operative environmental temperature and the preferred body temperature. This index fails to capture the statistical variance or spatial distribution of operative temperatures. Without controlling for environmental variance and spatial structure, the use of d_e to estimate energetic costs can lead to erroneous conclusions (Sears 2006). To comprehend the significance of spatial structure, picture an environment with a frequency distribution of operative temperatures, characterized by some value of d_e. This statistical distribution (and hence the d_e) tells us nothing about the spatial distribution of these temperatures. In fact, we can imagine many spatial distributions for any statistical distribution. A clumped distribution of operative temperatures poses a different challenge to an organism than does a dispersed distribution. When favorable temperatures are clumped, other resources would be either more or less likely to occur near preferred microclimates. Moreover, clumped distributions caused by large clearings or canopies should change more predictably as the sun moves across the sky. Yet, clumped and dispersed distributions can yield identical values of d_e! Because the cost of thermoregulation depends on both the spatial and statistical distributions of operative temperatures, one must control for spatial structure when testing Huey and Slatkin's model. Using computer simulations, we can estimate cost curves for specific thermal environments (Box 4.4); however, applying this approach requires extensive knowledge of the biology of the organism and the structure of the environment. Simply put, estimating the cost of thermoregulation is a complex task. This task becomes even more complex when we add the nonenergetic costs.

4.7 Nonenergetic costs of thermoregulation

4.7.1 Aggressive interactions with competitors

In a seminal paper, Magnuson and colleagues (1979) argued convincingly that competition for temperature could be viewed in the same way as competition for food. After all, preferred microclimates occur in space, and ecologists have long known that organisms compete directly for space (Connell 1961). Magnuson and colleagues characterized the thermal niche by statistically describing the temperatures selected in a thermal gradient (e.g., the median or range $\pm$ a percentage of the distribution). By adopting this view, we could characterize the use of available space in relation to its thermal quality (Huey 1991; Tracy and Christian 1986). When two species compete for thermal resources, the thermal niche of the subordinate competitor or both competitors could be altered by the interaction (Buckley and Roughgarden 2005).

Magnuson and his colleagues (Beitinger and Magnuson 1975; Medvick et al. 1981) experimentally verified the impact of intraspecific competition on thermoregulation. In a series of experiments, bluegill sunfish (*Lepomis macrochirus*) were placed individually or socially in a tank with two compartments: an optimal habitat (31°C) and a suboptimal habitat (a constant temperature between 21°C and 29.5°C). In isolation, fish preferred the optimal habitat over the suboptimal one (Fig. 4.23a). In pairs, the dominant fish always occupied the optimal habitat and the subordinate did not (Fig. 4.23b). In larger groups, all fish crowded into the optimal habitat, particularly when the suboptimal habitat was much colder than the optimal habitat (Fig. 4.23c–f). Observations of reptiles also support the view that intraspecific competition affects thermoregulatory behavior. According to Seebacher and Grigg (2000), crocodiles often emerge from water to bask, but dominant crocodiles chase subordinate crocodiles back into the water before these subordinates can reach their preferred body temperature. Moreover, subordinates do not return to bask as frequently after being chased. Using mathematical models, the body temperatures of subordinate crocodiles were estimated to have been 6°C below the body temperatures of dominant crocodiles.

Competition could cause rapidly reversible or developmentally fixed changes in thermoregulatory behavior. For example, Regal (1971) noticed a marked change in basking when a male lizard was exposed to another male and a rapid reversal of behavior when the competitor was removed. In a more rigorous study of the same phenomenon, lizards of a dominant species (*Podarcis sicula*) used warm microclimates more frequently when sharing an enclosure with a subordinate competitor (*P. melisellensis*) than when occupying a similar enclosure alone (Downes and Bauwens 2002). In contrast, individuals can express combinations of social and thermoregulatory behaviors in the absence of competitors; dominant, orange-colored males of the lizard *Pseudemoia entrecasteauxii* preferred a slightly higher mean temperature than subordinate, white-colored males, even when placed individually in a thermal gradient (Stapley 2006). Such combinations of social and thermoregulatory strategies could develop during early encounters with conspecifics and remain relatively fixed. Differences in thermoregulatory behavior between competitors, whether flexible or fixed, could minimize risks of stress, injury, or death.

4.7.2 Risk of predation or parasitism

When ectotherms thermoregulate, they often seek out areas that receive intense solar radiation. This behavior would place them in open habitats, where they would draw the attention of predators (Webb and Whiting 2005). Just about every environment offers some form of refuge from predation, but

Box 4.4 Estimating cost curves by computer simulation

Estimating the energetic cost of thermoregulation seems hopelessly difficult because this cost depends on the statistical and spatial distributions of operative temperatures, as well as physical and physiological properties of the organism. Still, we can estimate this cost through computer simulations of movement in virtual landscapes. Using such simulations, Sears and Angilletta (2007) calculated cost curves for hypothetical thermal environments. Simulated individuals were assigned properties of a small lizard ($\approx$ 10 g), such as *Sceloporus undulatus* (Angilletta 2001; Dzialowski and O'Connor 2004). These individuals moved through a 60 m × 60 m grid, with each cell (1 m^2) having an operative temperature drawn from a bimodal distribution (Fig. 4.21a). Holding this statistical distribution constant, we configured the thermal landscape in various ways; for example, Fig. 4.21 shows

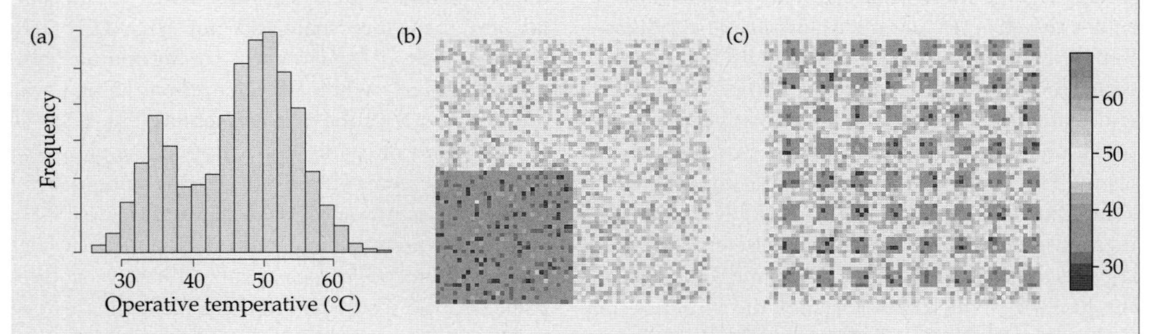

Figure 4.21 (a) A bimodal distribution of operative temperatures was used to simulate the energetic cost of behavioral thermoregulation. (b and c) Depending on the simulation, these temperatures were distributed in one large clump or in many small patches. See plate 2.

continues

Box 4.4 Continued

a clumped spatial distribution and a dispersed spatial distribution. For each spatial distribution, we simulated behavioral thermoregulation and estimated its cost.

Energetic costs of thermoregulation were estimated as the movement required to achieve a preferred body temperature. At the start of a simulation, an individual was placed in a randomly selected microenvironment and was assigned a body temperature equal to the operative temperature. From this initial position, the individual moved until its body temperature entered the set-point range (33–36°C). The angle and distance of each movement were randomly drawn from uniform distributions (angle = 0–360°; distance = 0–3 m). The change in body temperature at each interval was calculated from the thermal time constant of the individual (5 minutes) and the operative temperature of the microenvironment (Gates 1980). The distance moved during the simulation provides a simple proxy for the energetic cost of thermoregulation. By replicating this simulation 1000 times, we obtained costs of thermoregulation for a range of initial operative temperatures. The shape of the cost curve, as conceived by Huey and Slatkin (1976), was inferred by plotting the distance moved against the absolute deviation between the thermal optimum and the initial operative temperature ($|T_{opt} - T_e|$).

The spatial structure of the environment had a dramatic impact on the cost of thermoregulation (Fig. 4.22). In the clumped environment, more movement was necessary to locate a preferred microenvironment when the initial operative temperature deviated further from the set-point range. In the patchy environment, the initial operative temperature had little effect on the movement required to thermoregulate. These results make sense considering the spatial distributions used in the simulation. In a clumped environment, great distances might be required to reach a preferred microclimate if the initial microclimate deviates greatly from the preference (see Fig. 4.21b). In a patchy environment, an individual always resides near a preferred microclimate regardless of its position (see Fig.4.21c). On average, the energetic cost of thermoregulation in a clumped environment greatly exceeds that in a patchy environment.

This approach to estimating the cost of thermoregulation confers several advantages. First, we can consider properties of animals, such as perception, mobility, and memory. Second, we can account for both spatial and statistical distributions of operative temperatures when predicting thermoregulation. Finally, the explicit mapping of operative temperatures enables us to consider spatial covariations between temperature and other factors, such as food, water, and predators. These spatial covariations ultimately determine the costs of predation risk and missed opportunities.

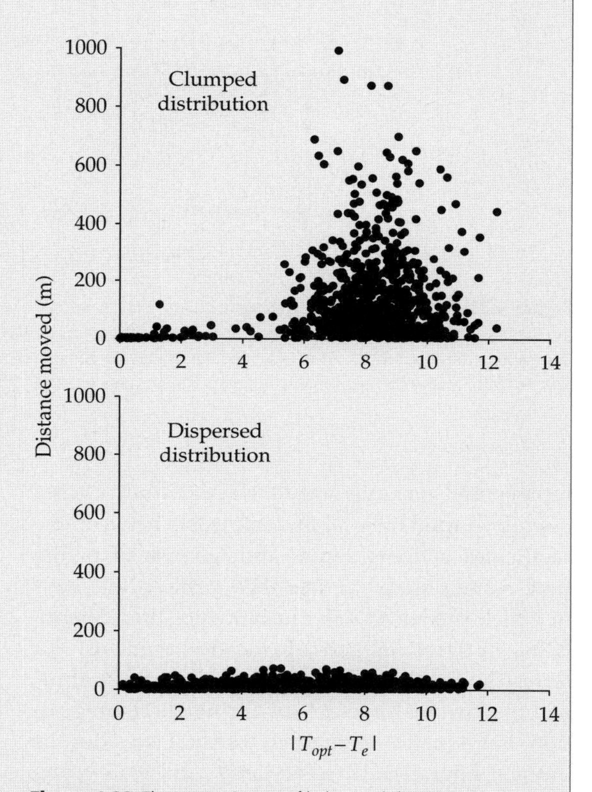

Figure 4.22 The energetic cost of behavioral thermoregulation depends on the spatial structure of the environment. In an environment where preferred microclimates occur in one large clump (positive spatial autocorrelation), simulated individuals had to move greater distances to thermoregulate as the operative temperature of their initial microenvironment (T_e) deviated further from the optimal temperature range. In an environment where preferred microclimates were dispersed (negative spatial autocorrelation), simulated individuals had to move only small distances to thermoregulate, regardless of their initial microenvironment. For calculations of $|T_{opt} - T_e|$, the thermal optimum (T_{opt}) was represented by either the lower or upper limit of the set-point range (33°C or 36°C).

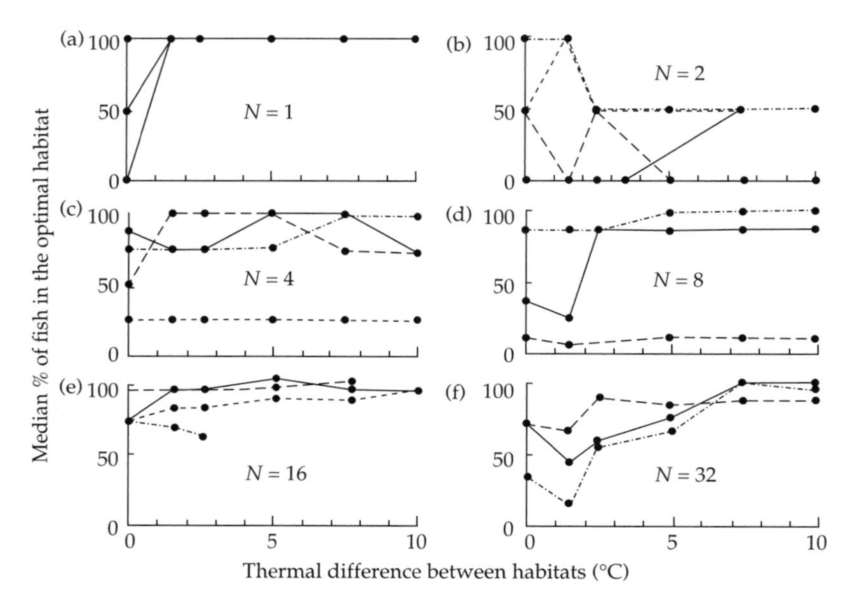

Figure 4.23 The percentage of fish using an optimal habitat instead of a suboptimal habitat depended on the number of competing fish (*N*) and the thermal difference between habitats. For any particular trial, a fish was either alone (a) or in the presence of some number of conspecifics (b–f). Competition forced some fish into the suboptimal habitat, except in cases when the difference in temperature between habitats was extreme. Adapted from Medvick et al. (1981) with permission from the American Society of Ichthyologists and Herpetologists.

refuges are generally insulated and cool. Consequently, a fundamental tradeoff exists between the regulation of temperature and the risk of predation. Recent modeling has shed light on the ways in which predation risk can influence thermoregulation in terrestrial and aquatic environments. For terrestrial ectotherms, we shall consider the optimal use of burrows or crevices during thermoregulation. For aquatic ectotherms, we shall consider the optimal migration between deep, cold waters and shallow, warm waters.

Imagine a terrestrial animal that basks to maintain its thermal optimum for performance, but this behavior attracts the attention of a predator. The animal can avoid predation by fleeing to a refuge; however, its body temperature will fall below its thermal optimum within the refuge. When should the animal emerge from the refuge to continue basking? To determine the optimal duration to spend within a refuge, one must define the costs and benefits of abandoning thermoregulation. Obviously, the benefit of staying in the refuge equals the increase in survivorship from avoiding the predator. The cost of staying in the refuge equals the loss of performance caused by the drop in body temperature (Amo et al. 2007a,b). Assuming the benefit and the cost can be expressed in the same currency (fitness), we can find the optimal emergence time.

Polo and colleagues (2005) modeled this problem with specific cost and benefit curves. They assumed attack by a predator imposes a risk of predation (R), but this risk decreases gradually over time (x):

$$R(x) = \frac{2}{e^{x/a} + e^{-x/a}}, \tag{4.9}$$

where a determines the rate at which the risk of predation decays over time. Although a refuge offers protection from predation, the cost of remaining in the refuge increases asymptotically over time:

$$C(x) = m(1 - e^{-x/n}), \tag{4.10}$$

where m and n define the shape of the cost curve. Based on this model, the time spent in a refuge

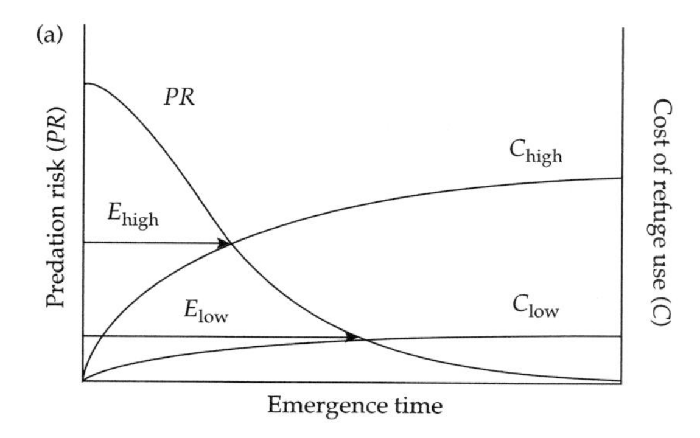

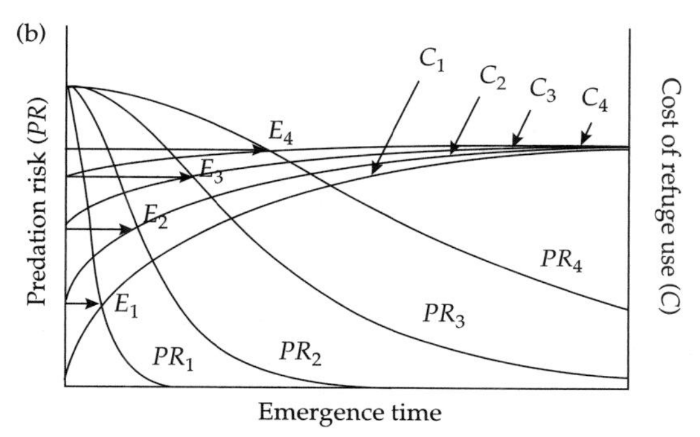

Figure 4.24 An optimality model predicts the time required to emerge from a cold refuge after an attack by a predator. (a) The optimal decision (*E*) depends on the change in predation risk over time (*PR*) and the energetic cost of remaining within the retreat (*C*). Arrows depict the optimal time of emergence for retreats imposing high and low costs. (b) The optimal decision changes with subsequent attacks if we assume an organism's predation risk increases or its energetic cost decreases with each new attack. Adapted from Polo et al. (2005) with permission from Evolutionary Ecology Ltd.

should decrease as the temperature of the refuge decreases (Fig. 4.24a). This result occurs because the loss of performance when hiding in a cold refuge exceeds the loss of performance when hiding in a warm refuge.

Polo and colleagues used their model to predict the behavior of alpine lizards in an artificial thermal arena. They set up a linear gradient of basking sites and placed a refuge in the coolest region. By adjusting the height of the heat sources, they could control the operative temperatures throughout the gradient. The operative temperature of the warmest region lay within the lizard's range of preferred body temperatures (31–34°C), whereas the operative temperature in the refuge equaled either 27°C (low-cost refuge) or 18°C (high-cost refuge). After allowing a lizard to habituate to the gradient, it was attacked by a simulated predator (a paint brush). As predicted, lizards provided with the warm refuge took about three times as long to emerge as did lizards provided with the cool refuge.

The researchers continued to attack each lizard after the initial emergence and noted that emergence time increased by about 1 minute with each successive attack. They conjectured this pattern reflects changes in both the perceived risk and the cost curve with repeated attacks. If each attack increases a lizard's perceived risk of predation, the risk curve, $R(x)$, becomes shallower during successive attacks (however, predation risk would actually increase with each attack if lizards tire progressively during repeated escapes). Additionally, a lizard would be cooler than it prefers if attacked when emerging from the refuge. Because the initial body temperature would be lower at the onset of the second attack, the lizard would experience a different cost curve, $C(x)$. The combined changes in these two curves should favor longer emergence times with repeated attacks (Fig. 4.24b). Results of a field experiment with the same species were fairly consistent with the model (Martin and Lopez 2001). Furthermore, the model also explained differences in the use of refuges between populations; lizards in a safer environment waited longer to flee and emerged sooner after hiding than did lizards in a risky environment (Diego-Rasilla 2003). Predation risk during basking can cause ineffective thermoregulation (Herczeg et al. 2008), which would ultimately slow growth and reproduction (Downes 2001).

The risk of predation also affects the thermoregulatory behaviors of aquatic organisms. Many species of zooplankton migrate vertically on a daily basis, spending days in cool, deep waters and nights in warm, shallow waters. Several lines of evidence suggest both energetic cost and predation risk modulate the degree of vertical migration (reviewed by Lampert 1989). The intensity of daily migration by water fleas (*Daphnia magna*) depended strongly on the concentration of fish kairomones (Loose and Dawidowicz 1994). In the presence of concentrated kairomones, individuals migrated to cool, deep water during the day. In the presence of dilute or no kairomones, the water fleas remained in warm surface waters at all times. Predation risk also seems to explain differences in the direction of migration (Lampert 1989); for example, an exceptional species, which spends the day in shallow waters and the night in deep waters, avoids predation by a larger species of zooplankton (Ohman et al. 1983). Similarly, snails avoided areas of a thermal gradient that were concentrated with the odor of crushed conspecifics. Upon encountering these areas, an individual withdrew immediately and often moved as far as possible from the source of the cue (Gerald and Spezzano 2005). Although snails usually favor very low temperatures, their movements in the presence of conspecific cues resulted in higher body temperatures.

Ectotherms may also alter their use of microclimates in response to the risk of parasitism. For example, certain ticks attach to animals as they pass through vegetation. Kerr and Bull (2006) argued this localized risk of parasitism affects the use of microenvironments by the sleepy lizard (*Tiliqua rugosa*). Lizards had the greatest risk of acquiring ticks when seeking thermal refuge in bushes. The risk of parasitism doubled or tripled as temperatures rose within bushes. Because the most extreme thermal conditions enhance both the need for a thermal refuge and the risk within that refuge, a tradeoff between thermoregulation and parasitism exists. As with the risk of predation, the risk of parasitism could impose a cost of thermoregulation that deserves the attention of both theorists and empiricists.

4.7.3 Risk of desiccation

Although evaporative water loss enables organisms to avoid overheating, dehydration causes a loss of performance that can ultimately lead to death. For this reason, we should expect the hydration of an organism to influence its thermoregulatory strategy. Since evaporation proceeds more rapidly at higher temperatures, we could reasonably predict that poorly hydrated organisms would behaviorally avoid high temperatures. Studies of animals and plants support this prediction. Dehydrated animals typically prefer lower temperatures than do hydrated animals (Crowley 1987; Dohm et al. 2001; Ladyman and Bradshaw 2003; Smith et al. 1999). Among 12 species of Australian frogs, preferred body temperature correlated positively with the water resistance of skin (Tracy and Christian 2005). In plants, the need for solar radiation inextricably links thermoregulation and photosynthesis. Plants that track the sun to absorb visible light must also use evaporation to avoid overheating (Ehleringer and Forseth 1980; Galen 2006). At the wilting point, leaves and flowers abandon solar tracking to avoid the resulting damage caused by high temperatures or poor hydration (Forseth and Ehleringer 1980). In natural environments, a shortage of water alters the pattern of activity (Lorenzon et al. 1999) and the use of microhabitats (Cohen and Alford 1996) in ways that affect the body temperatures of animals.

4.7.4 Missed opportunities for feeding or reproduction

Organisms do more than thermoregulate; they must also find food, defend territories, and attract mates. When thermoregulation conflicts with these other activities, an individual incurs a cost of missed opportunities. Because activities such as finding and handling prey determine the energy gain, the benefit of thermoregulation depends on the degree to which thermoregulatory behaviors cause missed opportunities. For example, a conflict between food and temperature drives the daily vertical migration of dogfish (*Scyliorhinus canicula*). In an artificial thermal gradient, these fish preferred cool water. Yet individuals in their natural environment spent the night within cool, deep waters and the day within warm, shallow waters, because shallower waters provided far greater densities of food (Sims et al. 2006). In some lakes, *Daphnia* can find more food in deep water, but grow more slowly there because of low temperatures. By manipulating vertical clines in artificial water towers, Kessler and Lampert (2004) showed that *Daphnia* migrate vertically in the water column to optimize the combination of food and temperature (Fig. 4.25). When presented with a steep thermal gradient, *Daphnia* spent most of their time at intermediate depths. As thermal gradient dampened, *Daphnia* spent more time in deep waters. Bluegill sunfish also chose a habitat for both its thermal properties and its prey density (Wildhaber 2001). When preferred temperatures and preferred

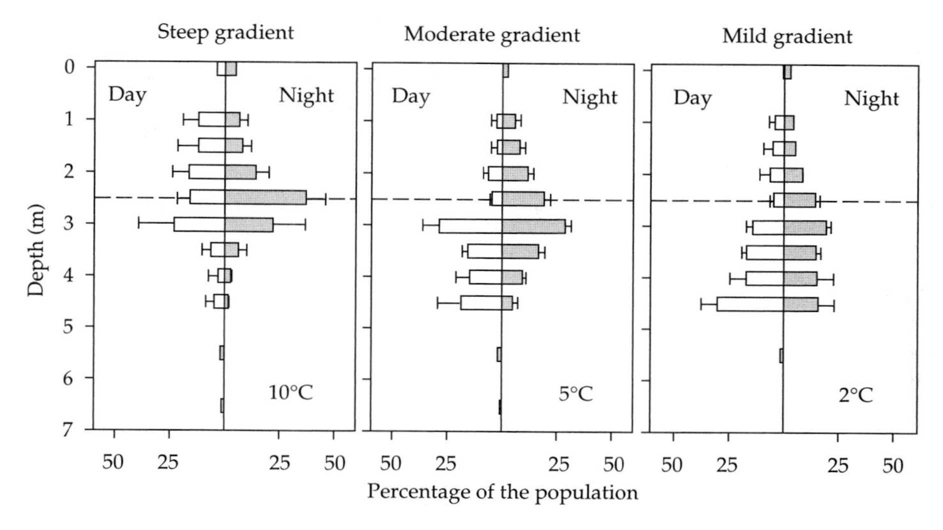

Figure 4.25 The vertical distribution of water fleas shifted from shallow water to deep water as the vertical gradient in temperature became milder. The horizontal bars in each plot show the mean percentage of the population (±SE) dwelling at the respective depths during the day (unfilled bars) and night (filled bars). The temperature of the top layer of water was always 20°C. The temperature differences between the top and bottom layers of water are given in the panels. The dashed line indicates the thermocline (the point at which the temperature changes rapidly with depth). Adapted from Kessler and Lampert (2004) with kind permission from Springer Science and Business Media. © Springer-Verlag 2004.

foods occur in different places, perfect thermoregulation can even decrease energy gain. Thus, we must consider both thermal and nutritional resources when predicting the use of microhabitats (e.g., Crowder and Magnuson 1983; Riechert and Tracy 1975; Wildhaber and Lamberson 2004).

Gvoždík (2002) investigated the tradeoff between basking and foraging for lizards (*Zootoca vivipara*) in four environments along an altitudinal gradient of 250–1450 m. A thermoregulator was assumed to maintain its temperature between lower and upper set points (T_{min} and T_{max}, respectively). Set points were estimated from body temperatures selected in an artificial thermal gradient (see Box 4.1). A lizard was hypothesized to initiate basking when its temperature fell to the lower set point and cease basking when its temperature rose to the upper set point. Using the biophysical model of Bakken and Gates (1975), Gvoždík estimated the duration of basking (t_{bask}) required to thermoregulate:

$$t_{bask} = -\tau_h \ln\left(\frac{T_{max} - T_e}{T_{min} - T_e}\right), \tag{4.11}$$

where τ_h equals the time constant for heating. Once an animal reached its upper set point, it could forage until its temperatures dropped to the lower set point. Thus, the time available for foraging (t_{forage}) was as follows:

$$t_{forage} = -\tau_c \ln\left(\frac{T_{min} - T_e}{T_{max} - T_e}\right), \tag{4.12}$$

where τ_c equals the time constant for cooling. To quantify the tradeoff between basking and foraging, Gvoždík assumed lizards cannot bask and forage simultaneously. Based on operative temperatures measured on clear days, he estimated the time required for basking at each elevation. At the highest elevation (1450 m), operative temperatures were mainly below the set-point range. Consequently, lizards at this elevation would have needed to spend about 50% of their time basking to maintain their temperature within the set-point range. Among the four sites, the estimated proportion of time required for basking was very highly correlated with the observed proportion of time spent basking. This greater duration of basking explains how lizards at high elevation were able to thermoregulate as accurately and effectively as lizards at lower elevations. If Gvoždík's assumptions were correct, lizards at the highest elevation might have gained less energy than lizards at lower elevations, despite their effective thermoregulation.

Opportunities for courtship and mating might also be compromised by thermoregulation. Males cannot bask uninterrupted while defending their territories. Long bouts of courtship or mating cause body temperatures of the participants to rise above or fall below the preferred range (Shine et al. 2000). Regulating conductive or convective exchange of heat can also interfere with the defense of resources and mates. Such tradeoffs between mating and thermoregulation were recently identified in male bees (Roberts 2005). Males of the solitary desert bee (*Centris pallida*) aggregate by the thousands in spring, attempting to mate with virgin females as they emerge from their nest. Bees either patrol the ground searching for mates or defend small territories by hovering along the edge of the aggregation. Like many insects, these bees alter the frequency of wing-beating and the flow of hemolymph to thermoregulate during flight. Despite these regulatory mechanisms, convective heat loss depends strongly on the specific manner of flight. Patrolling males beat their wings as rapidly as hovering males, yet patrollers were 1–3°C cooler than hoverers. Based on these observations, Roberts concluded that patrolling males suffered greater heat loss to convection than hovering males did. Because the reproductive benefits of these two strategies likely differ, a tradeoff exists between thermoregulation and mating. Although thermal biologists will need to investigate the generality of missed opportunities during thermoregulation, I suspect most organisms must choose between thermoregulation and competing activities.

Given a cost of missed opportunities, individuals should endeavor to spend as little time as possible to reach their preferred body temperatures. Belliure and Carrascal (2002) carried out an experimental test of this hypothesis. They placed lizards in an enclosure with two basking sites: one site that offered mainly radiative heat exchange and another site that offered mainly conductive heat exchange. Both sites provided the same operative temperature but differed in the rates of heating they enabled. Amazingly, lizards were able to detect the difference between these sites and selected the one that offered the faster rate of heating! This result confirms that, at least in one species, individuals strive to minimize missed opportunities during thermoregulation.

4.7.5 Interactions between different costs

Estimating the fitness consequence of a thermoregulatory strategy becomes further complicated when various costs interact with one another. For example, we know that animals avoid preferred microclimates when they occur in risky habitats (see Section 4.7.2). Although predation risk clearly deters individuals from making certain thermoregulatory decisions, the decision to avoid this risk depends on other factors, such as the density of conspecifics. Two experiments by Downes nicely illustrate this point. In the first experiment, Downes and Shine (1998) showed that antagonistic interactions between conspecifics could alter thermoregulation under the risk of predation. Geckos were placed in enclosures that offered two retreats that differed in temperature. Solitary geckos preferred to rest in the warm retreat, but shifted their preference to the cold retreat when the scent of a predator was applied to the warm retreat. The result differed when geckos were paired with larger conspecifics; the smaller member of the pair was forced to use the warm retreat even when this site had been marked by a predator (Fig. 4.26). In the second experiment, Downes and Hoefer (2004) showed that interactions among conspecifics can sometimes reduce the cost of thermoregulation imposed by predators. When alone or in small groups, sun lizards basked less in the presence of chemical cues from snakes. In larger groups, however, the same lizards basked frequently with or without cues from the predator. When animals find safety in numbers, the presence of many conspecifics dilutes the risk of predation during basking.

Dynamic optimality models can make predictions about thermoregulation when temperature, food, and risk vary among microenvironments (Box 4.5). Such models can easily include complex biological details and are best tailored for specific organisms. If these models include an explicit consideration of spatial structure, they could not only provide more precise estimates of energetic costs but could also incorporate the spatial covariation between temperature and the factors that influence nonenergetic costs. The distribution and abundance of other resources in space almost certainly determine costs of thermoregulation. Whether an organism views

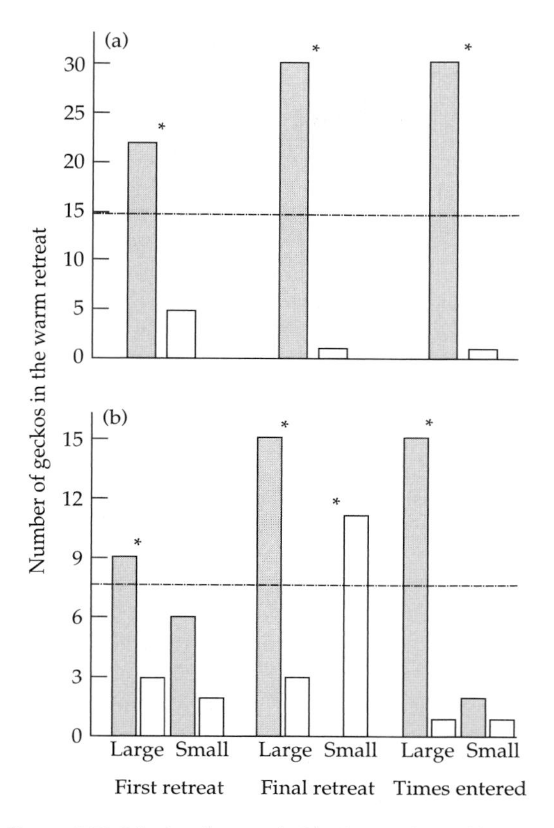

Figure 4.26 Selection of retreats by (a) solitary geckos and (b) paired geckos depended on the risk of predation. In each case, geckos were exposed to two conditions: a choice between cold and warm retreats without chemical cues of a predator (filled bars) and a choice between a cold retreat without chemical cues and a warm retreat with chemical cues (open bars). The dotted line shows the expected number of geckos in each retreat if no preference existed. An asterisk above a pair of bars indicates a significant difference in choice between the two conditions ($P < 0.05$). Adapted from Downes and Shine (1998) with permission from Elsevier. © 1998 Elsevier Ltd.

a microclimate as favorable or not depends on its proximity to competitors, predators, parasites, prey, water, and refuge. A spatially explicit model would enable us to determine how these factors interact with temperature to affect the optimal strategy of thermoregulation.

4.8 Endothermic thermoregulation

Endothermy—the regulation of body temperature via metabolic heat—occurs in mammals and birds, as well as certain insects, fish, and plants. Because many endotherms produce enough heat to maintain relatively high and stable temperatures (Clarke and Rothery 2008), they undoubtedly pay extraordinary energetic costs of thermoregulation. Indeed, the sheer energetic extravagance of endothermy probably explains why biologists have been obsessed with its explanation (Koteja 2004). Despite considerable efforts, the evolutionary origin of endothermy remains as mysterious today as it was several decades ago. Biologists have a much better understanding of the evolutionary maintenance of endothermy in extant populations. In this section, we shall explore both the origin and maintenance of endothermy. In doing so, we shall see how Huey and Slatkin's model of behavioral thermoregulation applies perfectly well to endothermic thermoregulation.

4.8.1 The evolutionary origins of endothermy

Two kinds of explanations for the evolution of endothermy have been offered: a response to direct selection for thermoregulatory capacity and an indirect response to selection for metabolic capacity (reviewed by Hayes and Garland 1995; Kemp 2006). Direct selection for thermoregulatory capacity would have required the benefit of thermoregulation to offset the cost of thermogenesis. By now, the benefits of thermoregulation should be obvious. Besides the physiological benefits of being hotter (Section 4.2.1), an organism might expand its temporal or spatial range of activity by offsetting heat loss with heat production (Crompton et al. 1978). Nonetheless, endotherms also require morphological and physiological mechanisms to store the heat produced by metabolism. Without ample size or insulation, the thermoregulatory benefits of endothermy lie beyond reach. For example, even a quadrupling of metabolic rate could not increase the body temperature of a large lizard by 1°C (Bennett et al. 2000). Without certain morphological features, the energetic cost of elevating body temperature seems too high.

Recently, Farmer (2000) proposed a twist on the original hypothesis for the evolution of endothermy. She proposed endothermy evolved because metabolic heat enables parents to warm their developing embryos. In this sense, she viewed

Box 4.5 Modeling optimal vertical movements in aquatic ectotherms

Alonzo and Mangel (2001) modeled the activity and growth of krill along a vertical cline in water temperature. The environment was structured into zones, comprising surface, shallow, and deep waters. Abundant prey resided in warm, surface water, but refuge from predators occurred in cool, deep water. Krill could adopt one of three behaviors: (1) spend both night and day at the surface, (2) spend the night at the surface but migrate to shallow or deep water during the day, or (3) spend both night and day in shallow or deep water. The energetic consequence of this decision was based on the thermal sensitivity of growth rate. Because the energetics of growth depended on body size, a complex function related food consumption (F) and metabolic costs (C) to the time of year (t), the length of the animal (L), and the temperature of its vertical zone (T). By adding a constant (K) that converts energy into a change in length, they could calculate the growth of an individual for each zone:

$$\Delta L = K\left[F(T, L, t) - C(T, L, t)\right]. \tag{4.13}$$

Because Alonzo and Mangel assumed that fecundity depended directly on body length, the use of warm water to speed growth would enhance reproduction. Empirical estimates of performance curves were used to choose reasonable rates of feeding and metabolism over a range of body sizes.

The presence of predators imposed a cost of using warm waters. Both baseline and size-specific probabilities of survivorship were considered. When defining the size-specific probability, predators were assumed to prefer larger individuals over smaller ones. Both baseline and size-specific probabilities could vary among vertical zones. Empirical estimates of age distributions in natural environments were used to choose reasonable survivorships over a range of body sizes.

To calculate the optimal behavior, Alonzo and Mangel had to choose a measure of fitness. Fecundity and survivorship were used to calculate the reproductive value—the sum of current and expected future reproduction. The optimal strategy maximizes the reproductive value of an individual at birth, thus maximizing the expected number of offspring produced during its lifetime. Because growth and survivorship depended on size, which changed over time, Alonzo and Mangel identified the optimal behavior through dynamic programming. This modeling approach enables one to track the influence of state variables (e.g., size) on the optimal strategy over the course of an individual's life (Mangel and Clark 1988).

The use of warm, surface waters depended strongly on the level of predation (Fig. 4.27). Without size-specific predation, the optimal behavior always involved spending the night on the surface. The optimal location during the day depended on the relative risk of predation. When the risk of predation was equal among zones, the optimal forager spent most of its day in surface water. When predation risk was greatest in surface water, the optimal forager spent most of its day in shallow water. Finally, when predation risk was great in both surface and shallow waters, the optimal forager spent most of its day in deep water. More complex patterns of optimal behavior were predicted when the risk of predation depended on size; specifically, optimal foragers split their diurnal activity between surface water and deeper waters when predation risk was concentrated. Although these qualitative predictions seem obvious in hindsight, Alonzo and Mangel's model provides a framework for generating quantitative predictions under many scenarios. Biologists could parameterize the model for other aquatic species or develop similar models for terrestrial species. Dynamic programming offers an efficient means of calculating the optimal thermoregulation under a variety of constraints.

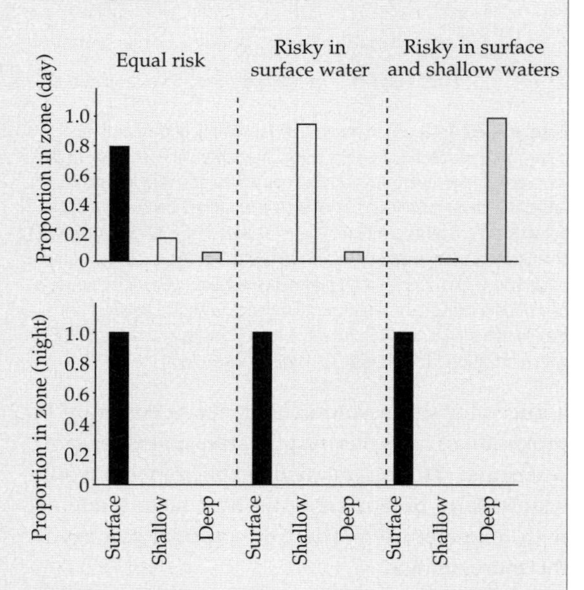

Figure 4.27 Dynamic optimization predicts that vertical migration by krill will depend on the risk of predation. The proportion of krill predicted in each vertical zone is shown for three scenarios of predation risk. Krill should spend their nights and days in warm surface water when equal predation risk exists at all depths. Concentrated predation risk in shallow or surface waters favors a shift to deeper water by day. Adapted from Alonzo and Mangel (2001) with permission from the Ecological Society of America.

endothermy as a form of parental care. A higher temperature would enable faster development and possibly cause greater survival of offspring. From the parent's perspective, the energetic cost of reproduction would increase, but the rapid development of offspring would provide additional time to acquire food or reproduce further. By quantifying the benefits of endothermic parental care and weighing them against the costs, we could determine the plausibility of Farmer's evolutionary scenario. Unfortunately, we lack the data needed to quantify the benefits and costs of endothermy for parental care (Angilletta and Sears 2003). Plus, Farmer's hypothesis does not get around the problem associated with the classic thermoregulatory hypothesis: ultimately, either size or insulation would need to evolve for the transition to homeothermy.

Alternatively, the initial selection of metabolic heat production might have had little (if anything) to do with thermoregulation. Instead, the initial target of selection might have been a greater maximal or sustained capacity for energy expenditure during activity, growth, or reproduction (Bennett and Ruben 1979; Koteja 2000). Proponents of this idea argued that selection of greater metabolic capacity would have caused a correlated increase in resting metabolic rate. According to Else and colleagues (2004), metabolic capacity and resting rates could be connected by the function of membranes, which appear to control rates of energy expenditure in cells. In this scenario, the heat produced by the first endotherms would have played an insignificant role in thermoregulation. Later, adaptation to conserve heat would have facilitated the transition to endothermic homeothermy.

Hypotheses based on an indirect response to selection rely entirely on the assumption that either pleiotropy or linkage exists between resting and maximal (or sustained) metabolic rates. Although these rates are often correlated (reviewed by Boily 2002; Hayes and Garland 1995), correlations can vary from strongly positive to strongly negative among closely related species (Gomes et al. 2004). We must also view these results cautiously because a phenotypic correlation does not necessarily possess the same magnitude or even the same sign as a genetic correlation (Hayes and Garland 1995). Recently, genetic correlations between resting and active metabolic rates have been quantified for certain mammals, either through breeding experiments or artificial selection. Sadowska and colleagues (2005) analyzed a multigenerational pedigree of bank voles to estimate the genetic correlation between metabolic rates while resting and swimming. A strong positive genetic correlation was observed, supporting the assumption of the aerobic capacity model. Ksiazek and colleagues (2004) artificially selected mice for either low or high basal metabolic rate. After 18 generations of selection, they had caused a divergence in basal metabolic rate of 2.3 standard deviations between selection lines. Contrary to the prediction of Bennett and Ruben's (1979) model, the maximal metabolic rate during swimming decreased as a correlated response to selection for a higher basal metabolic rate. Nevertheless, selection for high basal metabolic rate caused the evolution of greater food consumption and heavier internal organs; these results suggest that selection for high rates of sustained metabolism would have led to a correlated increase in basal metabolic rate, as suggested by Koteja (2000).

Evaluating a scenario for the evolution of endothermic thermoregulation requires three elements. First, we must ask whether metabolic heat could have caused the hypothetical increase in body temperature in the hypothetical environment. For a well-defined hypothesis, we could use mathematical models of heat transfer to answer this question, especially when we consider an ancestral organism unlike any living form (e.g., Phillips and Heath 2001). Second, we must argue convincingly that the benefit of thermogenesis would have outweighed the cost. Finally, we must consider whether the net benefit of endothermic thermoregulation would have exceeded the net benefit of ectothermic thermoregulation. Unfortunately, the lack of mathematical models describing the evolution of endothermy prevents a rigorous test of existing ideas. Recall from Chapter 1 that verbal models come with implicit assumptions and usually make imprecise predictions. For example, the idea that endothermy evolved as an indirect response to direct selection for aerobic capacity has generated a fixation on correlations between basal and maximal metabolic rates. But we know that genetic correlations change over time (Prasad and Shakarad

2004), such that any correlation in the ancestor of mammals or birds will not necessarily mirror the correlations existing today. How strong a correlation was necessary and for how long must it have persisted for endothermy to evolve? Once the incipient steps toward endothermy were taken, other developmental and genetic constraints might have kept organisms evolving towards homeothermic endothermy, even if genetic correlations between basal and maximal metabolic rates were lost in some lineages. Furthermore, direct selection for endothermy to expand the niche or care for offspring could have followed indirect selection for endothermy. Indeed, current hypotheses are not competing hypotheses in the sense that they mutually exclude one another. For this reason, biologists cannot conduct a critical experiment to distinguish among these hypotheses. Quantitative models of evolution will be necessary to evaluate the relative roles of hypothetical selective pressures leading to the origin of endothermy.

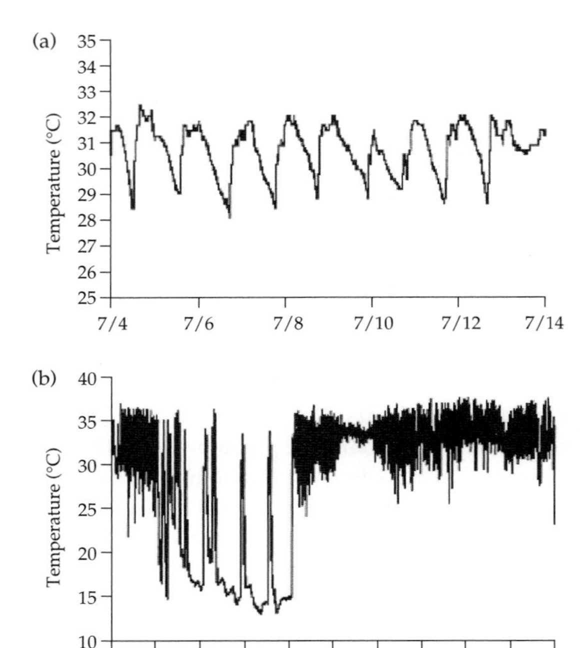

Figure 4.28 In echidnas, intermittent periods of hypothermia result from (a) daily torpor and (b) seasonal hibernation. Reproduced from Grigg et al. (2004, *Physiological and Biochemical Zoology*, University of Chicago Press). © 2004 by The University of Chicago.

4.8.2 Optimal thermoregulation by endotherms

Perfect homeotherms exist only in uniform environments, such as the depths of the ocean, and endothermic homeotherms are no exception. Endotherms generally exhibit bouts of hypothermia, ranging from a mild reduction in temperature during sleep to a drastic reduction in temperature during hibernation (Fig. 4.28). These hypothermic responses, commonly referred to as torpor, fall along a continuum; for example, the minimal temperature during hypothermia ranges from 4°C to 38°C among species of birds (McKechnie and Lovegrove 2002). Grigg and colleagues (2004) concluded that all degrees of torpor rely on similar physiological mechanisms related to ancestral features of physiology. Biologists agree that torpor was probably present in early mammals, but disagree on whether torpor was present in the first birds or evolved repeatedly within this group (Geiser 1998; Grigg 2004; McKechnie and Lovegrove 2002). If torpor was an ancestral property of mammals and birds, we could view endothermic homeothermy as a facultative strategy adopted by organisms when the energetic benefit outweighs the obvious cost. This view turns our original question on its head: instead of asking why endothermy evolved from ectothermy, we could ask why ectothermy remained an important strategy for organisms capable of endothermy. In doing so, we must consider how the respective costs and benefits of endothermy drive diel or seasonal shifts between alternative strategies.

Huey and Slatkin's model of optimal thermoregulation applies equally well to ectotherms and endotherms. Like an ectotherm, an endotherm would experience a decrement in performance if its body temperature were to deviate from its thermal optimum. Furthermore, the cost of thermoregulation should be a function of environmental temperature (Ames 1980). In fact, the benefit and cost curves for endotherms could share the same properties as those curves considered by Huey and Slatkin (see Fig. 4.11), except the implicit mechanism for thermoregulation would be physiological rather than behavioral. Despite this fact, biologists have only recently applied quantitative models of energetic costs and benefits to understand the timing and intensity of endothermy (but see Heinrich 1974). In this section, we shall consider optimal patterns of daily and seasonal torpor.

Do endothermic organisms maintain less precise temperatures when environmental conditions reduce the energetic benefits or raise the energetic costs? Based on the model, we should expect torpor to be more common when food is scarce or the environment is cold. The degree of torpor varies among species from different environments in ways that support this view. Species of ground squirrels express different degrees of metabolic depression, ranging from daily torpor in hot deserts to extreme hibernation in the cold Arctic (reviewed by Grigg 2004). Experimental manipulation of ambient temperature causes mammals to increase the use of torpor as temperature decreases (Ellison and Skinner 1992; Soriano et al. 2002). Similarly, food deprivation generally decreases the body temperatures of mammals (Piccione et al. 2002); in two experiments involving Siberian hamsters, minimal body temperature was positively related to food intake (Dark et al. 1996; Stamper et al. 1998). Injections of metabolic inhibitors (Westman and Geiser 2004) or murine leptin (Geiser et al. 1998) have experimentally linked the initiation of daily torpor to a lack of cellular energy. Birds, bees, and flowers (but not trees) also switch from endothermy to ectothermy when the net benefits of thermoregulation decline. Some birds enter torpor during periods of cold or starvation (reviewed by McKechnie and Lovegrove 2002). Bumblebees will use endothermy to sustain flight only when the environment contains sufficient nectar (reviewed by Heinrich 1974). Flowers of the skunk cabbage abandon endothermic homeothermy when air temperature drops below a certain level (Seymour and Blaylock 1999). Based on factorial experiments, both cold and starvation combine to cause a greater reduction in body temperature than does either factor alone (Westman and Geiser 2004).

Energetic status also dictates the timing and degree of seasonal torpor (or hibernation). Because hibernation involves much greater depressions of metabolism and temperature, costs of survival stem not only from predators but also from physiological damage to tissues and organs. The risk of physiological damage must be weighed against the risk of starvation. Humphries and colleagues (2003a) argued that hibernators would avoid deep and prolonged torpor if energetic resources were plentiful. Based on their reasoning, the intensity and duration of seasonal torpor should relate to the availability of food or the energy required for thermoregulation. Humphries and colleagues supported their idea with experimental evidence. Chipmunks that were supplementally fed prior to hibernation maintained higher temperatures during torpor and remained aroused for longer periods than did chipmunks belonging to an unfed control group (Humphries et al. 2003b). Furthermore, fat males tend to emerge from hibernation earlier than do skinny males and reproductive females (Dark 2005). Greater energy stores are less useful when they contribute indirectly to physiological damage. For example, an abundance of polyunsaturated fatty acids in membranes and fat bodies can lead to oxidative damage. The indirect cost associated with this energy source can offset the direct benefit. In support of this view, chipmunks fed a diet poor in polyunsaturated fatty acids underwent longer and deeper bouts of torpor than did chipmunks fed a diet rich in polyunsaturated fatty acids (Munro et al. 2005).

In addition to manipulations of environmental temperature and food intake, researchers have directly manipulated the energetic cost of hibernation. Kauffman and colleagues (2004) shaved ground squirrels 4 weeks into their hibernation and, as a control, similarly handled but did not shave other squirrels. Both shaved and control squirrels underwent multiday periods of torpor separated by homeothermic arousals lasting less than 24 hours. Shaved squirrels took longer to rewarm than control squirrels, presumably because of the greater loss of heat through their body surfaces. Surprisingly, shaved and control squirrels reached similar minimal body temperatures and remained in torpor or arousal for similar durations. Yet, the energetic cost of arousal did limit the duration of hibernation. Despite the fact that shaved squirrels consumed more food than control squirrels consumed during each arousal, shaved squirrels lost mass more rapidly and ceased hibernation several weeks earlier than control squirrels. Thus, the cost of rewarming forced shaved squirrels to abandon seasonal torpor earlier in search of food to replenish their energy stores.

Social interactions reduce the cost of daily or seasonal torpor, resulting in adaptive plasticity of endothermic thermoregulation. Huddling improves heat balance by minimizing convective heat loss and promoting radiative heat gain. Thus, a social animal would need to spend less energy to maintain a given body temperature than would a solitary animal. This observation leads to the prediction that social individuals should thermoregulate more precisely than solitary individuals. Consistent with this prediction, speckled mousebirds adjusted the depth of daily torpor according to the presence of conspecifics, as well as the abundance of food (McKechnie et al. 2006). Given surplus rations, birds in a communal roost maintained stable temperatures during the night, while birds roosting alone experienced steady declines in body temperature. When rations were restricted, both social and solitary birds cooled progressively during the night; however, solitary birds reached lower body temperatures than social birds. Plasticity of seasonal torpor also depends on social behavior. For example, skunks that wintered in isolation underwent more frequent and more intense bouts of torpor than did skunks that wintered in groups (Ten Hwang et al. 2007). Despite the difference in torpor, social skunks emerged from hibernation with more than twice the body fat possessed by solitary skunks.

As with behavioral thermoregulation, endothermic thermoregulation depends on the net benefits of nonenergetic currencies as well as energetic ones. Pravosudov and Lucas (2000) used stochastic dynamic programming to model the optimal degree of daily torpor when the energetic cost of endothermy trades off with its survival benefit. They reasoned that birds in torpor were unusually susceptible to predation. Therefore, an individual must balance the risk of running out of energy during thermoregulation with the risk of falling prey during torpor. Although additional feeding could provide the energy needed to avoid torpor, this behavior would impose a risk of predation. Based on their model, the interaction among environmental temperature, predation risk, and foraging success strongly affects the optimal strategy of torpor (Fig. 4.29). Given little

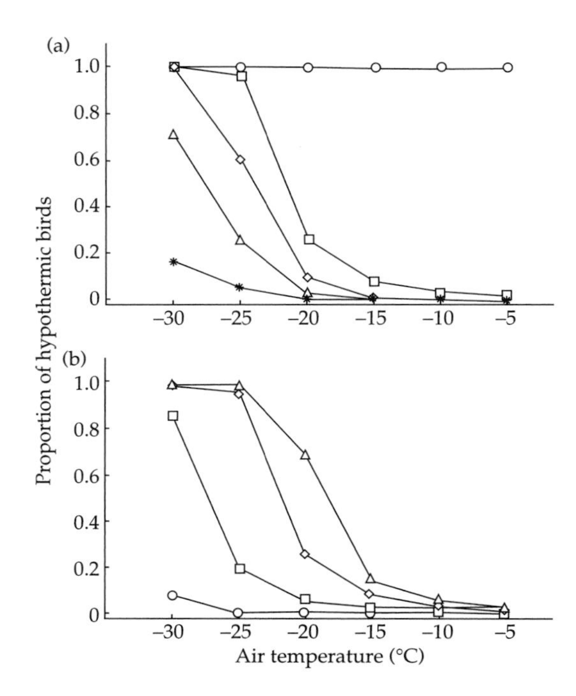

Figure 4.29 The interaction among environmental temperature, predation risk, and foraging success imposes a strong effect on the optimal strategy of torpor. These plots show the predictions of a stochastic dynamic model of optimal torpor in birds. (a) The optimal proportion of hypothermic birds at a particular air temperature depends on the risk of nocturnal predation (stars > triangles > diamonds > squares > circles). (b) The optimal proportion of hypothermic birds at a particular air temperature depends on the variance of foraging success (triangles > diamonds > squares > circles). Adapted from Pravosudov and Lucas (2000) with permission from Blackwell Publishing.

risk of predation, a bird should undergo daily torpor at any temperature. In a riskier environment, some torpor should occur at very low temperatures but no torpor should occur at higher temperatures. This result highlights a beautiful symmetry between models of ectothermic and endothermic thermoregulation: while ectotherms should forgo thermoregulation and its energetic benefits to avoid predation (Section 4.7.2), endotherms should adopt thermoregulation and its energetic costs to avoid predation. Nevertheless, the optimal strategy of torpor depends on energy availability; when foraging success varies greatly, the optimal strategy likely involves some torpor during cold nights in spite of predation risk. Unlike the daily torpor of birds, the seasonal torpor of mammals might reduce the risk of predation. Hibernation often begins before the environment restricts energy acquisition, suggesting that predation risk causes animals to abandon homeothermy before energetic costs do. In fact, echidnas hibernate even in mild climates where food occurs year round. Grigg and Beard (2000) argued this behavior enables echidnas to avoid predators before breeding, once sufficient energy reserves have accumulated. Their interpretation seems compelling because additional growth could occur if individuals were to delay hibernation. In light of these examples, we need more research to clarify the potentially contrasting roles of predation risk in the evolution of daily and seasonal torpor.

Based on the ubiquity of torpor among endotherms, Grigg and Beard (2000) hypothesized that the first endotherms were facultative homeotherms. These researchers also argued that we can learn much about the origin of endothermy by reflecting on the similarity and diversity of thermoregulatory strategies adopted by extant endotherms. Perhaps mathematical models tailored toward the costs and benefits of endothermic thermoregulation will ultimately help to answer questions about both the maintenance and the origin of endothermy.

4.9 Conclusions

Our evolutionary understanding of thermoregulation relies heavily on optimality models, ranging from Huey and Slatkin's classic model to the more recent models tailored to specific taxa. These models have guided several decades of research on the costs and benefits of behavioral thermoregulation, and have recently been applied to endothermic thermoregulation.

The theory has predicted the qualitative behaviors of animals extremely well when the costs of energy expenditure and mortality risk were manipulated in controlled environments. The theory also seems to explain the endothermic thermoregulation of birds and mammals in experiments conducted in artificial and natural environments. Overall, optimality models have been more successful at predicting thermoregulation than they have been at predicting thermal sensitivity (see Chapter 3). Perhaps for this reason, we know far less about the genetic constraints on thermoregulatory strategies than we do about genetic constraints on performance curves.

Despite the success of optimality models when applied to experimental systems, our understanding of behavioral thermoregulation in natural environments suffers from a failure to characterize thermal constraints and an inability to estimate diverse costs. Recent advances in mapping operative environments (Chapter 2) and simulating thermoregulatory costs (see Box. 4.4) should take us beyond the insights gained from experiments in artificial environments. These approaches will also enable us to consider the interactions among the costs of energy expenditure, mortality risk, and missed opportunities. Thus, we are poised to build on the progress of the past.

CHAPTER 5

Thermal Acclimation

5.1 Patterns of thermal acclimation

All organisms possess some capacity to modify their behavioral, physiological, or morphological characteristics in response to environmental temperature. Brief exposure to extreme heat or cold often causes greater tolerance of thermal extremes within hours (Hoffmann et al. 2003; Sinclair and Roberts 2005; Sinclair et al. 2003), a phenomenon called hardening. Prolonged exposures to moderate temperatures can trigger lasting changes in thermoregulation or thermosensitivity (Fig. 5.1). In this way, gradual shifts in the mean and variance of environmental temperature among seasons lead to responses that enhance performance (e.g., Hazel 2002; Packard et al. 2001; Sinclair 1997). Generally, both rapid and gradual responses to environmental temperature are reversible, but some responses remain fixed throughout the life of the organism (Johnston and Wilson 2006). Collectively, I refer to these examples of phenotypic plasticity as *thermal acclimation*.[11]

To be precise, I distinguish between two types of thermal acclimation: developmental acclimation and reversible acclimation. Developmental acclimation comprises irreversible responses to temperatures experienced during ontogeny, whereas reversible acclimation comprises regulated responses to diel or seasonal changes in temperature. As we shall see, this distinction becomes critical when discussing the evolution of acclimation.

Unfortunately, we cannot always distinguish between these forms of acclimation in practice. First, similar mechanisms could underlie the detection of and response to thermal heterogeneity in both

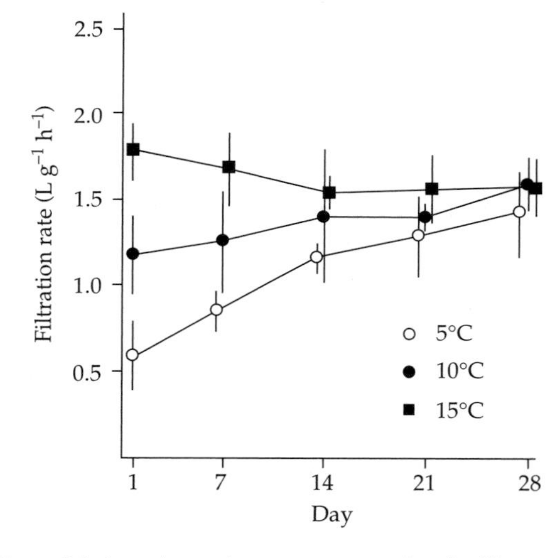

Figure 5.1 Acute changes in temperature caused predictable changes in the feeding performance of mussels (*Mytilus edulis*), but prolonged exposures to novel temperatures triggered physiological responses that gradually restored performance to its original level. Data are mean filtration rates of individuals exposed to 5°C, 10°C, or 15°C. Error bars are standard deviations. Adapted from Widdows and Bayne (1971) with permission from Cambridge University Press.

[11] Other researchers have defined thermal acclimation differently (Bowler 2005; Lagerspetz 2006; Wilson and Franklin 2002). Two features distinguish competing definitions. First, some researchers regard the period of thermal exposure (i.e., the treatment) as acclimation, whereas others regard the change in phenotype (i.e., the response) as acclimation. Second, some researchers who define acclimation as a change in phenotype also require that this response improves the fitness of the organism. I favor an intermediate position. In this book, I use thermal acclimation to refer to any phenotypic response to environmental temperature that alters performance and plausibly enhances fitness. Therefore, I do not require precise knowledge of the fitness consequences because such data rarely exist. Finally, I do not distinguish between responses to acute and chronic exposures to thermal extremes, and thus consider heat and cold hardening as forms of acclimation (for further discussion of this topic, see Loeschcke and Sorenson 2005).

processes, even though acclimation during early and late stages of ontogeny can differ dramatically (Johnsen et al. 2005; Terblanche and Chown 2006; Zeilstra and Fischer 2005). Second, we never know whether acclimation is truly irreversible; we only know that the time required for reversal exceeds the lifespan or that the alleles required to regulate reversal have not arisen. Researchers have compounded this problem by failing to design experiments that separate reversible acclimation from developmental acclimation (Wilson and Franklin 2002). For these reasons, some choose to disregard any categorization of acclimation responses. Indeed, we could place all patterns of acclimation along a continuum, defined by the rate of the response or the potential to reverse the response; for example, diel and seasonal changes bound a continuum defined by the rate of the response (Loeschcke and Sorenson 2005). At the very least, all acclimation responses share some common properties: the detection of environmental signals, the transduction of this signal into a cellular response, and the activation of the molecules (e.g., genes, polymerases, ribosomes, enzymes) that cause a change in phenotype (Angilletta et al. 2006a; Wilson and Franklin 2002).

In this chapter, we shall examine the acclimation of performance curves as an adaptation to thermal heterogeneity. In doing so, we must confront some challenging questions. Does acclimation to high temperatures impair performance at low temperatures, and vice versa? If not, what other costs do organisms incur during acclimation? Do most acclimation responses enhance the fitness of organisms or are these responses merely unavoidable consequences of thermal change? What kinds of thermal environments favor acclimation and what kinds favor alternative strategies? To address these questions, we need to consider the benefits and costs of acclimation in variable and unpredictable environments. Let us start by considering the benefits of acclimation.

5.2 The beneficial acclimation hypothesis

Although biologists have documented a broad diversity of acclimation responses in ectotherms and endotherms (Chown and Nicolson 2004; Hutchison and Maness 1979; McNab 2002), the fitness benefits of these responses have generally been assumed rather than demonstrated (Huey and Berrigan 1996). Imagine an organism that encounters a gradual decline in environmental temperature. This individual could decrease its thermal optimum for performance or modify structures needed to enhance thermoregulation (e.g., thicken its pelage or darken its surface). These changes seem obviously beneficial. Yet not all responses to thermal change enhance the performance of an organism. For example, the term deacclimation commonly refers to the reversal of a previous acclimation response. Because deacclimation often occurs rapidly (Kalberer et al. 2006), this response can be maladaptive when the environment fluctuates wildly. In the last decade, biologists have made a much greater effort to establish whether acclimation responses actually benefit organisms.

The assumption that all acclimation responses enhance fitness has come to be known as the *beneficial acclimation hypothesis*. According to this hypothesis, a change in the environment of an organism should elicit a change in the phenotype that enhances performance in the new environment (Leroi et al. 1994). The beneficial acclimation hypothesis has been tested many times, generally following a standard protocol. First, organisms are held under controlled environmental conditions for an initial period. Then the same individuals are shifted to one or more new environmental conditions for some period. Finally, the performance of each individual is assessed over a range of environmental conditions to determine whether performance in the new environment has been enhanced. Despite its intuitive appeal, the beneficial acclimation hypothesis makes predictions that often deviate from the results of experiments. To substantiate this claim, I must briefly review some studies of developmental and reversible acclimation.

5.2.1 Developmental acclimation

Two forms of developmental acclimation could benefit organisms in thermally heterogeneous environments. First, organisms could respond to temperatures during one stage of the life cycle to enhance performance during a subsequent stage (intragenerational response). Second, parents could respond to temperatures during their life to enhance the performance of their offspring (intergenerational response); this response requires a parent to supply its offspring with nutrients, hormones, mRNA, or other factors that alter the physiological state at birth. Both forms of developmental acclimation have been the focus of recent attempts to evaluate the beneficial acclimation hypothesis.

Plant physiologists have a long history of investigating developmental acclimation of photosynthetic rate, focusing primarily on intragenerational responses (reviewed by Berry and Bjorkman 1980). In most studies, researchers raised individuals at constant temperatures or constant diel cycles for some portion of development (e.g., from the germination stage to the seedling stage). After this period, net photosynthetic rates were measured over a range of temperatures that included the developmental temperatures. Because physiologists report photosynthetic rates for a given area or mass of leaves, differences in these rates between plants from different developmental environments reflect acclimation of physiology rather than plasticity of biomass (although the allocation of biomass also contributes to photosynthetic acclimation; Atkin et al. 2006a). Collectively, these experiments provide a wealth of evidence that pertains to the beneficial acclimation hypothesis. And most have generated patterns that differ markedly from those predicted by this hypothesis (Table 5.1). Quite often, plants that developed at a particular temperature performed best across a wide range of temperatures (Fig. 5.2). Interestingly, the diversity of performance curves produced within populations by developmental acclimation resembles that which has been produced among populations by genetic divergence (see Chapter 3).

Although fewer animal physiologists have studied the developmental acclimation of performance curves, their research further confirms the shortcomings of the beneficial acclimation hypothesis. Zamudio and colleagues (1995) compared the social dominance of fruit flies raised at 18°C and those raised at 25°C. Regardless of the temperature of their social environment, males raised at the higher temperature dominated males raised at the lower temperature. Rhen and Lang (1999) incubated chelonian embryos at 24°C, 26.5°C or 29°C and measured growth at 19°C and 28°C after hatching. At both temperatures, hatchlings from the colder environments grew faster than hatchlings from the warmest environment. Both studies revealed optimal temperatures for development rather than the beneficial acclimation of performance.

Escherichia coli and *Drosophila melanogaster* have been useful models for studies of intergenerational responses. In a highly influential experiment, Leroi and colleagues (1994) grew lines of *E. coli* at either 32°C or 41.5°C for 24 hours and then competed cells from these lines at the same temperatures. Regardless of the competitive environment, the descendents of bacteria grown at 32°C outcompeted the descendents of bacteria grown at 42°C. Nevertheless, the descendents of bacteria grown at 42°C survived longer when exposed to the stressful temperature of 50°C. Crill and colleagues (1996) used a factorial manipulation of developmental temperatures experienced by parents and offspring of *D. melanogaster*. They crossed males and females that were raised at either 18°C or 25°C and then raised some offspring at each temperature. The knockdown temperatures of offspring were unaffected by the father's developmental temperature and were only weakly affected by the mother's developmental temperature. However, offspring raised at 25°C tolerated higher temperatures than did offspring raised at 18°C. Similarly, the thermal sensitivity of walking speed was affected by the developmental temperature but not by parental temperature; flies raised at 25°C had a lower thermal optimum and a wider performance breadth than flies raised at 18°C. In contrast, a similar experiment revealed that the fitness of *D. melanogaster* was affected primarily by the developmental temperature of parents, with parents from warmer environments giving rise to offspring with greater fitness across a wide range of temperatures (Gilchrist and Huey 2001). Despite these inconsistencies, intergenerational studies also suggest an optimal temperature for development.

Based on the majority of evidence gathered to date, the beneficial acclimation hypothesis offers few insights about the physiological responses to developmental temperature. Instead, an optimal temperature for development seems to exist, at which organisms that develop at a certain temperature outperform those that develop at other

Table 5.1 Representative studies of the thermal acclimation of photosynthetic rate (adjusted for the area or mass of leaves). For each study, I note whether the results support the beneficial acclimation hypothesis (BAH).

Species	Acclimation temperatures	Test temperatures	Beneficial acclimation?	Comments	Source
Ambrosia psilostachya	1–2°C above ambient temperature vs. ambient temperature (control)	A range of 14°C centered near the ambient temperature	No		Zhou et al. (2007)
Aster ericoides	1–2°C above ambient temperature vs. ambient temperature (control)	A range of 14°C centered near the ambient temperature	No		Zhou et al. (2007)
Ballia callitricha	0, 5	0, 5, 10, 15, 20, 25	No	Algae grown at 0°C performed better at all temperatures.	Eggert and Wiencke (2000)
Cenchrus ciliaris	25/20, 35/30	5, 10, 15, 20, 25, 30, 35, 40	No	Plants grown at low temperatures outperformed plants grown at high temperatures.	Dwyer et al. (2007)
Colobanthus quitensis	7/7, 12/7, 20/7	2–30°C in various intervals	No	Plants grown at 12°C/7°C performed best over the range of 2–35°C.	Xiong et al. (2000)
Deschampsia antarctica	7/7, 12/7, 20/7	2–35°C in various intervals	No	Plants exposed to 12°C performed best at all temperatures.	Xiong et al. (2000)
Erigeron candensis	5, 25	5–50°C and 20–50°C in cold-acclimated and warm-acclimated plants, respectively	Yes		Dulai et al. (1998)
Flaveria bidentis	25/20, 35/30	5, 10, 15, 20, 25, 30, 35, 40	No	Plants grown at low temperatures outperformed plants grown at high temperatures.	Dwyer et al. (2007)
Gymnogongrus antarcticus	0, 5, 10	0, 5, 10, 15, 20, 25	No	Algae grown at 0°C performed as well as or better than algae grow at 5°C or 10°C.	Eggert and Wiencke (2000)
Helianthis mollis	1–2°C above ambient temperature vs. ambient temperature (control)	A range of 14°C centered near the ambient temperature	No	In summer, T_{opt} was greater for warmed plants than for control plants.	Zhou et al. (2007)
Kallymenia antarctica	0, 5	0, 5, 10, 15, 20, 25	No		Eggert and Wiencke (2000)
Panicum coloratum	25/20, 35/30	25, 28, 31, 34, 37, 40	No	Plants grown at low temperatures outperformed plants grown at high temperatures.	Dwyer et al. (2007)

Continued

Table 5.1 Continued

Species	Acclimation temperatures	Test temperatures	Beneficial acclimation?	Comments	Source
Pisum sativum	25/18, 35/18	25, 30, 35, 40, 45	No		Haldimann and Feller (2005)
Phyllophora ahnfeltioides	0, 5, 10	0, 5, 10, 15, 20, 25	No		Eggert and Wiencke (2000)
Ruppia drepanensis	10, 20	10, 20, 30	Partial	When measured at 30°C, algae grown at 30°C outperformed algae grown at 20°C.	Santamaría and Hoots-mans (1998)
Sorghastrum nutans	1–2°C above ambient temperature vs. ambient temperature (control)	A range of 14°C centered near the ambient temperature	No	In summer, T_{opt} was greater for warmed plants than for control plants.	Zhou et al. (2007)
Spinacia oleracea	30/25, 15/10	10–40°C in various intervals	Yes		Yamori et al. (2006)
Triticum aestivum	15, 25, 35	5, 10, 15, 20, 25, 30, 35, 40	Partial	Plants grown at 15°C and 25°C outperformed others at their respective developmental temperatures; however, leaves grown at 35°C performed poorly at all temperatures.	Yamasaki et al. (2002)
Valonia utricularis	15, 18, 20, 25	5, 10, 15, 20, 25, 30, 35	No	Algae grown at 18°C outperformed algae grown at all other temperatures.	Eggert et al. (2003b)
Xylia xylocarpa	25/20, 30/20, 35/20, 40/20	20, 25, 30, 35, 40, 45, 50	No	Seedlings grown at 40°C/20°C outperformed other seedlings at 0°C and 50°C.	Saelim and Zwiazek (2000)

temperatures (Huey and Berrigan 1996; Huey et al. 1999). Some researchers believe this trend results from the artificial conditions used in acclimation experiments; in other words, when an organism must undergo development at a constant temperature, the detrimental effect of development at an extreme temperature probably overwhelms any effect of thermal acclimation (Wilson and Franklin 2002). This argument gains some support when we contrast experiments conducted at constant temperatures with a recent experiment conducted at fluctuating temperatures. Bilcke and colleagues (2006) incubated lizard eggs at either warm or cool diel cycles and compared the feeding and digestive performances of hatchlings. For most measures of feeding performance, lizards that developed in the cooler environment performed better at 20°C, but the reverse was true at 28°C.

Bear in mind that these studies of developmental acclimation probably confounded irreversible and reversible responses to temperature. In each experiment, we do not know whether the organisms

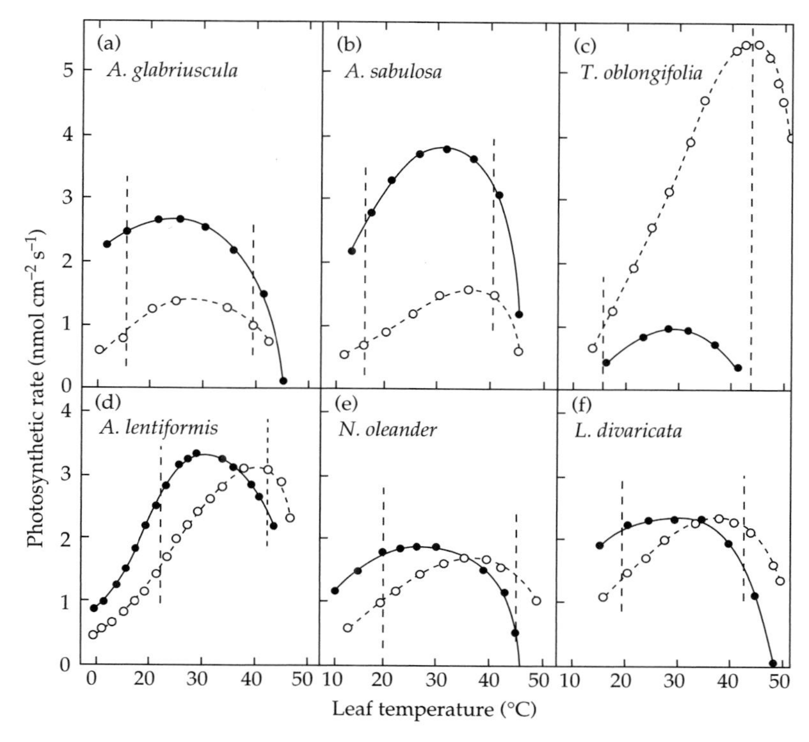

Figure 5.2 The developmental acclimation of net photosynthetic rate varied greatly among six species of plants (*Atriplex glabriuscula, Atriplex sabulosa, Tidestromia oblongifolia, Atriplex lentiformis, Nerium oleander*, and *Larrea divaricata*). Each panel shows performance curves of plants grown under cool and hot conditions (filled circles and open circles, respectively). Temperatures during development are indicated by the vertical dashed lines. Data in (d–f) support the beneficial acclimation hypothesis, but those in (a–c) do not. Reprinted, with permission, from the Annual Review of Plant Physiology (Berry and Bjorkman 1980), Volume 31. © 1980 by Annual Reviews (www.annualreviews.org).

could have reversed their responses to developmental temperature. By investigating the cellular mechanisms of change, one might find clues about the reversibility of these acclimation responses. Alternatively, one could simply attempt to reverse the effects of acclimation, but such attempts should be confined to a single ontogenetic stage to ensure comparable measures of performance during acclimation and deacclimation.

5.2.2 Reversible acclimation

The thermal acclimation of mature organisms has been a focus of physiologists for many decades. Presumably, changes in the capacity for performance during adulthood can be reversed if given enough time. Although this assumption has rarely been tested, I suspect most acclimation responses during adulthood reflect processes that occur during diel or seasonal change. By comparing the performance curves of adults exposed to different temperatures, we can infer whether acclimation enhances performance as described by the beneficial acclimation hypothesis.

The wealth of evidence clearly shows that acclimation enhances some performances in some species (Huey et al. 1999). Of the experiments summarized in Table 5.2, almost half revealed clear benefits of acclimation across a range of temperatures. Additional experiments revealed benefits of acclimation at some temperatures but not others. Nevertheless, acclimation was rarely perfect in the sense that thermal optima corresponded to environmental temperatures. Only in one exceptional case was perfect acclimation observed; after 28 days of exposure

Table 5.2 Representative studies of thermal acclimation in ectothermic animals. For each study, I note whether the results support the beneficial acclimation hypothesis (BAH).

Species	Performance	Acclimation temperatures	Test temperatures	Beneficial acclimation?	Comments	Source
Mollusks						
Ostrea edulis	Filtration	5, 10, 15, 20, 25	5, 10, 15, 20, 25, 30	No	T_{opt} was related to acclimation temperature.	Newell et al. (1977)
Tardigrades						
Macrobiotus harmsworthi	Walking	10, 25	10, 25	Yes		Li and Wang (2005)
Crustaceans						
Cherax destructor	Fighting success and chela force	20, 30	20, 30	Yes/No	Fighting success supported BAH, but chela force did not.	Seebacher and Wilson (2006)
Cherax dispar	Swimming and chela force	12, 20, 30	12, 20, 30	No		C. L. Bywater and R. S. Wilson, unpublished observations
Fishes						
Barbus barbus	Swimming	7, 25	7,13, 19, 25	Partial	Critical speed supported BAH, but voluntary and maximal speeds did not.	O'Steen and Bennett (2003)
Carassius auratus	Swimming	10, 35	5, 10, 15, 25, 30, 35, 40	Yes	T_{opt} was related to acclimation temperature.	Johnson and Bennett (1995)
Carassius auratus	Swimming	5, 15, 25, 35	5, 10, 15, 20, 25, 30, 35, 38	Yes	T_{opt} was related to acclimation temperature.	Fry and Hart (1948)
Cyprinus carpio	Swimming	8, 26	10, 20	Partial	Fish exposed to 8°C performed better at 10°C than fish exposed to 26°C, but both performed similarly at 20°C.	Rome et al. (1985)
Fundulus heteroclitus	Swimming	10, 35	5, 10, 15, 25, 30, 35	Yes	T_{opt} was related to acclimation temperature.	Johnson and Bennett (1995)
Gambusia holbrooki	Mating success	18, 30	18, 30	Yes		Wilson et al. (2007)
Gambusia holbrooki	Swimming	18, 30	18, 30	Yes/Partial	Low acclimation temperature had no effect in females.	Wilson et al. (2007)
Gasterosteus aculeatus	Swimming	18, 23	18, 22	Partial	Fish exposed to 23°C swam faster at high temperatures than fish exposed to 18°C in one of two seasons.	Guderley et al. (2001)

Continued

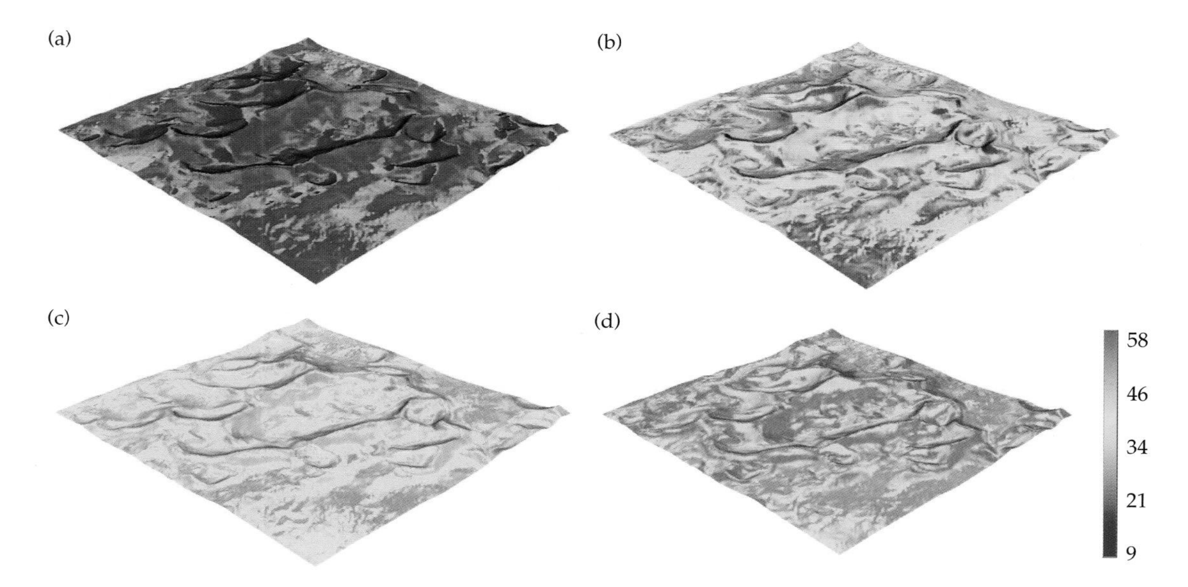

Plate 1 Operative temperatures available to small lizards in the Mescalaro sand dunes (New Mexico, USA) change dramatically throughout the morning on a summer's day (M. W. Sears, G. S. Bakken, M. J. Angilletta, and L. A. Fitzgerald, unpublished observations). (a–d) Each map depicts an operative thermal environment over an area of 460 × 320 m at a resolution of 0.5 m. (a) 7 a.m.; (b) 9 a.m.; (c) 11 a.m.; (d) 1 p.m. See Page 34.

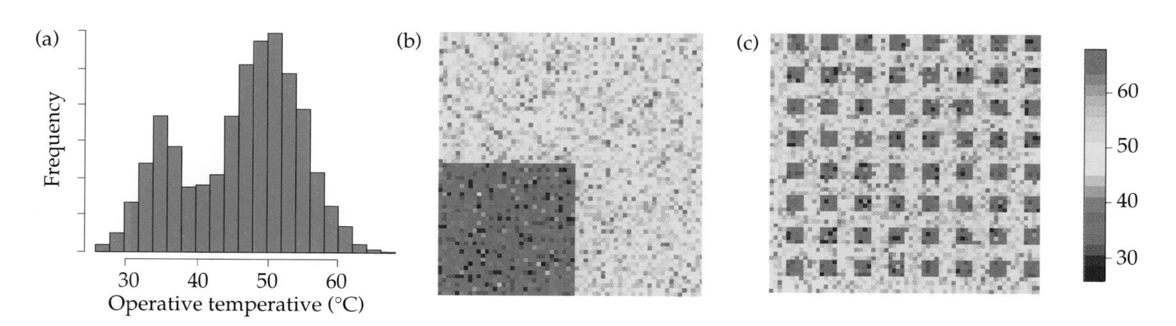

Plate 2 (a) A bimodal distribution of operative temperatures was used to simulate the energetic cost of behavioral thermoregulation. (b and c) Depending on the simulation, these temperatures were distributed in one large clump or in many small patches. See Page 112.

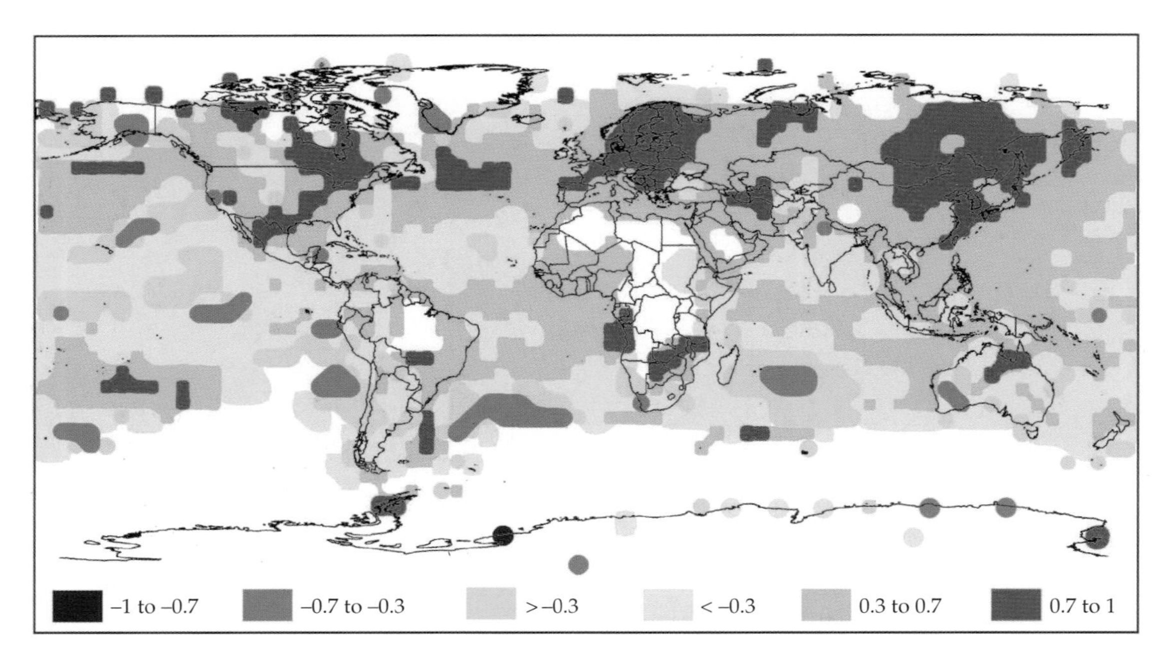

Plate 3 Environmental temperatures have changed at different rates throughout the world. This map shows rates of change (°C per decade) between 1976 and 2000, relative to normal temperatures for 1961–1990 (see Box 2.1). Reprinted by permission from Macmillan Publishers Ltd: *Nature* (Walther et al. 2002), © 2002. See Page 215.

Table 5.2 Continued

Species	Performance	Acclimation temperatures	Test temperatures	Beneficial acclimation?	Comments	Source
Micropterus dolomieui	Swimming	5, 10, 15, 20, 25, 30, 35	5, 10, 15, 20, 25, 30, 35	Partial	T_{opt} increased by 0.5°C per 1°C increment in acclimation temperature, over the range 5–25°C.	Larimore and Duever (1968)
Myoxocephalus scorpius	Swimming	5, 15	1, 5, 15, 20	Partial	Fish exposed to 15°C performed better at 15°C and 20°C than fish exposed to 5°C; however, treatments were conducted in different seasons.	Temple and Johnston (1998)
Oncorhynchus kisutch	Swimming	2, 5, 8, 11, 14, 17, 20, 23	2, 5, 8, 11, 14, 17, 20, 23	Yes	T_{opt} increased by 0.25°C per 1°C increment in acclimation temperature.	Griffiths and Alderdice (1972)
Pagothenia borchgrevinki	Swimming	−1, 4	−1, 2, 4, 6, 10	No		Wilson et al. (2001)
Pagothenia borchgrevincki	Swimming	−1, 4	−1, 2, 4, 6, 8	Yes		Seebacher et al. (2005)
Puntius schwanenfeldii	Swimming	17, 33	17, 20, 25, 30, 33	Partial	Critical speed supported BAH, but voluntary and maximal speeds did not.	O'Steen and Bennett (2003)
Taurulus bubalis	Swimming	5, 15	1, 5, 15, 20	No	Fish exposed to 5°C performed better at 1°C than fish exposed to 15°C, but both performed similarly at 5°C and 15°C; however, treatments were conducted in different seasons.	Temple and Johnston (1998)
Amphibians						
Bufo americanus	Jumping	5, 25	5, 10, 15, 20, 25	Partial	Toads exposed to 5°C jumped farther at 5°C than toads exposed to 25°C.	Renaud and Stevens (1983)
Limnodynastes peronii (tadpoles)	Swimming	10, 24	10, 24, 34	No	At all temperatures, tadpoles exposed to 10°C swam as fast as or faster than tadpoles exposed to 24°C.	Wilson and Franklin (1999)

Continued

Table 5.2 Continued

Species	Performance	Acclimation temperatures	Test temperatures	Beneficial acclimation?	Comments	Source
Limnodynastes peronii (adults)	Jumping	18, 30	8, 12, 18, 23, 30, 35	No	Frogs acclimated at 18°C would not jump at 35°C.	Wilson and Franklin (2000b)
Rana pipiens	Jumping	5, 25	5, 10, 15, 20, 25	Partial	Frogs exposed to 25°C jumped farther at 20°C and 25°C than frogs exposed to 5°C.	Renaud and Stevens (1983)
Triturus dobrogicus	Swimming/running	15, 25	10, 15, 20, 25, 30, 33	Partial	Newts exposed to 25°C ran faster at high temperatures than newts exposed to 15°C.	Gvoždík et al. (2007)
Xenopus laevis (tadpoles)	Swimming	12, 30	5, 12, 20, 30	Yes		Wilson et al. (2000b)
Xenopus laevis (adults)	Swimming	10, 25	5, 10, 15, 25, 30	Partial	Frogs exposed to 10°C swam faster at low temperatures than frogs exposed to 25°C.	Wilson et al. (2000b)
Reptiles						
Crocodylus porosus	Swimming	20, 29	17, 20, 25, 29, 32	Yes	T_{opt} matched the acclimation temperature.	Glanville and Seebacher (2006)
Xantusia vigilis	Sprinting	20, 30	12.5, 15, 20, 25, 30, 34, 37.5	No	Acclimation benefited performance at extreme temperatures.	Kaufmann and Bennett (1989)

Notes: T_{opt}, thermal optimum for performance.

to either 18°C or 29°C, crocodiles swam best at their respective acclimation temperature (Fig. 5.3a). More often, acclimation enhanced performance at one or more extreme temperatures without radically shifting the thermal optimum (Figs. 5.3b). And, in some cases, no beneficial acclimation was observed (Fig. 5.3c). This diversity of findings calls for a more complex view than the beneficial acclimation hypothesis.

Thermal limits to performance acclimate more readily than thermal optima. For example, Kaufmann and Bennett (1989) exposed adult lizards to either 20°C or 30°C for 50 days and then compared the thermal sensitivity of sprint speed between groups (Fig. 5.4). The mean performance curves for these groups were very similar over the range of 15°C to 34°C, but diverged at extreme temperatures (12.5°C and 37.5°C). Critical thermal limits had diverged in a manner consistent with the difference in sprinting performance. In fact, rapid and reversible acclimation of thermal tolerance occurs ubiquitously among animals. Acute exposure to extreme heat or cold results in greater tolerance upon subsequent exposure (see reviews by Chown and Nicolson 2004; Hoffmann et al. 2003; Sinclair and Roberts 2005). Interestingly, exposure to a moderate temperature can also alter the tolerance of extreme temperatures. In fish, the critical thermal limits scale linearly with water temperature over a range of 30°C (Fig. 5.5). Realistic diel cycles of temperature or light can induce rapid thermal hardening of fruit flies (Kelty and Lee 2001; Sorenson and Loeschcke 2002). Acclimation of thermal limits without exposure to extreme temperatures seems

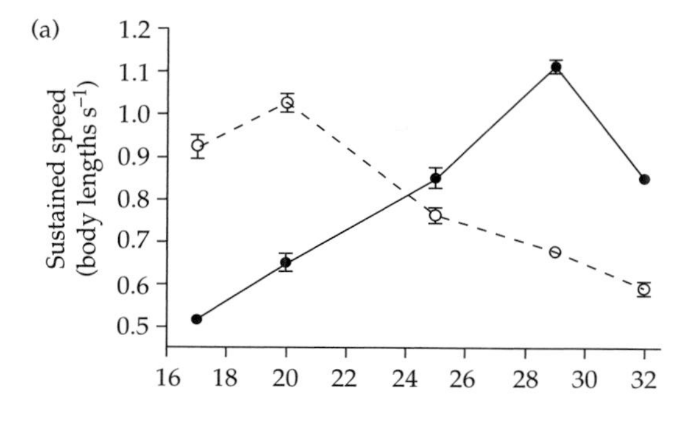

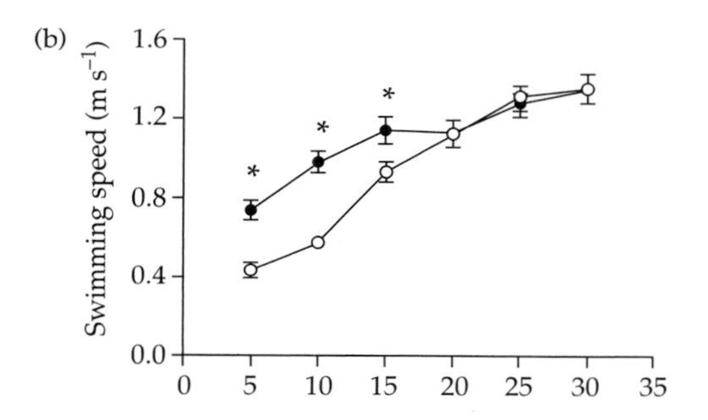

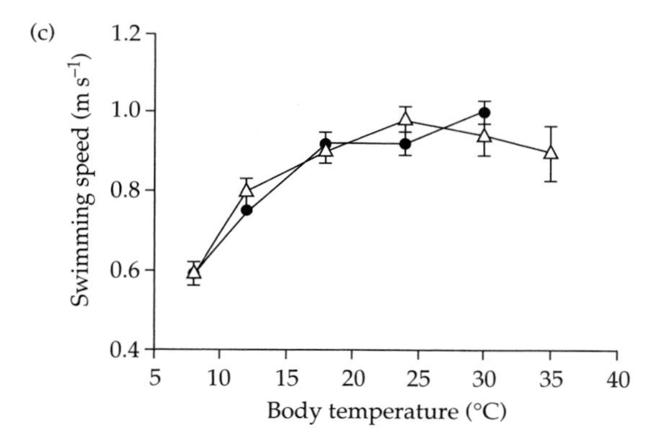

Figure 5.3 Acclimation responses of ectotherms range from perfect compensation to no compensation for a change in environmental temperature. (a) Estuarine crocodiles matched their thermal optimum for sustained swimming to their environmental temperature. Adapted from Glanville and Seebacher (2006) with permission from the Company of Biologists. (b) Clawed frogs altered their performance breadth for swimming speed in response to environmental temperature, but did not alter their thermal optimum. Adapted from Wilson et al. (2000) with kind permission from Springer Science and Business Media. © Springer-Verlag 2000. (c) Marsh frogs did not alter their thermal optimum or their performance breadth in response to environmental temperature Adapted from Wilson and Franklin (2000b) with permission from Elsevier. © 2000 Elsevier Ltd.

counterintuitive because acclimation of the thermal optimum would be more beneficial in moderate environments.

5.2.3 Beyond the beneficial acclimation hypothesis

If we consider the magnitude and unpredictability of thermal heterogeneity in natural environments, the discord between the predictions of the beneficial acclimation hypothesis and the observations of some researchers hardly seems surprising. The beneficial acclimation hypothesis is a naïve, cost-free view of nature. In other words, proponents of the beneficial acclimation hypothesis have overlooked the fact that acclimation must impose some cost to an organism (Hoffmann 1995), just as other forms of phenotypic plasticity impose costs (DeWitt et al. 1998). In the remainder of this chapter, we shall abandon the beneficial acclimation hypothesis in favor of mathematical models of optimal acclimation (Gabriel 1999; Gabriel and Lynch 1992). While the beneficial acclimation hypothesis implicitly assumes that acclimation imposes no cost, alternative models explicitly consider costs and benefits when predicting the optimal strategy of acclimation. Although optimality models of acclimation have been available for more than a decade, they have not been adopted by physiologists who study this phenomenon. Unfortunately, ignorance of these models has slowed the development of a quantitative theory of acclimation (see Huey and Berrigan 1996; Huey et al. 1999). By embracing these models, we can better understand why acclimation capacities have diverged among populations that experience different patterns of thermal heterogeneity.

5.3 Costs of thermal acclimation

To model the optimal strategy of acclimation, one must have some knowledge of the costs of plasticity. In Chapter 3, I introduced three classes of tradeoffs that constrain performance curves: acquisition tradeoffs, allocation tradeoffs, and specialist–generalist tradeoffs (see Box 3.5). These same tradeoffs apply not only to genetic variation in performance curves among individuals but also to phenotypic variation within individuals. In other words, when an organism alters its performance

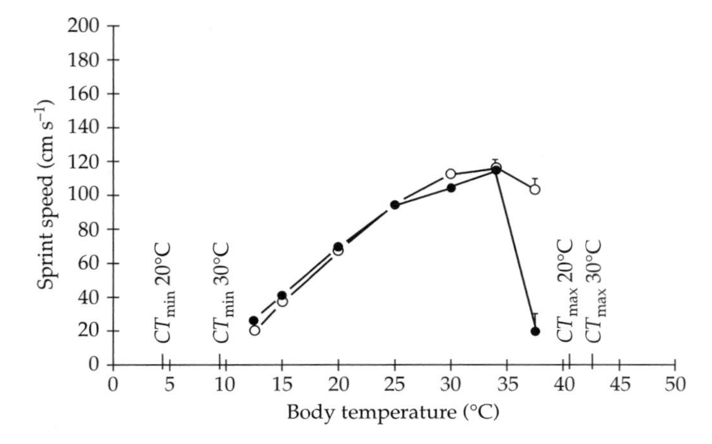

Figure 5.4 The performance of lizards (*Xantusia vigilis*) at extreme temperatures responded to acclimation at either 20°C (filled symbols) or 30°C (open symbols). Adapted from Kaufmann and Bennett (1989, *Physiological and Biochemical Zoology*, University of Chicago Press). © 1989 by The University of Chicago.

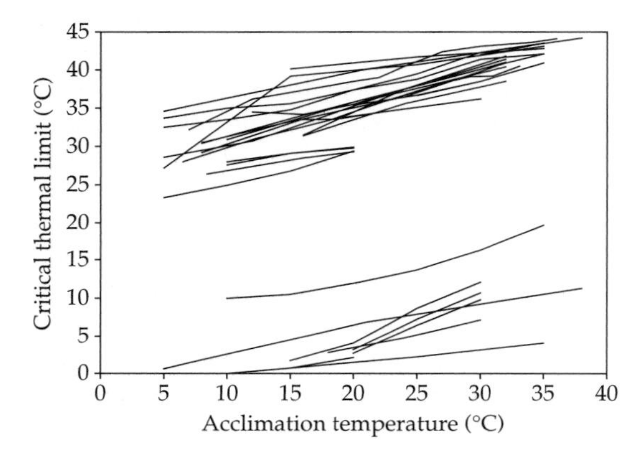

Figure 5.5 The critical thermal limits acclimate readily during prolonged exposure to a moderate temperature. Each line represents the acclimation response for a different species of fish. Data were summarized by Beitinger and colleagues (2000).

curve via acclimation, one or more of these tradeoffs can reduce the net benefit of this physiological response. Costs imposed by these tradeoffs follow from two facts. First, acclimation requires energy that an organism acquired at some risk and might otherwise use for a different function. Second, acclimation requires time for developmental or physiological changes to occur. Each of these requirements imposes costs in terms of survivorship or fecundity.

5.3.1 Costs of energetic demands

Energetic costs arise from the detection of and response to thermal change. Detection initiates a series of molecular responses, ultimately leading to modifications of cells and tissues. Cellular responses include the expression of new proteins (Schulte 2004; Somero and Hochachka 1971), the remodeling of cell membranes (Crockett 1998; Hazel 1995), and the operation of molecular chaperones (Feder and Hofmann 1999). From Chapter 3, recall that the functions of proteins and membranes define a performance curve. By changing these components of cells, an organism can alter its thermal sensitivity. For example, acclimation of photosynthetic capacity in plants involves the concentration of existing proteins and the expression of different isozymes (Yamasaki et al. 2002; Yamori et al. 2006). Similarly, acclimation of locomotor performance in animals involves changes in the quantities or qualities of metabolic enzymes (Rogers et al. 2004; Wilson and Franklin 1999). Changes in membrane fluidity during thermal acclimation occur in both unicellular and multicellular organisms (Chintalapati et al. 2004; Lee et al. 2006; Overgaard et al. 2005). As with the genetic divergence of performance curves, greater membrane fluidity enables better function at low temperatures (Fig. 5.6). Finally, acclimation to stressful temperatures usually involves antifreeze or heat-shock proteins (Lee and Costanzo 1998; Sorensen et al. 2003); for example, exposure to mild warming causes Antarctic fishes to produce fewer glycoproteins (Jin and DeVries 2006) and Antarctic grasses to produce more Hsp70 (Reyes et al. 2003). Heat-shock proteins consume energy during their function, leading to costs beyond those required for production (Macario and de Macario 2007).

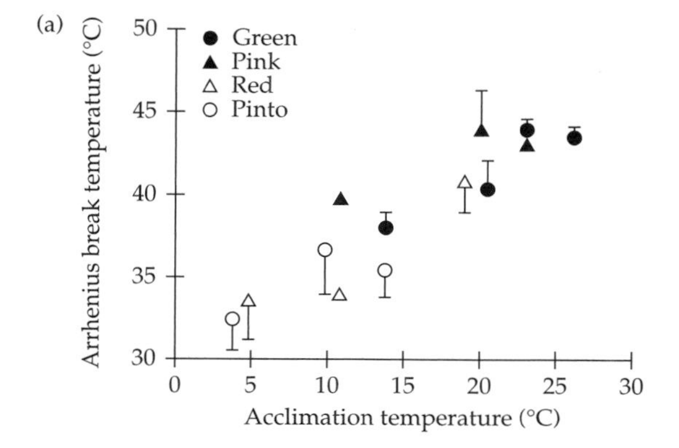

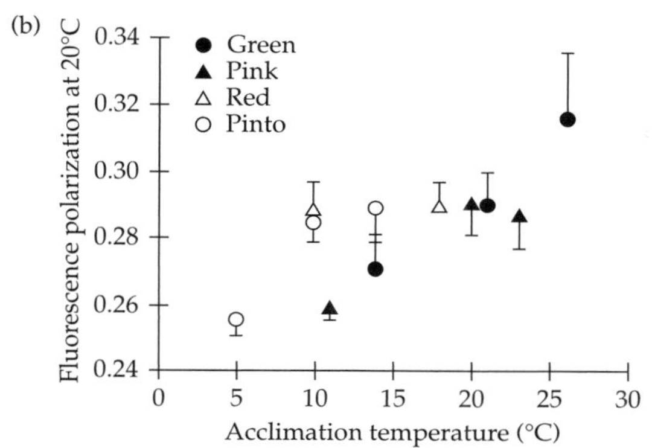

Figure 5.6 Changes in the fluidity of cellular membranes underlie the acclimation of physiological function. (a) For abalone of the genus *Haliotis*, acclimation alters the temperature at which mitochondrial respiration fails (the Arrhenius break temperature). (b) Acclimation also alters fluidity of the mitochondrial membrane, as estimated by fluorescence polarization. Abalone maintained at low temperatures possessed more fluid membranes (and hence less polarization). Symbols represent different species of abalone. Error bars are 1 standard error of the mean. Reproduced from Somero (2002) with permission from Oxford University Press.

The energetic costs of these cellular responses are difficult to quantify, but qualitative assessments suggest the costs are substantial (Somero 2002). Under certain conditions, these energetic costs will be manifested as specialist–generalist tradeoffs; for example, upregulation of new isozymes and downregulation of old ones would cause an increase in performance at some temperatures and a decrease in performance at others. Under other conditions, the energetic expense of acclimation would cause an acquisition or allocation tradeoff. In this way, acclimation to one range of temperature could occur without loss of performance at other temperatures. In fact, an organism could even gain some capacity for performance at temperatures to which they have not been exposed (e.g., an increase in cold resistance during exposure to heat). In such cases, the energetic cost of acclimation would be manifested as a decrease in survivorship or fecundity. In summary, physiological acclimation involves some of the same constraints faced during evolutionary adaptation.

Studies of phenotypic and genetic correlations associated with heat hardening have provided conflicting data about the costs of acclimation. In some experiments, heat hardening reduced fecundity (Krebs and Loeschcke 1994a) or longevity (Scott et al. 1997), but other experiments revealed no detrimental effects. For example, heat hardening of parasitic wasps (*Trichogramma carverae*) reduced the rate of parasitization in one study (Hercus et al. 2003), but enhanced parasitization without an obvious cost in another study (Thomson et al. 2001). This diversity of phenotypic correlations might reflect an underlying diversity of genetic correlations. Analyses of isofemale lines of *Drosophila melanogaster* revealed a negative relationship between heat hardening and survivorship (Krebs and Feder 1997), but artificial selection for heat hardening resulted in greater longevity (Norry and Loeschcke 2003). Moreover, isofemale lines of *D. buzzatii* exhibited a positive relationship between heat hardening and fecundity (Krebs and Loeschcke 1999). A recent study suggests these contrasting results might have stemmed from differences in the timing and intensity of heat stress. Repeated exposure of *D. melanogaster* to mild heat stress either enhanced or reduced its fecundity, depending on the age of the first exposure (Hercus et al. 2003).

Genetic engineering offers a more sophisticated approach to exploring potential costs of acclimation. Roberts and Feder (2000) engineered fruit flies with extra functional copies of *hsp70* and control flies with nonfunctional copies. Flies with extra genes survived better than control flies during moderate to severe heat stress, but the reverse was true in the absence of heat stress (Fig. 5.7). Other transgenic flies with extra copies of *hsp70* had lower fertility, primarily because their eggs hatched poorly (Silbermann and Tatar 2000). The poor survivorship and fecundity of transgenic flies might have resulted

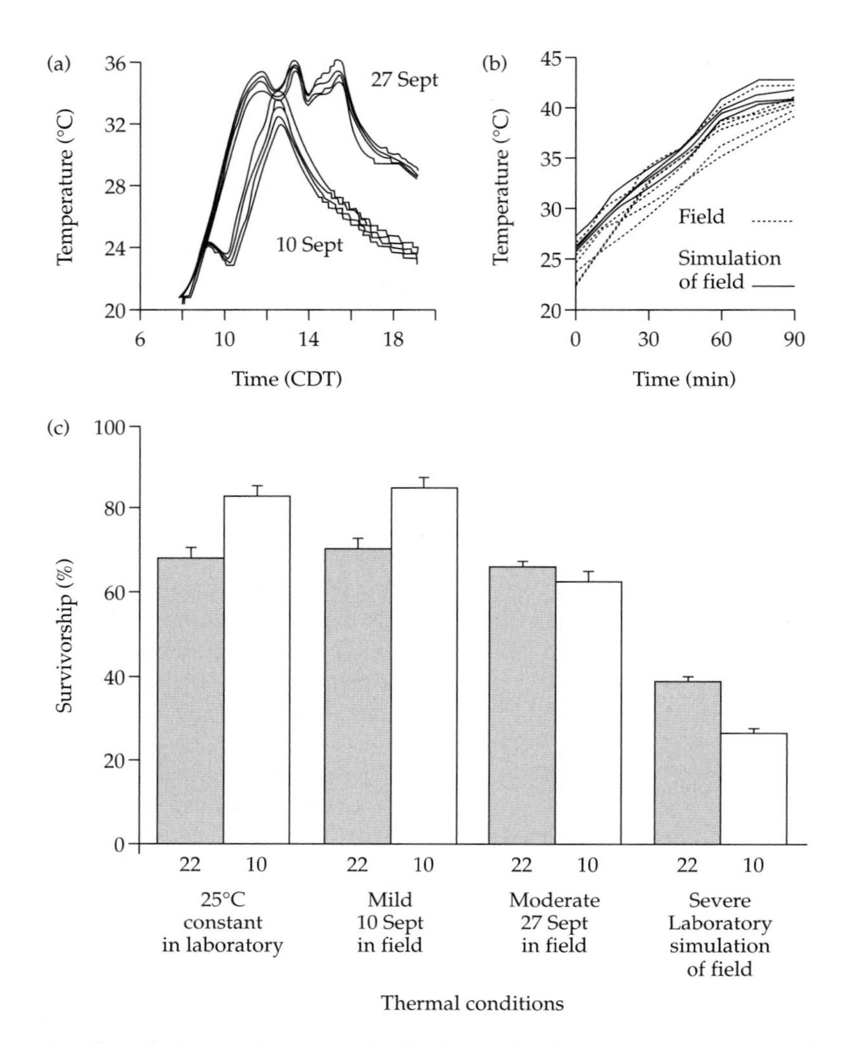

Figure 5.7 The expression of heat-shock proteins by transgenic fruit flies (*Drosophila melanogaster*) carries a cost under mild heat stress or no heat stress. (a) Temperatures experienced by larvae developing in rotting apples can rise above 32°C for varying durations. Temperatures are shown for days of mild and moderate heat stress (10 and 27 September, respectively; CDT, central daylight time).(b) Temperatures experienced during severe heat stress in apples were simulated in the laboratory. (c) Survivorship of transgenic flies (shaded bars) exceeded that of control flies (open bars) during moderate to severe heat stress, but the reverse was true in the absence of heat stress. Reproduced from Roberts and Feder (2000) with permission from Blackwell Publishing.

from energetic demands of heat-shock proteins or detrimental effects on development (Hoffmann et al. 2003). An elegant experiment by Krebs and Feder (1998) suggested that energetic costs of producing heat-shock proteins do not affect longevity. These researchers used lines of flies genetically engineered to express β-galactosidase instead of *hsp70*. Repeated heat shock of larvae decreased survivorship to adulthood, but the effect was greater at high rations than low rations; also, gene expression was unrelated to survivorship. Although one can interpret these findings as evidence that heat hardening costs little energy, energetic costs might have affected fecundity rather than survivorship. Furthermore, the function of heat-shock proteins could cost more than their production (Somero

2002), and Krebs and Feder did not quantify this cost in their experiment.

Despite the discrepancies among findings, acclimation clearly imposes costs of survivorship and fecundity in some cases. In other cases, perhaps a focus on animals raised under benign conditions makes tradeoffs less detectable. Although studies of heat hardening have given us an excellent start, we must devise creative ways to estimate the energetic cost of other acclimation responses. Undoubtedly, all forms of acclimation impose some energetic cost. The question for the future will be whether these costs outweigh others, such that acquisition and allocation tradeoffs greatly influence the evolution of acclimation. With this in mind, we now turn to a very different cost of acclimation: the cost arising from a time lag.

5.3.2 Costs of time lags

While scientists have mainly studied acclimation to constant temperatures, most natural environments are anything but constant. Stochastic variations over space and time leave many organisms with great doubt about the future of their environment (see Chapter 2). If acclimation occurs rapidly, an organism could continuously adjust its physiology to deal with its current environment. But if acclimation occurs slowly, the environment will likely change again before the organism can adjust to current conditions. Therefore, the net benefit of acclimation depends on the magnitude and predictability of diel, daily, and seasonal variations in temperature (see Kingsolver and Huey 1998).

Obviously, the impact of thermal stochasticity on the cost of acclimation depends strongly on the time required for acclimation. We can infer the existence of time lags from repeated measures of performance during thermal change. For example, Widdows and Bayne (1971) showed that mussels required at least 28 days to restore their rate of filtration after an abrupt change in temperature (see Fig. 5.1). In contrast, the acclimation of photosynthetic rate in some plants occurs within 2 weeks (Cunningham and Read 2003a). Tolerance of extreme temperatures can change throughout the course of a day, presumably because of the rapid cellular responses involving stress proteins (Buchner and Neuner 2003; Kelty and Lee 2001; Sorenson and Loeschcke 2002). Even when thermal acclimation occurs rapidly, the precise rate differs among closely related species (see, for example, Podrabsky and Somero 2006). Although available data afford crude comparisons of time lags among species, we know virtually nothing about the extent to which time lags vary among genotypes within species.

Given specialist–generalist tradeoffs, time lags will always impose some cost of acclimation. Specialization for the current temperature will lead to maladaptation when the environment changes more rapidly than the organism can respond. A generalist could avoid the problems associated with a time lag, but this strategy imposes other costs (see Chapter 3). The influence of time lags on the evolution of acclimation cannot be captured by simple ideas, such as the beneficial acclimation hypothesis. Instead, we must adopt mathematical models developed specifically for the study of phenotypic plasticity in stochastic environments.

5.3.3 Interaction between costs

In some cases, energetic costs and time lags depend on one another. As an example, consider the production of *hsp70* following a heat shock. An organism can use one of two strategies to regulate the expression of heat-shock proteins: feedforward and feedback. Using feedforward regulation, the organism would produce a certain quantity of *hsp70* immediately after heat shock. With this mechanism, new protein can function with minimal delay but uncertainty about the actual damage could lead to a wasteful overproduction of heat-shock proteins. Using feedback regulation, an organism would produce heat-shock proteins in direct response to denatured proteins. This mechanism ensures the production of *hsp70* will match the need, but the organism suffers a lag in protection.

Shudo and colleagues (2003) modeled the optimal combination of feedforward and feedback mechanisms of regulation. They assumed selection acts to minimize the sum of protein damage and production costs. Using this optimality criterion, an organism should always use feedforward regulation to some degree but should not necessarily use feedback regulation. Feedback regulation confers an advantage only when the organism cannot accurately predict the degree of damage caused by heat shock. In fact, an organism should rely entirely on a feedforward mechanism when the degree of heat damage corresponds to the degree of heat exposure or when expression of *hsp70* requires considerable time. Given poor predictably of damage and rapid upregulation of proteins, the relative degree of feedforward regulation versus feedback regulation depends on the actual damage caused by heat shock. Based on this model, the relative costs imposed by energetic demands and time lags should partly depend on the thermal environment, which determines the predictability and magnitude of heat stress.

5.4 Optimal acclimation of performance curves

After reviewing the benefits and costs of acclimation, we should understand the need to consider hypotheses about optimal acclimation. Obviously, a genotype that can assess its thermal environment and express the appropriate phenotype would perform better than a genotype that expressed a certain phenotype regardless of its environment. But the net benefit of acclimation depends on the energy required for change and the predictability of the environment. What if acclimation requires considerable energy, thereby negatively affecting other traits that contribute to fitness? And what if today's environment correlates imperfectly with tomorrow's environment? How do such constraints affect the optimal acclimation of performance curves? To answer these questions, we must turn to mathematical models plucked from the burgeoning literature on phenotypic plasticity. These models differ in their assumptions about environmental conditions and phenotypic strategies; therefore, let us briefly consider these assumptions before diving into some examples.

When modeling the evolution of phenotypic plasticity, theorists characterize environments as either fine-grained or coarse-grained. In a fine-grained environment, an individual experiences variation in environmental states during its life. In a coarse-grained environment, an individual experiences a single environmental state, but different individuals experience different environmental states. Coarse-grained environments arise in one of two ways: (1) the environment varies temporally, but changes on a scale that exceeds the longevity of an individual, or (2) the environment varies spatially, but each individual settles within a homogeneous patch. In either kind of environment, some form of acclimation could enhance fitness.

Theorists characterize phenotypes as being irreversible or reversible. Irreversible phenotypes become fixed early in ontogeny. The developmental environment provides cues about the future, including but not limited to temperature. These cues trigger the expression of a particular phenotype, which then remains constant thereafter. The cues provided by the developmental environment may or may not accurately reflect the temperatures of the selective environment. In contrast, reversible phenotypes can acclimate and deacclimate repeatedly. For real organisms, these phenotypic transitions occur gradually (see Fig. 5.1). For theoretical organisms, acclimation and deacclimation are more conveniently treated as abrupt transitions, subject to a time lag between phenotypic states. Obviously, reversible acclimation enables an organism to enhance its fitness by continuously tuning its physiology to its environment. In the models considered here, the phenotypes consist of parameters that define a performance curve.

Before considering any mathematics, our intuition should lead us to the following conclusions: (1) stable environments favor specialists without the capacity for acclimation; (2) environments that remain stable within generations but vary among generations favor developmental acclimation; and (3) environments that vary within generations favor reversible acclimation. Our intuition provides fewer insights about the optimal magnitude of acclimation, particularly for complex traits such as performance curves. First, any or all parameters of a performance curve might acclimate during ontogeny. Second, real environments vary spatially and temporally, and hence most organisms perceive their environment as both fine-grained and course-grained. Given these complications, we must supplement our intuition with insights from mathematical models.

5.4.1 Optimal developmental acclimation

Expanding on their previous work, Gabriel and Lynch (1992) modeled the optimal acclimation of a performance curve in an unpredictable environment. From Chapter 3, recall these researchers modeled the thermal sensitivity of survivorship with a Gaussian function (see Fig. 3.15). Two parameters—the thermal optimum (z_1) and performance breadth (z_2)—defined the specific shape of the curve (Fig. 5.8). Because the Gaussian function describes a probability distribution, the area under the curve remains constant for all values of z_1 and z_2. This function conveniently imposes specialist–generalist tradeoffs, in which an increase in performance at one temperature reduces performance at another temperature. As in their original model (Lynch and Gabriel 1987), values of z_1 and z_2 of an individual were determined by genotypic and environmental factors:

$$z_i = g_i + e_i, \tag{5.1}$$

where g_i equals the genetic value of the thermal optimum or performance breadth, and e_i represents developmental noise.

To model acclimation, Gabriel and Lynch added a term (r_i) to eqn 5.1 that represents an irreversible response to the developmental environment:

$$z_i = g_i + e_i + r_i. \tag{5.2}$$

If acclimation occurs without cost or constraint the optimal value of r_i equals

$$z_i^* - g_i, \tag{5.3}$$

where z_i^* equals the optimal phenotype. However, Gabriel and Lynch assumed the capacity for acclimation was constrained. Specifically, they assumed that small changes in the performance curve were more feasible than large changes, and very large changes were impossible. This constraint was implemented by choosing a bounded, nonlinear reaction norm for r_i (Fig. 5.9). Besides the addition of this acclimation response, the assumptions of this new model followed those of the earlier model (see Table 3.2); the most notable of these assumptions being that fitness depends on the product of performance over time.

In Gabriel and Lynch's model, developmental acclimation of the thermal optimum increases the fitness of an organism by reducing the need for a wide performance breadth. The optimal degree of acclimation depends on the spatial and temporal variance of environmental temperature. Selection favors a greater capacity for acclimation as the spatial variation within generations or the temporal variation among generations increases. In contrast, increasing the temporal variation within generations lessens the selective advantage of acclimation.

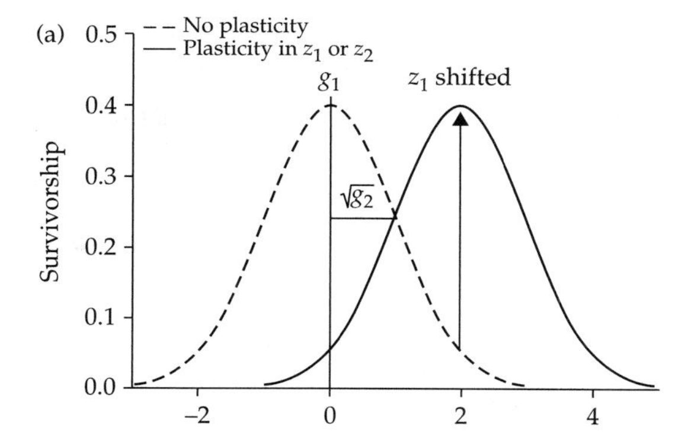

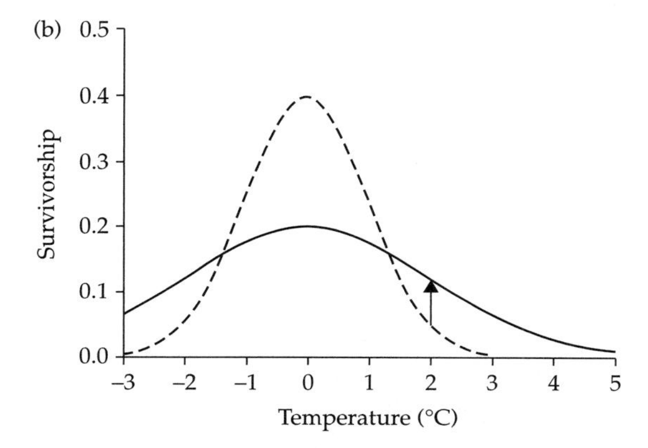

Figure 5.8 Developmental acclimation of the thermal optimum (a) or performance breadth (b) enhances the performance of an organism in a heterogeneous environment. Genotypic values of the thermal optimum and performance breadth (g_1 and g_2, respectively) interact with the environmental temperature to produce phenotypic values (z_1 and z_2, respectively). Environmental temperatures are scaled relative to the mean (i.e., mean = 0). In these examples, the developmental temperature exceeds the mean temperature by 2°C. The arrow in each plot shows the increase in survivorship derived from acclimation to the developmental temperature. Adapted from Gabriel (1999) in The Ecology and Evolution of Inducible Defenses (Harvell, C. D. and R. Tollrian, eds.) by permission from Princeton University Press. © 1999 Princeton University Press.

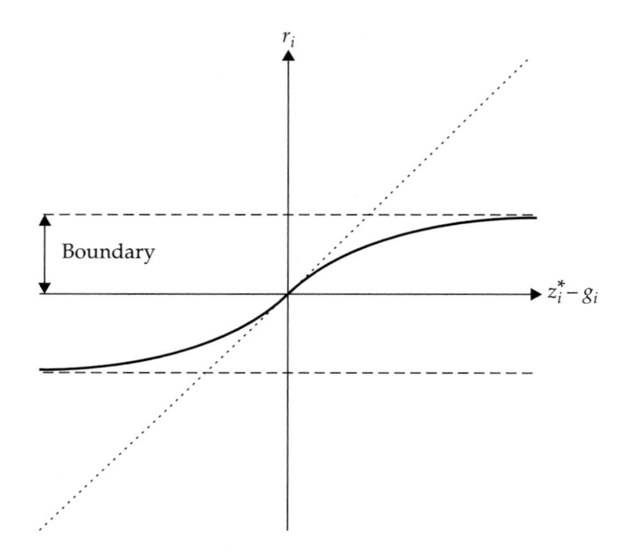

Figure 5.9 In the model by Gabriel and Lynch, a nonlinear reaction norm described a bounded relationship between the optimal shift in the phenotype ($z_i^* - g_i$) and the acclimation response (r_i). Adapted from Gabriel and Lynch (1992) with permission from Blackwell Publishing.

In a sense, the temporal variation experienced by an individual defines the stochasticity of the environment. A more stochastic (or less predictable) environment favors a relatively wide performance breadth instead of a developmental shift in the thermal optimum. Therefore, the greatest capacity for developmental acclimation should evolve in environments in which thermal heterogeneity among generations greatly exceeds that within generations.

Acclimation of the performance breadth yields little selective advantage over acclimation of the thermal optimum alone (Fig. 5.10). Given the potential for the thermal optimum to acclimate, the optimal performance breadth depends strictly on the variance of temperature within generations. A relatively constant (or predictable) environment favors the development of a specialist for the mean temperature. Even in the most heterogeneous environments, acclimation of the thermal optimum confers a greater selective advantage than acclimation of the performance breadth. This result follows from the assumption that the maximal performance of a specialist exceeds that of a generalist. If an organism can match its thermal optimum to its current environment, why accept the cost of being a generalist? Through developmental acclimation, an organism can perform over a wide range of temperatures without sacrificing performance at the optimal temperature. Unless some other constraint exists, the thermal optimum of performance should acclimate more readily than the performance breadth.

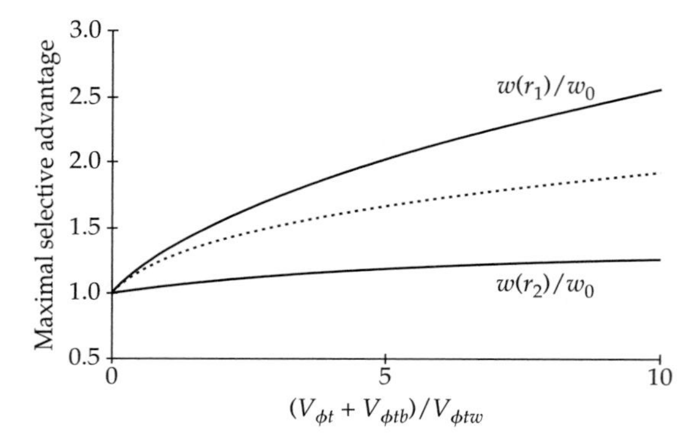

Figure 5.10 The selective advantage of developmental acclimation of the thermal optimum (z_1) or performance breadth (z_2) depends on several sources of thermal heterogeneity: spatial variation ($V_{\phi t}$), temporal variation within generations ($V_{\phi tw}$), and temporal variation among generations ($V_{\phi tb}$). Solid lines depict relative fitnesses of a plastic genotype to a fixed genotype for z_i: $w(r_i)/w_0$. The dotted line represents the selective advantage of a change in the thermal optimum without a correlated change in the performance breadth. The selective advantage increases with an increase in the ratio of the variation experienced among individuals to the variation experienced within individuals, $(V_{\phi t} + V_{\phi tb})/V_{\phi tw}$. These selective advantages are for unbounded reaction norms for z_1 and z_2. Adapted from Gabriel and Lynch (1992) with permission from Blackwell Publishing.

Gabriel and Lynch formulated their model under the assumption that acclimation does not cost energy or reduce survivorship. Therefore, let us briefly consider how such costs would alter the predictions of their model. Because developmental acclimation enables organisms to specialize for local temperatures, the fitness of this strategy must outweigh that of a generalist. Assuming generalization imposes a reduction in maximal performance (see Chapter 3), a cost of acclimation exists at which both acclimation and generalization yield equal fitness (DeWitt and Langerhans 2004). Below this cost, developmental acclimation would be favored over generalization. In this range, the qualitative results of Gabriel and Lynch's analysis hold unless the cost of acclimation depends on the thermal environment or the acclimation response. Given that generalization imposes a severe cost (Lynch and Gabriel 1987), we should expect useful insights to emerge from this model despite its assumption of no energetic cost.

Gabriel and Lynch also assumed that organisms receive accurate cues from their environment. But what if the temperature during development poorly predicts the temperature later in life? Would such unpredictability affect the evolution of developmental acclimation? Moran (1992) addressed this very question by modeling fitness landscapes in a simple, coarse-grained environment. She assumed an environment that comprised two kinds of patches, which we can think of as hot and cold. An individual disperses randomly after birth and experiences a single patch during its lifetime. A cold-adapted phenotype achieves greater fitness in a cold patch than does a heat-adapted phenotype; the opposite occurs when the two phenotypes encounter hot conditions. A plastic genotype could adopt either phenotype but would suffer a cost if the wrong phenotype were expressed. The optimal strategy depends strongly on the accuracy of environmental cues and the frequency of the two environments. If the accuracy of the cue exceeds the frequency of the more common environment, a genotype capable of developmental acclimation would outperform one that specializes on a single environment. If the cue is less accurate, selection favors a specialist for the more frequent of the two environments. In other words, plastic genotypes must express the correct phenotype more often than nonplastic genotypes would do so by chance alone. If one environment occurs much more frequently than the other, the evolution of developmental acclimation requires (1) extremely accurate cues, (2) a low cost of expressing the wrong phenotype in the more common environment, or (3) a high cost of expressing the wrong phenotype in the less common environment. Thus, developmental acclimation would likely evolve when the frequencies of different thermal patches equal one another, or when the fitness of one phenotype greatly exceeds that of the other phenotype in the rare patch.

5.4.2 Optimal reversible acclimation

Compared to developmental acclimation, reversible acclimation offers even greater potential to match physiology to the current environment. To explore this potential, Gabriel (1999) modeled the optimal strategy of reversible acclimation in a simple environment that switched between two states. Although Gabriel defined the environment in general terms, we can think of these states as operative environmental temperatures (T_1 and T_2); they could represent either diel or seasonal extremes depending on the temporal scale we choose.

As in previous analyses, the thermal sensitivity of survivorship was modeled as a Gaussian function, defined by z_1 and z_2 (see Section 5.4.1). The environment shifted abruptly between T_1 and T_2, each state lasting a certain period (t_s). Following the shift from T_1 to T_2, an organism adapted to T_1 would lose some ability to perform. Should the organism alter its performance curve in response to the change in temperature? The benefit of doing so seems obvious, but what about the costs?

Gabriel focused on the cost imposed by time lags. As discussed in Section 5.3.2, if a response to thermal change requires time, the temperature might change again before the response can occur. With this in mind, we must compare the performance of two genotypes: one whose performance curve acclimates reversibly (+) and one whose performance curve remains fixed (−). During a switch in temperature, the performance of both the acclimating and nonacclimating genotypes would decrease. After a time lag, however, the performance of the acclimating genotype would rebound. Once the environment switches back to its original temperature, the acclimating genotype suffers another temporary loss of performance while the nonacclimating genotype regains its original level of performance. The relative fitness of these genotypes (w_+/w_-) depends on the time required for an acclimation response (t_r) and the frequency and magnitude of environmental change:

$$\frac{w_+}{w_-} = e^{\left\{0.5(t_s - t_r)\frac{(T_2 - T_1)^2}{z_2}\right\}}. \qquad (5.4)$$

Decreasing either the frequency of environmental change ($1/t_s$) or the lag during acclimation (t_r) enhances the relative fitness of reversible acclimation. Similarly, increasing the magnitude of environmental change ($T_2 - T_1$) also favors acclimation. Specifically, the optimal shift in the thermal optimum (Δz_1^*) during acclimation would be

$$\Delta z_1^* = (T_2 - T_1)\left(1 - \frac{t_r}{2t_s}\right). \qquad (5.5)$$

Obviously, acclimation never benefits an organism when the time needed to acclimate (t_r) exceeds the time between thermal switches (t_s). Hence, reversible acclimation should benefit long-lived organisms, which experience seasonal changes in temperature during their life. Furthermore, reversible acclimation would also benefit organisms during diel variation in temperature if acclimation and deacclimation occur rapidly.

In contrast to developmental acclimation, reversible acclimation of the performance breadth can provide a significant benefit. Two conditions favor reversible acclimation of the performance breadth: (1) when acclimation occurs slowly but the environment changes frequently or (2) when the temperature shifts greatly between periods (Fig. 5.11). The benefits of a greater performance breadth under such conditions might seem obscure, but the key lies in thinking about performance during deacclimation. Acclimation of the thermal optimum causes maladaptation to T_1 in exchange for adaptation to T_2; however, acclimation of the performance breadth confers some performance at T_2 while reducing maladaptation to T_1 during deacclimation. This aspect of the model has direct implications for adaptation to diel and seasonal cycles (see Fig. 5.11). Diel fluctuations, which occur rapidly relative to physiological responses, should favor acclimation of the performance breadth. Seasonal fluctuations,

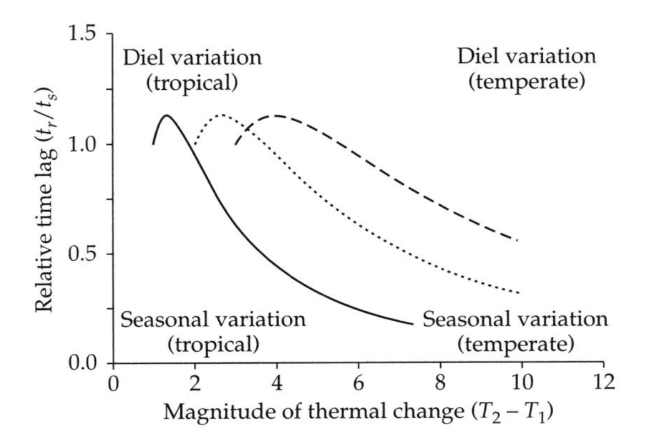

Figure 5.11 The selective advantage of reversible acclimation depends on the magnitude of thermal change ($T_2 - T_1$) and the relative time lag for acclimation (t_r/t_s). The lines depict isoclines of equal fitness resulting from acclimation of the thermal optimum (z_1) and acclimation of the performance breadth (z_2). Above the line, acclimation of the performance breadth yields greater fitness than acclimation of the thermal optimum; below the line, the reverse condition holds. From left to right, the three isoclines represent increasing performance breadths prior to acclimation (g_2 equals 1, 4, and 9, respectively). The parameter values at each of the four corners of the plot correspond to certain temporal patterns of variation in either temperate or tropical environments (see Chapter 2). Adapted from Gabriel (1999) in The Ecology and Evolution of Inducible Defenses (Harvell, C. D. and R. Tollrian, eds.) by permission from Princeton University Press. © 1999 Princeton University Press.

which are more prolonged than diel fluctuations, should favor acclimation of the thermal optimum. This prediction should hold regardless of the environment in which the organism has evolved (see Fig. 5.11).

As in his previous model (Section 5.4.1), Gabriel assumed reversible acclimation imposed no cost in terms of energy or survivorship. We can easily infer how such a cost would affect the predictions of his model. Figure 5.12 depicts the fitnesses of various strategies of reversible acclimation. Because acclimation imposes no cost in this analysis, the fitness of an acclimating genotype always exceeds that of a nonacclimating genotype. A direct cost of acclimation would shift the fitness curves for acclimating genotypes downward, without affecting the curve for the nonacclimating genotype. As a consequence, selection would favor a genotype that did not acclimate in relatively stable environments. All other things being equal, selection for

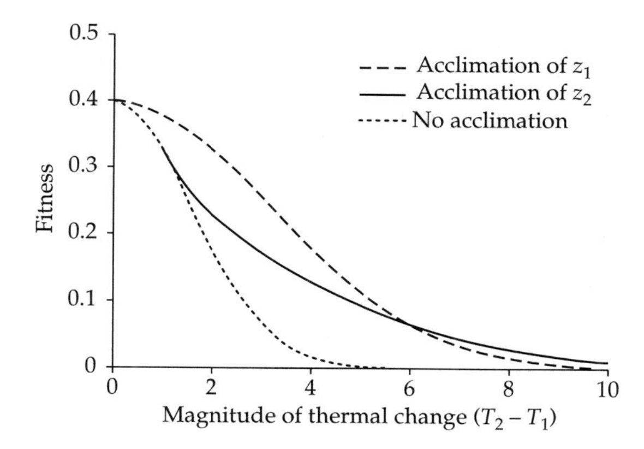

Figure 5.12 The selective advantage of reversible acclimation depends on the magnitude of thermal change ($T_2 - T_1$). Each line shows this dependence for a genotype that can change either its thermal optimum (z_1) or its performance breadth (z_2), or neither parameter. Other parameters were set as follows: $t_s = 0.4$, $t_r = 0.1$, and $g_2 = 1$. Adapted from Gabriel (1999) in The Ecology and Evolution of Inducible Defenses (Harvell, C. D. and R. Tollrian, eds.) by permission from Princeton University Press. © 1999 Princeton University Press.

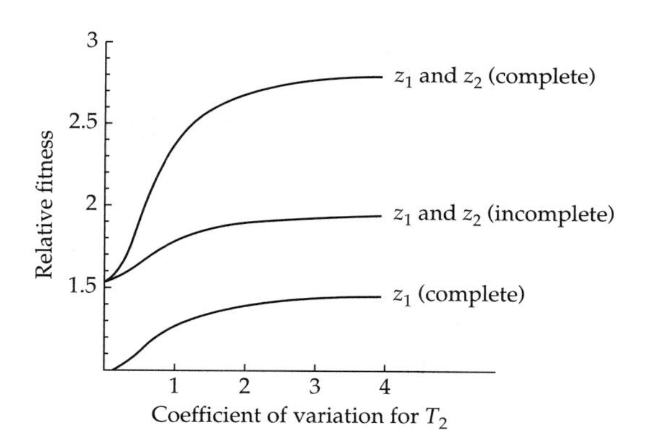

Figure 5.13 In a stochastic environment, the selective advantage of reversible acclimation of the thermal optimum (z_1) or the performance breadth (z_2) depends on whether the organism can determine the magnitude of thermal change (complete information) or only the occurrence of thermal change (incomplete information). Fitnesses are scaled relative to the fitness of a genotype whose thermal optimum acclimates with incomplete information. The three lines represent the following genotypes: (1) acclimation of the thermal optimum with complete information, (2) acclimation of the thermal optimum and the performance breadth with incomplete information, and (3) acclimation of the thermal optimum and the performance breadth with complete information. Adapted from Gabriel (2005) with permission from Blackwell Publishing.

costly reversible acclimation would require a greater magnitude of thermal change or a shorter time lag (Padilla and Adolph 1996).

The predictability of temperature also affects the optimal strategy of reversible acclimation. Gabriel (2005) extended his model of reversible acclimation to consider the effect of unpredictable magnitudes of thermal change. To introduce unpredictability, he assumed that T_2 was drawn from a distribution of temperatures; thus, the organism knows the environment has changed but does not know by how much. If the organism can detect a change in temperature but cannot predict the magnitude of this change, the optimal shift in the thermal optimum corresponds to the long-term mean of thermal change (i.e., mean T_2 − mean T_1). Acclimation of the performance breadth in addition to acclimation of the thermal optimum helps to compensate for the stochasticity of thermal change (Fig. 5.13). The optimal performance breadth after acclimation scales proportional to the variance of T_2. Even with the potential for acclimation, stochasticity reduces the fitness of a genotype in a heterogeneous environment.

5.4.3 Relaxing assumptions about fitness

Although the models presented in this chapter describe fitness as the product of performance over time, we could easily envision a scenario in which fitness depends on the sum of performance. In Chapter 3, we saw how the relationship between performance and fitness affects the evolution of thermal sensitivities. Recall that most patterns of thermal heterogeneity favor the evolution of thermal specialists when performance contributes additively to fitness; generalists are favored only when the variation among generations exceeds the variation within generations (Gilchrist 1995). No formal model of optimal acclimation has addressed this scenario, but we can intuit the behavior of such a model. Acclimation of the thermal optimum relaxes the selective pressure imposed by thermal heterogeneity among generations. Therefore, acclimation would always favor specialists when performance contributes additively to fitness. Consequently, the selective advantage of developmental acclimation would not depend on the thermal heterogeneity within generations. Similarly, reversible acclimation of the performance breadth would not benefit an organism. Of course, both of these conclusions hinge on the existence of specialist–generalist tradeoffs.

5.5 Evidence of optimal acclimation

The models described in the preceding section provide testable predictions about the acclimation of performance curves. First, thermal optima should acclimate more often than performance breadths, particularly during prolonged changes in temperature (i.e., seasonal or intergenerational changes). Conversely, performance breadth should acclimate primarily in response to rapid changes in temperature, such as diel cycles.

Second, organisms in variable environments should acclimate more readily than organisms in stable environments. In this section, I review the experimental evidence for each of these hypotheses. Keep in mind that most of the experiments were not designed explicitly to test the theory of optimal acclimation.

5.5.1 Does the thermal optimum acclimate more than the performance breadth?

Although the theory suggests the thermal optimum of performance should acclimate readily in variable environments, studies of plants and animals indicate that thermal optima rarely respond to changes in environmental temperature (see Tables 5.1 and 5.2). As discussed in Section 5.2, plants generally exhibit an optimal temperature for development, rather than plasticity of their thermal optima (but see Mooney 1980; Slatyer 1977; Smith and Hadley 1974). Some species of animals undergo acclimation of locomotor performance consistent with the theory. For example, a few species of fish (Fry and Hart 1948; Griffiths and Alderdice 1972; Johnson and Bennett 1995; Larimore and Duever 1968; Seebacher et al. 2005) and one crocodilian (Glanville and Seebacher 2006) adjusted their thermal optima for swimming following abrupt changes in temperature. Similarly, the thermal optimum of filtration by oysters scaled proportional to acclimation temperature (Newell et al. 1977). These examples merely prove exceptions to the rule. The majority of studies revealed no evidence for the optimal acclimation of thermal optima (see Table 5.2).

5.5.2 Do organisms from variable environments acclimate more than organisms from stable environments?

In principle, we could answer this question through an experimental or a comparative approach. A powerful experimental approach involves natural selection in the laboratory (see Section 1.5.2). One could establish replicated populations and force them to evolve in constant and fluctuating environments. After some period, one would compare the acclimation capacities of individuals from these populations. Unfortunately, no researcher has conducted a selection experiment designed to address the evolution of acclimation. For this reason, we must rely solely on comparisons of acclimation capacities among natural populations. An ideal comparative approach involves a quantitative analysis of acclimation capacity in a phylogenetic context (see Section 1.5.3). The populations or species in this analysis would span a range of thermal environments, perhaps distributed over latitudinal or altitudinal gradients. Unfortunately, comparative studies of acclimation generally lack the replication required for quantitative phylogenetic analysis. Therefore, we must rely on qualitative comparisons of acclimation capacity between genotypes from different thermal environments.

Comparative studies of developmental acclimation provide virtually no support for the notion that developmental acclimation evolves according to thermal heterogeneity. Most of the relevant data come from studies of photosynthetic rates (Table 5.3). In an early review of this subject, Berry and Björkman (1980) concluded the acclimation capacity of a plant species corresponds to its selective environment. Among six species they considered, three patterns were observed: (1) two species from cold environments performed poorly after developing at high temperatures, (2) one species from a hot environment performed poorly after developing at low temperatures, and (3) three species from variable environments acclimated well to high and low temperatures (see Fig 5.2). More

Table 5.3 Comparative studies of the thermal acclimation of photosynthetic rate (adjusted for the area or mass of leaves). For each study, I note patterns that are relevant to current optimality models. I also note whether the experiment revealed evidence of an optimal developmental temperature (ODT).

Species	Cline	Acclimation temperatures	Test temperatures	Pattern	ODT	Source
Acer saccharum	Latitude	27/18, 31/18	11, 16, 21, 26, 31, 36	Plants performed better when they had developed at higher temperatures.	Yes	Gunderson et al. (2000)
Atriplex lentiformis	Desert vs. coastal	23/18, 43/30	10–46°C in 2°C increments	T_{opt} was related to acclimation temperature in desert plants, but development at low temperature resulted in maximal performance of coastal plants.	Yes/no	Pearcy (1977)
Australian trees	Latitude	22°C:14°C and a stochastic cycle with the same mean temperature	12, 17, 22, 27, 32	Temperate species underwent acclimation of performance breadth, whereas tropical species did not; however, the thermal breadth of photosynthesis was greater in the less variable and more predictable treatment.	No	Cunningham and Read (2003b)
Australian trees	Latitude	14/6, 19/11, 22/14, 25/17, 30/22	10, 14, 18, 20, 22, 24, 26, 30	Plasticity of the thermal optimum was greater in tropical species than in temperature species; performance breadth did not acclimate.	Yes	Cunningham and Read (2002)
Eucalyptus pauciflorac	Altitude	8/4, 15/10, 21/16, 27/22, 33/28	10, 15, 20, 25, 30, 35, 40	T_{opt} was related to acclimation temperature in all populations.	No	Slatyer (1977)
Heliotropium curassavicum	Desert vs. coastal	25/15, 40/26	20, 25, 30, 35, 40, 45	T_{opt} was related to acclimation temperature in both desert and coastal plants.	No	Mooney (1980)
Ledum groenlandicum	Latitude and longitude	15/10, 30/25	15, 20, 25, 30, 35	T_{opt} was related to acclimation temperature in all four populations.	No	Smith and Hadley (1974)
Oxyria digyna	Altitude	12/4, 21/10, 32/21	10, 21, 32, 38, 43	T_{opt} was related to acclimation temperature in alpine populations, but Arctic populations performed best when they had developed at low temperatures.	Yes/no	Billings et al. (1971)

Continued

Table 5.3 Continued

Species	Cline	Acclimation temperatures	Test temperatures	Pattern	ODT	Source
Plantago major, P. euryphylla, and P. lanceolata	Altitude	13, 20, 27	6, 13, 20, 27, 34	Alpine plants performed better at low temperatures (13–20°C), regardless of their developmental temperature. Lowland plants that developed at 13°C performed better at low temperatures (6–13°C) than plants that developed at 20°C or 27°C.	Yes	Atkin et al. (2006b)
Valonia utricularis	Latitude	15, 18, 20, 25	5, 10, 15, 20, 25, 30, 35	T_{opt} was related to acclimation temperature in algae from the Indian Ocean, but T_{opt} did not acclimate in Mediterranean algae.	Yes	Eggert et al. (2003b)
Valonia utricularis	Latitude	$\leq$ 20, 25, 30	$\leq$ 20, 25, 30	Under chronic exposure to low temperatures, temperate algae maintained high rates of photosynthesis but tropical algae did not. The opposite pattern was observed when algae were exposed to high temperatures	Yes	Eggert et al. (2006)

Notes: T_{opt}, thermal optimum for performance.

rigorous comparisons suggest a very different view: little correspondence exists between thermal heterogeneity and acclimation capacity. For several species, populations exhibited strong acclimation of thermal optima regardless of the thermal heterogeneity of their native environment (Mooney 1980; Slatyer 1977; Smith and Hadley 1974). Moreover, genotypes from variable environments sometimes exhibited less capacity for acclimation than genotypes from stable environments.

In a comprehensive case study, Cunningham and Read (2002) compared the acclimation of photosynthetic rate in tropical and temperate trees of Australia. These investigators raised eight species of trees in five fluctuating environments and compared their photosynthetic performance. Contrary to the theory, the thermal optimum acclimated more in tropical species than in temperature species. The performance breadth did not acclimate in any species. These investigators also grew the same eight species under predictable and unpredictable thermal cycles; both treatments had a mean diel cycle of 22°C : 14°C, but one treatment varied randomly between 17°C : 9°C and 27°C : 19°C on a daily basis.

Consistent with the theory, temperate species had wider performance breadths and underwent greater acclimation of performance breadths than did tropical species. But narrower performance breadths were expressed in the more variable environment! Thermal optima did not acclimate but were 2–3°C higher in tropical species than in temperate species. Despite the weak support for current models, at least one key assumption appears valid for these trees: a wider performance breadth resulted in a lower maximal performance.

Although rare, comparative studies of reversible acclimation have also produced patterns that contradict the theory. Two examples are particularly relevant. In the first example, Cunningham and Read (2003b) followed up on their study of developmental acclimation (see above) by exposing temperate and tropical seedlings to progressively lower temperatures; they started at 30°C and ended at 14°C, giving seedlings 14 days to acclimate before each decrease in temperature. In most species, the thermal optimum for photosynthesis decreased in response to the decrease in temperature. However, both tropical and temperate species acclimated similarly. In the second example, Deere and Chown (2006) exposed five species of mites—three marine and two terrestrial—to a range of constant temperatures for 7 days. After this period, they compared the thermal sensitivities of walking speed among mites from different thermal treatments. For four of the five species, thermal optima and performance breadths shifted little during acclimation. Instead, thermal treatments tended to enhance or impair performance across a wide range of temperatures, leading to variation in mean performance among groups. Clearly, both examples contradict the prediction that reversible acclimation will correspond to the expected change in environmental temperature.

Finally, most comparative studies of thermal hardening fail to support the current theory (Table 5.4). Although heat and cold tolerances acclimate readily in many species, the thermal heterogeneity of the selective environment seems to bear little on the capacity for acclimation. A striking exception was documented for *D. buzzatii* along an altitudinal cline (Sorenson and Loeschcke 2002). Flies from high and low altitudes exhibited distinct diel cycles of heat tolerance. A subsequent experiment confirmed that photoperiod served as a cue for the timing of acclimation; by artificially coupling photoperiod and temperature, the researchers reversed the diel cycle of heat tolerance in flies from a low altitude, creating flies that tolerated heat better during a cool photophase instead of a warm scotophase.

5.6 Constraints on the evolution of acclimation

In general, current models of optimal acclimation predict observed patterns poorly. First, the thermal optimum for performance rarely acclimates, whereas thermal tolerances often acclimate during exposure to extreme temperatures. Second, the capacity for acclimation rarely corresponds to the thermal heterogeneity of the selective environment. Why do current optimality models yield such poor predictions about thermal physiology? As with any optimality model, the models described in Section 5.4 focus on functional constraints and ignore genetic constraints, such as additive genetic variance, pleiotropy between traits, and gene flow among populations. In Chapter 3, we saw how genetic variances and gene flow influence the evolution of performance curves. Now, we shall draw on quantitative genetic models to see how these factors might influence the evolution of thermal acclimation.

5.6.1 Genetic variance and covariance

To explore the potential impacts of additive genetic variance and covariance on the evolution of thermal acclimation, we can use simple quantitative genetic models of phenotypic plasticity. In Chapter 1, we discussed how quantitative genetics models help us to understand constraints on the evolution of reaction norms. If we treat thermal acclimation as a form of phenotypic plasticity, these models can define constraints on the evolution of acclimation responses. For example, imagine a reaction norm describing how the mean environmental temperature relates to the thermal optimum of performance. The expression of the thermal optimum in multiple environments could be modeled as if it were multiple traits evolving in a single environment

Table 5.4 Comparative studies of the acclimation of thermal tolerance in ectotherms

Species	Thermal cline	Acclimation temperatures	Acclimation period	Measure of tolerance	Effect of treatment	Source
Crustacea	Tidal depth	18–32	30–60 min	CT_{max}	CT_{max} of all species increased in winter; CT_{max} of most species did not acclimate in summer.	Hopkin et al. (2006)
Drosophila buzzatii	Altitude	39	1 hour	Knockdown time	No acclimation	Sorensen et al. (2001)
Drosophila buzzatii	Latitude/other	38	60 min	Survival at 42°C for 100 min	All populations exhibited similar heat hardening.	Krebs and Loeschcke (1995)
Drosophila melanogaster	Latitude	11	3–4 weeks	LT_{50} at 0°C	LT_{50} of all populations increased sevenfold.	Bubliy et al. (2002)
Drosophila melanogaster	Latitude	12, 21, 31	Larval period	Recovery from 16 hours at 0°C	Both temperate and tropical populations exhibited strong acclimation, but plasticity was slightly greater for temperature populations.	Ayrinhac et al. (2004)
Drosophila melanogaster	Latitude	4	2 hours	Survival at −2°C for 23 hours and −5°C for 50 min	Temperate and tropical populations acclimated similarly.	Hoffmann and Watson (1993)
Drosophila melanogaster	Latitude	18, 25	Larval period	Survival at −2°C for 135 min and −5°C for 45 min	Temperate and tropical populations acclimated similarly.	Hoffmann and Watson (1993)
Drosophila melanogaster	Latitude	13, 25, 29	4 days	Survival at 39°C for 28 min	Temperate and tropical populations acclimated similarly.	Hoffmann and Watson (1993)
Drosophila similans	Latitude	13, 25, 29	4 days	Survival at 39°C for 14 min	Temperate and tropical populations acclimated similarly.	Hoffmann and Watson (1993)
Drosophila spp.	Latitude	35–38	30 min	Heat-shock recovery	Flies from low latitude underwent heat hardening, but flies from high latitude were detrimentally affected.	Garbuz et al. (2003)
Fundulus heterclitus	Latitude	2, 7, 13, 22, 27, 32, 34	≥ 21 days	CT_{min}/ CT_{max}	Both populations acclimated equally.	Fangue et al. (2006)
Halozetes spp. and *Podacarus auberti*	Marine vs. terrestrial	0, 5, 10, 15	7 days	CT_{min}/ CT_{max}	Both CT_{min} and CT_{max} acclimated more in marine species, but only changes in CT_{max} were in the predicted direction.	Deere et al. (2006)
Orchestia gammarellus	Latitude	5, 10, 15, 20	10 days	CT_{max}	CT_{max} of both populations increased with increasing acclimation temperature.	Gaston and Spicer (1998)
Petrolisthes spp.	Latitude	sea temperature vs. 10°C above sea temperature		CT_{max} (heart failure)	CT_{min} acclimated more and CT_{max} acclimated less in crabs from lower latitudes.	Stillman (2003)

Continued

Table 5.4 Continued

Species	Thermal cline	Acclimation temperatures	Acclimation period	Measure of tolerance	Effect of treatment	Source
Sarcophagidae	Latitude	0, 25, 40	2 hours	Survival at −10°C or 45°C until adulthood	Acclimation of heat and cold tolerances was greater for flies from higher latitudes.	Chen et al. (1990)
Sceloporus spp.	Latitude	35/16 at 12L:12D, 35/10 at 8L:16D	1 month	CT_{min}	CT_{min} of all populations decreased in both treatments.	Tsuji (1988)
Tegula spp.	Tidal depth	14, 22	15–19 days	CT_{min}/ CT_{max} (heart failure)	Acclimation was greater for subtidal species than for intertidal species.	Stenseng et al. (2005)

Notes: T_{opt}, thermal optimum for performance; CT_{min}, critical thermal minimum; CT_{max}, critical thermal maximum; LT_{50}, the time required for 50% of the individuals in a sample to die from exposure to a temperature.

(see Section 1.4.2). A lack of genetic variance and the presence of genetic covariances will slow adaptation of the reaction norm. Nevertheless, the optimal strategy of acclimation should evolve eventually as long as the population meets two conditions (van Tienderen 1991; Via 1987, 1994). First, some additive genetic variance for acclimation must exist. Second, the acclimation response to one temperature cannot be perfectly correlated (genetically) with the acclimation response to another temperature. If responses at different temperatures were perfectly correlated, the shape of the reaction norm would not be free to evolve toward the optimum.

Very little information exists about the genetic variances of acclimation responses (reviewed by Kingsolver and Huey 1998). What we do know pertains mainly to heat hardening by drosophilids. The capacity for heat hardening varies among isofemale lines in some populations (Krebs and Feder 1997) but not in others (Krebs and Loeschcke 1994b; Loeschcke et al. 1994). In *Drosophila buzzatii*, artificial selection led to an increase in heat hardening over 10 generations (Fig. 5.14). From Chapter 1, recall that the magnitude of this response to selection reflects the heritability of heat hardening in this population. With respect to performance curves, we know virtually nothing about the genetic variance of thermal acclimation.

Genetic correlations resulting from pleiotropy would limit thermal adaptation in heterogeneous

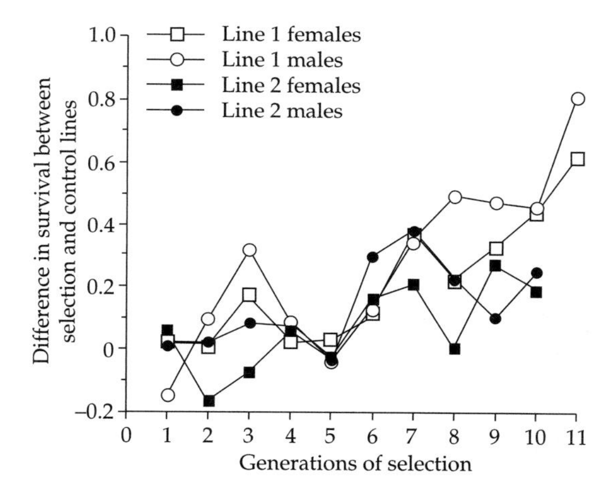

Figure 5.14 Artificial selection led to an increase in heat hardening over 10 generations. In each generation, flies (*Drosophila buzzatii*) were exposed to 39°C for 90 min after a previous exposure to heat. Individuals of selection lines were mated based on their ability to resist heat after acclimation, whereas those from the control lines were mated randomly with respect to this trait. Reproduced from Krebs and Loeschcke (1996) with permission from the Genetics Society of America.

environments. Consider this problem in the context of Gabriel and Lynch's model of developmental acclimation (Section 5.4.1). The thermal optimum for performance depended on a genetic value (g_1), an environmental value (e_1), and an acclimation response (r_1). The acclimation response (r_1) was limited by a boundary (b). Now imagine the boundary for acclimation depends on the genetic value for the thermal optimum. Such a relationship constitutes a genetic correlation that constrains the evolution of acclimation. Although genetic correlations of this nature have not been documented, Stillman (2003) discovered a phenotypic correlation consistent with this idea. Among species of porcelain crabs, species that possessed higher critical thermal minima and lower critical thermal maxima exhibited greater capacities for acclimation (Fig. 5.15). This pattern might reflect an underlying genetic correlation between thermal sensitivity and thermal acclimation. Alternatively, the phenomenon could reflect adaptive divergence between crabs in warm, stable environments of the tropics and those in cool, variable environments of the temperate zone.

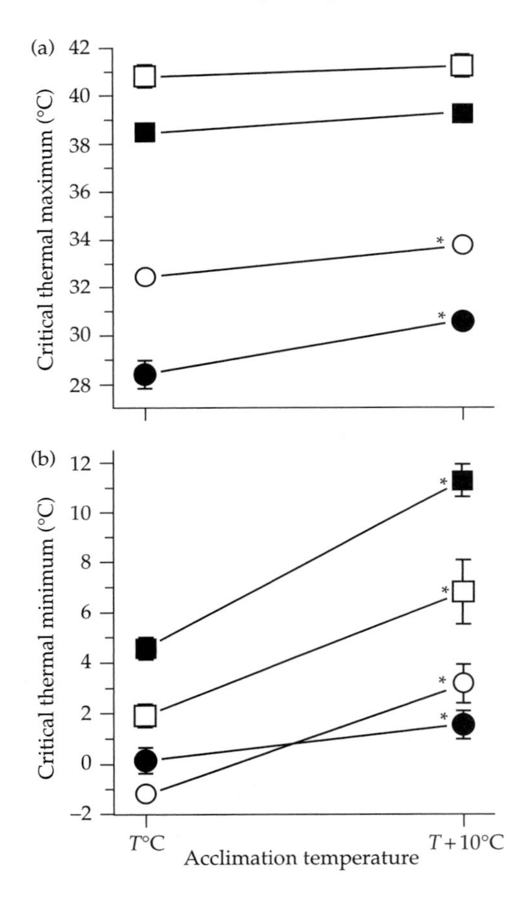

Figure 5.15 The acclimation of critical thermal limits for cardiac function in porcelain crabs (*Petrolisthes* spp.) depended on the value of the thermal limit prior to acclimation. Lines connect critical thermal maxima (a) or critical thermal minima (b) expressed in thermal treatments separated by 10°C (*T*°C and *T*+10°C). Data are for *P. eriomerus* (filled circles), *P. cinctipes* (open circles), *P. hirtipes* (filled squares), and *P. gracilis* (open squares). From Stillman (2003). Reprinted with permission from AAAS.

5.6.2 Gene flow

When temperature varies spatially, gene flow among populations can constrain the evolution of thermal acclimation. At first, this statement might seem counterintuitive. After all, acclimation enables an organism to adjust its thermal physiology according to its current environment; therefore, migration within a spatially heterogeneous environment should facilitate the evolution of acclimation rather than constrain it (see Section 5.4.1). But when dispersal occurs in the middle of ontogeny, the developmental environment no longer reflects the selective environment, increasing the unpredictability of future environmental conditions. To demonstrate this effect, de Jong (1999) modeled the evolution of plasticity for several types of dispersal. She envisioned a plastic trait under stabilizing selection, whose reaction norm could take on any shape defined by a linear or quadratic function. This phenotype equates to a reaction norm for the thermal optimum of performance (e.g., $T_{opt} = aT + b$). Irreversible acclimation of the phenotype occurred during the zygotic stage but affected the performance of the organism during the adult stage. Three cases of dispersal were considered (Fig. 5.16). In the first case, zygotes dispersed randomly and remained in their environment until reproduction (as in Gavrilets and Scheiner 1993; Via and Lande 1985). Because individuals could predict their selective environment under these conditions, the reaction norm in each population converged on the global optimum (i.e., the thermal optimum expressed by a genotype would equal the temperature of its environment). In the second case, zygotes dispersed prior to acclimation and then dispersed again as adults; the dispersal of zygotes determined the developmental environment, but the dispersal of adults determined the selective environment (as in Scheiner 1998). Under these conditions, the reaction norm that evolved in each population differed from the global reaction norm (Fig. 5.17). In the third case, zygotes developed in the same environment as their parents and then dispersed as adults. Again, dispersal after

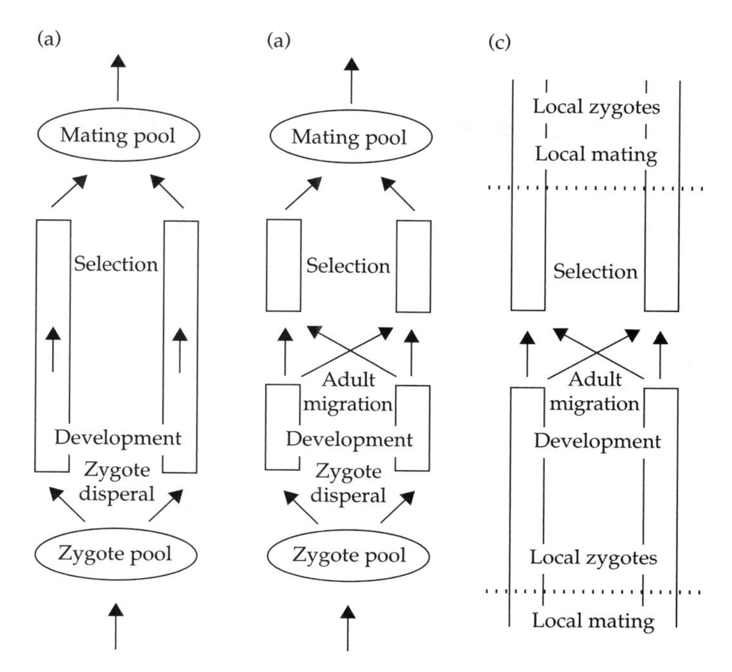

Figure 5.16 Three distinct life cycles used to model selection and migration. (a) Random dispersal of zygotes, development and selection in the same habitat, and random mating among habitats. (b) Random dispersal of zygotes, development to adulthood, migration among habitats, selection in the adult habitat, and random mating among habitats. (c) No dispersal of zygotes, development to adulthood, migration among habitats, selection in the adult habitat, random mating within each habitat. Reproduced from de Jong (1999) with permission from Blackwell Publishing.

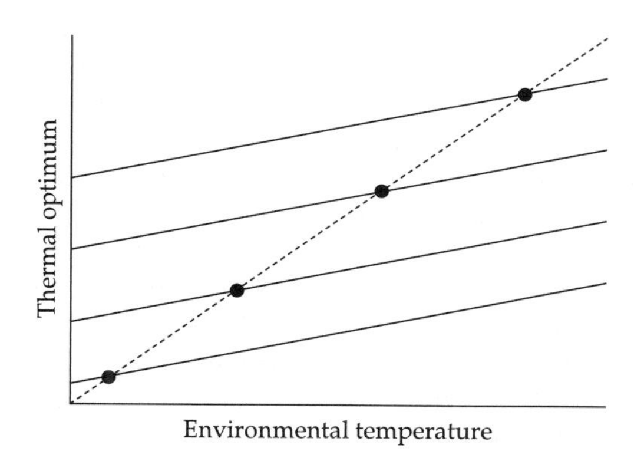

Figure 5.17 When the dispersal of zygotes and the dispersal of adults disconnect the developmental environment from the selective environment (Fig. 5.16b), the evolved reaction norms (solid lines) deviate from the globally optimal reaction norm (dashed line). A circle marks the temperature of the local environment in which each reaction norm evolved. Each reaction norm intersects the globally optimal reaction norm at the temperature of the genotype's local environment. Adapted from de Jong (1999) with permission from Blackwell Publishing.

development reduced the predictability of the selective environment. And once again, reaction norms deviated from the global optimum. Because these results hold only under weak selection, we must assume the performances breadth exceeds a certain width to apply this model to the evolution of thermal acclimation.

Gene flow also affects the evolution of acclimation when temporal heterogeneity varies geographically. Imagine a species consisting of populations distributed along a latitudinal cline. Without gene flow, the optimal strategy of developmental or reversible acclimation depends on the thermal heterogeneity experienced by a population, not that experienced by the species as a whole (see Section 5.4). To apply current optimality models to metapopulations, we must assume individuals in all localities experience the same temporal variation in temperature. Despite this simplifying assumption, seasonality clearly differs radically across the range of a widespread species. Based on intuition, the optimal strategy of acclimation should depend on both geographic variation in seasonality and the pattern of migration among sites. Unfortunately, no one has formally modeled this phenomenon. By generalizing the models discussed in Section 3.9, we could gain valuable insights about the effects of gene flow on the evolution of thermal acclimation.

5.7 Toward ecological relevance

In part, our poor understanding of thermal acclimation stems from a common disregard for the complexity of natural environments. Consider Gabriel's model of optimal acclimation (Section 5.4.2). In this model, the mean environmental temperature switches abruptly between two states. Moreover, consider the typical experimental design for the study of acclimation (see, for example, Fig. 5.1), in which a researcher exposes organisms to an abrupt shift from one constant temperature to another and subsequently records changes in thermal physiology. This standard design ignores natural variation in environmental temperature. In contrast to theoretical and experimental environments, natural environments change both continuously and unpredictably.

Continuous variation in temperature ameliorates the loss of performance during time lags. For example, seasonal change occurs gradually and follows a predictable cycle. Although heat waves and cold snaps add a stochastic element to this cycle, variation among seasons generally exceeds variation within seasons. Two conclusions must follow. First, current models overestimate the loss of performance during acclimation. In other words, the gradual change of natural environments affords a gradual response by an organism. Under gradual thermal change, a given time lag would prove less costly than it would under abrupt thermal change. Second, experimenters have generally exposed organisms to unnatural environmental cues for acclimation. When one manipulates temperature while holding other environmental variables constant, one breaks the natural covariation that organisms rely on to trigger beneficial acclimation. But this natural covariation plays an important role during acclimation. As mentioned above, Sorenson and Loeschcke (2002) showed that flies (*D. buzzatii*) used the diel cycle of light to coordinate expression of heat-shock proteins with thermal stress. In this species, light truly supersedes temperature as a cue. Similarly, photoperiod serves as a more reliable cue for seasonal changes in temperature, given the stochasticity of environmental temperature within seasons (see Chapter 2). Because patterns of thermal acclimation depend on other environmental variables (e.g., Hernandez et al. 2006; Licht 1968), the separation of thermal and photoperiodic cues during an experiment could introduce artifacts in the data, leading to erroneous conclusions about adaptation.

To conduct ecologically relevant studies of thermal acclimation, researchers must mimic the variability and stochasticity of natural environments. Only two experiments have included such treatments. Schaefer and Ryan (2006) raised zebra fish (*Danio rerio*) in constant, variable, and stochastic environments and then compared their critical thermal maxima. Mean temperature was either 24°C or 28°C, with a daily range of 0°C, 4°C, or 6°C. Not surprisingly, fish from hotter or more variable environments tolerated higher temperatures. However, fish that experienced stochastic variation tolerated heat as well as those that experienced predictable variation. This result suggests that the absolute range of temperatures during development influences acclimation more than the predictability of this range. In another experiment, Niehaus and colleagues exposed field crickets (*Gryllus pennsylvanicus*) to predictable and stochastic changes in temperature (A. C. Niehaus, R. S. Wilson, J. J. Storm, and M. J. Angilletta; unpublished observations). Temperatures of the predictable treatments decreased linearly over time, while those of the stochastic treatment decreased chaotically. The stochastic treatment mimicked the seasonal decline in the temperature of the natural environment (Fig. 5.18). Interestingly, thermal sensitivities of feeding and locomotion failed to acclimate during the experiment, regardless of the thermal treatment. Possibly, the gradual change in temperature provided less information about the benefits of acclimation than an abrupt change in temperature would have provided. Unfortunately, the photoperiod remained constant during the course of the experiment, eliminating a more reliable cue for the seasonal decline in temperature. Although simulating stochastic environments

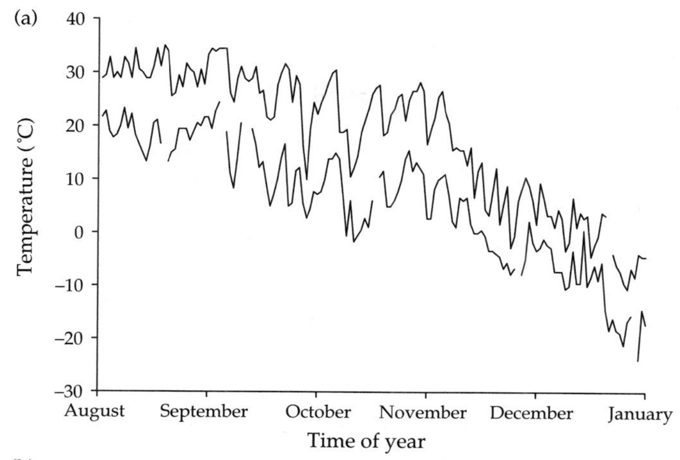

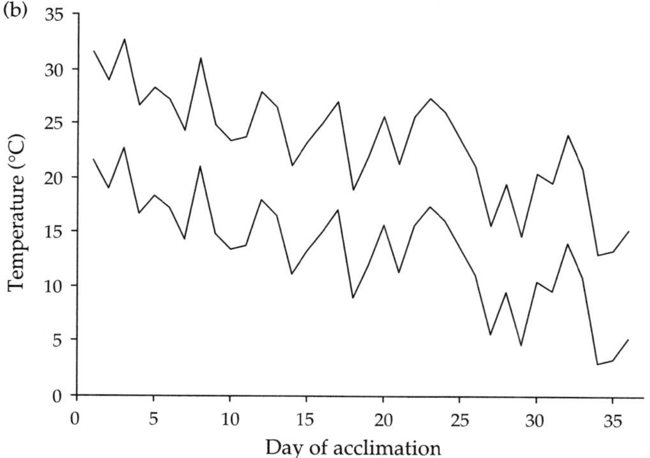

Figure 5.18 Natural temperatures experienced by field crickets (a) were used to design a stochastic thermal treatment (b) in an acclimation study (A.C. Niehaus, R.S. Wilson, J.J. Storm, and M.J. Angilletta, unpublished observations). In both plots, upper and lower lines represent maximal and minimal temperatures for each day. Data for natural temperatures were sampled from the year 2000 in Terre Haute, IN, USA.

requires more effort than simulating constant ones, these kinds of experiments will likely yield new insights and raise important questions.

5.8 Conclusions

A coherent theory of thermal acclimation only recently emerged from efforts to understand the evolution of phenotypic plasticity. This theory consists primarily of optimality models of developmental or reversible acclimation in fluctuating environments. These models represent logical extensions of Lynch and Gabriel's classic model of the evolution of performance curves, discussed in Chapter 3. While Lynch and Gabriel's model has had some success in explaining thermal sensitivities, extensions of this model have had very little success in explaining the acclimation of thermal sensitivities. Contrary to the theory, acclimation of the thermal optimum rarely occurs in laboratory experiments. Additionally, a genotype's capacity for acclimation rarely correlates with the magnitude or predictability of thermal heterogeneity in its natural environment. Rather, acclimation usually produces variation in performance curves within populations that resembles that produced by evolutionary divergence among populations. As with the evolution of thermal sensitivity, thermal acclimation can lead to phenotypes best described as masters of all temperatures. From this observation, we can draw two conclusions: (1)

some factor, not considered by the theory, limits the evolution of thermal acclimation and (2) constraints on thermal acclimation have resulted in optimal temperatures for development.

The failure of recent optimality models to account for the observed patterns of acclimation likely reflects several problems. First, these models hinge on the assumption that specialist–generalist trade-offs impose a universal constraint on evolution. We have already discussed the danger of this view in Chapter 3. Second, these models assume sufficient genetic variation for adaptation, yet we know virtually nothing about the genetic variances and covariances of acclimation responses. By reflecting back on the previous chapters, we should realize that optimality models have helped us to understand thermoregulatory behaviors far more than they have helped us to understand thermal sensitivities and their acclimation; the differential success of optimality models might stem from the relative strength of genetic constraints on behavior and physiology. Finally, the theory should ultimately be extended to deal with the complexity of natural environments, such as the interaction between gradual seasonal changes in temperature and sudden stochastic changes. Even with these shortcomings, current optimality models provide a major advance over verbal models such as the beneficial acclimation hypothesis. These quantitative models provide a clear statement of assumptions leading to objective and testable predictions about the evolution of acclimation in a wide range of environments. The explicit nature of their assumptions enables one to generalize these models and advance the theory.

CHAPTER 6

Temperature and the Life History

6.1 The link between performance and the life history

In Chapter 3, we needed to translate performance into fitness to predict the evolution of thermal sensitivity. In doing so, we assumed a semelparous organism (i.e., reproducing only once before death), in which either survivorship or fecundity scaled proportional to performance. This choice of life history, whether explicit or not, imposes a particular fitness function. For example, Lynch and Gabriel (1987) used the following function to define fitness (w) when they modeled optimal performance curves:

$$w = l(\alpha)m(\alpha), \tag{6.1}$$

where α equals the age at maturity, $l(\alpha)$ equals the probability of surviving to maturation, and $m(\alpha)$ equals the fecundity at maturation. Performance (P) affected fitness by determining the survivorship:

$$l(\alpha) = \prod_{x=0}^{\alpha} P(x). \tag{6.2}$$

Alternatively, Lynch and Gabriel might have chosen a fitness function that enables iteroparous reproduction, such as Euler's equation:

$$1 = \int_{x=\alpha}^{\infty} e^{-rx} l(x)m(x)\, dx. \tag{6.3}$$

In this case, the Malthusian parameter (r) would have represented the spread of a genotype in the population, which is a reasonable estimate of fitness in a growing population (see Box 3.3).

No biological consideration justified the use of the simpler function. In fact, by choosing eqn 6.1, Lynch and Gabriel constrained the life history considerably. Still, if they had chosen eqn 6.3, they would have needed to specify the age at maturity and the age-specific fecundities. In other words, they would still have constrained the life history substantially. Modeling thermal adaptation of behavior or physiology for specific life histories serves an important purpose: it yields theoretical insights while keeping the mathematics tractable. But we could just as easily wonder how traits such as age at maturity or age-specific fecundity evolve in response to thermal heterogeneity. In this chapter, we shall flip the coin and ask the following: assuming a specific performance curve, how would natural selection shape the life history?

Thermal heterogeneity favors the evolution of plastic life histories. In Chapter 3, we saw that temperature sets the capacity for physiological performance. Performances such as energy assimilation constrain the evolution of life-history traits, including the age and the size at maturity, the frequency and magnitude of reproduction, and the size and the number of offspring. These life-history traits define the parameters of the fitness function—$l(x)$ and $m(x)$. Therefore, reaction norms for life-history traits should evolve as a consequence of the thermal sensitivity of performance. We shall focus on two fundamental questions about the evolution of life-histories. First, how do thermal sensitivities of growth and survivorship affect the optimal reaction norms for age and size at maturity? Second, should organisms in cold environments make many small offspring or a few large ones, relative to organisms in warm environments? To answer each of these questions, we can draw on a rich store of models that has accumulated in recent decades. These models will help us to understand the thermal adaptation of life-history traits in ectotherms.

To understand thermal adaptation of the life-history, we must consider tradeoffs that constrain the allocation of energy. Indeed, such tradeoffs have formed the core of life-history theory since its inception (reviewed by Roff 2002; Stearns 1992). Individuals allocate their energy to the competing demands of maintenance, growth, and reproduction. Maturation generally marks a pronounced shift in the allocation of energy from maintenance or growth to reproduction. Presumably, natural selection shapes the relative investment in each of these functions, and these allocations produce the life-history traits that we observe (Dunham et al. 1989). In practice, theorists rarely focus on the entire life history at once because the mathematics would become very complex. Rather, they partition the life history into interesting yet manageable chunks. For instance, if we wanted to model the evolution age and size at maturity, we would likely focus on the tradeoff between growth and reproduction. The allocation of energy to growth would benefit future reproduction at the expense of current reproduction. Alternatively, if we wanted to understand the evolution of reproductive traits, we should focus on the tradeoff between the size and number of offspring. The allocation of energy to each offspring reduces the potential number of offspring. In either case, the optimal phenotype emerges from the tradeoffs that dictate the phenotypic function (see Chapter 1). Temperature modulates the fitness consequences of these tradeoffs through its influence on performance. This theme permeates the remainder of this chapter.

6.2 General patterns of age and size at maturity

6.2.1 Thermal plasticity of age and size at maturity

At least a decade before the proliferation of life-history theory, Ray (1960) discovered that many ectotherms respond similarly to variation in their developmental temperature. By raising 13 species of plants and animals at different temperatures, he observed the same pattern in almost every species: individuals at low temperatures grew relatively slowly, but delayed maturation long enough to outgrow individuals at high temperatures. More than 30 years later, Atkinson further generalized Ray's

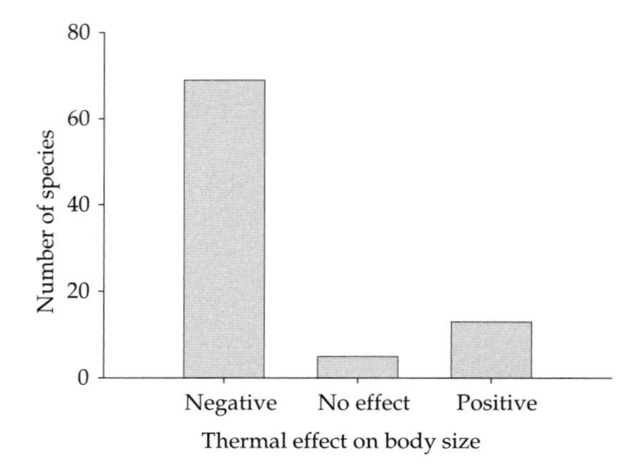

Figure 6.1 The majority of ectothermic species studied in the laboratory attained larger sizes when raised at lower temperatures. Data were summarized by Atkinson (1994).

observation. In a review of 109 experiments, Atkinson (1994) concluded Ray's relationship between temperature and size was strikingly widespread; 80% of species[12]—including animals, plants, protists, and one bacterium—reached larger sizes at lower temperatures (Fig. 6.1). Atkinson went as far as to suggest a "near-universal relationship," which he later called the temperature–size rule (Atkinson 1996a). Subsequent experiments have supported the generality of this relationship (Fischer and Fiedler 2000, 2001a; Møller et al. 1989; Stillwell and Fox 2005), despite the discovery of additional exceptions (e.g., De Block and Stoks 2003). Most recently, Atkinson and colleagues (2003) confirmed the generality of the temperature–size rule among protists. Without a doubt, the temperature–size rule ranks among the most taxonomically widespread "rules" in biology.

[12] Although Atkinson reviewed 109 experiments, some of these experiments involved the same species (e.g., he included seven studies of *Drosophila melanogaster*). Treating these experiments as independent would ignore the evolutionary history shared by populations of a species (see Chapter 1 and Box 6.1). Therefore, I collapsed these results into a single result for each species, as the results of these experiments generally agreed with one another. For two species, a negative effect of temperature was observed in one or more experiments and mixed effects were observed in another experiment. For these species, I assumed some investigators failed to raise organisms over a wide enough range of temperature to detect a maximal body size.

Despite the tendency described as the temperature–size rule, reaction norms for stage-specific size do vary within and among populations. Two degrees of variation call for our attention. First, the strength of the relationship between temperature and size varies among genotypes, even when all genotypes follow the temperature–size rule. This phenomenon has been documented among isogenetic lines of drosophilids (Loeschcke et al. 1999) and among distinct populations of other insects (Stillwell and Fox 2005). Second, the very nature of the temperature–size relationship can differ among genotypes. For example, three populations of snails diverged in their reaction norms for size at maturity; genotypes in one population followed the temperature–size rule while those in the other populations did not respond to developmental temperature (Dybdahl and Kane 2005). A similar divergence occurred between two populations of butterflies in less than 150 years (Fig. 6.2).

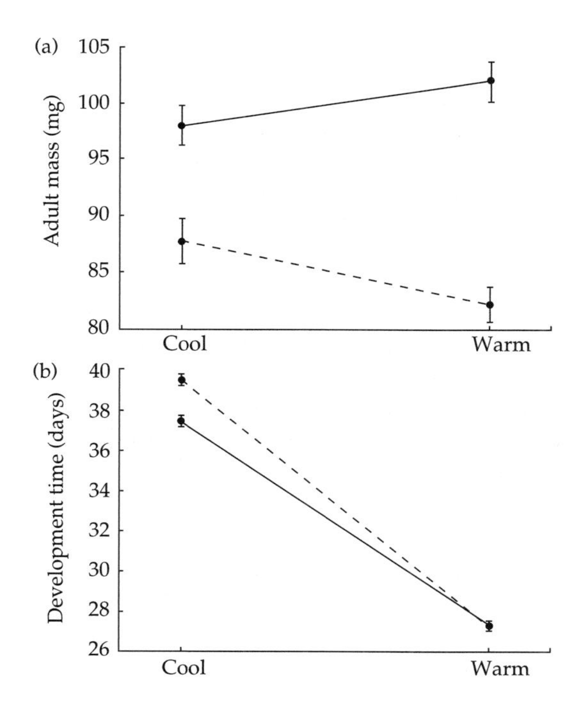

Figure 6.2 Thermal reaction norms for adult mass (a) and development time (b) diverged between populations of butterflies (*Pieris rapae*) from North Carolina (solid line) and Washington (dashed line) Error bars represent standard errors. Adapted from Kingsolver (2007) with permission from Blackwell Publishing.

Likely, a nonlinear function better describes the true form of these thermal reaction norms, such that smaller size results from development at extremely high and low temperatures. For example, several species of drosophilids attained their largest size when developing between 15°C and 20°C (David et al. 2006; Petavy et al. 1997; Yadav and Singh 2005). Similarly, *Daphnia pulex* attained its largest length at maturity when developing at an intermediate temperature (Dufresne and Hebert 1998). Moreover, the effect of developmental temperature on the life history depends on other environmental conditions, such as the abundance of food or the risk of predation (Gebhardt and Stearns 1993; Newman 1998; Weetman and Atkinson 2002). Despite these exceptions, the temperature–size rule still describes the thermal plasticity of many genotypes over a wide range of temperatures. Thus, even when reaction norms appear nonlinear, we should wonder why organisms that grow very rapidly would cease growing at a relatively small size.

6.2.2 Thermal clines in age and size at maturity

Species distributed over broad geographic ranges often exhibit thermal clines in body size. Recent phylogenetic comparative analyses indicate the majority of tetrapods reach larger adult sizes in colder environments (Box 6.1), a phenomenon widely referred to as Bergmann's rule (Blackburn et al. 1999). The patterns described by Bergmann's rule parallel those described by the temperature–size rule. Although Bergmann originally considered variation in size among endotherms, the size clines that bear his name seem to occur in many ectotherms as well. Among tetrapods, the majority of amphibians and turtles exhibit Bergmann's clines along latitudinal or altitudinal gradients (Ashton 2002a; Ashton and Feldman 2003; Measey and Van Dongen 2006; Morrison and Hero 2003b), whereas most squamates exhibit opposing clines (i.e., smaller sizes in colder environments). Among fishes, Bergmann's clines rarely occur in freshwater species, but occur commonly in anadromous and marine species (Belk and Houston 2002; Blanckenhorn et al. 2006; Gilligan 1991).

Box 6.1 Phylogenetic comparative analyses of Bergmann's clines

How did researchers assess the generality of Bergmann's clines? In general, this task requires a meta-analysis, or an analysis of analyses (see Gurevitch and Hedges 1993). Two meta-analytical approaches were taken: vote counting and correlational analysis. For the vote-counting approach, researchers treated a size cline as discrete trait, assigning a "plus" to all species exhibiting a Bergmann's cline and a "minus" to all other species. Then, they used a statistical test to see whether the proportion of species with a Bergmann's cline exceeded the expected proportion of 0.5. When vote counting, some investigators counted only statistically significant Bergmann's clines (e.g., Meiri and Dayan 2003) while others counted all trends, regardless of statistical significance (e.g., Ashton et al. 2000). For the correlational approach, researchers calculated the coefficient of the correlation between latitude (or temperature) and size for each species. Then, these coefficients were used to estimate a mean latitudinal (or thermal) effect, weighted for the sample size of each species (see Gurevitch and Hedges 1993).

With either approach, one should worry about pseudoreplication resulting from the evolutionary relationships among species (see Chapter 1). This problem compromises a statistical inference when the data possess a strong phylogenetic signal. Some novel efforts have been made to detect phylogenetic signal or account for its influence on Bergmann's clines. Ashton (2002b) computed independent contrasts of correlation coefficients and their associated variance. Then, he randomized correlation coefficients within the phylogeny and recalculated the variance of the contrasts; repeating this procedure 1000 times resulted in a null distribution for his test statistic. Because the real variances always fell well within the distribution of simulated variances, Ashton ruled out the presence of a strong phylogenetic signal. To control for phylogeny during meta-analysis, Ashton and Feldman (2003) simulated the evolution of correlation coefficients within a phylogeny, assuming equal branch lengths and an evolutionary process equivalent to Brownian motion. For each simulation, they calculated the proportion of extant species that followed Bergmann's rule (vote-counting approach) and the mean correlation coefficient (correlational approach). After 1000 simulations, they asked whether the observed latitudinal effect was more extreme than the expected latitudinal effect caused by their simulated evolution. Keep in mind, these analyses focused on the evolutionary relationships among species but ignored the relationships among populations within species. Addressing the latter issue requires a phylogenetic hypothesis for each species (see, for example, Angilletta et al. 2004b).

Phylogenetic analyses of Bergmann's clines appear robust to several potentially confounding variables, such as the mean body sizes, mean latitudes, or latitudinal ranges of the species. To check for systematic bias in the data, Ashton (2004) computed phylogenetic contrasts of the correlation coefficients and regressed these contrasts against potentially confounding variables. The direction and magnitude of a size cline was unrelated to the mean body size, mean latitude, or latitudinal range. However, larger samples yielded stronger clines in body size, suggesting that statistical power limited the resolution of size clines for some species. Overall, these analyses underscore the reality of size clines and justify our search for adaptive explanations.

de Quieroz and Ashton (2004) concluded that a Bergmann's cline was a deeply ancestral characteristic of tetrapods. They came to this conclusion by mapping Bergmann's clines onto a phylogenetic hypothesis for the Tetrapoda. Although de Quieroz and Ashton's argument seems interesting, I am skeptical of their interpretation. The problem lies in equating a Bergmann's cline with a heritable trait. An organism cannot pass on a geographic cline to its offspring; it can only pass on genes that affect a thermal reaction norm. By mapping size clines as traits, de Quieroz and Ashton implicitly assumed each size cline reflected a reaction norm, passed from ancestor to descendant. In which case, the size clines must have resulted from thermal plasticity rather than genetic divergence among populations. But if size clines resulted from genetic divergence among populations, this genetic structure would not necessarily pass to descendents during speciation. Because body size adapts readily to local environments (Fig. 6.2; Partridge and Coyne 1997), this logical flaw likely undermines the main conclusion reached by de Quieroz and Ashton.

In contrast to vertebrates, arthropods seem to violate Bergmann's rule more than they follow it, and sometimes they exhibit more complex patterns. In a recent survey of latitudinal clines, only nine of 33 species exhibited larger sizes at higher latitudes (Blanckenhorn and Demont 2004). A survey of altitudinal clines in insects generated more equivocal results: eight species followed Bergmann's rule, 14 species opposed the rule, 11 species showed no pattern, and two species showed a complex pattern (Dillon et al. 2006). Complex relationships between latitude and size often resemble a saw-tooth pattern, such as the one seen in crickets (Fig. 6.3); complex relationships of a slightly different form have arisen in some vertebrates, such as the common frog (*Rana temporaria*) in Scandinavia (Laugen et al. 2005). Granted, we have no idea what size clines look like for the vast majority of arthropods, but available evidence indicates that geographic clines in this group contrast the thermal plasticity observed in the laboratory (see Section 6.2.1).

Obviously, geographic clines stem from factors other than thermal plasticity. In the eastern fence lizard (*Sceloporus undulatus*), females from cold environments mature at larger sizes than females from warm environments; nevertheless, body size appears more closely related to latitude than to temperature (Angilletta et al. 2004b). As with most phenotypic phenomena, geographic clines likely depend on the interplay of genetic and environmental factors. In some species, we know that genetic factors contribute to size clines because variation among populations persisted in common garden experiments (Blanckenhorn and Demont 2004; Partridge and Coyne 1997; Schutze and Clarke 2008). Natural selection seems a likely cause of genetic divergence along thermal clines because the direction of genetic differences often parallels the direction of thermal plasticity (i.e., animals from colder environments attain larger sizes). Environmental factors can also enhance or diminish thermal clines in body size. This point was nicely illustrated by Warren and colleagues (2006), who studied *Drosophila* in an outdoor mesocosm. Hot regions of the environment caused greater mortality of larvae, which in turn reduced competition among surviving larvae. The direct effect of temperature on development and the indirect effect on competition counteracted one another perfectly; larvae in hot microenvironments grew to the same size as those in cold microenvironments. Given the multitude of factors that vary geographically, it is no wonder ectotherms more commonly follow the temperature–size rule in artificial environments than they do in natural environments.

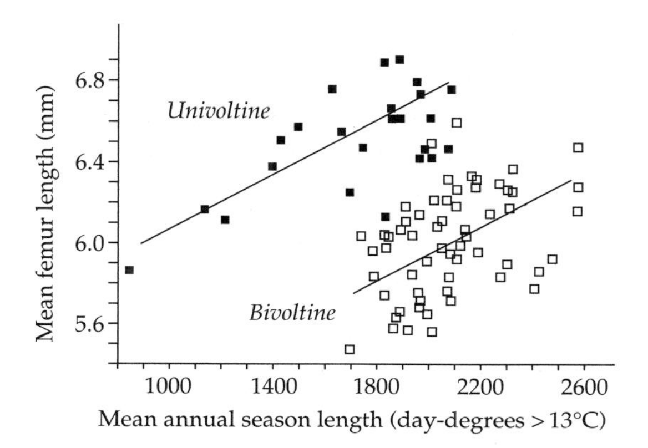

Figure 6.3 Mean body sizes of ground crickets increased with increasing season length over portions of the range, but decreased sharply at the thermal boundary between populations of univoltine and bivoltine individuals. Adapted from Mousseau (1997) with permission from Blackwell Publishing.

6.2.3 Experimental evolution of age and size at maturity

If thermal reaction norms for age and size at maturity resulted from selection, similar patterns should evolve in the controlled environments of research laboratories. Two selection experiments have adressed this point. In the first experiment, Partridge and colleagues (1994) maintained replicated populations of *Drosophila melanogaster* at either 25°C or 16.5°C. After 5 years, populations at 16.5°C had evolved a larger mean size at maturity than had populations at 25°C. Santos and colleagues (2006) conducted a similar experiment with *Drosophila subobscura*, but obtained a different result. After 3.5 years, wing size had not diverged among populations maintained at 13°C, 18°C and 22°C; nevertheless, rapid development evolved within the lines kept at 18°C and 22°C. Thus, only age at maturity evolved in a way that parallels patterns of thermal plasticity. From these experiments, we might conclude that natural selection maintains the temperature–size rule, if we assume *D. subobscura* lacked sufficient genetic variation to respond to selection.

Although the selection experiments by Santos and colleagues failed to trigger the evolution of body size, an "experiment" in nature has done so. In the late 1970s, *D. subobscura* spread from Europe to North and South America. In Europe, this species tends to have larger wings in colder environments (Misra and Reeve 1964). Since invading the Americas, *D. subobscura* has evolved parallel clines in the wing sizes of females (Huey et al. 2000), although the cellular basis of clinal variation differs among continents (Gilchrist et al. 2004). Larger wings in cold environments might have evolved to reduce wing loading during flight or as a correlated response to selection for larger bodies; available data support the former hypothesis for European flies and the latter hypothesis for South American flies (Gilchrist and Huey 2004). The specific benefits of large size in a cold environment remain unclear for *D. subobscura*, but might be inferred from recent studies of *D. melanogaster*. Like its congener, *D. melanogaster* evolved Bergmann's clines on more than one continent. When raised in a common environment, both South American and Australian flies from high latitudes developed faster and matured at larger sizes than their counterparts from low latitudes (James et al. 1995, 1997). At least for this species, temperature appears to play a role in convergent evolution because selection experiments led to the evolution of similar variation in size (see above). Furthermore, researchers have compared components of fitness between flies adapted to low and high temperatures (Bochdanovits and de Jong 2003; McCabe and Partridge 1997; Reeve et al. 2000). Flies that evolved at low temperature enjoyed greater longevity and fecundity in a cold environment than did flies that evolved at high temperature. Interestingly, cold-adapted flies enjoyed no noticeable benefits in a warm environment.

The convergent evolution of body size during selection experiments and biological invasions favors an explanation based on natural selection over one based on genetic drift. Combined with the abundant evidence for the temperature–size rule, these size clines have motivated some researchers to seek simple explanations based on physical constraints (van der Have 2002). Nevertheless, the existence of genetic variation within and among populations demands an adaptive explanation for these patterns. With this in mind, let us now ponder the obvious question: why does natural selection in cold environments often favor delayed maturation at a large size?

6.3 Optimal reaction norms for age and size at maturity

Although thermal clines and thermal plasticity generally take on a common form, we can always identify exceptions (see Fig. 6.2). Hence the term "rule" must be taken with a grain of salt. Importantly, references to the temperature–size rule or Bergmann's rule focus attention on common phenomena, but provide no real explanations for these phenomena. We should not think that the temperature–size rule predicts or explains anything; it merely describes the existence of a pattern. To understand *why* this pattern exists, we must turn to quantitative models of evolution.

Optimality modeling provides a tool for understanding thermal reaction norms for age and size at maturity. Two very different types of models exist. The first type constrains the organism's rate of growth, such that body size follows a particular trajectory. The second type constrains the energy assimilated by an organism, such that body size results from the allocation of energy to growth. Both types of models share several key features. First, the models focus on tradeoffs between current and future fecundity, mediated by strategies of growth. Second, they assume temperature remains constant throughout the life of an organism; this assumption eliminates the need to worry about the asymmetry of performance curves. Finally, these models assume plasticity occurs without cost or constraint (i.e., complete plasticity, sensu Sibly and Atkinson 1994). In this section, we shall consider an example of each type of optimality model and explore the conditions under which it can explain the temperature–size rule. In doing so, we shall see that the optimal life history depends on the thermal sensitivities of juvenile survivorship, maximal size, and population growth.

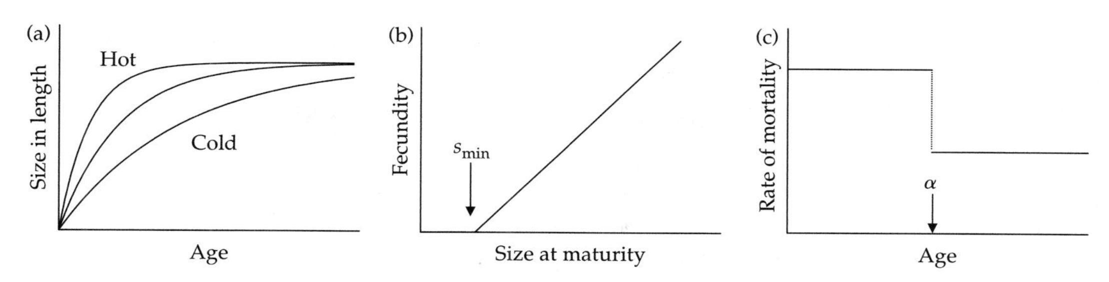

Figure 6.4 Hypothetical relationships used to model optimal reaction norms for age and size at maturity (based on Berrigan and Koella 1994). (a) Growth follows the Bertalanffy function, such that initial growth rate increases as the parameter k increases. (b) Fecundity increases linearly with increasing size once the organism reaches a minimal size (s_{min}). (c) The rate of mortality remains constant during the juvenile and adult stages, but can differ between stages; the example shown here depicts a greater rate of mortality for juveniles.

6.3.1 A comparison of two modeling approaches

Berrigan and Koella (1994) modeled optimal reaction norms relating growth rate to age and size at maturity. Because growth rate depends on temperature, their model nicely illustrates the selective pressures that thermal heterogeneity can impose on the life history.

Berrigan and Koella started by defining the fitness of an organism as the net reproductive rate:

$$R_0 = \int_{\alpha}^{\infty} l(x)m(x)\,dx. \tag{6.4}$$

In doing so, they assumed an asexual, iteroparous organism whose population remains stable during evolution (see Box 3.3). When density dependence acts early during ontogeny, eqn 6.4 also equals the reproductive value at birth (Kozłowski et al. 2004), a measure that many researchers consider an ideal estimate of fitness (Stearns 1992).

Their hypothetical organism grows according to an asymptotic function, first described by Bertalanffy (1960):

$$s(x) = s_{\infty}(1 - Be^{-kx}), \tag{6.5}$$

where $s(x)$ equals size in length at age x, s_{∞} equals the asymptotic size, B equals $1-s(0)/s_{\infty}$, and k determines the rate of growth between birth and maturation. For our purposes, assume temperature affects growth by changing the value of k. Our hypothetical relationship between temperature and k resembles a typical performance curve (see Chapter 3), such that a thermal optimum exists for k. For simplicity, we shall restrict our attention to an organism that ceases to grow at maturation; this restriction will not influence our qualitative predictions about reaction norms (Berrigan and Koella 1994).

Berrigan and Koella assumed vital rates depended directly on size or age (Fig. 6.4). Once the organism passes a minimal size (s_{min}), fecundity increases linearly with an increase in size at maturity:

$$m(\alpha) = f[s(\alpha) - s_{min}], \tag{6.6}$$

where f determines the effect of size on fecundity. Survivorship was calculated by assuming constant rates of mortality for juveniles (μ_j) and adults (μ_a). Under these strict assumptions, the net reproductive rate equals

$$R_0 = \int_{\alpha}^{\infty} l(x)m(x)dx$$

$$= e^{-\mu_j\alpha} \int_{\alpha}^{\infty} e^{-\mu_a(x-\alpha)} f\left[s_{\infty}\left(1 - Be^{-kx}\right) - s_{min}\right] dx. \tag{6.7}$$

By differentiating eqn 6.7 with respect to α and setting the resulting equation to zero, Berrigan and Koella found the optimal age at maturity (α^*):

$$\alpha^* = \frac{\ln\left[\frac{f s_{\infty} B(k+\mu_j)}{\mu_j f(s_{\infty} - s_{min})}\right]}{k}. \tag{6.8}$$

The optimal size at maturity was then calculated as follows:

$$s(\alpha^*) = s_{\infty}(1 - Be^{-k\alpha^*}) \tag{6.9}$$

Using eqns 6.8 and 6.9, Berrigan and Koella explored the effects of growth and survivorship on the optimal age and size at maturity. By solving the optimal ages and sizes for a range of parameter values, they constructed optimal reaction norms. In doing so, they implicitly assumed plasticity imposes no cost.

Instead of assuming a fixed growth trajectory, other modelers have determined the optimal age and size at maturity for organisms that allocate energy freely between growth and reproduction (reviewed by Kozłowski 1992; Kozłowski et al. 2004; Perrin and Sibly 1993). Although organisms also use energy for maintenance and activity, most researchers have determined how organisms should use energy for production after meeting other demands. Temperature influences the optimal life history by constraining the rate of assimilation and the cost of maintenance. Prior to maturation, an organism allocates 100% of its surplus energy to growth. The diversion of some energy to reproduction marks the onset of maturation; however, body size can continue to increase after maturation, producing a pattern referred to as indeterminate growth.

For comparison with the model of Berrigan and Koella, we shall focus on a simple model of optimal energy allocation described by Kozłowski and colleagues (2004). In this model, growth and fecundity are limited by the organism's rate of production (or P, since this is a measure of performance). Production depends on the rates of anabolism (A) and catabolism (C), measured in units of energy per time:

$$P = A(M) - C(M) = aM^c - bM^d, \qquad (6.10)$$

where M equals body size in units of energy, a and b determine rates of anabolism and catabolism, and c and d define the allometries of anabolism and catabolism. The organism grows and reproduces according to its allocation of surplus energy:

$$\frac{dM}{dx} = p(x)P(M) \qquad (6.11)$$

and

$$m(x) = \frac{(1 - p(x))P(M)}{n}, \qquad (6.12)$$

where $p(x)$ equals the proportion of energy allocated to growth at age x, and n equals the energy required to produce each offspring. As in Berrigan and Koella's model, let us assume the mortality rates of juveniles (μ_j) and adults (μ_a) remain constant (for optimal strategies under size-specific mortality, see Kozłowski et al. 2004).

An optimal age and size at maturity exist as long as the increase in production rate (dP/dM) slows during growth (Kozłowski et al. 2004; Perrin and Sibly 1993); this condition occurs when the allometry of catabolism exceeds the allometry of anabolism (i.e., $d > c$). In a constant environment, the optimal strategy is quite simple: allocate all surplus energy to growth ($p = 1$) up to a certain age, and allocate all surplus energy to reproduction ($p = 0$) thereafter. But how big should an organism grow before switching to reproduction? An organism reaches its optimal size at maturity when the energetic benefit of delaying maturation equals the survival cost:

$$\frac{dP(M)}{dM} = \mu_j \qquad (6.13)$$

Using this result, Kozłowski and colleagues could solve for the optimal age and size at maturity over a range of environmental conditions. By linking together the optimal solutions, they arrived at optimal reaction norms for age and size at maturity. Like Berrigan and Koella, Kozłowski and colleagues assumed plasticity imposes no cost.

6.3.2 Thermal effects on juvenile mortality

Using the models described above, let us consider the optimal reaction norms for three scenarios: (1) juvenile mortality remains constant as the growth rate increases, (2) juvenile mortality decreases as the growth rate increases, and (3) juvenile mortality increases as the growth rate increases. Each of these scenarios corresponds to particular assumptions about the thermal performance curves for survivorship and growth (Fig. 6.5). Which of these scenarios (if any) might explain the temperature–size rule? For Berrigan and Koella's model, we answer this question by making juvenile mortality (μ_j) a function of the growth parameter (k). When juvenile mortality remains constant, an increase in the growth parameter favors earlier maturation at a *larger* size (Fig. 6.6a), which differs from the pattern observed in most ectotherms (see Section 6.2). When juvenile mortality decreases as the growth

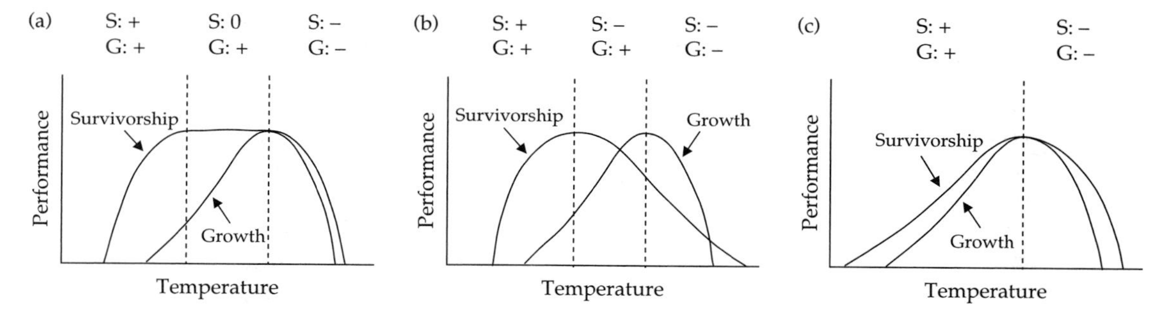

Figure 6.5 Hypothetical thermal sensitivities of survivorship (S) and growth (G) depict certain assumptions in the context of thermal performance curves. Survivorship can either increase (+) or decrease (−) with increasing growth rate. The dashed vertical lines mark temperatures at which the covariation between survivorship and growth changes qualitatively.

parameter increases, the prediction does not differ qualitatively from the first scenario; in fact, an increase in both growth rate and survivorship should lead to an even larger size at maturity than would an increase in growth rate alone (Fig. 6.6b). Either scenario 1 or 2 might explain exceptions to the temperature–size rule, but scenario 3 predicts a pattern that follows this rule. When an increase in the growth parameter causes a substantial increase in juvenile mortality, an organism should mature earlier at a *smaller* size (Fig. 6.6c). Relaxing assumptions about growth and fecundity does not alter the main conclusion: higher temperatures must increase juvenile mortality for the optimal reaction norm to follow the temperature–size rule (Sibly and Atkinson 1994). The model by Kozłowski and colleagues can also explain the temperature–size rule when high temperatures simultaneously raise rates of production and mortality (Fig. 6.7). For both models, the cost of greater mortality at higher temperatures must overwhelm the benefit of fecundity associated with rapid growth; otherwise, higher temperatures should lead to delayed maturation at a larger size.

Although the two models make similar predictions, the hypothetical mechanism differs greatly (Fig. 6.8). In Berrigan and Koella's model, the growth of an organism must decelerate with age, even though production clearly accelerates with age (see Fig. 6.4). In Kozłowski et al.'s model, a change in the capacity for production with age favors the diversion of energy from growth to reproduction. Despite very different assumptions about energetics, both models identify a plausible explanation for the temperature–size rule: higher temperatures reduce the survivorship of juveniles. A similar result emerges when higher temperatures limit the lifespan of an organism (Kozłowski et al. 2004; Perrin and Rubin 1990). Still, do higher temperatures actually decrease the survivorship of real organisms? And, if so, is the thermal sensitivity of survivorship sufficient to explain the temperature–size rule?

For a simple model of optimal allocation, my colleagues and I calculated the minimal thermal sensitivity of survivorship needed to explain the temperature–size rule (Angilletta et al. 2004a). This thermal sensitivity depends on two aspects of an organism's energetics: (1) the thermal sensitivities of anabolism and catabolism (Q_{10}s of a and b, respectively) and (2) the allometries of anabolism and catabolism (c and d, respectively). For a realistic range of parameters, we asked whether observed thermal sensitivities of survivorship exceed the minimal values required by the model. Two kinds of data bear on this issue. Data from artificial environments include physiological sources of mortality but exclude ecological sources (e.g., predation). Data from natural environments include both sources, but mortality results from processes occurring at fluctuating temperatures. Based on more than 100 experimental studies in artificial environments, we concluded temperature has no general physiological effect on survivorship. Even when a higher temperature decreased survivorship, the effect was well below that required by the model (Fig. 6.9). Comparative studies in natural environments generally

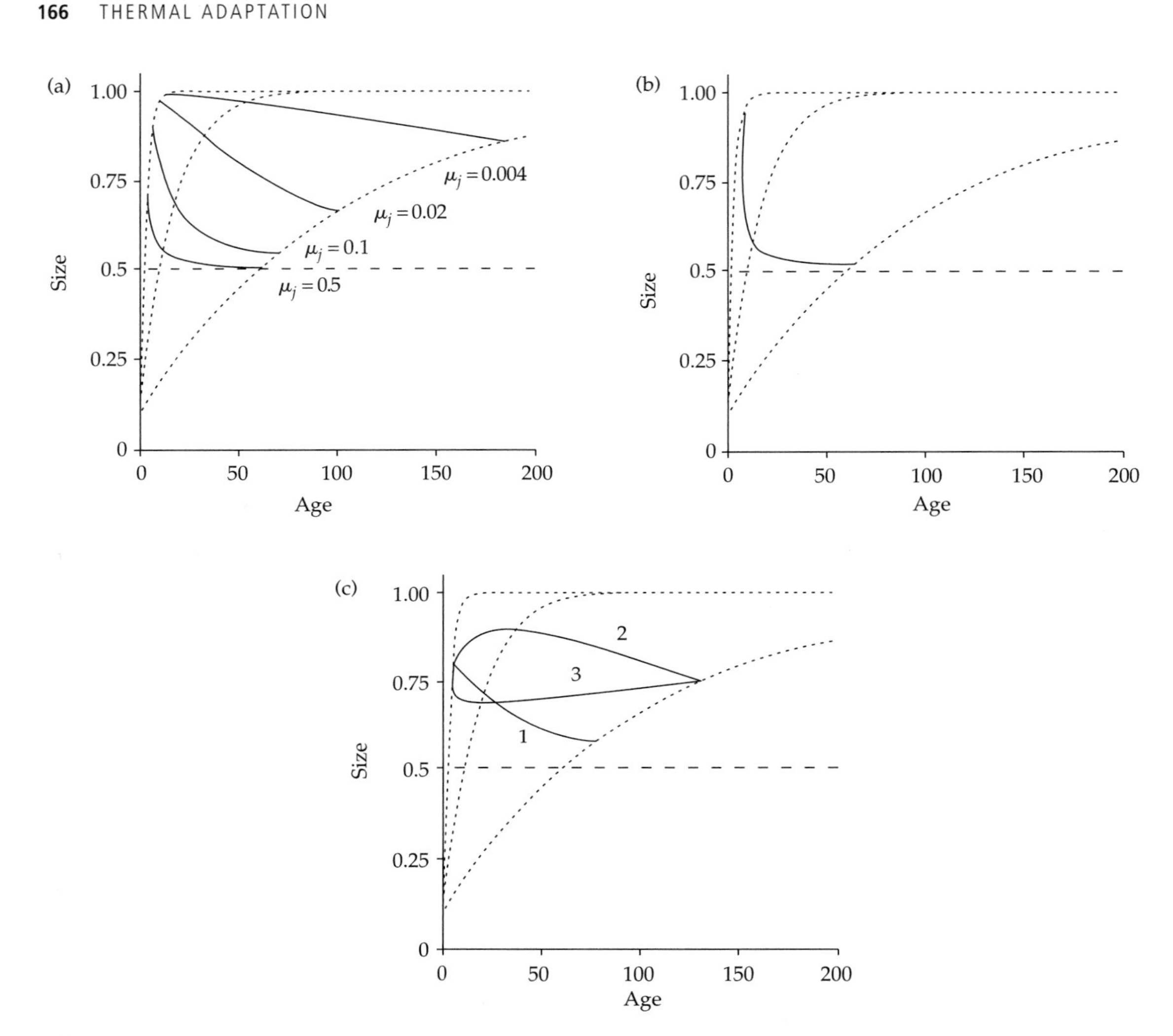

Figure 6.6 Optimal reaction norms for age and size at maturity depend on the covariation between growth rate and juvenile mortality. These plots show optimal reaction norms for three patterns of covariation: (a) juvenile mortality remains constant as growth rate increases, (b) juvenile mortality decreases as growth rate increases, and (c) juvenile mortality increases as growth rate increases. In each plot, the dotted lines depict growth curves and the solid line defines the corresponding optimal reaction norms for age and size at maturity in response to variation in growth rate. The dashed line marks the minimal size at maturity. In (a), reaction norms for several rates of juvenile mortality (μ_j) are shown. In (c), optimal reaction norms for three different relationships between growth rate and juvenile mortality are shown (1–3). The optimal reaction norm follows the temperature–size rule only when juvenile mortality increases appreciably with increasing growth rate (c, line 3). Adapted from Berrigan and Koellas (1994) with permission from Blackwell Publishing.

support the idea that higher temperatures decrease the survivorship of animals; however, the thermal sensitivities still fell short of the critical values (reviewed by Angilletta et al. 2004a). Taken together, these data suggest a complete explanation for the temperature–size rule will require at least one additional mechanism.

6.3.3 Thermal constraints on maximal body size

Physiological constraints can also generate optimal reaction norms consistent with the temperature–size rule (Atkinson 1996a; Berrigan and Charnov 1994; Kindlmann et al. 2001). Although warm

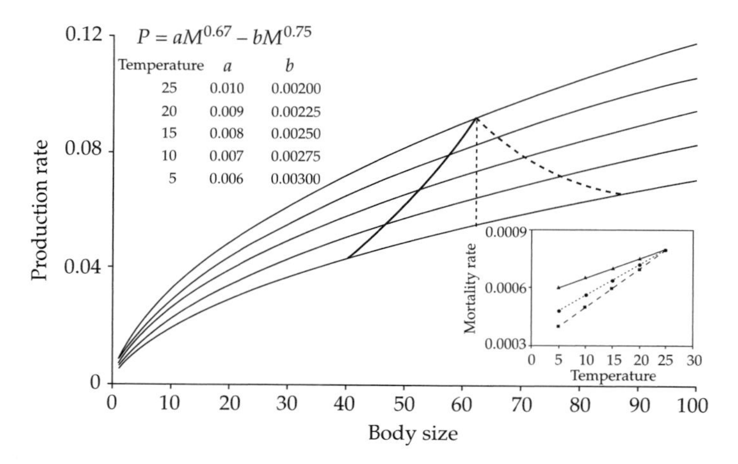

Figure 6.7 The optimal reaction norm might follow the temperature–size rule if high temperatures simultaneously raise rates of production and mortality. To illustrate this point, optimal reaction norms are shown for three different thermal sensitivities of mortality rate. The lines connecting the five production curves represent optimal reaction norms for size at maturity. In all cases, production occurs more rapidly at higher temperatures; parameter values used to calculate production rate (P) are listed in the upper left corner. Reproduced from Kozłowski et al. (2004) with permission from Oxford University Press.

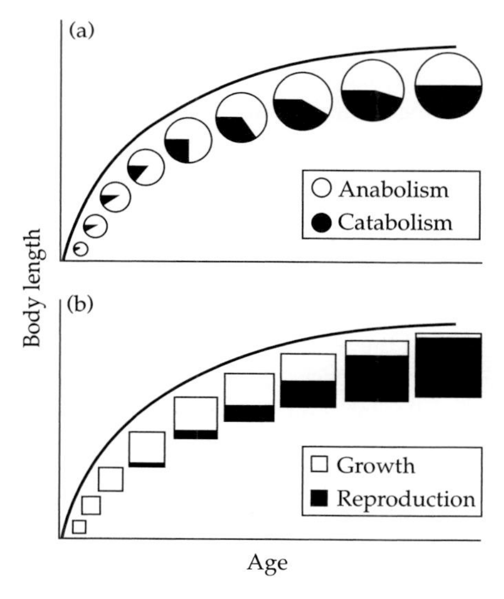

Figure 6.8 Models of optimal age and size at maturity differ fundamentally in their assumptions about energetics. (a) Berrigan and Koella (1994) assumed an organism allocates an increasing fraction of its energy to catabolism as it ages (represented by the pie diagrams). Consequently, growth must cease at some age because catabolism equals anabolism. (b) In contrast, Kozłowski and colleagues (2004) assumed an organism allocates an increasing fraction of its energy to reproduction as it ages. Consequently, growth ceases even though production continues to increase with age. Reproduced from Czarnołeski and Kozłowski (1998) with permission from Blackwell Publishing.

environments favor rapid growth early in life, they might limit growth later in life. Such constraints could arise when an organism must absorb some limiting resource, such as oxygen, through its body surface. When temperature constrains the maximal size of an organism, the benefit of delayed maturation disappears. When high temperatures constrain the maximal size more severely than low temperatures do, the optimal reaction norm conforms to the temperature–size rule.

In a model of energy allocation, we can impose thermal constraints on maximal size by making rates of anabolism and catabolism depend on temperature:

$$P = a(T)M^{c(T)} - b(T)M^{d(T)}. \tag{6.14}$$

The optimal life history follows the temperature size-rule under two conditions. First, temperature can affect catabolism more than it affects anabolism; specifically, the Q_{10} of b must exceed the Q_{10} of a (Perrin 1995). Second, temperature can affect the allometries of anabolism and catabolism in opposite ways; higher temperatures must decrease c while increasing d (Strong and Daborn 1980). In fact, both conditions can work together to favor smaller sizes at higher temperatures. Do these hypothetical

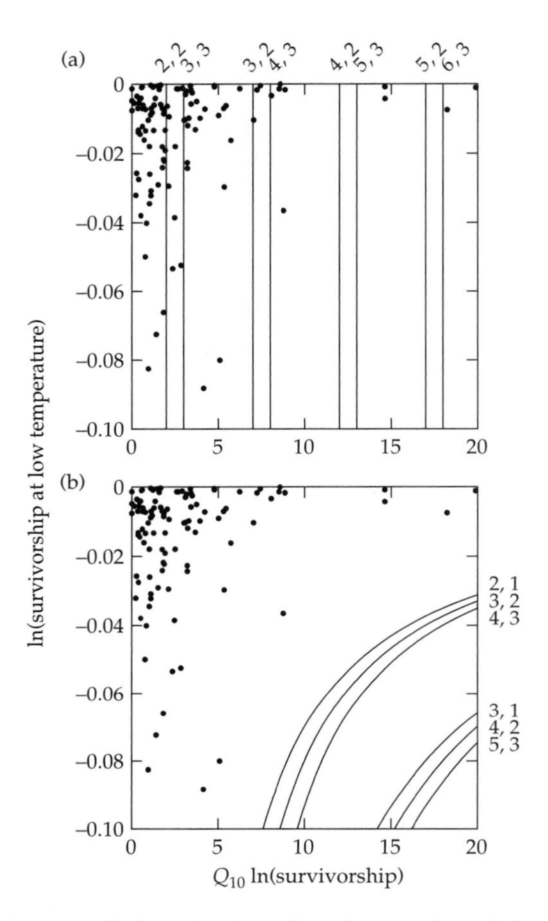

Figure 6.9 The thermal sensitivity of juvenile survivorship needed to explain the temperature–size rule depends on the rates of anabolism and catabolism. (a) The case in which the allometries of anabolism and catabolism are identical (i.e., $c = d$). (b) The case in which anabolism scales allometrically with body mass ($c < 1$) and catabolism scales isometrically ($d = 1$). The lines are isoclines for realistic combinations of thermal sensitivities of anabolism and catabolism, which are listed in the margin as [Q_{10} of anabolism, Q_{10} of catabolism]. To the right of each isocline, the optimal size at maturity decreases with increasing temperature in accord with the temperature–size rule. The points are thermal sensitivities of survivorship for 130 populations of 114 species of ectotherms. Reproduced from Angilletta et al. (2004a) with permission from Oxford University Press.

constraints on maximal size help to explain patterns of age and size at maturity?

Empirical evidence strongly refutes the hypothesis that temperature speeds catabolism more than anabolism. If this condition applies, organisms must grow less efficiently in warmer environments (Box 6.2). This prediction can be widely tested because physiologists have measured the thermal sensitivity of growth efficiency for many species of arthropods, mollusks, and fish. In a survey of more than 50 species, Angilletta and Dunham (2003) concluded that ectotherms usually grow more rapidly *and* more efficiently in warmer environments (Fig. 6.10). Even when data for a particular species do support this hypothesis, the range of temperatures over which this mechanism can explain the temperature–size rule rarely exceeds 5°C (but see Karl and Fischer 2008). Therefore, this explanation applies only to a narrow range of temperatures in the minority of ectotherms studied to date.

In contrast, available evidence strongly supports the hypothesis that allometries of anabolism and catabolism respond differently to temperature. In a study of isopods (*Idotea baltica*), ingestion scaled almost isometrically ($c = 0.94$) at a low temperature but allometrically ($c = 0.71$) at a high temperature, whereas respiration scaled allometrically ($d = 0.68$) at a low temperature and isometrically ($d = 1.00$) at a high temperature. Consequently, growth decelerated with age in the warm environment but accelerated with age in the cold environment. Similar trends have been documented for several other species (see Angilletta and Dunham 2003). Still, the true generality and physiological basis of this phenomenon remain unknown.

To predict the temperature–size rule, these models require production to decelerate as an organism ages. Kindlmann and colleagues (2001) assumed that production accelerated early in ontogeny but decelerated late in ontogeny because of senescence. Under these conditions, the growth of juveniles accelerates with age while the fecundity of adults decelerates with age. This senescence of production can result from a decrease in resource acquisition or an increase in required maintenance. The optimal age and size at maturity depends on the rate of senescence; if senescence proceeds more rapidly at higher temperatures, early maturation enables reproduction before senescence takes a major toll on fecundity. Kindlmann and colleagues claimed their model likely explains the general effect of temperature on body size. However, their conclusion depends on the validity of the Malthusian parameter

Box 6.2 Growth efficiency and the temperature-size rule

Recently, Dunham and I derived an implicit assumption of Perrin's (1995) explanation for the temperature–size rule: an organism must grow less efficiently in a warmer environment (Angilletta and Dunham 2003). We started by redefining Bertalanffy's equation in terms of growth efficiency. Bertalanffy (1960) defined the rates of anabolism and catabolism as $aM^{2/3}$ and bM^1, respectively. These rates determine what physiologists refer to as the net growth efficiency (K_2):

$$K_2 = \frac{A - C}{A} = \frac{aM^{2/3} - bM^1}{aM^{2/3}}. \quad (6.15)$$

To explore the influence of body temperature on the net growth efficiency, we made a and b functions of temperature, as assumed by Perrin (1995):

$$K_2 = \frac{a(T_b)M^{2/3} - b(T_b)M^1}{a(T_b)M^{2/3}}, \quad (6.16)$$

which simplifies to

$$K_2 = 1 - \frac{b(T_b)}{a(T_b)}M^{1/3}. \quad (6.17)$$

Taking the derivative of K_2 with respect to body temperature yields the following:

$$\frac{dK_2}{dT_b} = -M^{1/3}\left[\frac{b'(T_b)a(T_b) - b(T_b)a'(T_b)}{a(T_b)^2}\right]. \quad (6.18)$$

Note the derivative of K_2 with respect to body temperature is negative when the following condition is met:

$$\frac{a'(T_b)}{b'(T_b)} < \frac{a(T_b)}{b(T_b)}. \quad (6.19)$$

Perrin's model requires this condition by assuming $b'(T_b)$ exceeds $a'(T_b)$ and $a(T_b)$ exceeds $b(T_b)$ for a growing organism. Therefore, K_2 must decrease with increasing temperature if Perrin's model accurately describes growth.

as a criterion of fitness, the time-course of senescence, and the thermal sensitivities of assimilation, gonadogenesis, and senescence. As such, we cannot infer the generality of this hypothesis at the present time.

6.3.4 Thermai effects on population growth

A positive effect of temperature on the growth of a population can also favor the temperature–size rule. When density dependence prevents the growth of a population, the timing of reproduction has no direct bearing on the fitness of an organism. In other words, two offspring produced late in life benefit an organism as much as two offspring produced early in life. In a growing population, the situation differs greatly; all else being equal, an organism that reproduces early in life enjoys greater fitness than one that reproduces late in life (see Box 3.3). Therefore, early maturation at a small size might reflect a greater potential for population growth in warm environments (Sibly and Atkinson 1994). In laboratory experiments, the Malthusian parameter depends on temperature in much the same way that other performances do (Huey and Berrigan 2001). Therefore, this hypothesis provides a plausible explanation for the temperature–size rule when similar temperatures maximize the growth of an individual and the growth of its population.

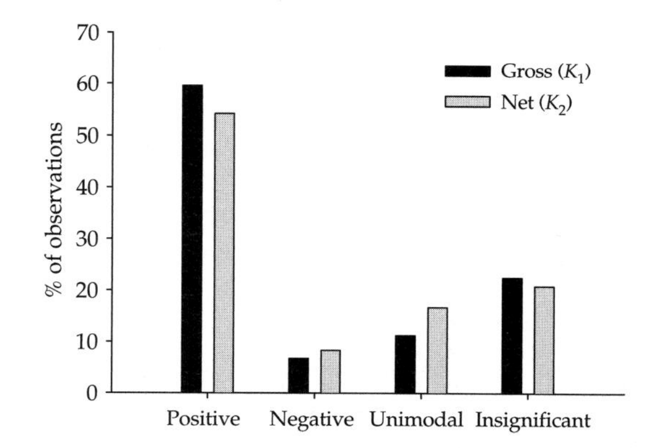

Figure 6.10 An increase in temperature usually enhances the growth efficiency of an ectotherm. Relationships between temperature and growth efficiency (gross or net) were characterized as positive, negative, unimodal, or statistically insignificant. These gross and net growth efficiencies represent 89 populations of 53 species and 24 populations of 20 species, respectively. Data are from Angilletta and Dunham (2003). Reproduced from Angilletta et al. (2004) with permission from Oxford University Press.

Unfortunately, thermal sensitivities of population growth have generally been estimated under artificial conditions, which excluded many sources of density dependence such as competition and predation. Until we have more information about the thermal sensitivity of population growth in natural environments, we cannot confirm the generality of this hypothesis.

Geographic clines might also stem from thermal effects on population growth. Lower latitudes and altitudes offer longer seasons that enable more generations per year. Even if a population remains stable over the long run, natural selection favors early maturation when genotypes can complete multiple generations between periods of density dependence. This situation exists in seasonal environments, where reproduction occurs during part of the year and density-dependent mortality culls the population during the remainder of the year. In such environments, the optimal life history might conform to the temperature–size rule. Some support for this hypothesis exists among populations of butterflies (*Lycaena hippothoe*). Genotypes complete either one or two generations per year depending on their environment. Genotypes from a population that completes two generations per year responded more strongly to developmental temperature than those from populations that complete one generation per year (Fischer and Fiedler 2002). Do not confuse this variation in thermal plasticity among populations with the variation in body size associated with multivoltinism in insects. In the latter case, genotypes evolve small size at maturity in environments where they can achieve two or more generations per year (Masaki 1967), but the phenomenon does not require plasticity.

6.3.5 A synergy of evolutionary mechanisms

In summary, any one of three mechanisms can cause the evolution of reaction norms consistent with the temperature–size rule: thermal constraints on juvenile mortality, maximal size, and population growth (Table 6.1). Support for each of these hypotheses varies greatly among species, even though the phenomenon we seek to explain seems nearly universal. Before we abandon hope of making sense of this situation, let us briefly consider the potential synergism among these hypothetical mechanisms.

The overwhelming generality of the temperature–size rule has had an unfortunate effect on the way many researchers have sought to understand thermal reaction norms. Atkinson (1996a) recognized this effect when he stated the following:

> The discovery of a widespread relationship such as a temperature–size rule tends to direct research towards general explanations which apply throughout ectotherms, and away from those specific to particular populations, species or groups of species. Simple, rather than complex explanations are usually sought.

Table 6.1 Hypothetical mechanisms that can generate optimal reaction norms consistent with the temperature–size rule. These mechanisms are identified as extrinsic or intrinsic factors

Hypothetical mechanism	Predicted relationship	Source
Extrinsic factors		
Juvenile mortality increases with temperature	Negative	Kozłowski et al. (2004); Sibly and Atkinson (1994); Berrigan and Koella (1994)
Adult mortality decreases with temperature	Negative	Charnov and Gillooly (2004); Sibly and Atkinson (1994)
Rate of population growth increases with temperature	Negative	Sibly and Atkinson (1994)
Seasonal constraints on growth and development	Saw-toothed	Roff (1980)
Intrinsic factors		
Growth efficiency decreases with temperature	Negative	Atkinson (1996b); Berrigan and Charnov (1994)
Lifespan decreases with temperature	Dome-shaped	Perrin (1988)
Rate of gonadogenesis increases with temperature	Negative	Kindlmann et al. (2001)

On the one hand, simple explanations appeal intuitively to our human brains (despite how the world repeatedly demonstrates its complexity). On the other hand, simple explanations usually fall very short of offering a meaningful understanding of broad patterns of life-history variation (Angilletta and Dunham 2003; Angilletta et al. 2004a). Rather than accept any of these simple explanations, we should expect a multitude of factors to drive the evolution of thermal reaction norms. Indeed, all three of the mechanisms discussed in this section could operate in synergy. Sibly and Atkinson (1994) noted the conditions for the evolution of large adults in cold environments were more likely to occur when temperature affected both juvenile mortality and population growth. Likewise, Kozłowski and colleagues (2004) modeled a synergism between the thermal sensitivities of juvenile mortality and maximal size. Although complex explanations often fail to win the hearts of biologists, the points made by these researchers should not be ignored. Ultimately, we will need to integrate our knowledge of physiology and ecology to arrive at a general explanation for the thermal plasticity of age and size at maturity.

6.4 General patterns of reproductive allocation

6.4.1 Thermal plasticity of offspring size

Temperature appears to exert a general effect on reproductive allocation as well as age and size at maturity. Atkinson and colleagues (2001) reviewed 36 experiments in which researchers measured the effects of temperature on offspring size (or egg size). More than 70% of these experiments revealed a decrease in offspring size with increasing temperature; in fact, only 3% reported the opposite relationship (Fig. 6.11). The same trend was observed for relative offspring size (offspring size/maternal size), suggesting the pattern was not merely a consequence of the temperature–size rule. More recent experiments yielded additional support for this conclusion; lower temperatures caused females to produce larger offspring (Fischer et al. 2003a; Karlsson and Van Dyck 2005; Stelzer 2002) or caused no clear change in offspring size (Du et al. 2005; Tveiten et al. 2001), but did not lead to smaller offspring.

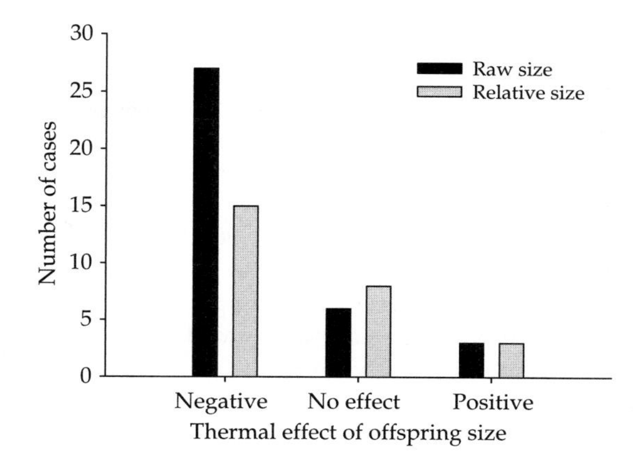

Figure 6.11 In ectotherms, mothers experiencing higher temperatures generally produce smaller offspring. Only a minority of populations studied in the laboratory consisted of genotypes that deviated from this pattern. This conclusion also held when the effect of maternal size was removed by computing a relative offspring size (offspring size/maternal size). Data were summarized by Atkinson et al. (2001).

For some species, we know females can adjust their egg size very rapidly in response to a change in temperature. For example, the yellow dung fly (*Scathophaga stercoraria*) typically produces large eggs when developing at a low temperature in natural or artificial environments (Blanckenhorn 2000). To determine whether females could respond rapidly to environmental temperature, Blanckenhorn (2000) took females that had developed at high temperatures and shifted them to a low temperature after they had produced their first clutch. As he expected, females produced larger eggs after the switch to a lower temperature. The phenotypic plasticity that Blanckenhorn observed in the laboratory mirrored a seasonal change he had documented in the natural environment. At least for dung flies, thermal plasticity of egg size can enhance fitness during thermal changes within generations.

6.4.2 Thermal clines in offspring size

Offspring size (or egg size) also varies greatly along geographic clines. Much of this variation accords with the patterns of thermal plasticity observed in controlled environments. Thermal clines in egg size have been documented among arthropods (Fischer and Fiedler 2001b; Fox and Czesak 2000;

Parry et al. 2001; Wilhelm and Schindler 2000), amphibians (Morrison and Hero 2003b), and fish (Laptikhovsky 2006). In most cases, individuals in colder environments produced larger eggs (but see Fleming and Gross 1990). In *D. melanogaster*, thermal clines evolved independently in Australia and South America (Azevedo et al. 1996). An interesting "experiment" in nature supported temperature as the cause of local variation in egg size. Meffe (1990) compared eggs of mosquitofish (*Gambusia holbrooki*) in a natural pond with those of fish in a polluted pond; the pollution consisted of thermal effluent, which periodically raised the temperature of the pond to as high as 45°C. Fish in the polluted pond produced smaller offspring throughout the year than did fish in the unpolluted pond. If we consider these patterns in the context of thermal plasticity, we might conclude that a model of thermal adaptation could explain geographic variation in egg size.

Although thermal clines in egg size resemble thermal reaction norms, not all geographic variation can be attributed to plasticity. For a few species, researchers have documented the contributions of genetic and environmental factors by raising populations in the laboratory or by transplanting populations to new environments. For example, Armbruster and colleagues (2001) raised pitcher-plant mosquitoes from six populations in a common environment. All mosquitoes experienced ideal summer conditions, yet mosquitoes from higher latitudes laid larger eggs. The variation in egg size among populations diminished when mosquitoes were raised at high densities, suggesting intraspecific competition plays some role in the reproductive decision. Given the common environment, patterns at high and low densities likely reflect the genetic divergence of egg size among populations; unfortunately, patterns of thermal plasticity within populations were not quantified. Parallel clines in the egg size of *D. melanogaster* also have a genetic basis; flies from higher latitudes laid larger eggs, even after flies from all populations were raised at constant temperatures for more than a year (Azevedo et al. 1996). In contrast, a transplant experiment revealed plasticity as the major cause of a thermal cline in the land snail, *Arianta arbustorum* (Baur and Raboud 1988). This species occurs at altitudes up to 2700 m, where the relative egg size exceeds that observed at lower altitudes. When snails were transplanted from five sites between 1200 and 2600 m to a site at 500 m, the majority of snails laid eggs similar in size to those laid in their natal environment. However, snails transplanted from the highest altitude laid smaller eggs than did snails that remained at this altitude. In this case, the thermal cline in egg size resulted from the combined effects of genetic and environmental factors. These experiments highlight the need to identify the sources of variation in offspring size before devising theories to explain thermal clines.

6.4.3 Experimental evolution of offspring size

Three types of experimental evidence prompt an adaptive interpretation of the thermal plasticity of offspring size: (1) selection gradients at different temperatures, (2) additive genetic variances of reaction norms, and (3) responses to selection in controlled environments.

Only two experiments have been conducted to identify the selective benefits of egg size over a range of temperatures. Blanckenhorn (2000) raised dung flies (*Scathophaga stercoraria*) at 19°C and divided mature siblings among four thermal environments (11°C, 15°C, 19°C, and 23°C). Eggs produced at each temperature were then divided among the same thermal environments to document variation in rates of mortality, development, and growth. Surprisingly, both large eggs produced at 11°C and small eggs produced at 23°C survived better than the intermediate eggs produced at 15°C or 19°C. Moreover, hatchlings from eggs produced at 23°C survived the best at all temperatures except 23°C! Rates of development and growth by offspring depended primarily on their thermal environment. Since temperature had little effect on fecundity, the patterns uncovered by Blanckenhorn's experiment seem very puzzling. A similar experiment by Fischer and colleagues (2003b) focused on the selective advantage of egg size for butterflies in cold and hot environments. Females of *Bicyclus anynana* alter the size of their eggs rapidly in response to environmental temperature (Fig. 6.12). After raising families of butterflies at 23°C, these researchers divided sisters between 20°C and 27°C to obtain eggs of different sizes. Then, they reciprocally transferred eggs

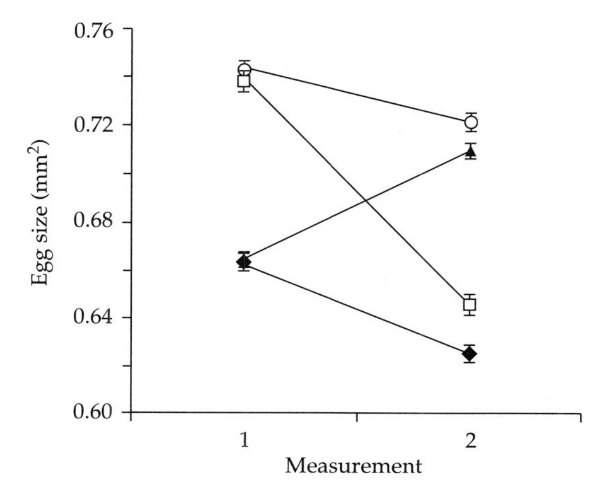

Figure 6.12 Egg size of the butterfly *Bicyclus anynana* responded rapidly to changes in environmental temperature. Females were raised at one temperature and either were maintained at this temperature as adults or were switched to a different temperature. Measurements of egg size occurred just after development at the first temperature (1) and 10 days after being switched to the second temperature (2). Symbols are as follows: circles, 20–20°C; squares, 20–27°C; triangles, 27–20°C; and diamonds 27–27°C. Adapted from Fischer et al. (2003a) with permission from the Royal Society.

produced at each temperature to see whether the selective advantage of egg size depended on temperature. Females produced small eggs at 27°C, but these eggs suffered relatively high rates of mortality at 20°C. Consequently, females that laid fewer, larger eggs would expect more of their offspring to reach maturation at 20°C. Yet the benefit of fecundity associated with small eggs outweighed the cost of survivorship at 27°C. In a subsequent experiment, these researchers found that eggs produced at 20°C and 27°C suffered equally during starvation and desiccation (Fischer et al. 2003c). Thus, the mechanism by which large size enhances the survivorship of offspring in cold environments remains elusive.

Using a half-sibling design, Steigenga and colleagues (2005) estimated the additive genetic variance of thermal reaction norms for egg size in *B. anynana*. Genetic correlations between environmental temperatures were positive but significantly less than 1.0, indicating a potential for thermal plasticity to respond to natural selection. Unfortunately, the experimental design confounded environmental temperature and laying date, creating doubt about the true magnitude of thermal plasticity. Interestingly, parent–offspring and full sibling analyses of the same species did not detect genetic variation in reaction norms (Fischer et al. 2004), despite clear plasticity of egg size in response to temperature. If we assume no effects of dominance and epistasis, maternal effects on egg size far exceeded genetic effects (see Box 1.2).

Azevedo and colleagues (1996) exposed replicated lines of *D. melanogaster* to constant temperatures. Originally, selection lines were initiated at 16.5°C and 25°C, but an additional set of lines were established at 29°C by splitting the lines at 25°C. After 9 years of selection at either 16.5°C or 25°C and 4 years of selection at 29°C, flies from all lines were raised at 25°C and their eggs were compared. Flies selected at 16.5°C laid larger eggs than flies selected at either 25°C or 29°C, but flies from these warmer selective environments laid eggs of similar sizes. These researchers also compared the thermal reaction norms of flies evolving at 16.5°C and 25°C. Females from these selective environments were raised at 25°C and then oviposited sequentially at 16.5°C and 25°C. Selection had not altered the thermal plasticity of egg size, but eggs laid by cold-adapted flies were 5–10% larger than eggs laid by warm-adapted flies (Fig. 6.13).

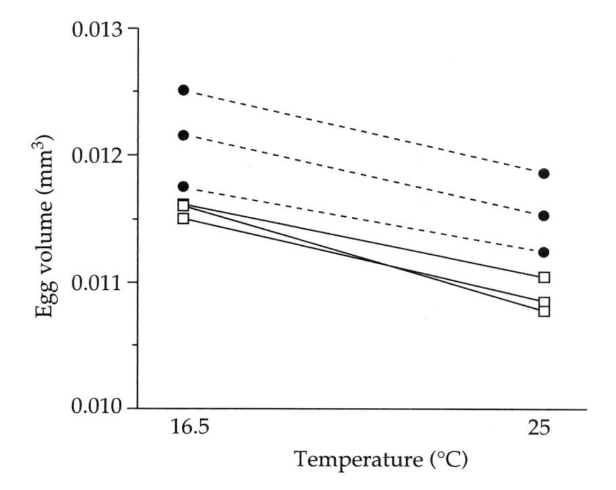

Figure 6.13 Experimental evolution led to a divergence in mean egg size between populations of *Drosophila melanogaster*, but the thermal plasticity of egg size remained similar among selection lines. Dashed and solid lines represent reaction norms for populations evolving at 16.5°C and 25°C, respectively. Adapted from Azevedo et al. (1996) with permission from Blackwell Publishing.

This body of evidence suggests three points. First, selection favors particular reaction norms for egg size. Second, these reaction norms persist despite heritable variation. Finally, changes in environmental temperature elicit genetic responses in egg size that parallel plastic responses. Given these points, we now turn to models of optimal reproductive allocation to identify possible causes for the thermal plasticity of egg size.

6.5 Optimal size and number of offspring

The theory of optimal reproductive allocation attempts to define the environmental causes of variation in reproductive traits, such as the size and number of offspring. Since the seminal model of Smith and Fretwell (1974), the theory has grown around a set of core assumptions: (1) reproductive decisions evolve to maximize the product of the number of offspring and their probability of surviving; (2) the fitness of offspring increases asymptotically as their share of resources increases; and (3) mothers have a fixed quantity of resources available for reproduction. Under these conditions, a single optimal size should dictate the allocation of resources among offspring (Smith and Fretwell 1974). Subsequent modeling (McGinley et al. 1987) supported the existence of an optimal offspring size, except under rare conditions when a mixture of sizes confers greater fitness (see Section 6.6). In this section, we shall see how thermal heterogeneity affects the optimal size and number of offspring—be they seeds, eggs, or babies.

6.5.1 Direct effect of temperature on the optimal offspring size

Changes in temperature over space and time should favor variation in offspring size. Among ectotherms, the performance of offspring decreases as temperatures rise above or fall below the thermal optimum (see Chapter 3). The optimal offspring size depends on several performances of offspring. Specifically, the optimal size decreases as (1) the growth rate of offspring increases, (2) the mortality rate of offspring decreases, and (3) the growth rate of the population increases (Sibly and Calow 1983; Taylor and Williams 1984). Because temperature potentially affects all three of these conditions, ectotherms should exhibit thermal plasticity of offspring size (Perrin 1988).

To understand the effects of temperature on the optimal offspring size, we can use a simple model described by Yampolsky and Scheiner (1996). Like previous modelers, Yampolsky and Scheiner assumed the survivorship of offspring increased asymptotically with increasing size. Furthermore, the survivorship of offspring continued to increase as they approached maturity. To model this phenomenon, they let the rate of juvenile mortality (j) decrease exponentially as offspring grew in size (s):

$$j(s) = j_0 e^{-ks}, \qquad (6.20)$$

where j_0 is the rate of juvenile mortality when $s = 1$, and k determines the size dependence of mortality. Importantly, they also assumed the rate of mortality did not depend on temperature. Offspring grew according to a linear function:

$$s(x) = s_0 + \frac{s_\alpha - s_0}{\alpha} x, \qquad (6.21)$$

where s_0 and s_α equal sizes at birth and maturation, respectively. Thus, Yampolsky and Scheiner implicitly assumed that energy acquisition and allocation remained constant during the juvenile stage.

With these assumptions in place, they solved for the fitness of a genotype in a growing population (r). This calculation became greatly simplified when the genotype reproduces semelparously. In this special case, fitness equals

$$r = \frac{\ln l(\alpha) + \ln m(\alpha)}{\alpha}. \qquad (6.22)$$

By sequentially substituting eqns 6.20 and 6.21 into eqn 6.22, they arrived at a function relating fitness to several life-history traits, including size at birth:

$$r = \frac{j_0 \left[e^{-ks_\alpha} - e^{-ks_0} \right]}{k(s_\alpha - s_0)} + \frac{\ln m(\alpha)}{\alpha} \qquad (6.23)$$

From this equation, we can see that r depends strongly on the age at maturity (α) and the rate of juvenile mortality (as influenced by j_0 and k).

Yampolsky and Scheiner used eqn 6.23 to infer the selective pressures imposed by changes in temperature. Two effects stand out. First, a genotype that matures at a younger age (i.e., a lower α) would enjoy a greater chance of reaching adulthood and hence would experience a greater r. This effect is reflected by an increase in the right-hand term of eqn 6.23 with a decrease in α. Second, a genotype that suffers less mortality (i.e., a lower j_0) would also experience a greater r. This effect is reflected by an increase in the left-hand term of eqn 6.23 with a decrease in j_0 (note that $e^{-ks_\alpha} - e^{-ks_0} < 0$). Both effects cause selection for smaller offspring because of the concomitant increase in fecundity. From Fig. 6.14, we can see that selection favors small offspring whenever they would mature at an early age or experience little risk of mortality (see also Taylor and Williams 1984).

Because ectotherms tend to grow faster and mature earlier in warmer environments (see Section 6.2), Yampolsky and Scheiner reasoned that individuals in warm environments should produce small offspring. Unfortunately, we cannot generalize the effect of environmental temperature on juvenile survivorship (see Section 6.3.2). If juveniles suffer greater mortality in warmer environments, mothers should produce relatively large offspring. Thus, opposing selective pressures could result from thermal sensitivities of growth and development and the thermal sensitivity of juvenile survivorship.

The predictions of this model depend greatly on the underlying assumption about the relationship between the size and performance of offspring. As with most models of optimal reproductive allocation, this one assumes large offspring always outperform small offspring. The model breaks down when the relationship between size and performance differs among thermal environments, either through acclimation or through interactions between temperature and other factors (see, for

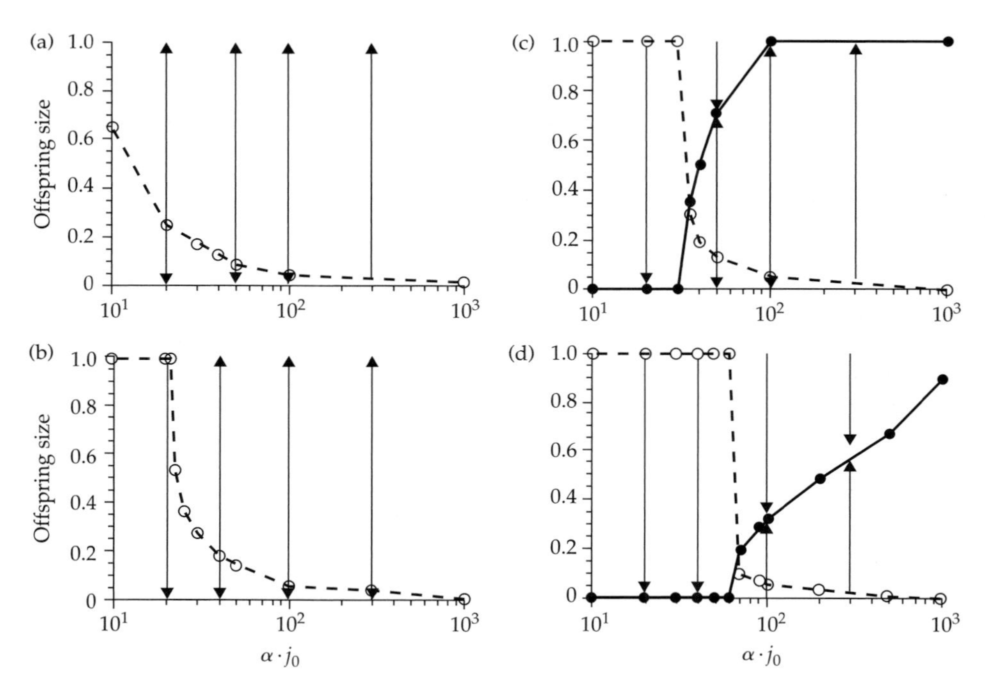

Figure 6.14 The optimal offspring size increases as either the age at maturity (α) or the rate of mortality (j_0) increases. (a–d) Equilibrial reaction norms given for four relationships between offspring size and juvenile mortality (i.e., k increases from plot a to plot d; see eqn 6.20). Both stable equilibria (solid lines, filled circles) and unstable equilibria (dashed lines, open circles) are shown. Stable equilibria occur only when an increase in offspring size results in a large decrease in juvenile mortality (c and d). Under these conditions, we should expect selection to favor an increase in egg size with decreasing environmental temperature. Adapted from Yampolsky and Scheiner (1996, *The American Naturalist*, University of Chicago Press). © 1996 by The University of Chicago.

example, Gagliano et al. 2007). One can envision potential disadvantages of large size in both aquatic and terrestrial environments. For example, large embryos could fail to acquire sufficient oxygen from warm waters, as oxygen delivery scales disproportionally to body size (Woods 1999). Additionally, large eggs might attract the attention of predators because of their great energetic value (Stephens and Krebs 1986). In reality, we do not know the true relationship between the size and performance of offspring for most species in most environments (but see Sinervo et al. 1992). Therefore, we should cautiously recognize that thermal plasticity of egg size could be caused by something other than thermal sensitivities of growth and development.

6.5.2 Indirect effects of temperature on the optimal offspring size

The optimal reproductive strategy depends not only on the thermal sensitivity of offspring performance but also on the thermal sensitivity of the maternal phenotype. Adults in warm environments differ dramatically from adults in cold environments. As discussed in Section 6.2, a mother in a cold environment would likely be older and larger than a mother in a warm environment. These differences would affect the relative capacities for energy acquisition and parental care—capacities that undoubtedly influence reproductive decisions. Large mothers would also be more capable of passing large offspring during birth or oviposition. Therefore, the thermal plasticity of age and size at maturity mediates indirect effects on the optimal reproductive decision.

The capacities of large mothers to acquire and store more energy greatly affect the optimal size and number of offspring. Parker and Begon (1986) modeled a capital breeder that acquired energy at a constant rate until reaching a maximal level. They assumed a large female acquired energy more rapidly and could store energy more readily than a small female. Should a large female use her extra energy to produce more offspring, larger offspring, or both? The answer depends on the relationship between offspring size and fitness. If we assume the survivorship of offspring increases asymptotically with size (as in most models), all mothers should produce offspring of the same size. A large female should use her additional energy to produce more offspring because the benefit of greater fecundity exceeds the benefit of producing offspring beyond a certain size. The situation changes when the survivorship of offspring depends on more than just their size. Imagine that offspring compete with one another, such that an increase in the number of offspring heightens the intensity of competition. Then a large female that produced many offspring would have her increase in fecundity offset by a decrease in her offspring's performance. Under this condition, a large female should use her additional energy to produce larger offspring. When the survivorship of offspring depends on both their size and their density, larger mothers should adopt a mixed strategy, producing both more and larger offspring.

The care provided by a mother also affects the optimal size of her offspring. Conceivably, larger mothers construct better nests, deter more predators, or provide more food for their offspring. Sargent and colleagues (1987) modeled the effect of parental care on the optimal offspring size. These researchers separated the survivorship of offspring into two stages: embryonic and juvenile stages. Assume that embryos inside larger eggs (or seeds) take longer to develop or suffer greater rates of mortality. In either case, the probability of an offspring reaching the juvenile stage (and ultimately adulthood) decreases with increasing size. In contrast, the rate of juvenile mortality decreases with increasing size. Under these conditions, a mother should produce larger eggs if her care can improve the survivorship of her offspring during their embryonic stage. Parental care offsets the decrement in embryonic survivorship associated with large eggs, such that mothers reap the benefits of providing offspring with the energy needed to become large juveniles. Thus, a large mother should produce fewer, larger eggs than a small mother if she provides better parental care. Hendry and colleagues (2001) predicted the same phenomenon from a very different relationship between the size and survivorship of offspring. Specifically, they assumed that offspring of an intermediate size enjoyed the greatest survivorship. Yet through some form of parental care, a large mother could ameliorate the disadvantages of extremely large offspring. The two models essentially portray the same idea despite containing different details.

Finally, large mothers escape the constraints on offspring size imposed by morphology. Most animals pass their eggs or babies through apertures in their internal or external skeletons. The maximal size of the aperture determines the maximal size of the offspring. Consequently, small mothers can be forced to produce smaller offspring than would be optimal for larger mothers (Congdon and Gibbons 1987). This nonadaptive mechanism would generate a unique pattern of variation: offspring size would increase with increasing maternal size up to a maximum that approximates the optimal size in the absence of a morphological constraint (Oufiero et al. 2007).

6.5.3 Teasing apart direct and indirect effects on reproductive allocation

From the models described in the preceding section, we can safely conclude that temperature should have direct and indirect effects on reproductive decisions (Table 6.2). Direct effects result from the thermal sensitivities of offspring performances. Indirect effects are mediated by maternal size and its consequences for energy acquisition, parental care, and morphological constraints. Because two or more of these effects can occur simultaneously, we should try to distinguish their relative contribution to variation in reproductive allocation.

A strong inference approach can tease apart the direct and indirect effects of environmental temperature on the evolution of reproductive traits (Angilletta et al. 2006b). We can use the predictions of optimality models to construct statistical models relating environmental temperature, maternal size, and reproductive traits. A simple path model accounts for all of the hypothetical mechanisms linking environmental temperature to reproductive traits (Fig. 6.15a). This statistical model can be evaluated with phylogenetic comparative data, which describe the correlated evolution of two or more traits (e.g., independent contrasts). Because any or all of these mechanisms could operate simultaneously, we can use Akaike's information criterion (see Box 3.1) to select the model that best fits the available data. Support for a particular statistical model implies support for its corresponding optimality model(s). Angilletta and colleagues (2006) used this approach to infer the likely causes of reproductive decisions in the eastern fence lizard (*Sceloporus undulatus*). We found considerable support for the hypothesis derived from Parker and Begon's model (1986), in which (1) the size of a mother enhances her energy acquisition and (2) the survivorship of offspring depends on their size and their density (Fig. 6.15b). Other statistical approaches can help distinguish between direct and indirect effects. For

Table 6.2 Models of optimal reproductive allocation predict specific relationships between temperature and reproductive traits under sets of hypothetical conditions (reproduced from Angilletta et al. 2006b)

Hypothetical conditions	Thermal effect on egg size	Thermal effect on clutch size	Source
1. Offspring grow faster in warmer environments	Decrease	Increase	Perrin (1988); Yampolsky and Scheiner (1996)
2. (a) Mothers are smaller in warmer environments, (b) smaller mothers acquire less energy, and (c) offspring survival increases as egg size increases	None	Decrease	Parker and Begon (1986)
3. (a) Mothers are smaller in warmer environments, (b) smaller mothers acquire less energy, and (c) offspring survival decreases as density increases	Decrease	None	Parker and Begon (1986)
4. (a) Mothers are smaller in warmer environments and (b) offspring survival increases as maternal size increases	Decrease	Increase	Hendry et al. (2001); Sargent et al. (1987)

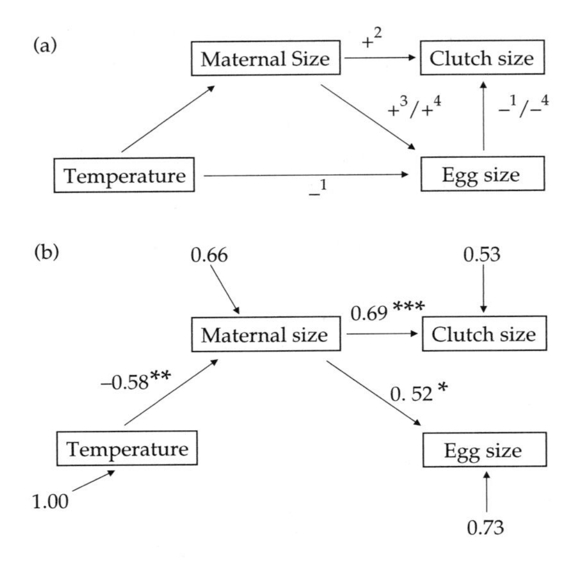

Figure 6.15 An analysis of the direct and indirect effects of environmental temperature on reproductive traits can reveal how multiple factors simultaneously affect reproductive allocation. (a) A path model depicts the predicted relationships among environmental temperature, maternal size, and reproductive traits under different conditions. These relationships were derived from the optimality models summarized in Table 6.2. The predicted relationships (positive or negative) are linked to specific conditions by superscripts: 1, Thermal sensitivity of growth; 2, survival depends on size; 3, survival depends on density; 4, survival depends on maternal size. (b) A submodel best described the phenotypic covariation among populations of the eastern fence lizard (*Sceloporus undulatus*). Data were independent contrasts generated from an intraspecific phylogenetic hypothesis. Adapted from Angilletta et al. (2006b, *The American Naturalist*, University of Chicago Press). © 2006 by The University of Chicago.

example, Perrin (1988) used an analysis of covariance to partition the effects of maternal size and environmental temperature on the egg size of *Simocephalus vetulus*; his analysis clearly showed that the thermal effect on egg size exceeded the maternal effect.

6.6 Optimal variation in offspring size

Thermal heterogeneity can also select for females that produce a mixture of offspring sizes. The models described in the preceding section assume that a mother must produce offspring of one size at each reproductive event. Now consider the possibility that a mother could produce two or more sizes of offspring at one time. Under what thermal conditions would this strategy outperform a single offspring size?

We can use a model by McGinley and colleagues (1987) to infer the conditions that favor a mixture of offspring sizes. These researchers modeled the production of offspring in an environment that varied either spatially or temporally. To simplify their analysis, they assumed the environment consisted of good and bad patches, instead of a continuum of patch qualities. For our purpose, good and bad patches would provide optimal and suboptimal temperatures, respectively. Offspring can disperse randomly among patches or they can be directed to particular patches by their parents. Given our previous considerations, we should expect good patches to favor small offspring and bad patches to favor large offspring. But should mothers ever produce a combination of large and small offspring?

According to the model, most conditions favor a single offspring size over a mixture of sizes. The optimal offspring size depends on the dispersal of offspring, the frequency of patches, and the degree of density dependence. If offspring disperse randomly among patches, a single offspring size always returns greater fitness than does a mixture of two sizes. Selection for small offspring would occur when patches differ greatly in temperature or when the environment consists mainly of good patches. Otherwise, a mother should produce offspring of the minimal viable size in a bad patch. If a mother can direct her offspring to a good patch, she should produce small offspring; however, anything less than a perfect direction of offspring to good patches favors larger offspring. Thus, spatial heterogeneity in temperature should alter the mean offspring size without adding to its variance.

Extreme variation in temperature among generations can select for variation in offspring size (McGinley et al. 1987). A mixture of offspring sizes enhances fecundity while guarding against extinction during generations of extreme temperatures. A mixed strategy should evolve when temperatures differ greatly between good and bad generations but favorable temperatures occur frequently. Otherwise, mothers should produce offspring large enough to survive in a poor thermal environment. Not surprisingly, the condition favoring a mixture of offspring sizes also favors a thermal generalist over a thermal specialist (see Chapter 3). Therefore, selection for variation in offspring size likely occurs in short-lived organisms that have multiple generations per year, because such organisms would experience the greatest variation in temperature among generations.

6.7 Conclusions

Optimality models have provided a wealth of competing hypotheses for the thermal plasticities of life-history traits. Reaction norms for the age and size at maturity should evolve according to the thermal sensitivities of juvenile survivorship, maximal size, and population growth. Reaction norms for the size and number of offspring should evolve according to the thermal sensitivity of juvenile performance and the thermal plasticity of maternal size. Although this theory provides potential explanations for general patterns of life-history variation, the critical assumptions of most models remain to be validated. Existing data cast doubt on the assumptions required to explain the temperature–size rule, such as strong thermal effects on juvenile survivorship or maximal size. Insufficient data exist to verify assumptions pertaining to optimal offspring size, such as the relationships among the size, density, and fitness of offspring.

More importantly, all of the models implicitly assume that the problems of optimizing adult size and offspring size can be solved independently. Yet, we know that environmental temperature influences the optimal offspring size indirectly via changes in maternal size. Stelzer (2002) turned this observation around by arguing that thermal effects on egg size contribute to temperature–size relationships. By independently manipulating the temperatures of reproducing rotifers and their growing offspring, he found that the thermal plasticity of egg size contributed substantially to variation in size at maturity. When modeling the optimal size at maturity, researchers have assumed temperature does not affect the size at birth. The fact that temperature *does* affect the size at birth prompts an important question. Does evolution by natural selection produce a coadapted suite of life-history phenotypes?

Comparative and experimental studies of *D. melanogaster* suggest several life-history traits coevolve in response to temperature. In South America and Australia, flies from higher latitudes developed faster and grew more efficiently as larvae (James and Partridge 1995; Robinson and Partridge 2001; van't Land et al. 1999), reached larger sizes at maturity (James et al. 1995, 1997), and produced larger eggs (Azevedo et al. 1996). Temperature appears to be a major selective pressure because the same suite of phenotypes arose during experimental evolution in the laboratory (Azevedo et al. 1996; James and Partridge 1995; Neat et al. 1995; Partridge et al. 1994). Phenotypes that evolved at low temperature conferred greater longevity and fecundity in a cold environment, but conferred no noticeable benefits in a warm environment (Bochdanovits and de Jong 2003; McCabe and Partridge 1997; Reeve et al. 2000). Contrary to Stelzer's idea, egg size did not consistently affect size at maturity, but large eggs did enhance rates of development and growth (Azevedo et al. 1997). Thermal adaptation produced a suite of life-history phenotypes, apparently involving some functional linkage.

Despite the evidence that life-history traits coevolve, current theory cannot predict how the entire life history will vary over thermal clines. This shortfall stems from a focus on the evolution of individual traits (or pairs of traits) rather than on the integration of these traits. For example, some models address the effect of size-specific or stage-specific mortality on the optimal rate of growth and development, but they do not consider how growth should affect the evolution of reproductive strategies (Cichoń and Kozłowski 2000; Conover 1998; Wieser 1994). Alternatively, models of the optimal size at maturity explore the consequences of tradeoffs between current and future reproduction, but they do not link strategies of maturation to the allocation of energy among offspring (Berrigan and Charnov 1994; Kozłowski et al. 2004; Sibly and Atkinson 1994). Finally, models of reproductive allocation consider the tradeoffs between offspring quality and quantity, but they do not link reproductive strategies to the evolution of juvenile growth and development (Perrin 1988; Yampolsky and Scheiner 1996). Although much progress has been made by studying specific tradeoffs and isolated traits, major strides could be made by developing theories that explore the integration of traits throughout ontogeny.

To transition from current models that atomize the life history to new models that take a holistic approach, we must embrace the concept of phenotypic integration. As Pigliucci (2003) suggests, think of phenotypic integration as the genetic correlations among traits driven by their functional relationships. Traits such as offspring size and growth rate should be tightly integrated, while those such as offspring size and adult size might be less integrated. For many ectotherms, we could describe the integration of life-history phenotypes as follows: individuals in colder environments grow slower, mature later at a larger size, and produce fewer, larger offspring. This pattern of phenotypic integration describes covariation among individuals resulting from either genetic divergence or thermal plasticity. Although the pattern does not apply to all species, it occurs frequently enough to warrant our attention. We can interpret this phenotypic integration as a product of constraint or a product of adaptation.

If natural selection causes patterns of phenotypic integration, we are faced with the challenge of developing a theory that describes the coadaptation of phenotypes. Such a theory must identify the behavioral and physiological processes that span ontogenetic stages and the tradeoffs caused by these processes. The degree to which adaptation will involve changes in growth and developmental rates, age and size at maturity, or offspring size and number should depend on the relative costs of a change in each trait. But coadaptation involves more than just life-history traits; strategies of thermosensitivity, thermoregulation, and thermal acclimation must also coevolve with the life history. In the next chapter, we shall directly tackle the problem of thermal coadaptation.

CHAPTER 7

Thermal Coadaptation

7.1 Traits interact to determine fitness

In the preceding chapters, we covered some major issues in evolutionary thermal biology, namely adaptations of thermal sensitivity, thermoregulatory behavior, thermal acclimation, and life history. As would most evolutionary biologists, we addressed these issues by atomizing the phenotype into mathematically tractable pieces. This approach assumes that natural selection shapes each trait independently. In Chapter 3, we considered the evolution of thermal sensitivity, but assumed organisms do not thermoregulate (or do so independently of their thermal sensitivities). In Chapter 4, we considered the evolution of thermoregulation while assuming organisms have a fixed thermal sensitivity. In Chapter 5, we examined the evolution of thermal acclimation, again assuming no thermoregulatory behavior. Finally, in Chapter 6, we examined the evolution of the life history by assuming fixed thermal sensitivities of survivorship, growth, and development. In each case, these simplifying assumptions enabled us to derive predictions with relative ease. But these predictions should be viewed with caution because real organisms do not evolve under such constraints.

Real organisms possess a suite of strategies to cope with thermal heterogeneity, and these strategies undoubtedly evolve according to their combined influence on fitness. In other words, the behavior, physiology, and life history of a species coadapt to the environment. The concept of coadaptation applies perfectly well to both the genotype and the phenotype. For example, a set of genes will coevolve if their interaction confers greater fitness than other possible sets of genes (reviewed by Templeton 1986; Whitlock et al. 1995). Alternatively, one can focus on the coadaptation of traits controlled by these genes; two or more traits will coevolve if they enhance fitness through an interaction (Lande 1984; Wolf and Brodie 1998). Two examples of the coadaptation of traits include behavioral drive and behavioral inertia (Huey et al. 2003). In behavioral drive, the plasticity or evolution of behavior drives subsequent evolution of physiology, morphology, or life history. In behavioral inertia, the behavior of an organism ameliorates the stress of thermal heterogeneity, thereby limiting the evolution of other traits. Phenomena such as these suggest that evolutionary biologists are looking at the integrated superstrategies of organisms, whether they care to recognize it or not.

Why think in terms of coadaptation when modeling evolution? By constraining some traits and not others, we will miss selective pressures for the correlated evolution of interacting traits. For example, the thermal sensitivity of performance evolves according to the body temperatures experienced by organisms (Chapter 3). An organism can use behavior and physiology to regulate its body temperature, but the benefit of thermoregulation depends on the thermal sensitivity of performance (Chapter 4). Hence, selection should favor some combination of thermal sensitivity and thermoregulatory behavior (Huey and Bennett 1987). An organism that thermoregulates very precisely would gain little advantage from a wide performance breadth. On the other hand, an organism that thermoregulates poorly might need a wide performance breadth to survive and reproduce. Other combinations of phenotypes should also confer particularly high fitness when compared with alternatives. As we shall see, the correlated evolution of these traits can differ qualitatively from the evolution of each trait in isolation. Therefore, a general theory of thermal adaptation must be a theory of thermal coadaptation. In this chapter, I hope to convey some of the more interesting insights that have emerged from this perspective, starting with the coadaptation of pairs of traits and ending with the coadaptation of thermoregulatory behavior, thermal physiology, and life history.

7.2 Coadaptation of thermal sensitivity and thermal acclimation

In previous chapters, we studied the evolution of performance curves and their capacity to acclimate to thermal heterogeneity. To better understand the evolution of performance curves, we applied an optimality model developed by Lynch and Gabriel (1987). This model describes the fitness landscape for the thermal optimum and performance breadth, but assumes these parameters cannot acclimate to a changing environment. In Chapter 5, we moved on to consider a model of thermal acclimation developed by Gabriel (1999). This model defines the optimal shifts in the thermal optimum and performance breadth, given an initial performance curve and the expected thermal change. Given the interaction between the thermal sensitivity and its acclimation capacity, these two traits surely evolve in synchrony. In this section, we shall use a more general model by Gabriel (2005) to determine the optimal combination of thermal sensitivity and thermal acclimation. As we shall see, this model of coadaptation provides unique insights about the evolution of thermal physiology.

As in his previous work, Gabriel (2005) modeled a performance curve for survivorship. This curve was determined by a Gaussian function, with the parameters z_1 and z_2 defining the thermal optimum and performance breadth, respectively (see eqn 3.5). The fitness of the organism depends on the product of performance throughout its life. Thus, an organism dies if its temperature falls outside the range of temperatures encompassed by its performance curve. Given these considerations, what should the performance curve look like in a changing environment? Should the organism possess a wide performance breadth, to tolerate the full range of body temperatures experienced during its life? Or should it possess a narrower performance breadth and shift its thermal optimum as its temperature changes? The answer depends on both the variability and predictability of the environment.

Imagine an environment that switches between two operative temperatures: T_1 and T_2. At each transition, the organism knows its body temperature will change, but does not necessarily know the exact magnitude of this change; in other words, the organism has incomplete information. The pattern of thermal change and the phenotypic consequences of acclimation follow those of Gabriel's original model (Fig. 7.1). The environment switches from T_1 to T_2 after a period of time (a). Yet, the organism requires some time (b) to acclimate to the thermal change. After some additional time at T_2 (c), the environment switches back from T_2 to T_1. Following this reversal in temperature, deacclimation requires some time (d). During the time lags (b and d), the performance of the organism suffers. We can interpret this model as a single cycle of thermal change during the lifespan or multiple cycles throughout the lifespan. For either interpretation, we need to know only the total time that an organism spends in each of the four phases: t_a, t_b, t_c, and t_d.

If we assume all genotypes have the same fecundity, fitness (w) scales proportional to the product of

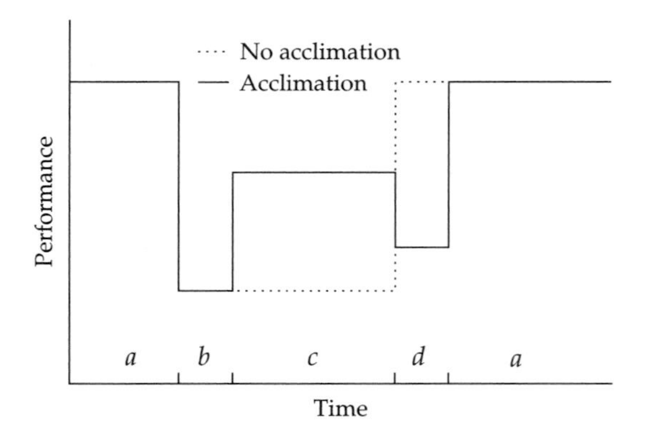

Figure 7.1 Acclimation improves performance in a fluctuating environment. After some period (a), an abrupt shift in environmental temperature cues acclimation, but an organism requires time to respond (b). After some additional period (c), the environment switches back to its original temperature. However, deacclimation also requires time (d). The performance of an acclimating genotype suffers during the time lags (b and d), but the performance of a nonacclimating genotype suffers longer during thermal change (b and c). Adapted from Gabriel (1999) in The Ecology and Evolution of Inducible Defenses (Harvell, C. D. and R. Tollrian, eds.) by permission from Princeton University Press. © 1999 Princeton University Press.

performance during all phases:

$$w \propto P^{t_a} \cdot P^{t_b} \cdot P^{t_c} \cdot P^{t_d}. \qquad (7.1)$$

Because performance depends on the operative temperature and the thermal sensitivity, we can rewrite eqn 7.1 as

$$w \propto P(T_1, z_{1_{ab}}, z_{2_{ab}})^{t_a} \cdot P(T_2, z_{1_{ab}}, z_{2_{ab}})^{t_b} \cdot P(T_2, z_{1_{cd}}, z_{2_{cd}})^{t_c} \cdot P(T_1, z_{1_{cd}}, z_{2_{cd}})^{t_d} \qquad (7.2)$$

where $z_{1_{ab}}$ and $z_{2_{ab}}$ equal the thermal optimum and performance breadth during t_a and t_b, and $z_{1_{cd}}$ and $z_{2_{cd}}$ equal the thermal optimum and performance breadth during t_c and t_d. Using eqn 7.2, we can determine the thermal optima and performance breadths that maximize fitness, given time lags and environmental stochasticity. The conclusions drawn here follow from Gabriel's analysis (Gabriel 2005); additional insights can be found in other analyses of this model (Gabriel 2006; Gabriel et al. 2005).

The capacity for acclimation alters the optimal thermal sensitivity throughout the life of the organism. With no capacity for acclimation, selection favors a thermal optimum equal to the expected temperature and a performance breadth proportional to the thermal variance (see Chapter 3). With an unlimited capacity for acclimation, the thermal optimum should instantaneously switch between T_1 and T_2 as the environment changes; with such capacity, the performance breadth should be as small as physically possible to minimize the impact of a specialist–generalist tradeoff. In reality, time lags for acclimation and deacclimation prevent this ideal response to thermal change. A time lag for acclimation creates a tradeoff between performances during t_a and t_b. If the thermal optimum equaled T_1 during t_a, the organism would maximize its performance during t_a but suffer poor performance during t_b. Alternatively, if the thermal optimum equaled T_2 during t_a, the organism would suffer poor performance during t_a but maximize its performance during t_b. A similar tradeoff between performances during t_c and t_d occurs because of the time lag for deacclimation.

When time lags exist, the optimal strategy represents a compromise imposed by these tradeoffs. The thermal optimum should always lie somewhere between T_1 and T_2, such that the optimal genotype never maximizes its performance in the current environment but always reduces its loss of performance in the future environment. When the operative temperature equals T_1, the thermal optimum that maximizes fitness ($z^*_{1_{ab}}$) can be calculated as follows:

$$z^*_{1_{ab}} = T_1 + \left[(T_2 - T_1)\left(\frac{t_b}{t_a + t_b}\right)\right]. \qquad (7.3)$$

Intuitively, an increase in the time required for acclimation (and thus t_b) should move the thermal optimum during t_a closer to the temperature of the future environment, T_2. Likewise, an increase in the time required for deacclimation (and thus t_d) should move the thermal optimum during t_c closer to T_1:

$$z^*_{1_{cd}} = T_1 + \left[(T_2 - T_1)\left(\frac{t_c}{t_c + t_d}\right)\right]. \qquad (7.4)$$

Time lags also favor the evolution of wider performance breadths, which enhance performance during acclimation and deacclimation. The optimal performance breadths ($z^*_{2_{ab}}$ and $z^*_{2_{cd}}$) scale proportional to the magnitude of thermal change ($T_2 - T_1$).

If time lags become sufficiently long, an interesting phenomenon emerges. When the time spent acclimating to thermal change exceeds the period of benefit (i.e., $t_b > t_c$ and $t_d > t_a$), selection favors a thermal optimum closer to the future environmental temperature than the current one! Figure 7.2 illustrates this phenomenon for genotypes with incomplete and complete information about the magnitude of change. In the rare event that the periods of acclimation and deacclimation equal the periods of benefit (i.e., $t_a = t_b = t_c = t_d$), selection favors no acclimation of the performance curve. In this special case, the optimal performance curve corresponds to that predicted by a model of evolution without acclimation (e.g., Lynch and Gabriel 1987): the thermal optimum should equal the expected temperature and the performance breadth should scale proportional to the variance of temperature. In Gabriel's model of coadaptation, only this special case favors a perfect match between the thermal optimum and the expected temperature. All other conditions favor a constant mismatch between the two, which differs qualitatively from the predictions of models that ignore coadaptation (see Chapters 3 and 5).

Coadaptation also depends on the predictability of thermal change. We can reasonably assume that

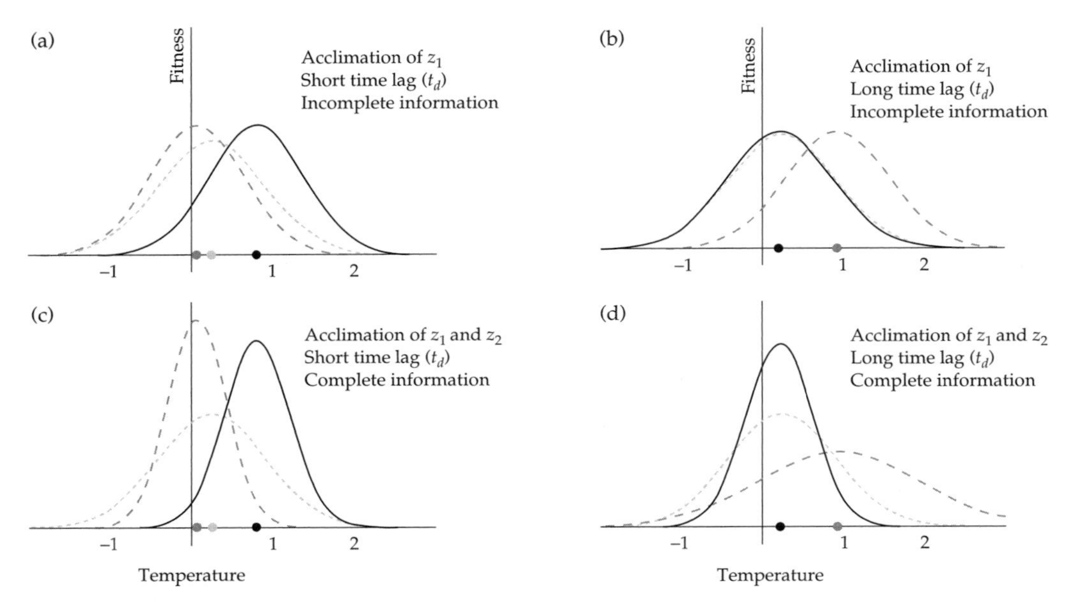

Figure 7.2 The optimal acclimation of performance curves depends on the time lag for physiological change. In each plot, the optimal performance curves at T_1 and T_2 are depicted by the long-dashed line and the solid line, respectively. For comparison, the short-dashed line depicts the optimal phenotype without acclimation. The thermal axis is relative, with zero being equal to T_1. Optimal strategies are shown for conditions in which individuals have either incomplete or complete information about the thermal change (a and b vs. c and d). In all cases, the duration of T_1 was three times as long as the duration of T_2 ($t_a + t_d = 0.75$). Therefore, when deacclimation requires a very long time (b and d), the optimal strategy of acclimation prepares the organism for the future temperature rather than maximizes performance at the current temperature (i.e., the thermal optimum during T_1 lies closer to T_2 and the thermal optimum during T_2 lies closer to T_1). Adapted from Gabriel (2005) with permission from Blackwell Publishing.

organisms know of a change in temperature without knowing the exact magnitude of the change. In natural environments, stochasticity of diel and seasonal changes seems more the rule than the exception (see Chapter 2). To incorporate this stochasticity, Gabriel characterized T_1 and T_2 by their means and variances. Two results emerged from this modification of the model. Stochasticity should not affect the thermal optimum at any phase of the organism's life. For any degree of uncertainty, the thermal optimum shifts according to the long-term mean of thermal change (see also Chapter 5). Nonetheless, unpredictable values of T_1 and T_2 do affect the optimal performance breadths. As stochasticity increases, the performance breadth should increase. In fact, the selective advantage of thermal acclimation diminishes as the variance of temperature increases (Fig. 7.3). In the rare situation where the environment cycles predictably, the optimal performance breadths before and after acclimation depend only on time lags for shifting the thermal optimum.

The unique predictions offered by Gabriel's model may help us to better understand the diversity of performance curves observed in nature. Although we currently know little about the time lags for acclimation (see Chapter 5), such constraints could explain why certain species exhibit thermal optima that deviate from their mean body temperatures. For example, Smith and Hadley (1974) repeatedly observed an imperfect match between the environmental temperature and the thermal optimum for photosynthesis in four populations of the Labrador tea (*Ledum groenlandicum*). After raising plants in a common environment for several years, these investigators exposed the plants to repeated shifts in environmental temperature. Initially, some plants were exposed to a diel cycle of 15°C light–10°C dark, while others were exposed to 30°C light–25°C dark. After 12–16 weeks, the environments of these groups were switched; plants in the cool environment were shifted to the warm environment, and vice versa. Two weeks later, both groups were switched back

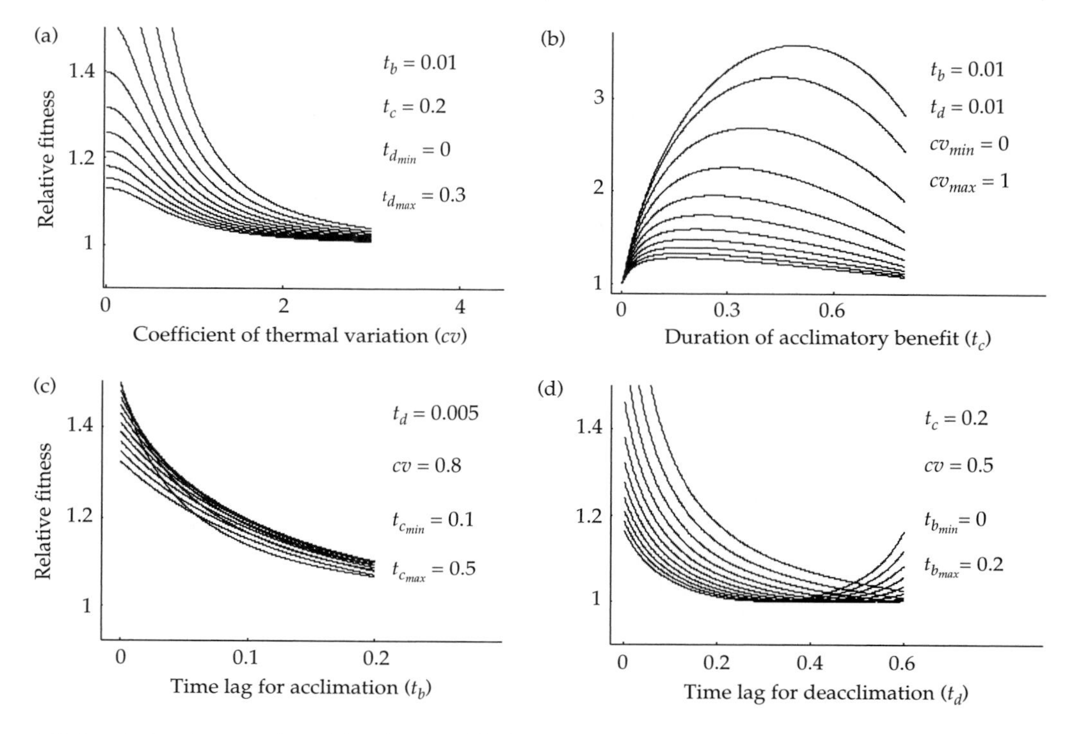

Figure 7.3 The selective advantage of reversible acclimation of the thermal optimum depends on properties of the environment and the organism. In each plot, lines depict the selective advantage when a particular parameter was varied in constant steps between minimal (*min*) and maximal (*max*) values. The fitness of an acclimating phenotype was scaled relative to that of a fixed phenotype. Relative fitness depends on the following conditions: (a) the unpredictability of thermal change; (b) the duration of acclimatory benefit (t_c), assuming a brief time lag for acclimation (t_b); (c) the time lag for acclimation (t_b), assuming a brief time lag for deacclimation (t_d); and (d) the time lag for deacclimation (t_d). Adapted from Gabriel (2005) with permission from Blackwell Publishing.

to their initial environments. Prior to each shift, the investigators measured the thermal sensitivity of the net photosynthetic rate. Based on Gabriel's model, acclimation and deacclimation should have resulted in thermal optima that lay somewhere between the two diurnal temperatures (15°C and 30°C). Consistent with this prediction, thermal optima of all populations shifted from 25°C in the warm environment to 20°C in the cool environment (Fig. 7.4). When exposed to the reverse order of temperatures, only two of the four populations followed this pattern; the other populations exhibited a thermal optimum of 25°C in both environments. Possibly, these mismatches between the thermal optimum and the environmental temperature stemmed from an insufficient period for acclimation. But this explanation seems unlikely because 12–16 weeks of acclimation to the same thermal environments did not produce a closer match between thermal physiology and environmental temperature.

Coadaptation would also create geographic variation in performance curves. Environments at high latitudes are colder and vary more than environments at low latitudes. Because of differences in seasonality, temperatures during the winter differ far more along latitudinal clines than do temperatures during the summer (see Fig. 2.8). Simple models of thermal sensitivity predict a decrease in the thermal optimum and an increase in the performance breadth with increasing latitude (see Chapter 3). However, thermal acclimation would enable an organism to specialize for the mean temperature of each season. Latitudinal clines in the seasonality and stochasticity of temperature should affect patterns of thermal physiology. At high latitudes, we should find genotypes that shift their

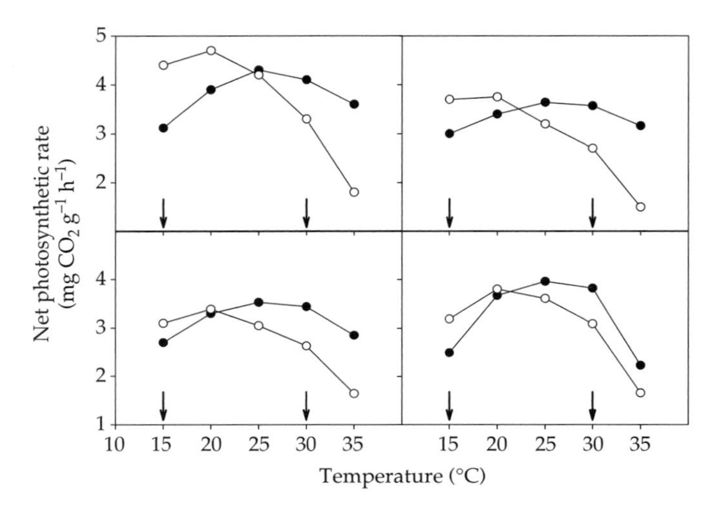

Figure 7.4 Thermal sensitivities of net photosynthetic rate acclimated reversibly to cool environments (open symbols) and warm environments (filled symbols). Each plot shows data for a population of *Ledum groenlandicum*. The arrows mark the temperatures of the two acclimation environments during photophase. Adapted from Smith and Hadley (1974) with permission from The Institute of Arctic and Alpine Research.

thermal optimum between seasonal extremes but retain relatively wide performance breadths in each season. At low latitudes, we should find genotypes with relatively constant thermal optima and narrow performance breadths. If we examined performance curves in the winter, we should see a strong latitudinal cline in the thermal optimum for performance. However, the same cline should be much weaker in the summer. One way to evaluate Gabriel's model of coadaptation would be to see whether latitudinal clines in thermal physiology actually differ between seasons.

7.3 Coadaptation of thermal physiology and thermoregulatory behavior

Given the ideas about coadaptation laid out in the preceding section, we can now go one step further and ask how the interaction of thermal physiology and thermoregulatory behavior influences adaptation. In Chapter 4, we saw that the optimal body temperature depends strongly on the shape of the performance curve. At the same time, the optimal performance curve depends on the body temperatures experienced by an organism. The union of these two results demands the correlated evolution of thermal sensitivity and thermal preference; an evolutionary shift in the thermal optimum should provide selective pressure for a subsequent shift in the preferred temperature, and vice versa (Huey and Bennett 1987). A patchy environment can even cause the evolution of a genetic correlation between physiology and behavior (Nosil et al. 2006). This coadaptation should produce a strong correlation between the thermal optimum and the preferred temperature within and among populations. Evidence of the coadaptation of physiology and behavior comes from two sources: covariations of phenotypes within individuals and among species (reviewed by Angilletta et al. 2002a).

For a coadapted genotype, a parallel shift in thermoregulatory behavior should accompany acclimation of a performance curve. For example, when acclimation to a cold environment involves a decrease in the thermal optimum, we should expect a corresponding decrease in the preferred temperature. Only a few teams have investigated the joint acclimation of thermal physiology and thermoregulatory behavior, but their work has generated a range of findings. Crocodiles maintained at 20°C underwent a perfect shift in their thermal optimum for swimming and a small parallel shift in their preferred body temperature (see Fig. 5.3a; Glanville and Seebacher 2006). In contrast, newts held at 25°C shifted their thermal optimum for swimming upward and their preferred body temperature downward (Fig. 7.5). Acclimation sometimes involves parallel shifts in the critical thermal limits of performance and the preferred body temperature. Prawns (*Macrobrachium acanthurus*) maintained at 20°C and those maintained at 30°C differed in their

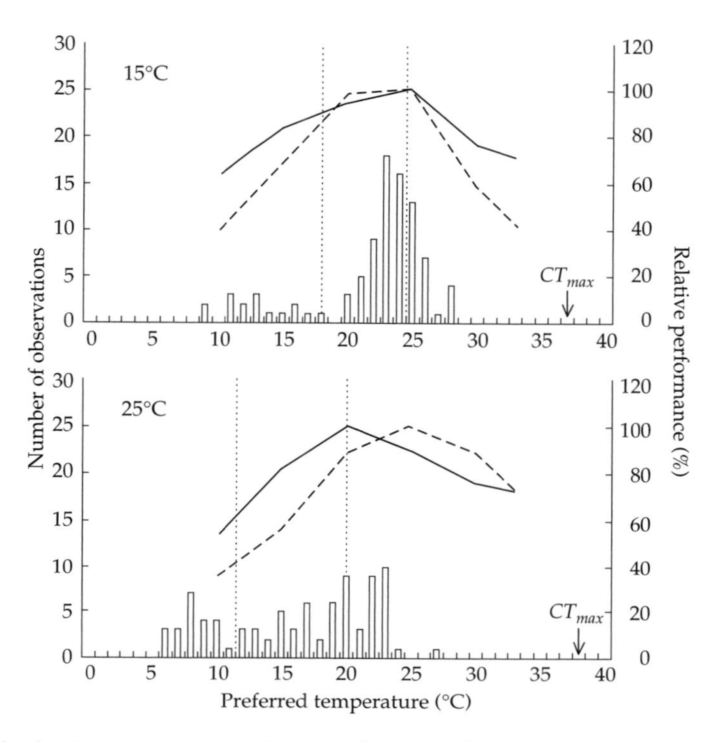

Figure 7.5 Acclimation of preferred body temperature (bars) and thermal sensitivities for swimming (solid lines) and running (dashed lines) did not support the hypothesis of coadaptation. After exposure to 15°C, the mean preferred temperature and the thermal optima for locomotor performances were approximately 25° C. After exposure to 25°C, the preferred temperature was much lower than the thermal optima. Vertical dotted lines denote the lower and upper quartiles of preferred body temperature. Adapted from Gvoždík et al. (2007) with permission from Blackwell Publishing.

preferred body temperatures by 7°C, but only differed in their critical thermal minima and maxima by 4°C (Diaz et al. 2002). The preferred body temperatures and critical thermal limits of grass lizards were positively correlated with acclimation temperature, but acclimation responses were small relative to thermal change; specifically, the preferred body temperature, critical thermal minimum, and critical thermal maximum of lizards kept at 35°C exceeded those of lizards kept at 20°C by only 3°C (Yang et al. 2008). Acclimation of angelfish to constant temperatures ranging from 20°C to 32°C affected the critical thermal maximum, but did not affect the preferred body temperature (Pérez et al. 2003). The diversity of responses observed in just these few studies justifies further development of the theory.

Interspecific comparisons have also provided mixed evidence of coadaptation. Dawson (1975) first examined interspecific correlations between preferred body temperatures and the thermal optima for physiological functions, including heart rate, muscle contraction, and enzyme activity. His review was restricted to reptiles, but established a qualitative match between behavior and physiology in many species (see, for example, Fig. 7.6). Beitinger and Fitzpatrick (1979) concluded the same was true for fish. Since these seminal reviews, two teams have conducted more rigorous tests of coadaptation at the interspecific level. Not only did their studies include quantitative analyses of correlated evolution, they also included efforts to control for phylogenetic effects. Both studies were focused on the locomotor performance of lizards; this focus was likely motivated by the observation that many active lizards maintain body temperatures that enable near maximal running speeds (Hertz et al. 1988).

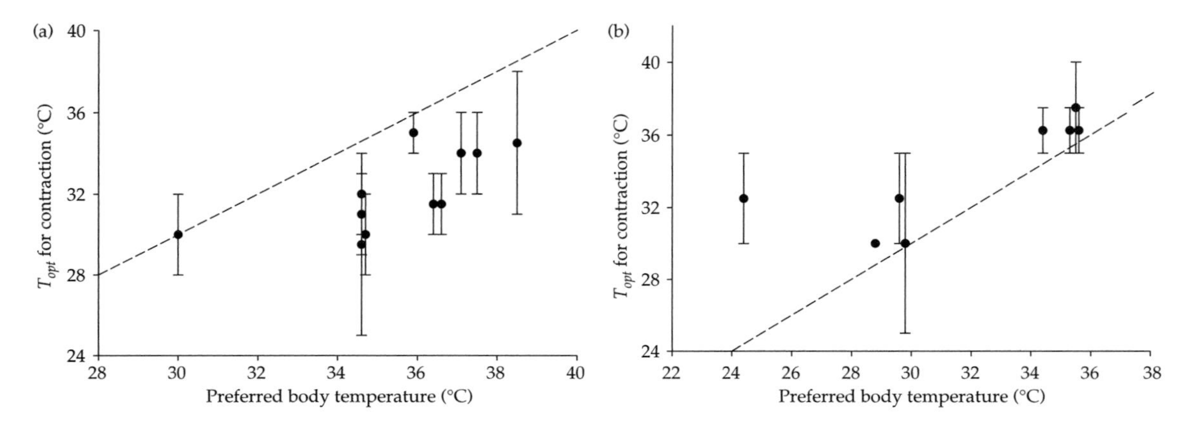

Figure 7.6 Among species of lizards, the thermal optimum for muscle contraction correlated with the preferred body temperature. Data are isometric twitch tensions (a, Dawson 1975) and tetanic tensions (b, John-Alder and Bennett 1987). Error bars show the optimal range of temperatures for contraction. Dashed lines show where the points should lie if the preferred temperatures perfectly matched the thermal optima.

Huey and Bennett (1987) conducted the first phylogenetic comparative analysis of coadaptation. They examined the correlated evolution of the preferred body temperature and the thermal sensitivity of sprinting in 12 species of Australian skinks. Because they did not know the timing of divergence among taxa, they chose a minimum-evolution approach in which ancestral phenotypes were estimated by averaging extant phenotypes. Then, they regressed hypothetical evolutionary changes in the thermal optimum, critical thermal minimum, and critical thermal maximum against changes in the preferred body temperature. This analysis supported the conclusion that the thermal optimum and the critical thermal maximum evolved in synchrony with the preferred body temperature. An analysis of raw variables also supported this conclusion. But this story has an interesting twist. Just a few years later, the emergence of a new phylogenetic hypothesis prompted Huey and Bennett to reanalyze their data. With the help of Garland, they used the new phylogeny to conduct a variety of phylogenetic analyses, including correlations of independent contrasts (Garland et al. 1991). The new analyses suggested only partial coadaptation had occurred. The preferred body temperature appears to have evolved in synchrony with the critical thermal maximum for sprinting, but did not evolve in synchrony with the critical thermal minimum or the thermal optimum (Fig. 7.7).

A similar study by Bauwens and colleagues (1995) provided stronger evidence of coadaptation. These investigators measured the preferred body temperatures and the thermal sensitivities of sprinting in 13 species of lacertid lizards. Analyses of independent contrasts showed that preferred body temperature covaried positively with the critical thermal maximum and the thermal optimum (Fig. 7.8). Moreover, the precision of thermoregulation and the breadth of performance coevolved very tightly, but in a way that contradicts logic: species with narrower performance breadths thermoregulated less precisely! As in Australian skinks, thermoregulatory behavior and thermal physiology seems only partially coadapted in lacertids.

These phylogenetic analyses tell us whether behavior and physiology have evolved in the same direction, but they do not tell whether traits have coadapted perfectly. This problem stems from an inherent limitation of the phylogenetic method. Independent contrasts define the deviation in the value of a trait between two taxa since they diverged from a common ancestor (see Box 1.5). For instance, we would calculate an independent contrast for the thermal optimum by subtracting the thermal optimum for one species from that of its sister species. Unfortunately, this contrast retains no information about the absolute value of the thermal optimum in each species. The same would be true of independent contrasts for preferred body temperature.

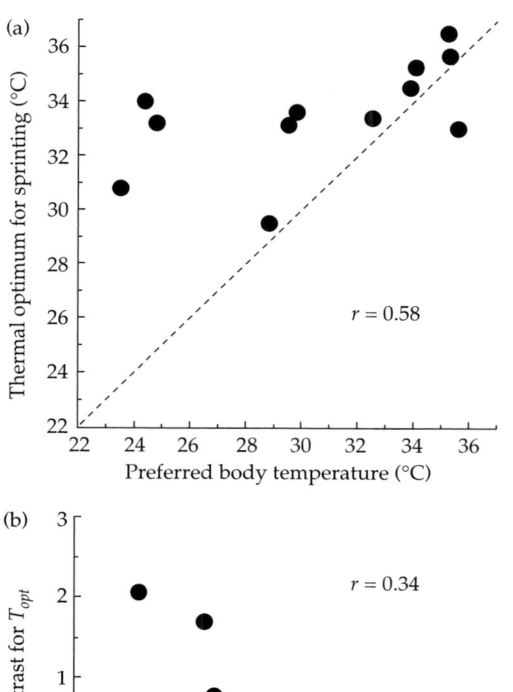

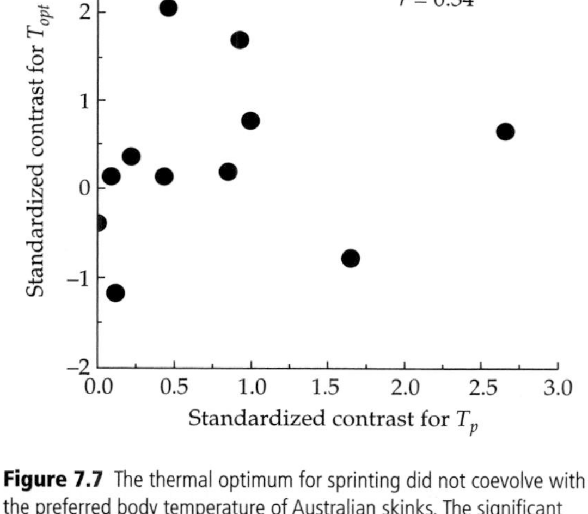

Figure 7.7 The thermal optimum for sprinting did not coevolve with the preferred body temperature of Australian skinks. The significant correlation between raw variables (a) indicates coadaptation, but the absence of a correlation between independent contrasts (b) suggests the relationship between raw variables resulted from phylogenetic pseudoreplication. Data are from Garland et al. (1991).

Therefore, a correlation between contrasts cannot tell us whether the preferred body temperature matches the thermal optimum for performance. To determine the degree of coadaptation, we must examine the correlation between raw variables. In doing so, we discover something interesting: the thermal optimum for sprinting usually exceeds the preferred body temperature (Fig. 7.7a; Bauwens et al. 1995; Huey et al. 1989). This observation has

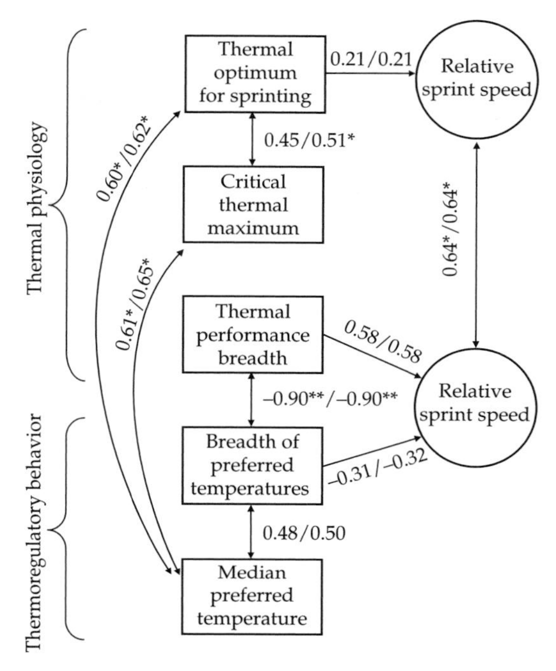

Figure 7.8 Relationships among aspects of thermal physiology and thermoregulatory behavior in lacertid lizards support the hypothesis of coadaptation. Correlation coefficients (double-headed arrows) or standardized regression coefficients (single-headed arrows) were calculated using independent contrasts; two values are reported for each relationship because two plausible phylogenetic hypotheses were available. Asterisks denote statistically significant coefficients. Adapted from Bauwens et al. (1995) with permission from Blackwell Publishing.

two possible interpretations. Either some unidentified mechanism constrains coadaptation, or perfect coadaptation does not necessarily imply the preferred temperature should equal the thermal optimum. As we shall see, recent modeling points toward the latter interpretation.

7.3.1 Mechanisms favoring a mismatch between preferred temperatures and thermal optima

Despite the current dogma, coadaptation should not always produce a perfect match between the preferred temperature and the thermal optimum. Two distinct conditions can favor a mismatch. First, when individuals cannot thermoregulate with perfect accuracy, the thermal optimum should exceed the preferred temperature. Second, when warm-adapted organisms outperform cold-adapted organisms, the thermal optimum should also exceed the preferred temperature. Let us briefly examine the reasons for these conclusions.

Martin and Huey (2008) developed a simple model to show how the accuracy of thermoregulation affects the course of coadaptation. If an organism thermoregulates with perfect accuracy, the greatest performance obviously results from maintaining a body temperature that equals the thermal optimum. But what mean body temperature maximizes performance when an organism thermoregulates imperfectly? The answer to this question depends not only on the variation in body temperature but also on the shape of the performance curve. For a symmetrical performance curve, unusually high body temperatures and unusually low body temperatures have equal impacts on performance (Fig. 7.9a). For an asymmetrical performance curve, unusually high body temperatures impair performance more than unusually low temperatures do (Fig. 7.9b). Because performance curves tend to be asymmetrical, inaccurate thermoregulation should affect the optimal preferred temperature.

To illustrate this point, Martin and Huey found the mean body temperature that maximized the following fitness function:

$$w = \sum_{T_b=CT_{\min}}^{CT_{\max}} f(T_b) \cdot P(T_b), \tag{7.5}$$

where $f(T_b)$ equals the frequency distribution of body temperatures. In biological terms, fitness equals the sum of performance over time, such as the case when performance directly affects fecundity (see Box 3.3). Both symmetrical and asymmetrical functions were examined for the frequency distribution, $f(T_b)$, and the performance curve, $P(T_b)$. For an asymmetrical performance curve, the thermal optimum exceeded the mean body temperature that maximized fitness (Fig. 7.10). The more an organism's temperature varied, the greater the optimal mismatch between the preferred temperature and the thermal optimum.[13] This effect was particularly exaggerated for thermal specialists because deviations from the thermal optimum imposed severe reductions in fitness (Fig. 7.10). However, both specialists and generalists should prefer temperatures below their thermal optima given extreme asymmetry of the performance curve.

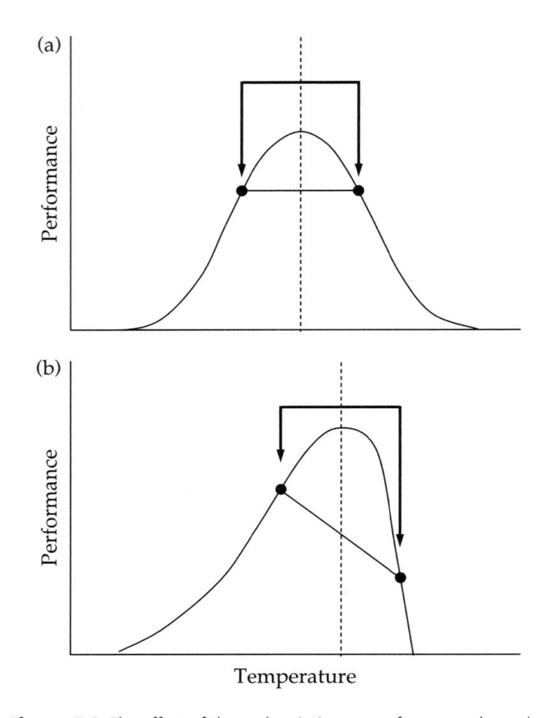

Figure 7.9 The effect of thermal variation on performance depends on the shape of the performance curve. For symmetric curves (a), body temperatures greater than and less than the thermal optimum lead to equal performances. For asymmetric curves, a body temperature below the thermal optimum enables better performance than does a body temperature above the thermal optimum. Based on Martin and Huey (2008).

To evaluate the predictions of their model, Martin and Huey conducted a phylogenetic comparative analysis of coadaptation. For 52 species of lizards, they calculated the difference between the mean preferred temperature and the thermal optimum for sprinting. The asymmetry (A) of each performance curve was characterized by

[13] The same phenomenon was predicted by earlier models of thermal adaptation (Beuchat and Ellner 1987; Gilchrist 1995). In Fig. 3.18, compare the optimal performance curves predicted by the levels of thermal heterogeneity. As thermal variation increases among generations, selection favors a greater difference between the mean environmental temperature and the thermal optimum.

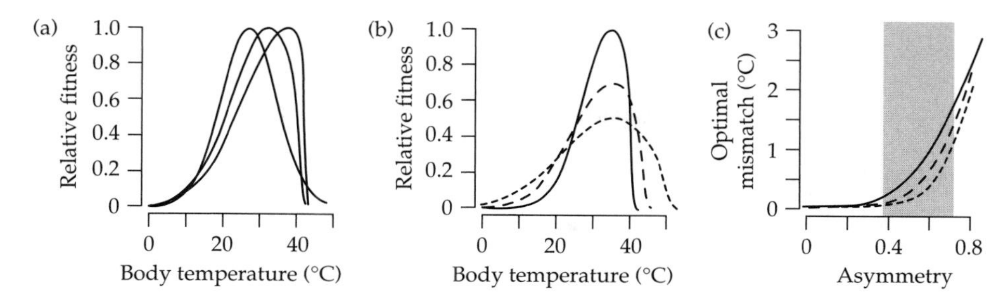

Figure 7.10 The optimal mismatch between the preferred temperature and the thermal optimum depends on the asymmetry and breadth of the performance curve. (a and b) Examples of symmetries and breadths used to model performance curves. (c) The three lines correspond to the breadths shown in (b). The optimal mismatch increases with increasing asymmetry but decreases with increasing breadth. Adapted from Martin and Huey (2008, *The American Naturalist*, University of Chicago Press). © 2008 by The University of Chicago.

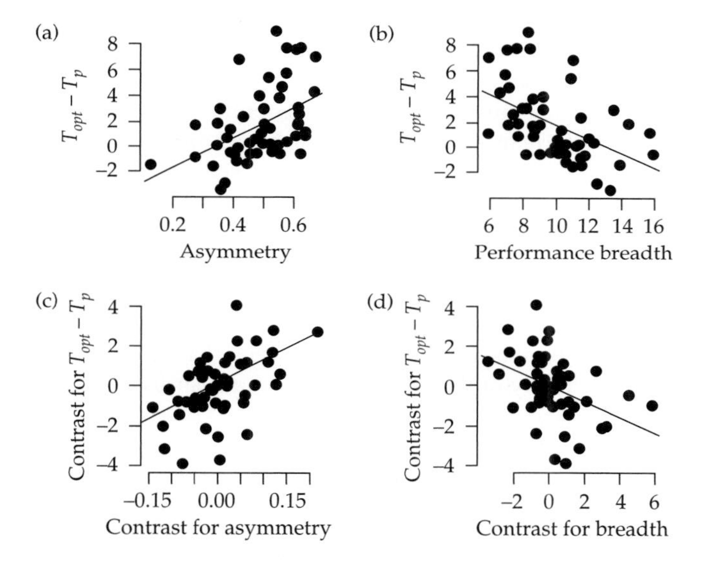

Figure 7.11 The observed mismatch between the preferred body temperature and the thermal optimum for sprinting depends on the asymmetry and breadth of the performance curve. As predicted, the mismatch increases as asymmetry increases and decreases as breadth increases. (a and b) Raw data. (c and d) Independent contrasts. Adapted from Martin and Huey (2008, *The American Naturalist*, University of Chicago Press). © 2008 by The University of Chicago.

the following function:

$$A = \frac{2T_{opt} - CT_{\max} - CT_{\min}}{CT_{\max} - CT_{\min}} \tag{7.6}$$

The empirical data were qualitatively consistent with the model (Fig. 7.11). First, the observed mismatch between the preferred temperature and the thermal optimum increased as the asymmetry of the performance curve increased. Second, thermal specialists exhibited a greater mismatch than did thermal generalists. Analyses of raw data and independent contrasts showed similar patterns. However, close inspection of the raw data reveals that the observed mismatches greatly exceed those predicted by the model (compare Figs 7.10c and 7.11a). Therefore, some of this variation likely reflects the evolutionary consequences of other processes. A possible explanation relates to the fitness function used in the model. By calculating fitness as the sum of performance, Martin and Huey assume that the body temperature of the organism never exceeds its thermal limit for survival. Relaxing this assumption would favor a greater mismatch between the thermal optimum and the preferred temperature (Martin and Huey 2008). But some of this mismatch might stem from a mechanism that has nothing to do with the asymmetry of performance curves.

A second mechanism that promotes a mismatch between the preferred temperature and the thermal optimum involves thermodynamic constraints on

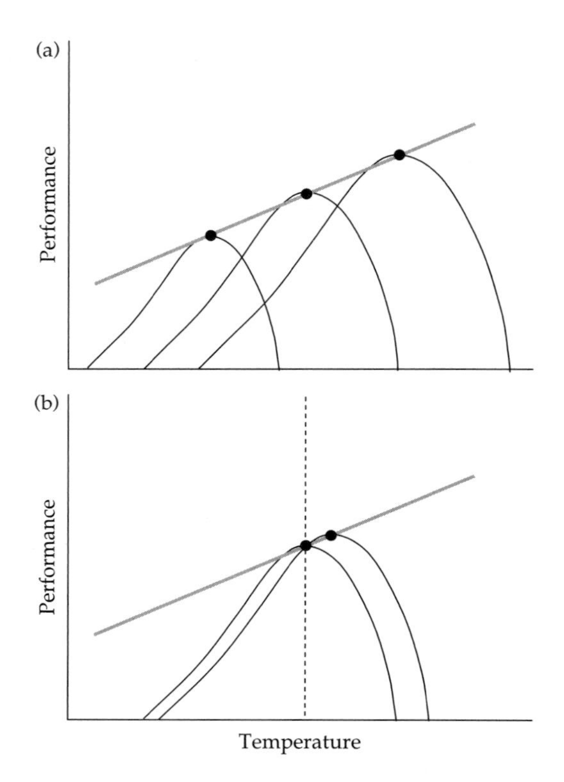

Figure 7.12 (a) The hypothesis that "hotter is better" describes a thermodynamic effect on the performance of an organism. The mean or maximal performance over a range of temperatures increases as the thermal optimum increases. (b) When hotter is better, selection favors a thermal optimum for performance that exceeds the mean body temperature (dashed line). This shift enhances performance at high body temperatures without impairing performance at the mean body temperature.

performance. From Chapter 4, recall that the performance of warm-adapted organisms usually exceeds that of cold-adapted organisms. In other words, hotter is better (Fig. 7.12a). Despite this common observation, models of thermal adaptation assume that adaptation to cold environments yields the same mean performance as adaptation to warm environments. Which predictions, if any, would change if we included a thermodynamic effect on performance? To answer this question, Asbury and I modeled the optimal performance curve under the assumption that hotter is better (D. Asbury and M. J. Angilletta, unpublished observations). We used a model very similar to the one developed by Gilchrist (1995; see Chapter 3), except we assumed the mean performance of the organism depended on its thermal optimum:

$$w = \sum_{T_b=CT_{\min}}^{CT_{\max}} f_g(T_b) \cdot P(T_b, T_{opt}), \qquad (7.7)$$

where $f_g(T_b)$ equals the frequency distribution of body temperatures in the current generation (compare with eqn 3.13). The thermodynamic effect on performance was modeled according to metabolic theory (Gillooly et al. 2001, 2002; Savage et al. 2004):

$$P \propto e^{-E/kT_{opt}}, \qquad (7.8)$$

where E equals the mean activation energy of rate-limiting reactions, and k equals Boltzmann's constant.

Because a higher thermal optimum enhances performance, an organism achieves maximal fitness when its thermal optimum exceeds its mean body temperature (Fig. 7.12b). An increase in the variance of body temperature intensifies this effect (Fig. 7.13). Thermodynamic constraints and inaccurate thermoregulation independently promote a mismatch between physiology and behavior, as long as the thermodynamic effect does not change the asymmetry of the performance curve (Martin and Huey 2008).

These models assume the thermal sensitivity of a single performance determines the fitness of an organism, but many physiological processes occur simultaneously. These processes likely differ in their thermal sensitivity and their impact on fitness. Variations in thermal optima and performance breadths among processes should influence coadaptation. In some species, very different performances proceed maximally at a single body temperature (Du et al. 2000; Huey 1982; Ji et al. 1995; Luo et al. 2006; Stevenson et al. 1985). In other species, thermal optima differ among physiological performances (Chen et al. 2003; Ji et al. 1996; Van Damme et al. 1991). An organism can solve this problem by varying its body temperature such that all processes proceed at a maximal rate for some duration. Alternatively, an organism can maintain a body temperature that permits all processes to proceed at a moderate rate. When adopting the latter strategy, the optimal body temperature depends on the relative impact of each process on fitness (see Chapter 4).

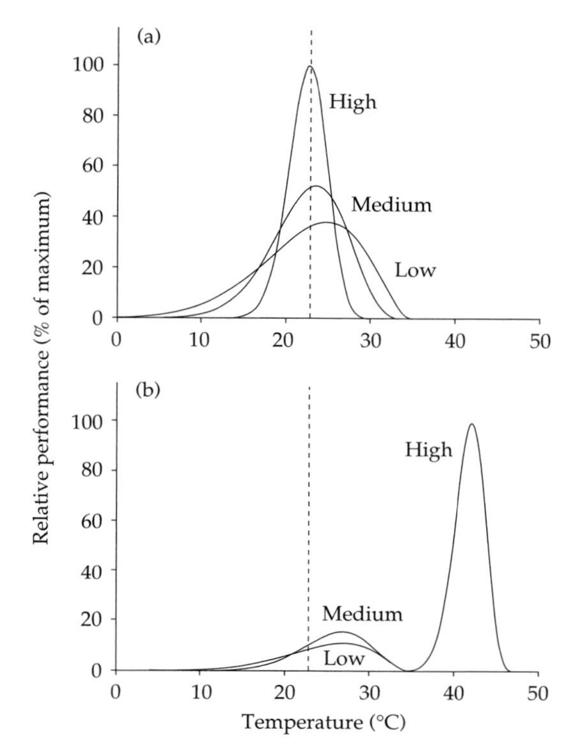

Figure 7.13 When hotter is better, optimal performance curves deviate from those described in Chapter 3 (D. Asbury and M. J. Angilletta, unpublished observations). Here, optimal performance curves without a thermodynamic constraint (a) are compared with those when hotter is better (b). For each curve, variation in body temperature among generations was high, while variation within generations was low, medium, or high. Variations within and among generations correspond to those shown in Fig. 3.17. As a reference, the dashed line in each plot approximates the mean environmental temperature. When body temperature varies within generations, a thermodynamic effect on performance favors a thermal optimum that exceeds the mean body temperature.

If performance breadths vary among physiological processes, thermoregulatory behavior should coevolve with the most thermally sensitive performances. For example, the preferred temperature of *Sceloporus undulatus* appears more closely related to the thermal sensitivity of energy assimilation than the thermal sensitivities of sprinting and endurance (Angilletta et al. 2002b). When performances differ in their thermal sensitivity, detecting coadaptation will require extensive knowledge of the relationships between performance and fitness. Keep in mind that empirical studies of coadaptation have focused on one or two physiological performances, whose links to fitness remain unclear.

7.3.2 Predicting coadapted phenotypes

The current theory falls short of providing a full understanding of coadaptation. First, let us review what the theory does address. The theory clearly predicts covariation between thermal physiology and thermoregulatory behavior. We expect the preferred temperature to correlate with the thermal optimum for performance, although variations in the precision of thermoregulation and the sensitivity to high temperatures would weaken this correlation. We also expect the preferred range of temperatures to correlate with the thermal breadth of performance, particularly when performance determines survivorship. These predictions come from the informal integration of models that focus on either thermal physiology or thermoregulatory behavior. Nevertheless, this informal integration cannot predict which combination of strategies should occur in a particular environment. Why do some organisms thermoregulate precisely and tolerate a very narrow range of temperatures? Why do other organisms tolerate a broad range of temperatures and let their body temperature fluctuate with the environmental temperature? To answer these questions, we need a very different theory of coadaptation. This theory must evaluate the costs and benefits of all combinations of thermal sensitivity and thermoregulatory behavior in a range of environments.

In the absence of a formal theory, how do we know which combination of strategies confers the greatest fitness? At one extreme, an organism can regulate its body temperature precisely and specialize for physiological function at its target temperature. At the other extreme, the organism can let its temperature fluctuate and evolve a performance curve and an acclimation capacity that suit the variation in body temperature. Somewhere in the middle lie countless combinations of thermoregulatory behavior and thermal physiology. The course of adaptation depends on the costs and benefits of these strategies in the current environment. If the net benefit of a pure behavioral strategy exceeds that of a pure physiological strategy, then some degree of thermoregulation should evolve (and vice versa). Given the many factors that determine the costs and benefits of these strategies (see Chapters 3–5), coadaptation could be a very complex process to understand. Yet, I hold some hope that we can simplify the problem.

Coadaptation would be a much simpler process if one strategy conferred an inherently greater benefit or lesser cost than did other strategies. Recently, my colleagues and I drew some conclusions about coadaptation from the belief that the benefit of thermoregulation inherently exceeds the benefits of thermal generalization and thermal acclimation (Angilletta et al. 2006a). This belief rests on two assumptions: (1) a jack of all temperatures is a master of none (see Chapter 3) and (2) hotter is better (see Chapter 4). If a jack of all temperatures really is a master of none, a thermal specialist with a constant body temperature would have a higher fitness than a thermal generalist with any distribution of body temperatures (Gilchrist 1995; Lynch and Gabriel 1987). If hotter is truly better, a specialist adapted to a high body temperature would enjoy greater fitness than a specialist adapted to a low body temperature (Frazier et al. 2006). Combining these assumptions, we proposed that the cost of thermoregulation drives coadaptation (Fig. 7.14). Environments that pose little cost of thermoregulation favor specialists that thermoregulate precisely. The optimal body temperature depends not only on the cost of thermoregulation but also on the exact magnitude of the thermodynamic effect and physical constraints on molecular stability. Alternatively, more costly environments favor genotypes that thermoregulate imprecisely or even not at all. The optimal thermal sensitivity of these genotypes depends on the thermal heterogeneity within and among generations and the mathematical relationship between performance and fitness. If fitness depends on the sum of performance over time, a specialist should evolve when body temperature varies relatively little among generations (Gilchrist 1995). A generalist should evolve if body temperature varies greatly among generations (Gilchrist 1995) or if fitness depends on the product of performance over time (Lynch and Gabriel 1987). Importantly, selection might favor generalization through developmental or reversible acclimation (Gabriel 2005; Gabriel and Lynch 1992), which would appear as specialization on a short temporal scale.

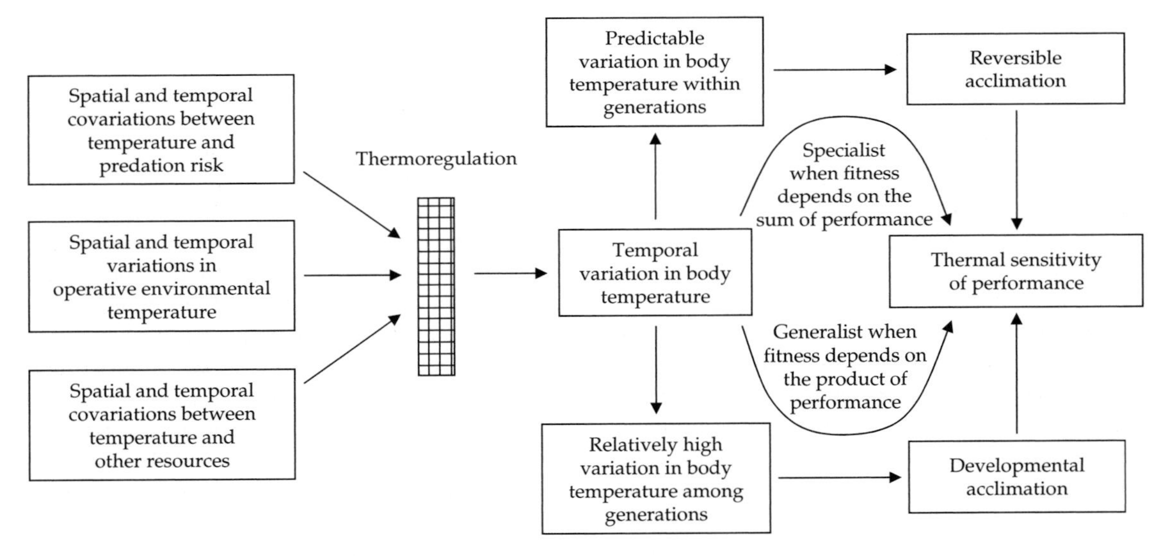

Figure 7.14 The evolution of thermoregulatory behavior should drive the thermal coadaptation of traits. Both abiotic and biotic aspects of the environment determine the optimal degree of thermoregulation, which in turn defines the variation in body temperature. Variations in body temperature within and among generations define the selective pressures on thermal sensitivity and its acclimation. This view implicitly assumes that (1) specialists outperform generalists and (2) warm-adapted genotypes outperform cold-adapted genotypes. Adapted from Angilletta et al. (2006a, *Physiological and Biochemical Zoology*, University of Chicago Press). © 2008 by The University of Chicago.

This simplified view of coadaptation relies on the validity of our two critical assumptions. Current evidence supports the existence of a thermodynamic effect on performance. Indeed, warm-adapted genotypes usually outperform cold-adapted genotypes (i.e., hotter is better; see Chapter 4). Less convincing evidence exists for a universal tradeoff between specialization and generalization; a jack of all temperatures can be a master of all (see Chapter 3). Yet, our current understanding of biochemical adaptation indicates this tradeoff must occur at the biochemical level, even if it can be masked at the organismal level (see Angilletta et al. 2003). If these assumptions ultimately prove incorrect, the process of coadaptation will differ from the one depicted in Fig. 7.14. The absence of thermodynamic constraints or specialist–generalist tradeoffs would weaken selection for thermoregulation and strengthen selection for generalization (see, for example, Gilchrist 1995). Ultimately, biologists may have no choice but to develop more complex models of coadaptation.

7.4 Coadaptation of thermoregulatory behavior, thermal physiology, and life history

To achieve either precise thermoregulation or thermal generalization, an organism must invest resources which then become unavailable for other functions. Regulating body temperature in a changing environment requires time or energy and exposes organisms to risks they might otherwise avoid. Tolerating extreme temperatures requires energy to maintain a suite of proteins or to turn over existing proteins during thermal change. Decisions to make these investments undoubtedly shape the survivorship and fecundity of an organism in multiple ways. Therefore, we should expect the thermoregulatory behavior, thermal physiology, and life history of an organism to interact during evolution.

Biologists use the concept of a tradeoff to study the interactions between physiological performances and life-history traits (Dunham et al. 1989; Stearns 1992). An acquisition tradeoff occurs whenever an organism incurs some risk of mortality associated with acquiring resources (e.g., energy, mates, or even microclimates). An allocation tradeoff occurs whenever an organism divides a finite supply of resources among competing demands (e.g., growth vs. reproduction). The influx and division of resources determine life-history traits, such as age-specific body size, survivorship, and fecundity. Most models of life-history evolution consider relatively simple problems, such as the division of energy between growth and reproduction (see Chapter 6). But organisms must divert incoming energy to many other functions, including maintenance, locomotion, and development. The thermal sensitivity of these performances will depend on the same factors thought to influence life-history traits.

We need consider only the most basic conclusions of life-history theory to forge a connection with thermal adaptation. Classic models have shown the optimal life history depends strongly on age-specific rates of mortality (Gadgil and Bossert 1970; Hirshfield and Tinkle 1975; Law 1979; Michod 1979). These models distinguish between two types of mortality risk: intrinsic and extrinsic. The intrinsic risk of mortality depends on the organism's phenotype, whereas the extrinsic risk does not. For example, an animal that thermoregulates behaviorally may draw the attention of predators and hence have a greater *intrinsic* risk of mortality. Yet, even an inactive animal can fall prey to a predator, which suggests some *extrinsic* risk also exists. This background risk of mortality should drive the evolution of behavior, physiology, and life history. When the environment poses a high extrinsic risk of mortality, an individual faces low odds of surviving regardless of its phenotype. Hence, natural selection favors individuals that reproduce earlier and more often, even if doing so involves risky behaviors to acquire the necessary resources (Stearns 2000). Let us think about this result in the context of thermal adaptation. As with any performance, the rate of reproduction depends on body temperature. More precise thermoregulation should enhance reproductive performance. Thus, higher extrinsic mortality should cause individuals to accept more risk when seeking preferred microclimates.

Extrinsic mortality should also affect the evolution of performance curves. Direct effects stem from the allocation of resources between competing functions (see Fig. 3.30a). For example, mean growth rate over a range of temperatures depends on the allocation of energy between growth and reproduction. If extrinsic mortality increases, the optimal allocation of energy shifts from growth toward reproduction. Consequently, natural selection will decrease the mean growth rate and increase the mean reproductive rate. Size-specific mortality can have the opposite effect. When an individual must reach a certain size to survive a stressful period, energy should be allocated to growth at the expense of other performances. In this case, natural selection will increase the mean growth rate, resulting in patterns such as countergradient variation (Conover 1992). Indirect effects stem from the coadaptation of thermoregulatory behavior and thermal sensitivity. Because thermoregulation determines the mean and variance of body temperature, high extrinsic mortality not only favors precise thermoregulation but also favors a corresponding change in the thermal optimum and performance breadth (see Section 7.3). Overall, we should expect the evolution of precise thermoregulation, thermal specialization, early maturation, and greater reproduction when the environment poses a high extrinsic risk of mortality.

7.5 Constraints on coadaptation

Optimality models define the fitness landscape, but the dynamics of coadaptation depend on genetic constraints as well as selective pressures (see Chapter 1). Because traits interact to determine fitness, a genetic constraint for one trait can alter the natural selection of other traits (Warner 1980). Some biologists have suggested the primary response to thermal change involves thermoregulatory behavior (Hertz 1981). This behavioral response would then impose selective pressure on thermal physiology (Huey and Bennett 1987), a process referred to as behavioral drive (Huey et al. 2003). Yet little evidence exists to confirm the belief that behavioral traits evolve more readily than physiological or life-historical traits (Artacho et al. 2005; Carlson and Seamons 2008). In fact, far more information exists about the heritability of thermal physiology and life history than exists about the heritability of thermoregulatory behavior. That said, we really do not know all that much about the heritability of thermal physiology (see Chapter 3). Therefore, we must await further research before drawing general conclusions about the limits to coadaptation imposed by genetic variance.

Genetic correlations between traits can either prohibit or facilitate coadaptation (Wolf and Brodie 1998). For example, if genotypes that prefer high temperatures tend to perform best at low temperatures, selection to increase the preferred temperature will decrease the thermal optimum. This negative genetic correlation would prevent coadaptation. Alternatively, a positive genetic correlation would facilitate coadaptation (Box 7.1). Such genetic correlations result from pleiotropic genes, physical linkage, or linkage disequilibrium. Therefore, mutation, recombination, and selection can eliminate existing genetic correlations and generate new ones. Indeed, coadaptation itself can result in the evolution of genetic correlations that facilitate future coadaptation. For example, imagine a metapopulation in which local environments cause a different suite of phenotypes to evolve in each population. Migration among populations will introduce genotypes whose suite of phenotypes differs from the local suite. Even if no physical linkage exists between coadapted alleles, this combination of adaptation and migration would cause a genetic correlation by linkage disequilibrium. The strength of the genetic correlation increases in direct proportion to the rate of migration among populations (Nosil et al. 2006). In this way, selected individuals tend to pass on alleles that promote coadaptation.

7.6 Conclusions

In this chapter, a fundamental tenet of the theory of thermal adaptation has been called into question. Simple models either predict or assume a close match between an organism's mean body temperature and its thermal optimum. Two evolutionary processes can lead to mismatches between the body temperature and the thermal optimum. First, time lags for acclimation and deacclimation favor a mismatch to reduce the loss of performance during thermal change. Second, imprecise thermoregulation favors a mismatch to prevent the loss of performance at high temperatures. These conclusions could have been inferred from models that ignore coadaptation (Gabriel 1999; Gilchrist 1995), but an

Box 7.1 Evolution of preferred body temperature by indirect selection

Experimental evolution of *Drosophila melanogaster* provided indirect evidence of a genetic correlation between thermoregulatory behavior and thermal physiology. Good (1993) maintained 10 populations of flies at each of three temperatures (25°C, 27°C, and 30°C) for 15 generations. After 10 generations of selection, flies from all lines were raised at 25°C and their preferred body temperatures were measured in a thermal gradient. Despite the fact that all generations experienced a uniform environment, preferred body temperatures had diverged genetically among selection lines. Specifically, flies that evolved at a higher temperature preferred a higher temperature (Fig. 7.15). Because no direct selection for thermoregulation could have occurred without thermal heterogeneity, only two processes can explain the evolution of preferred body temperature: genetic drift caused by a finite population or a correlated response to selection of thermal physiology. The latter explanation seems plausible considering that other selection experiments with *D. melanogaster* documented the evolution of thermal physiology (see Chapter 3). Unfortunately, Good did not determine whether the thermal optimum for any performance had diverged among selection lines. A positive genetic correlation between the preferred temperature and the thermal optimum would facilitate coadaptation, except in cases where selection favors a greater mismatch between these traits (see Section 7.3.1).

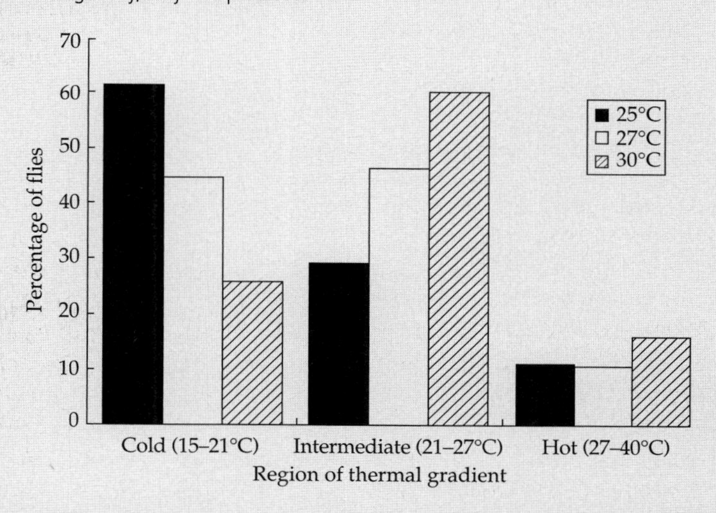

Figure 7.15 The distributions of flies (*Drosophila melanogaster*) in a thermal gradient differed among lines evolving at constant temperatures (25°C, 27°C, and 30°C). Flies that evolved at 25°C preferred the cool region of the gradient, while flies that evolved at 30°C preferred the intermediate region. Flies from all lines avoided the hot region. Adapted from Good (1993) with permission from Elsevier. © 1993 Elsevier Ltd.

explicit focus on the interactions among traits makes certain processes more transparent.

A formal theory of coadaptation would enable significant advances in evolutionary thermal biology. Models that ignore coadaptation predict thermoregulatory behavior, thermal physiology, or life history, but cannot predict the optimal combination of phenotypes. Two constraints on thermal adaptation will facilitate our transition to a general theory: (1) specialist–generalist tradeoffs and (2) thermodynamic effects. Given these constraints, coadaptation should proceed according to the costs of thermoregulation. Efforts should continue to establish the nature of these constraints in a wide range of organisms. At the same time, we should also develop mathematical models of coadaptation that will enable us to relax these constraints. Finally, we need to exploit a larger set of empirical tools for testing hypotheses about coadaptation. Multivariate selection gradients can reveal interactions between traits that influence fitness (Brodie et al. 1995), such as the interaction between the preferred temperature and the thermal optimum. Physical or genetic manipulations of thermoregulatory behavior and thermal physiology (e.g., Krochmal and Bakken 2003; Sack et al. 2005) can also be used to assess the fitness consequences of decoupling coadapted traits. Experimental evolution in the laboratory can reveal the correlated evolution of behavior, physiology, and life history. If we fully embrace a coevolutionary perspective, the theory of thermal adaptation may change radically in the coming years.

CHAPTER 8

Thermal Games

8.1 Filling the ecological vacuum

In previous chapters, we implicitly assumed that the thermal adaptation of a species does not depend on the density or frequency of other species. In other words, we have dreamed of evolution in an ecological vacuum. This incredibly naïve approach is slowly being supplanted by a more realistic one: evolutionary games. The theory of evolutionary games deals with the evolution of strategies whose fitness depends on the strategies of other organisms. The concept of an evolutionary game was borrowed from economics and popularized in biology by John Maynard Smith (Maynard Smith 1972, 1976, 1982; Maynard Smith and Price 1973). Unfortunately, the seeds sown by Maynard Smith have grown slowly within certain biological disciplines. Indeed, thermal biology seems untouched by game theory. In this chapter, I seek to stimulate progress by promoting a theory of *thermal games*.

By thermal games, I do not mean the Olympics of temperature; rather, I mean thermal adaptation in an environment where the optimal strategy depends on the strategies of competitors, predators, prey, parasites, and mutualists. Biologists refer to this process as frequency-dependent selection (Nowak and Sigmund 2004). Although *game theory* usually calls to mind analytical models that define evolutionarily stable strategies (Maynard Smith 1976), no single approach defines all of game theory. The current theory includes any model where the fitness of a strategy depends on its frequency and the frequencies of other strategies. Therefore, optimality, quantitative genetic, and allelic models constitute the modern theory of evolutionary games (for a history of this theory see Sigmund 2005).

Why must we consider models of thermal games? Numerous case studies have shown that organisms adjust their phenotype in response to the presence of other organisms. Recall the thermoregulatory behaviors of individuals in the presence and absence of others. Bluegill sunfish occupy their preferred thermal habitat in isolation, but shift to a less preferable habitat when exposed to a larger conspecific (Medvick et al. 1981). Similarly, we know zooplankton avoid warm shallow waters when they detect the presence of a predatory fish (Loose and Dawidowicz 1994). Finally, locusts bask more frequently when infected by a fungus than they do in the absence of this pathogen (Ouedraogo et al. 2004). Results such as these expose the need for a theory of thermal adaptation that accounts for the interactions among individuals. But how should we incorporate these interactions into our theory? One approach would be to assume certain behaviors impose costs that arise from the interaction with another species; for instance, we might assume that predators impose an elevated risk of mortality in certain habitats but not others. The current theory can easily be extended to include such assumptions and some models explicitly consider interactions among species (in Chapter 3, see the model by Polo et al. 2005). Still, these models treat other species as inert entities, which cannot alter their own phenotypes (Lima 2002). Game theory avoids this unrealistic assumption, letting us consider the coevolution of species that results from any type of ecological interaction. In this chapter, I focus more on introducing a theory of thermal games than on evaluating its performance. Rigorous tests must wait until more thermal biologists adopt this theory and design experiments to test its predictions.

8.2 Approaches to the study of frequency-dependent selection

The classic approach to understanding evolutionary games rests on the concept of an evolutionarily stable strategy. First, we would translate our assumptions about the game into a set of equations that link the phenotypes of two or more organisms to their respective fitnesses. Next, we would use these equations to find a strategy that can invade a population when rare, but cannot be invaded by another strategy once it becomes common (Maynard Smith 1982). In other words, we find an evolutionarily stable strategy. Such a strategy would ultimately evolve if a population was adapting to a stable set of abiotic and biotic interactions and possessed sufficient additive genetic variance in the phenotype of interest (see Chapter 1).

An evolutionarily stable strategy represents a special case of a more general kind of equilibrium, called a Nash equilibrium. A Nash equilibrium equals a set of strategies for which no other strategy would enhance an individual's fitness, assuming other individuals maintain their strategies (Riechert and Hammerstein 1983). However, a population of individuals exhibiting a Nash equilibrium can be infiltrated by individuals exhibiting an alternative strategy that confers equal fitness; thus, the possibility of genetic drift requires a distinction between a Nash equilibrium and an evolutionarily stable strategy. For a Nash equilibrium to qualify as an evolutionarily stable strategy, either no alternative strategy can confer as much fitness as the Nash equilibrium, or alternative strategies become less fit than the Nash equilibrium as they increase in frequency. Some models predict no evolutionarily stable strategy and others predict several.

In practice, an evolutionarily stable strategy could be pure or mixed. Pure strategies exist when all individuals of a population should exhibit the same phenotype. Mixed strategies exist when either (1) proportions of a population adopt different strategies or (2) an individual possesses a set of strategies adopted with certain probabilities (Dawkins 1980). Usually, the evolutionarily stable strategy differs from the optimal phenotype in the absence of frequency-dependent selection.

Complementary approaches to the search for an evolutionarily stable strategy involve the use of analytical models and numerical simulations, including individual-based models (DeAngelis and Mooij 2005). For an analytical model, one must first identify whether a rare genotype could invade a large population of other genotypes. Successful invasion suggests the mutant can become established in the population, whereupon it would coexist with or exclude other genotypes. To determine a genotype's potential for invasion, one calculates the limit of the growth rate of a genotype as its density approaches zero; this limit obviously depends on the resident genotypes and other environmental factors. After establishing a genotype's ability to invade a population when rare, one must also establish the genotype's ability to resist invasion when abundant. Thus, an evolutionarily stable strategy prevents the limit of all other genotypes from exceeding the critical value for population growth (Proulx and Day 2001). For more complex models, one cannot solve analytically for an evolutionarily stable strategy. Instead, one must use an individual-based model to simulate the invasion of a rare mutant in a population dominated by another genotype. In such simulations, phenotypes can evolve under constraints imposed by quantitative or multilocus genetics (see Chapter 1). The details of the genetic system and the assumptions about the selection gradients determine the dynamics of coevolution. Individual-based models of evolutionary games can yield stable equilibria that are analogous to the optimal phenotype in a model of frequency-independent selection. In this chapter, we shall consider analytical and individual-based models of thermal games.

8.3 Optimal thermoregulation in an evolutionary game

Considering the examples provided in Section 8.1, we should gain considerable insight by analyzing thermoregulatory strategies in the context of an evolutionary game. The simplest of games occurs between thermoregulators of a single species, which would be described as a game between competitors. This simple model could then be generalized to explore the properties of more complex games, such as the evolutionary interaction between a group of thermoregulators and their predators. In either case, think of behavioral thermoregulation as a process of choosing among potential microhabitats, which differ not only in their operative temperatures but also in their densities of food, competitors, and predators. Adopting this point of view enables us to draw on the rich body of theory that was developed for analyzing the optimal use of habitats in an evolutionary game.

Game models of habitat choice have steadily accumulated since the early 1970s (reviewed by Sih 1998). All of these models describe a heterogeneous environment as a mixture of two or more patches that differ in resources; here, we shall explicitly consider thermal resources. In the simplest case, we can envision a matrix of warm and cool patches. If we were ambitious, we might extend this analysis to a continuum of thermal patches occurring in equal or unequal frequencies. We might also envision a scenario in which movements among patches impose costs. Here, we shall explore the properties of very simple models to see how novel hypotheses emerge from the study of thermal games.

8.3.1 Competition during thermoregulation

To see how a game between competitors can influence thermoregulatory behavior, we shall adapt a simple model of habitat choice described by Sih (1998). Imagine an environment consisting of two patches that differ in their mean operative temperatures (T_1 and T_2) and their densities of food (F_1 and F_2). In this simple scenario, thermoregulation consists of choosing between these alternative microhabitats on the basis of temperature. Naturally, this choice has repercussions for the performance of an individual. In this model, we can think of performance as the ability to consume and process food; specifically, performance equals the rate of energy assimilation per unit of food available in a patch. The expected performance of an individual in patch i equals $P(T_i)$. Therefore, an individual's rate of energy gain (E) differs between patches because of food and temperature: $E(P(T_1), F_1) \neq E(P(T_2), F_2)$. As with other models (e.g., Huey and Slatkin 1976), the optimal strategy of thermoregulation maximizes the net rate of energy gain (assuming an individual expends no energy when moving between patches).

Because we desire a thermal game, we must introduce some form of competition among individuals. The actual mechanism of competition does not matter too much; either exploitative competition for food or interference competition for basking sites would yield similar results. In both cases, competition reduces the energy gain of an individual. Mathematically, we can introduce competition by making energy gain a function of the density of competitors as well as food and temperature (Sih 1998):

$$E = \frac{P_i(T_i)F_i}{N_i^x}, \tag{8.1}$$

where N_i equals the density of individuals in patch i. This function enables some flexibility regarding the intensity of competition. When $x = 1$, individuals compete so severely that a doubling in their density cuts each individual's performance in half. When $x < 1$, individuals compete less severely. By setting $x = 0$, we can also find the optimal strategy in the absence of competition.

How should competitors distribute themselves between thermal patches? To answer this question, we shall invoke the concept of an ideal free distribution (Fretwell and Lucas 1969). When individuals conform to an ideal free distribution, they occupy patches such that all receive the same rate of energy gain. If any individual changes its behavior, its rate of energy gain would decrease. This loss of energy should lead to further changes in behavior that restore the ideal free distribution. Think of this distribution as the outcome of flexible behaviors shaped by generations of natural selection. To find the ideal free distribution, we must set $E(P(T_1), F_1, N_1)$ equal to $E(P(T_2), F_2, N_2)$ and solve for N_1/N_2. In a population of competitors, the ideal free distribution depends on the disparity of operative temperatures and the intensity of competition:

$$\frac{N_1}{N_2} = \left(\frac{P(T_1)F_1}{P(T_2)F_2}\right)^{1/x}. \tag{8.2}$$

As we can see from this equation, individuals should distribute themselves according to their relative rate of energy gain in the two patches. During severe competition ($x = 1$), we should expect twice the number of individuals in the warmer patch if this patch enables twice the rate of energy gain (Fig. 8.1). As competition becomes less severe (i.e., $x \to 0$), the proportion of individuals occupying the warm patch should increase. Nevertheless, we

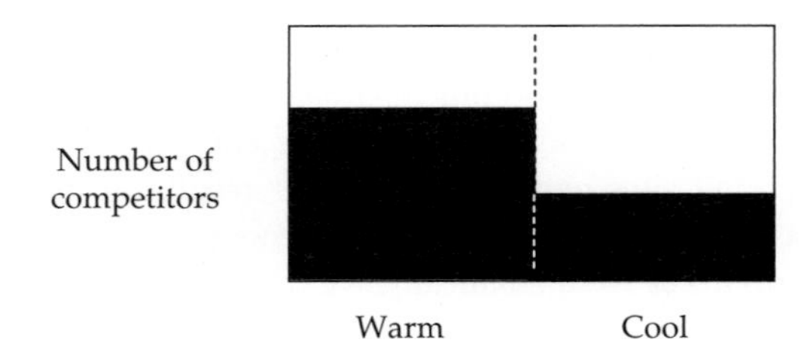

Figure 8.1 In an environment with two patches, the ideal free distribution of competitors should match the distribution of thermal resources. If individuals in the warm patch gain energy twice as fast as individuals in the cool patch, competitors should spend twice as much time in the warm patch.

should expect some individuals to occupy the cooler patch as long as this patch enables some performance. Only in the absence of competition (i.e., $x = 0$) should all individuals congregate in the warmer patch. But without competition, the model no longer represents a thermal game.

The ideal free distribution becomes particularly interesting when food and temperature interact to determine the quality of a patch. Hughes and Grand (2000) explored this phenomenon by modeling the optimal distribution of fish between two thermal patches. From Chapter 3, recall that food availability affects the thermal sensitivity of growth rate. In fish, high temperatures maximize growth when food is abundant, but low temperatures maximize growth when food is scarce (see Fig 3.3 and Huey 1982). When many fish compete for food, this interaction between food and temperature devalues a warm patch beyond the simple effect of competition (Fig. 8.2). At low densities of fish, all or most fish should congregate in the warmer patch, which enables rapid growth. As the number of fish increase in the warm patch, each fish not only receives less food but also grows less efficiently (i.e., grows less per unit of food). At some density, a new fish entering the warm patch would be better off entering the cool patch instead, because the higher growth efficiency in this patch more than compensates for the lower food intake. When many competitors coexist, most fish will congregate in the cool patch because the limited food available to each fish causes very poor growth in the warm patch. If the strength of the interaction depends on body size, individuals should segregate between patches according to their

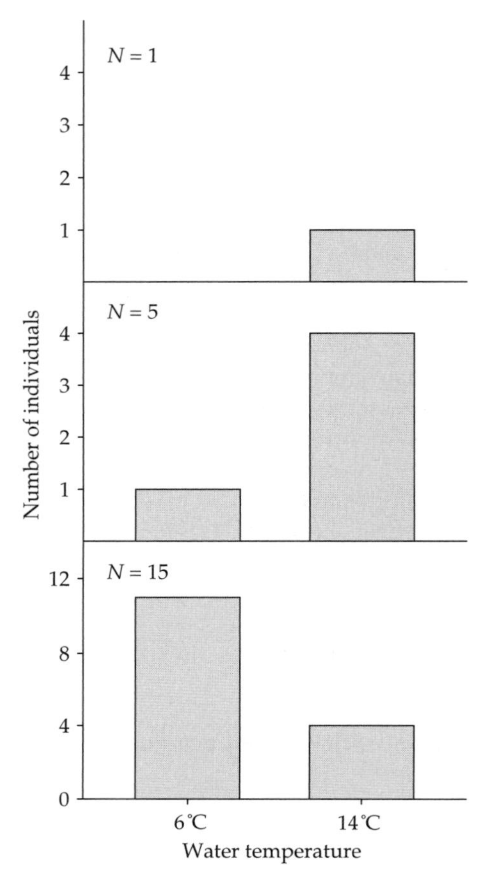

Figure 8.2 The interaction between food and temperature affects the ideal free distribution of fish. At low densities ($N = 1$ or 5), fish should spend most of their time in the warm patch. Yet, competition at a higher density ($N = 15$) reduces the feeding rate and growth efficiency of fish in the warm patch, causing fish to spend more time in the cool patch. The renewal rates of food (5000 mg day^{-1}) were equal for both patches. Energetics of growth were based on empirical data for brown trout (*Salmo trutta*). Based on Hughes and Grand (2000).

size; all else being equal, large fish should prefer the cooler patch because they grow less than small fish given the same food intake and body temperature (Hughes 1998). If this interaction represents the rule rather than the exception, the behavior of organisms in nature should deviate from a simple matching of either food or temperature.

An elegant experiment conducted by Lampert and colleagues (2003) illustrates the power of game theory to predict thermoregulatory behaviors. These researchers studied the vertical distribution of water fleas (*Daphnia pulicaria*) in a controlled environment. To manipulate distributions of food and temperature precisely, they erected water towers stretching more than 11 m in height! The surface and depths were heated and cooled, respectively, to mimic the thermal stratification of a natural lake. The warm epilimnion facilitated growth and development, but most of the food (algae) occurred in the hypolimnion. A model of energetics predicted that the fastest growth would occur at an intermediate depth, just below the thermocline. To achieve an ideal free distribution, water fleas should match their relative potential for growth; hence, the peak abundance of water fleas should occur just below the thermocline. By differentially heating the regions of each tower, the researchers were able to vary the intensity of the thermal stratification, creating conditions under which the ideal free distribution would change. Consistent with the prediction of game theory, the distribution of water fleas shifted toward the food-rich hypolimnion as the steepness of the thermocline weakened (Fig. 8.3). In a subsequent experiment, Lampert (2005) showed that another species of *Daphnia* spreads out toward the food-rich hypolimnion under increasing density. Large individuals altered their distribution the most in response to density, which accords with the model by Hughes and Grand (2000), in which temperature and food interact to determine growth rate.

8.3.2 Predation during thermoregulation

A more complex and certainly more realistic game occurs when we consider optimal thermoregulation in the presence of predators. Again, we can adapt a simple model by Sih (1998) in which the thermoregulators must balance their use of the superior patch against the risk of predation (see also Hugie and Dill 1994). In mathematical terms, a thermoregulator should maximize the ratio of energy gain to predation risk. Let us assume predators do not concern themselves with behavioral thermoregulation, either because they rely primarily on endothermy or because they perform well over a broad range of temperatures. Thus, the predators care only about maximizing the rate at which they consume their prey (f). This rate increases with the density of prey in patch i (N_i), as represented by this linear function:

$$f = cN_i, \tag{8.3}$$

where c equals the rate at which predators capture prey. Given ideal free distributions of predators and prey, (1) predators in both patches will gain energy at equal rates and (2) prey in both patches will experience the same ratio of energy gain to predation risk (Box 8.1).

Some counterintuitive insights emerge from the ideal free distributions of predators and their prey. If the prey congregate within the superior patch, they become predictable to predators. For this reason,

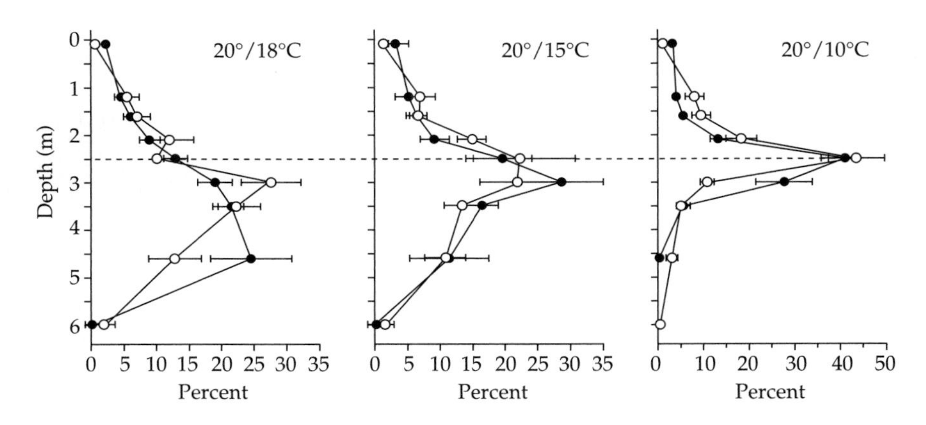

Figure 8.3 The vertical distribution of water fleas (*Daphnia pulicaria*) in experimental water columns shifted in response to a change in the thermal gradient. In all three cases, the temperature of the epilimnion equaled 20°C, while that of the hypolimnion was 18°C, 15°C, or 10°C. The dashed line marks the depth of the thermocline. Open and filled circles define the distributions during the day and night, respectively. Error bars represent 1 standard deviation. Adapted from Lampert et al. (2003) with permission from the Royal Society.

Box 8.1 Solving for the double ideal free distribution

The evolutionarily stable strategy for a system of predators and prey in an environment with two patches must meet two conditions (Sih 1998). First, no individual can increase its performance by switching patches. Second, all individuals must perform equally regardless of which patch they inhabit. For predators, performance equals the rate of energy gain. For prey, performance equals the ratio of energy gain to predation risk. Traditionally, biologists refer to the joint distribution of predators and prey at the evolutionarily stable strategy as a double ideal free distribution.

The ideal free distribution of prey equalizes the feeding rates of predators in the two patches:

$$c_1 N_1 = c_2 N_2, \tag{8.4}$$

where c_i equals the rate at which predators capture prey. If we assume $c_1 = c_2$, this equation simplifies to yield

$$N_1 = N_2. \tag{8.5}$$

From eqn 8.4, we learn that a uniform distribution of prey equalizes the feeding rates of predators.

Likewise, the ideal free distribution of predators must equalize the prey's ratio of energy gain to predation risk in the two patches:

$$\frac{P(T_1)F_1/N_1^X}{c_1 \rho_1} = \frac{P(T_2)F_2/N_2^X}{c_2 \rho_2} \tag{8.6}$$

where ρ_i equals the density of predators in patch i. Because $N_1 = N_2$ at the equilibrium (and $c_1 = c_2$), eqn 8.5 reduces to

$$\frac{\rho_1}{\rho_2} = \frac{P(T_1)F_1}{P(T_2)F_2}. \tag{8.7}$$

Hence, the evolutionarily stable strategy for this system occurs when the predators match the distribution of their prey's resources while the prey spread evenly throughout the environment. These distributions are stable because any movement of prey from patch 1 to patch 2 would cause predators to shift to patch 2 as well. This response would increase the risk of predation in patch 2 and favor a redistribution of prey. Ultimately, the system would settle back to equal densities of prey in the two patches.

the prey should not prefer either patch, regardless of the thermal benefits. Instead, prey should divide themselves evenly between the patches ($N_1 = N_2$). Surprisingly, the predators should distribute themselves according to operative temperatures, even though the predators gain no thermal benefits! The ideal free distribution of predators resembles that of the prey in the absence of predators (see eqn 8.2):

$$\frac{\rho_1}{\rho_2} = \frac{P(T_1)F_1}{P(T_2)F_2}, \tag{8.8}$$

where ρ_i equals the density of predators in patch i. Sih (1998) referred to this phenomenon as the leap-frog effect because the distribution of predators matches the distribution of their prey's resources, instead of the distribution of prey (Fig. 8.4). This surprising result occurs because (1) predators can obtain equal fitness in both patches only when prey are equally abundant in both patches and (2) prey can obtain equal fitness in both patches only when a greater predation risk offsets the greater energy gain in the superior patch. The leap-frog effect disappears if we relax the assumption that predators do not compete for prey. To introduce competition, we

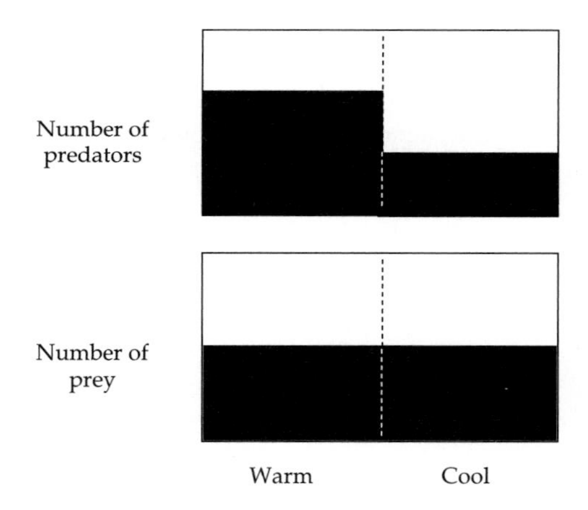

Figure 8.4 In an environment with two patches, the ideal free distributions of prey and their predators become counterintuitive. If prey in the warm patch gain energy twice as fast as prey in the cool patch, predators should spend twice as much time in the warm patch whereas prey should spread evenly between the patches.

can make the feeding rate of predators a function of their density:

$$f = \frac{cN_i}{\rho_i}. \tag{8.9}$$

In this case, both predators and prey should distribute themselves according to the prey's resources (Sih 1998).

Finally, we can imagine that one of the patches provides some refuge from predation. For instance, this patch could be a burrow or some other shelter. We can model the effect of the refuge on predation risk by introducing a coefficient to eqn 8.9 (Sih 1998):

$$f = k_i \left(\frac{cN_i}{\rho_i} \right), \tag{8.10}$$

where $k_1 = 1$ makes patch 1 risky and $k_2 < 1$ makes patch 2 safer. Realistically, the operative temperature and food density of the risky patch would exceed those of the refuge (i.e., $T_1 > T_2$ and $F_1 > F_2$). Assuming both competitors and predators compete, the ideal free distribution of prey becomes

$$\frac{N_1}{N_2} = k_2 \left(\frac{P(T_1)F_1}{P(T_2)F_2} \right), \tag{8.11}$$

and the ideal free distribution of predators becomes

$$\frac{\rho_1}{\rho_2} = \left(\frac{P(T_1)F_1}{P(T_2)F_2} \right). \tag{8.12}$$

Therefore, prey should underutilize the warmer patch to take advantage of the refuge. Still, the majority of predators should occupy the warmer patch, even though their feeding rate does not depend on temperature and fewer prey occupy this patch. This result seems counterintuitive but makes sense after some reflection. If fewer predators were to occupy the warmer patch, then prey would shift to this patch and enjoy an increase in energy gain. This shift in prey would draw predators back into the warm patch, causing some prey to leave this patch. Thus, the majority of predators must occupy the warm patch to achieve an evolutionarily stable equilibrium.

8.3.3 Relaxing assumptions of simple models

The models described in the previous section could be generalized in two ways. First, one could relax assumptions about the biotic environment. For example, additional trophic levels will change the optimal behavior for thermoregulators and their predators, because the predators must then worry about their own mortality. Second, we could relax assumptions about the abiotic environment. For example, the course dissection of an environment into two patches represents an implicit spatial landscape. A more realistic model would account for the spatial relationship among patches and the cost of travel within the environment. Before turning to other kinds of thermal games, let us briefly consider the evolutionary predictions of more complex models of habitat choice.

In a community with more than two species, a behavioral cascade can emerge from a thermal game. A behavioral cascade involves an indirect effect of a predator on the behavior of its prey's prey. For example, consider the simple system in which large fish feed on small fish, which in turn feed on zooplankton (Romare and Hansson 2003). In the absence of large fish, we have a predator–prey interaction that might loosely conform to one of the models described above. Specifically, we could characterize the environment by two kinds of patches: warm, shallow water and cool, deep water. Assuming equal densities of food in these patches, zooplankton should spread evenly between shallow and deep waters while small fish tend to occupy the shallow water (see Box 8.1). But what happens when we add a secondary predator to the system? Then, the evolutionarily stable strategy becomes a triple ideal free distribution. With no competition among predators, the triple ideal free distribution resembles the double ideal free distribution with a twist; small fish should spread evenly between shallow and deep waters while large fish and zooplankton tend to occupy the shallow water (Rosenheim 2004). In other words, thermoregulators tend to occupy the patch containing superior resources, primary predators spread their risk between patches, and secondary predators tend to occupy the patch containing superior resources for the thermoregulators (Fig. 8.5a). Thus, the prediction of this model appears similar to but fundamentally differs from the leap-frog effect (see Fig. 8.4), particularly with respect to the behavior of the lower trophic levels. Adding yet another trophic level to the system restores the leap-frog effect (Fig. 8.5b). Thus, the conflict between exploiting thermal resources

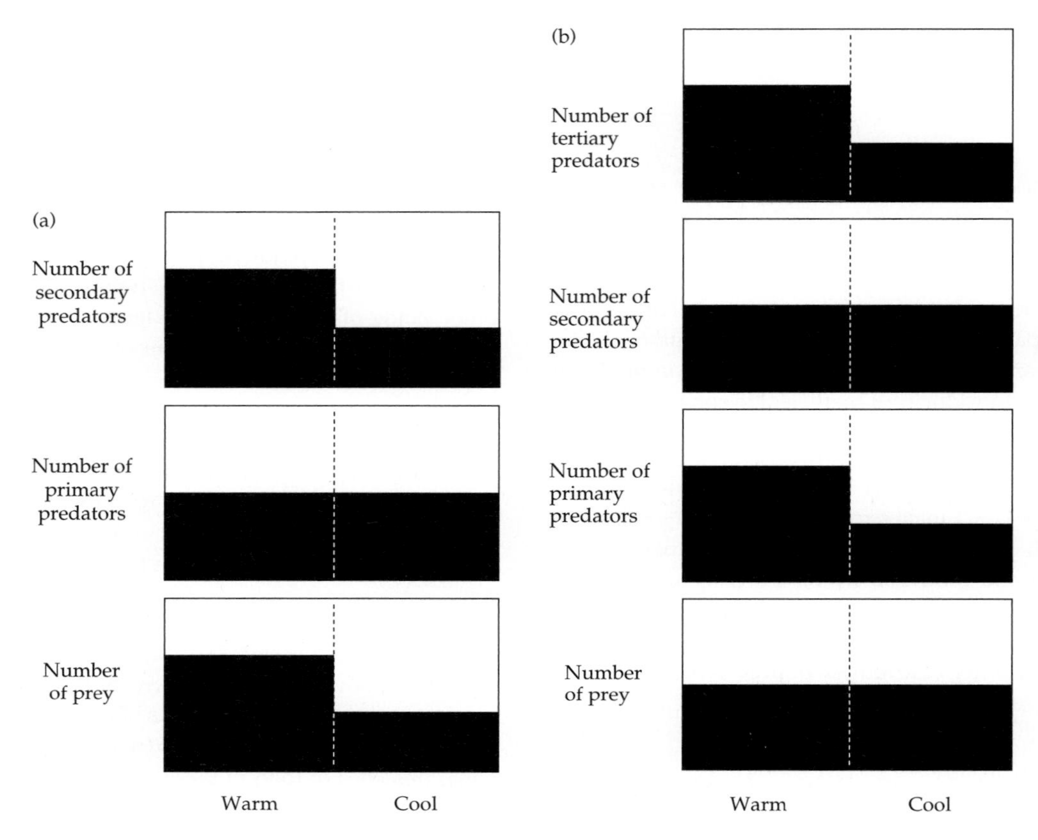

Figure 8.5 Adding a secondary predator causes a shift in the optimal use of patches by the primary predator and its prey—a process referred to as a behavioral cascade. As in Fig. 8.4, assume prey in the warm patch gain energy twice as fast as prey in the cool patch. (a) In the presence of secondary predators, primary predators spread evenly between the patches, while their prey use the patches in proportion to the relative energetic benefits. (b) In the presence of secondary predators and tertiary predators, primary predators use the patches in proportion to the relative energetic benefits for prey, while prey spread evenly between the patches.

and avoiding predation risk exists for organisms in complex food chains, as well as simple ones.

A more realistic model also results from an explicit description of the spatial environment. Alonzo and colleagues (2003) modeled the optimal movements of penguins and krill among waters of a continental shelf as a consequence of their trophic interactions and thermal physiologies. During this period, penguins leave their breeding grounds periodically to forage at sea. Figure 8.6 depicts the habitats of penguins and krill as characterized by the model. Penguins can move horizontally between inshore or offshore waters and vertically among depths, whereas krill can only move vertically. At deeper depths, krill escape penguins more easily but encounter less food and experience lower temperatures. These low temperatures reduce rates of energy assimilation and expenditure. Krill also escape penguins more easily during the night than they do during the day. Consequently, diel shifts among depths enable krill to exploit their resources while minimizing predation risk; however, these movements impose an energetic cost. How does the distribution of water temperatures affect the optimal behavior of these organisms?

The vertical cline in water temperature directly affects the optimal behavior of krill and thus indirectly affects the optimal behavior penguins. The optimal distributions of the two species depend on whether penguins seek to maximize their daily energy gain or minimize their daily foraging time. If penguins maximize their daily energy gain,

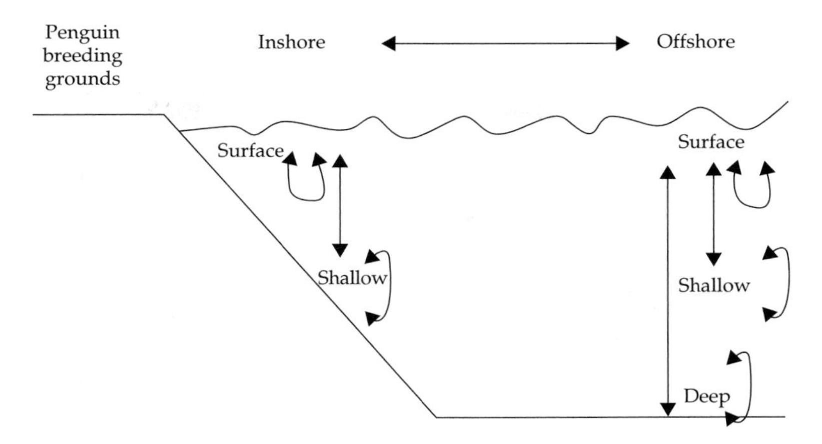

Figure 8.6 Alonzo and colleagues modeled the optimal use of microhabitats by krill and penguins. Penguins could forage inshore or offshore at any depth, whereas krill could move only among depths. Reproduced from Alonzo et al. (2003) with permission from the Ecological Society of America.

individuals should predominately forage inshore, with some individuals moving offshore at night. The optimal behavior of krill depends on their size, as penguins prefer to eat large krill. During the day, large krill should mainly occupy safe habitats: shallow inshore and deep offshore waters. At night, these krill should move to the surface to feed. Small krill should remain at the surface while medium krill should retreat only to intermediate depths. If penguins minimize the foraging time required to meet their metabolic demands, all penguins should forage inshore (assuming a sufficient density of krill exists there). In this case, krill in offshore waters should remain at the surface. When surface waters are much warmer than shallow and deep waters, more krill spend the day in cool, deep waters to avoid predation and minimize energy expenditure. Thus, warm surface water directly enhances the growth of krill but indirectly reduces the growth of penguins. This model illustrates the complex behaviors that result from a modest increment in the spatial structure of the thermal environment.

8.4 Optimal performance curves in an evolutionary game

In Chapter 3, we examined the factors that drive the adaptation of performance curves in heterogeneous environments. In doing so, we concerned ourselves mainly with the nature of thermal heterogeneity, the fitness consequences of performance, and the genetic variance of thermal sensitivity. Now, let us expand this focus to include a critical factor: the interactions within and among species. As with thermoregulatory decisions, the evolution of thermal sensitivity depends strongly on interactions with competitors, predators, and prey. The predictions emerging from game theory differ qualitatively from those discussed in Chapter 3.

8.4.1 The coevolution of thermal optima between species

In a pair of papers, Gavrilets and Kopp developed a game theory of adaptation that applies very nicely to our concept of a performance curve (Gavrilets 1997; Kopp and Gavrilets 2006). Here, we shall focus on the basic components of their theory and the qualitative predictions for the coevolution of two species.

Assume species x possesses a performance curve that we can approximate as a Gaussian function (see Chapters 3, 5, and 7 for models based on the same assumption). This particular function has convenient properties for modeling natural selection: as the thermal optimum for performance deviates from the mean temperature of the organism, fitness

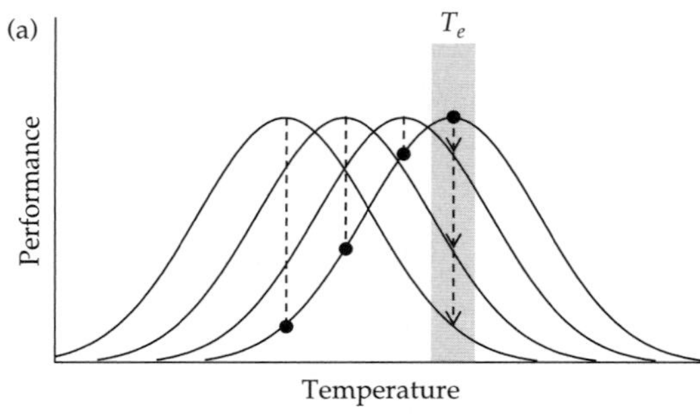

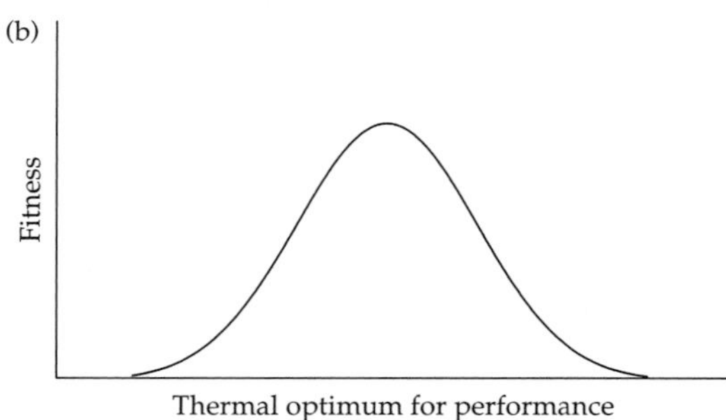

Figure 8.7 To apply existing models to thermal games, we must approximate performance curves with a Gaussian function. (a) Deviations between the thermal optimum for performance (T_{opt}) and the operative environmental temperature (T_e) lead to decreases in the expected performance of individuals in their environment. The solid lines represent performance curves for four genotypes. The gray bar highlights the mean operative temperature. The dashed arrows and lines associated with each performance curve quantify the loss of performance caused by the mismatch between the operative temperature and the thermal optimum. The gain in performance as the thermal optimum approaches the operative temperature mirrors the Gaussian form of the performance curve. (b) The phenomenon described in (a) leads to stabilizing selection of the thermal optimum.

declines according to a Gaussian function that mirrors the performance curve (Fig. 8.7). In other words, a Gaussian performance curve leads to stabilizing selection of the thermal optimum. If the thermal sensitivity of performance were the only selective pressure, the thermal optimum would evolve to match the mean temperature experienced by the species (see Chapter 3). However, species x participates in a thermal game with a second species, y. The interaction between species x and species y imposes directional selection of their thermal optima. The contributions of stabilizing and directional selection to the fitness of each species (w_i) can be characterized as follows:

$$w_x = \exp\left[-s_x(z_x - \phi_x)^2\right] \cdot \sum_{z_y} \exp\left[-\gamma_x(z_x - z_y)^2\right] \cdot f_y(z_y) \qquad (8.13a)$$

and

$$w_y = \exp\left[-s_y(z_y - \phi_y)^2\right] \cdot \sum_{z_x} \exp\left[-\gamma_y(z_y - z_x)^2\right] \cdot f_x(z_x). \qquad (8.13b)$$

The left-hand term of each function represents stabilizing selection imposed by the thermal sensitivity of performance; for species i, z_i equals the thermal optimum, ϕ_i equals the mean body temperature, and s_i scales inversely proportional to the performance breadth. The right-hand term represents directional selection imposed by the interaction between species; γ_i equals the effect of the interaction on species i, and $f_i(z_i)$ equals the frequency of individuals with phenotypes z_i.

Because the signs of γ_x and γ_y determine the nature of the interaction, we can use this general model to understand the dynamics of several thermal games. When both γ_x and γ_y are positive, the species engage in a mutualistic interaction. When both γ_x and γ_y are negative, the species engage in a competitive interaction. When γ_x is positive but γ_y is negative, species x and y are predator and prey (or parasite and host), respectively. Regardless of the ecological interaction, the strength of directional selection depends on the difference between the thermal optima of the two species. Assuming individuals choose microhabitats that maximize their fitness, the deviation between thermal optima determines the degree to which the species can (or must) segregate over space and time. Such segregation between competitors (Cerda et al. 1998) or between predators and prey (Joern et al. 2006) occurs in natural environments (but see Kronfeld-Schor and Dayan 2003).

Two different approaches were used to model genetic constraints during coevolution. In the first approach (Gavrilets 1997), the thermal optimum represents a quantitative genetic trait, whose

Box 8.2 Modeling the genetics of a performance curve

Kopp and Gavrilets (2006) supplemented their analysis of the weak-selection approximation with an analysis of coevolution under more realistic genetic constraints. In these simulations, the continuous traits that we are calling thermal optima (x and y) were affected by L loci, each having two alleles. Alleles at these loci affect the phenotype in an additive fashion. Alleles at each locus were coded as 1 or 0, yielding phenotypic effects of $+\alpha/2$ or $-\alpha/2$, respectively. Values of α were drawn from a Gaussian distribution with a mean of zero; therefore, most alleles had very small positive or negative effects on the thermal optimum while a few had large effects. By defining the mean phenotypic value of x and y as m_x and m_y, respectively, they could calculate the range of phenotypes:

$$m_x - \sum_{i=1}^{L} \alpha_i < z_x < m_x + \sum_{i=1}^{L} \alpha_i \tag{8.14a}$$

and

$$m_y - \sum_{i=1}^{L} \alpha_i < z_y < m_y + \sum_{i=1}^{L} \alpha_i. \tag{8.14b}$$

Evolutionary dynamics were modeled by assuming discrete, non-overlapping generations. In each generation, the populations underwent selection, random mating, recombination, and mutation. Mutation converted an allele coded as 0 to an allele coded as 1, or vice versa. Given the complexity of this model, Kopp and Gavrilets resorted to computer simulations to find evolutionarily stable strategies. Fortunately, the behavior of this complex allelic model supported the conclusions drawn from a quantitative genetic model (Gavrilets 1997).

evolution depends on a genetic variance. In the second approach (Kopp and Gavrilets 2006), the contribution of alleles at many loci was modeled explicitly. Because allelic models require far more analytical complexity, the modelers had to make some simplifying assumptions to explore the coevolutionary dynamics. Analytical results were computed for the special case of weak selection and no genetic drift or mutation (the weak selection approximation). More general scenarios demanded computer simulations to define the model's behavior (Box 8.2; see also Kopp and Gavrilets 2006; Kopp and Hermisson 2006).

The course of coevolution depends on the nature and strength of the biotic interaction, but most scenarios lead to predictions that differ qualitatively from those of traditional optimality models. For the weak selection approximation, the difference between the thermal optima of the species at the equilibrium is given by:

$$\frac{|\bar{z}_x - \bar{z}_y|}{|\phi_x - \phi_y|} = \frac{1}{|1 + \gamma_x/s_x + \gamma_y/s_y|}. \tag{8.15}$$

Although this equation cannot be defined when the two species experience the same mean body temperatures (i.e., when $N_x = N_y$), this condition should never exist because each species will have a unique operative temperature in the same microhabitat (see Chapter 2).

Using eqn 8.15, we can predict the thermal optima of two species engaged in a competitive, mutualistic, or an exploitative interaction. Recall that γ_x and γ_y determine the nature of the interaction between the species. For competitors, both γ_x and γ_y are negative, which causes their thermal optima to diverge (Fig. 8.8a); consequently, neither species would exhibit maximal performance at the mean operative temperature of its environment. For mutualists, both γ_x and γ_y are positive, which causes the thermal optima of the two species to converge (Fig. 8.8b). For predator and prey, γ_x is negative (prey) and γ_y is positive (predator), which could lead to one of several outcomes depending on the relative strengths of directional and stabilizing selection in the two species (Fig. 8.8b–f). When prey experience strong stabilizing selection, the predators converge on the same thermal optimum as the prey. When prey experience strong directional selection or weak stabilizing selection, the prey escape predation by evolving an extreme thermal optimum. When the genetic variance of the prey limits their escape from the predators, disruptive selection produces two distinct thermal optima in the prey and an intermediate thermal optimum in the predators. Finally, the

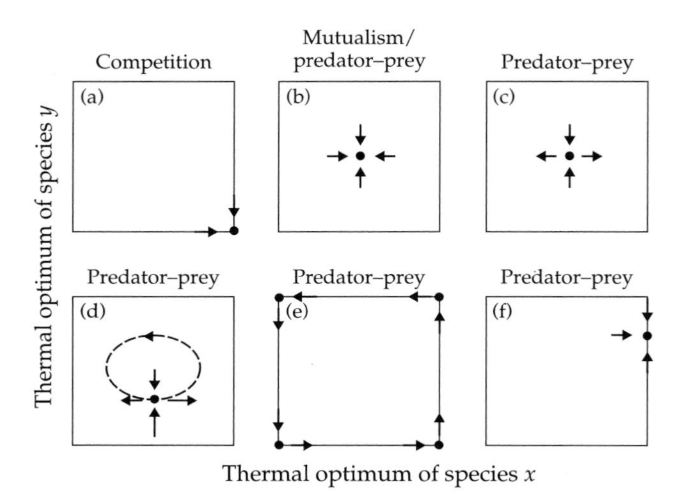

Figure 8.8 The coevolution of thermal optima between two species depends on the nature of their interaction, their breadths of performance, and the additive genetic variances. (a) The thermal optima of competitors should diverge from the mean operative temperature to reduce the intensity of competition. (b) Stabilizing selection should cause the thermal optima of mutualists to converge on the same value. (c) Disruptive selection can cause distinct thermal optima to evolve within a population of prey, while an intermediate thermal optimum evolves in the population of predators. (d and e) The thermal optima of predators and prey can cycle repeatedly as both experience selection in one direction imposed by their interaction, followed by selection in the opposite direction imposed by the thermal environment. (f) Prey can evolve an extreme thermal optimum while stabilizing selection prevents the predators from matching the thermal optimum of the prey. Adapted from Kopp and Gavrilets (2006) with permission from Blackwell Publishing.

thermal optima of predators and prey cycle indefinitely when the prey have a stronger incentive to win ($|\gamma_x| > \gamma_y$) or have greater genetic variance and experience stronger stabilizing selection ($s_x > s_y$).

8.4.2 The coevolution of thermal breadths between species

Although frequency-dependent selection clearly affects the evolution of thermal optima, performance breadths can also coevolve. Ackermann and Doebeli (2004) developed a model that offers stimulating insights on the evolution of performance breadths in a thermal game. The model focuses on genotypes that compete for a continuously distributed resource, which we shall think of as temperature. Because the genotypes reproduce asexually, they could represent genetic diversity within a single species or between two competing species.

The performance curve was modeled as a variant of the Gaussian function:

$$P(T_b) = \frac{e^{-cz_2}}{\sqrt{2\pi} \cdot z_2} \exp\left[-(z_1 - T_b)^2/2z_2^2\right], \quad (8.16)$$

where T_b equals the temperature of the organism, and z_1 and z_2 define the thermal optimum and performance breadth. This function strongly resembles the one used to model the evolution of thermal sensitivity in Chapter 3, with the distinction that a new parameter (c) defines the benefit of specialization. If $c = 0$, the performance curve evolves according to the typical constraint imposed by specialist–generalist tradeoffs: the integral of the performance curve remains constant. If $c > 0$, specialist–generalist tradeoffs become more severe, such that widening the performance breadth reduces the integral of the curve. If $c < 0$, the reverse becomes true, and generalists have an advantage over specialists. Because the model considers a single organismal performance affected by a single environmental resource, the constraints imposed when $c \neq 0$ cannot be interpreted as acquisition or allocation tradeoffs (see Box 3.5). Instead, these constraints must stem from some unspecified physical process associated with specialization or generalization. Whether or not we believe such constraints exist will become an important consideration once we have examined the behavior of the model.

To apply this model, we must assume individuals occupy thermal niches according to their performance curve.[14] Consequently, most individuals of a species will occupy microhabitats that provide them with their optimal temperature, whereas fewer individuals will occupy microhabitats that provide a higher or lower temperature. No individual can occupy a microhabitat in which the operative temperature falls outside its performance curve. In the absence of any organisms, a Gaussian function

[14] The distribution of individuals assumed by this model equals the ideal free distribution when individuals of the same species compete for resources.

(a) Frequency distribution of temperatures

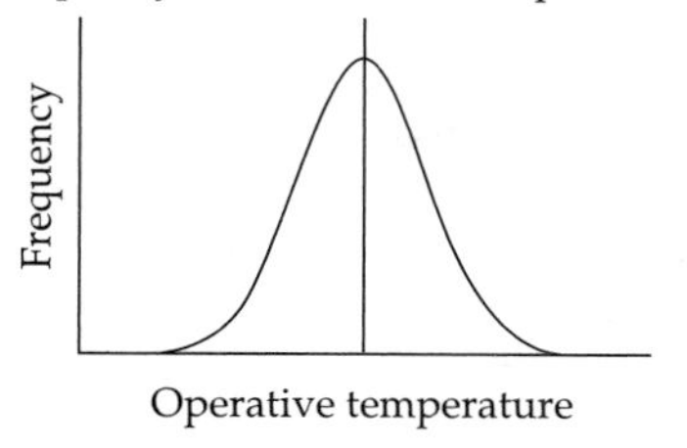

(b) Performance curve of resident genotype

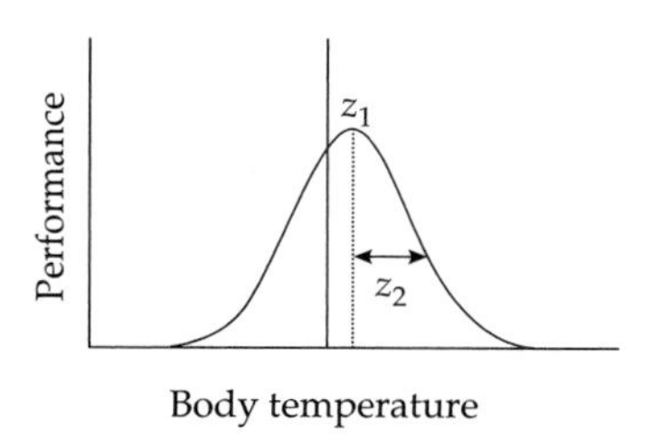

(c) Frequency distribution of temperatures available to an invading genotype

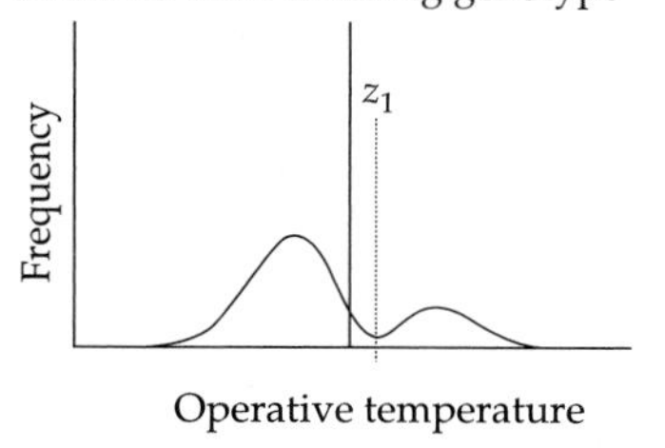

Figure 8.9 Thermal resources can be exploited by a resident genotype, which depletes the availability of operative temperatures (and associated resources) for other genotypes. In this example, the resident genotype uses thermal resources (a) according to its thermal performance curve (b), causing the greatest reduction in thermal resources near the resident's thermal optimum (c). The parameters z_1 and z_2 determine the thermal optimum and performance breadth, respectively. Adapted from Ackermann and Doebeli (2004) with permission from Blackwell Publishing.

defines the availability of operative temperatures (Fig. 8.9a). These temperatures are distributed in space such that any individual can access any microhabitat without paying a cost. Nevertheless, the presence of an individual in a microhabitat prevents another individual from effectively using this site, thus altering the frequency of available temperatures (Fig. 8.9b and c). To be clear, Ackermann and Doebeli envisioned the environmental axis as a biotic resource, which grew logistically in response to consumption, but we can apply their model to an abiotic resource without too much imagination. We need only assume that individuals occupy microhabitats, thereby depleting the availability of certain operative temperatures to other individuals.

Both the coexistence of competing genotypes and the evolution of their performance curves depend on the strength of the constraint imposed by specialist–generalist tradeoffs (Fig. 8.10). When these tradeoffs impose a weak or modest constraint ($c \leq 0$), a common genotype evolves. This genotype has a thermal optimum equal to the mean operative temperature and a performance breadth just wide enough to exclude all other genotypes. This performance breadth greatly exceeds the one predicted by a model of frequency-independent selection (Gilchrist 1995; see Chapter 3). When specialist–generalist tradeoffs impose a strong constraint ($c > 0$), distinct genotypes evolve by disruptive selection. As the cost of generalization increases, narrower performance breadths should evolve and more genotypes should coexist (compare Fig. 8.10b and Fig. 8.10c). Under this severe constraint, an important prediction of the model emerges: no phenotype should have a thermal optimum equal to the mean operative temperature.

8.4.3 Gene flow and the coevolution of thermal optima

From Chapter 3, recall that gene flow among disparate thermal environments can either facilitate or impede the adaptation of performance curves (see Section 3.9). Here, we shall use a model developed by Day (2000) to explore a thermal game between two populations that experience disparate thermal environments but exchange genes via migration. Imagine two environments that both have a Gaussian distribution of operative temperatures but have different mean temperatures. Each generation, individuals migrate between environments in proportion to the sizes of their populations. In the absence of migration, the divergence of thermal optima within environments would reduce the impact of competition. Would migration between populations enhance or impede this disruptive selection?

The answer depends on the symmetry of performance between environments. Environmental symmetry means that locally adapted organisms in each

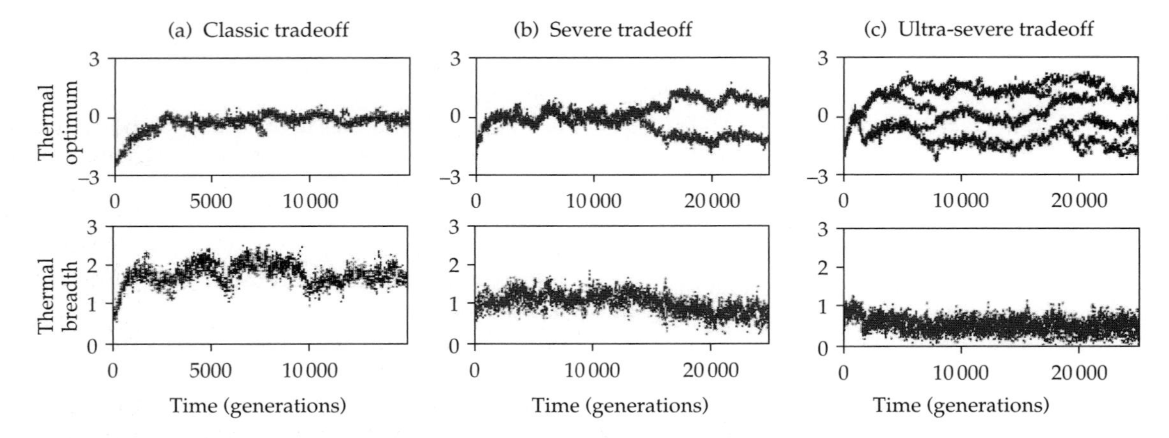

Figure 8.10 Computer simulations show that the evolution of thermal performance breadth depends on the constraint imposed by specialist–generalist tradeoffs. (a) The typical constraint, in which the integral of the performance curve remains constant ($c = 0$), leads to the evolution of a wide performance breadth. (b and c) More severe tradeoffs cause the evolution of distinct thermal optima and relatively narrow performance breadths. Values of c were 0.6 and 1.25 for simulations shown in (b) and (c), respectively. Thermal optima are plotted relative to the mean operative temperature. Adapted from Ackermann and Doebeli (2004) with permission from Blackwell Publishing.

environment achieve the same rates of energy gain, whereas environmental asymmetry means that local adaptation enables a greater rate of energy gain in one of the two environments. Asymmetry of performance can result from either of two mechanisms. First, one environment may contain more food, enabling individuals to gain energy faster than they could in the other environment. Second, a warm environment might enable greater performance because of thermodynamic constraints (i.e., hotter is better; see Chapter 4). When environmental symmetry exists, migration between environments favors a single thermal optimum that lies in between the mean operative temperatures of the two environments. The critical rate of migration depends on the mean performance in each environment; if locally adapted individuals perform very well, migration must occur more frequently to favor an intermediate thermal optimum. Thus, greater productivity or warmer conditions tend to favor disruptive selection of the thermal optimum. When environmental asymmetry exists, stabilizing selection should produce a single thermal optimum. The value of this thermal optimum depends on the frequency of migration; infrequent migration favors a thermal optimum closer to the mean temperature of the better environment, whereas frequent migration favors a thermal optimum closer to the mean temperature of the poorer environment. In general, thermal heterogeneity coupled with gene flow inhibits the divergence of thermal optima between competitors.

8.5 Life-history evolution in a thermal game

The concept of a thermal game applies not only to the evolution of thermoregulation and thermosensitivity but also to the evolution of the life history. To illustrate this point, let us consider how the predictions of a simple model of optimal reproductive allocation depend on the frequency dependence of selection.

From Chapter 6, recall that we used a model by McGinley and colleagues (1987) to determine the optimal size(s) of offspring in a patchy environment. For example, assume warm patches enhance the performance of offspring, whereas cold patches impair their performance. When parents can direct their offspring to a warm patch, the model predicts that mothers should produce relatively small offspring. But when the performance of offspring depends on their density, not all parents can reap the benefits of a warm patch. If enough parents were to direct their offspring to warm patches, the quality of these patches would fall below that of the cold patches. At that point, other parents would do better to direct their offspring to cold patches. Obviously, each parent should allocate its resources based on the environment they expect their offspring to experience (see Chapter 6). But the quality of the environment depends on the reproductive decisions of other mothers. Thus, the optimal strategy becomes the evolutionarily stable strategy.

The addition of frequency-dependent selection to the model of optimal offspring size revealed a new condition under which variation in offspring size might evolve. Through computer simulations, McGinley and colleagues (1987) calculated a range of evolutionarily stable strategies, varying the size of small offspring, the ability to direct offspring to the correct habitat, the intensity of density dependence, and the difference in quality between habitats. Based on their simulations, we can draw several conclusions about reproductive decisions in a thermal game. Under random dispersal, the evolutionarily stable strategy always results in a single offspring size: either the optimal size for a warm patch or the optimal size for a cold patch. Even when parents can direct their offspring perfectly to a warm or cold patch, the evolutionarily stable strategy often involves a single offspring size. A mixture of offspring sizes is favored only when very strong density dependence exists, such that a rare genotype directing large offspring to a cold patch fares better than a common genotype directing small offspring to a warm patch. The critical level of density dependence required to favor a mixture of offspring sizes increases as the quality of the patches becomes more disparate. Therefore, strong density dependence coupled with moderate thermal heterogeneity favors a mixed reproductive strategy.

8.6 Conclusions

Although the models summarized in this chapter are far too simple to be realistic, we should still come away with a sobering conclusion: a thermal game can generate patterns of adaptation that differ drastically from the patterns generated by traditional optimality models. Natural selection might favor a very low accuracy of thermoregulation when an organism must deal with competitors or predators. The same ecological interactions can shape the evolution of performance curves in surprising ways. Competition causes the thermal optimum to deviate from the mean operative temperature of the environment. Predation can cause endless cycles of adaptation in which the thermal optima of predators and prey rarely match their mean operative temperatures. Equally novel insights will likely emerge from the analysis of other thermal games. For example, the decision to mature early or late should depend on the number of conspecifics who make a similar decision. What effect will competition or predation have on the evolution of this life-history decision? How might these ecological interactions influence the coadaptation of behavior, physiology, and life history within species? Given what we know already, I am eager to learn the full extent to which game theory will change our perspective of thermal adaptation.

CHAPTER 9

Adaptation to Anthropogenic Climate Change

As I type these words, former Vice President Al Gore and members of the Intergovernmental Panel on Climate Change (IPCC) head to Stockholm to collect a Nobel Peace Prize. The prize was awarded "for their efforts to build up and disseminate greater knowledge about man-made climate change, and to lay the foundations for the measures that are needed to counteract such change." These efforts include multiple syntheses of scientific data and the recent documentary *An Inconvenient Truth*—all of which brought the issue of climate change into the public spotlight. While politicians, reporters, and laypeople continue to question the causes of the recent global warming, few scientists doubt the influence of human activities. The mean global temperature increased by 0.6°C during the last century, and reasonable models predict an additional increase of 2–4°C by the end of this century (IPCC 2007). Such scenarios could have startling consequences for ecosystems and biodiversity. But make no mistake. Adaptation will play some role in determining these consequences.

Throughout this book, we have explored the ways in which populations can adapt to thermal heterogeneity. Current patterns of phenotypic variation were produced by evolution during countless generations of thermal change. The direction and rate of change have certainly varied regionally over space and globally over time. But at this moment, many ecosystems are warming faster than they have for thousands of years (IPCC 2007). The biological impacts of this environmental perturbation are already under way. Indeed, some would say we are in the midst of a mass extinction caused by the advance of one species: *Homo sapiens*. Understanding the role that global warming will play in this drama has become one the greatest challenges facing ecologists. In this final chapter, we shall see how ecologists currently model the impacts of climate change and consider how evolutionary biologists can help to improve these models.

9.1 Recent patterns of climate change

9.1.1 Global change

According to the IPCC (2007), our planet warmed by 0.6°C during the last century and continues to warm in the present century. The case for global warming is open and shut. Of the last 12 years, 11 were among the 12 warmest years in recorded history (1850–2006). Diel variation has become smaller because nighttime temperatures have increased more than daytime temperatures. Seasonal variation has also become smaller because winter temperatures have increased more than summer temperatures. Heat waves have become more frequent, and cold snaps have become less frequent. The indirect effects should startle even a layperson. Some regions, such as the Mediterranean, southern Africa, and southern Asia, receive less precipitation than in previous decades. Other regions receive more precipitation concentrated into fewer events, leading to greater runoff. The melting of ice in the Arctic has contributed to an increase in sea levels. To make matters worse, global warming has accelerated over time; the rate of warming during the last decade exceeded the rate during the last 50 years!

Identifying the primary cause of global warming has enabled scientists to make projections about the future. Few scientists doubt the mechanistic link between the accelerating emissions of greenhouse gases and recent patterns of environmental warming (IPCC 2007). Greenhouse gases (e.g., carbon dioxide) influence the absorption, scattering, and emission of radiation, leading to an increase in surface and air temperatures. Because these emissions stem from the industries that supply energy, transportation, and other products, we can expect current trends to continue unless we reduce or cease these activities. Even if we stopped emitting greenhouse gases immediately, the globe would likely warm by 0.5°C during the next century (IPCC 2007). A more realistic scenario includes an increase in the annual emission of carbon dioxide by 40–110% between 2000 and 2030, causing the earth to warm 0.6°C during this period alone (IPCC 2000, 2007). If emissions continue to accelerate, the mean surface temperature of the earth might increase by as much as 6°C by 2100 (IPCC 2007). Although predictions on the scales of decades to centuries carry great uncertainty, the prediction offered by the IPCC in 1990 accords well with the subsequent rate of warming (predicted rate, 0.15–0.3°C per decade; observed rate, 0.2°C per decade).

9.1.2 Regional change

Although the pattern of global warming appears relatively straightforward, regional changes in temperature have been far more complex. The most rapid warming has occurred at high northern latitudes, with the Arctic having warmed twice as fast as the global average. Not surprisingly, the continents have warmed more rapidly than the oceans. Among oceanic environments, warming has occurred most slowly in the Southern Ocean surrounding Antarctica. Even these patterns obscure heterogeneity of thermal change on smaller scales (Walther et al. 2002). A detailed look reveals that some regions have warmed by as much as 1°C per decade while others have cooled at a similar rate (Fig. 9.1). Clearly, these regional patterns matter more to organisms and populations than do global patterns.

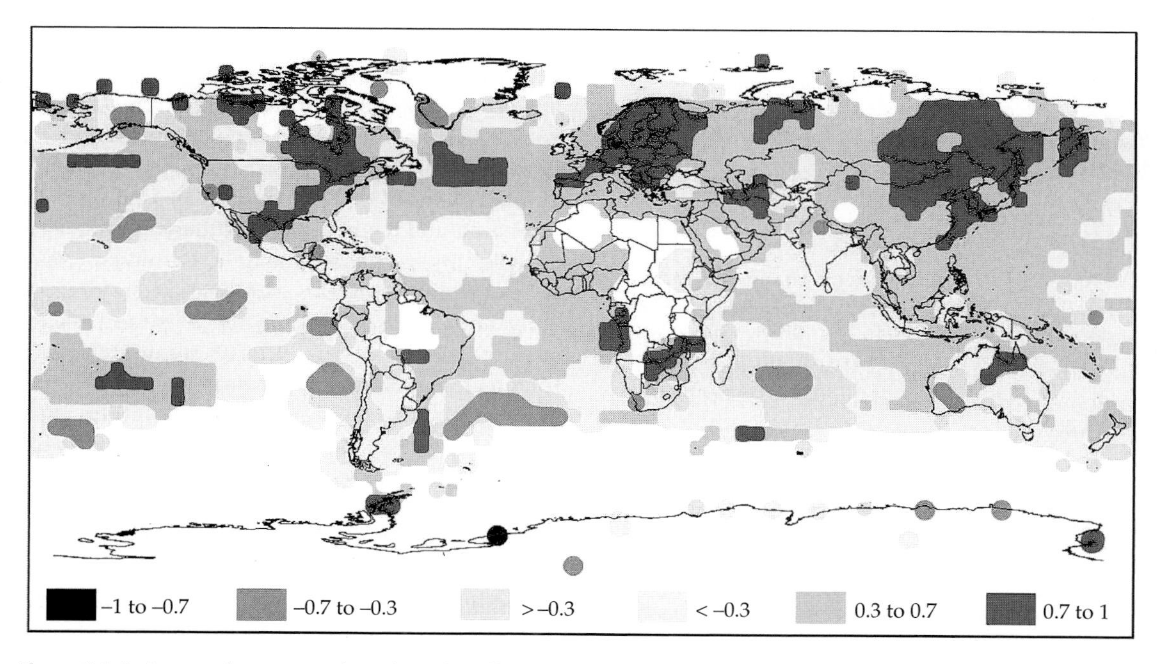

Figure 9.1 Environmental temperatures have changed at different rates throughout the world. This map shows rates of change (°C per decade) between 1976 and 2000, relative to normal temperatures for 1961–1990 (see Box 2.1). Reprinted by permission from Macmillan Publishers Ltd: *Nature* (Walther et al. 2002), © 2002. See plate 3.

9.1.3 Local change

On local scales, the activities of humans have added to the thermal change caused by greenhouse gases. Both the destruction of natural habitats and construction of human dwellings have generated marked changes in the thermal landscape. Devegetation and urbanization have raised local temperatures of both terrestrial and aquatic habitats (Grimm et al. 2008; Nelson and Palmer 2007; Roth 2002). Geographers refer to the elevations of air and surface temperatures in cities as urban heat islands. The magnitude of this thermal effect scales logarithmically with the population of a city (Oke 1973). In very large cities, urban warming has matched or exceeded the rate of global warming. Urban heat islands possess complex spatial and temporal structures (Arnfield 2003; Weng et al. 2004). Spatial variation depends on the historical factors that shape the development and growth of each city. Temporal variation appears less idiosyncratic; within a particular city, urban temperatures exceed rural temperatures more during the summer than any other time of year.[15] Because urban warming and global warming have proceeded independently (Parker 2004), the two processes combine to magnify potential impacts on organisms.

9.2 Observed responses to recent thermal change

In the past decade, scientists have made considerable progress toward assessing the biological consequences of recent climate change. A mixture of detailed case studies and comprehensive meta-analyses uncovered four major trends: (1) shifts in phenology, (2) shifts in geographic ranges, (3) disruption of ecological interactions, and (4) changes in primary productivity (see reviews by Hughes 2000; Parmesan 2006; Parmesan and Galbraith 2004; Walther et al. 2002). Although these trends could stem from factors other than anthropogenic climate change (Forchhammer and Post 2000), some effort has been made to control for other factors in these analyses. Below, we shall examine these four trends in some detail and consider their potential impact on the future of biological systems.

[15] The time of day at which urban heat islands reach their greatest intensity depends on whether one characterizes these heat islands by surface or air temperatures. The difference between urban and rural air temperatures peaks during the night, whereas the difference between urban and rural surface temperatures peaks during the day (reviewed by Arnfield 2003; Weng et al. 2004).

9.2.1 Shifts in phenology

By far, the most widely documented impacts of global warming involve phenology. Because temperature determines the performance of most organisms, environmental warming should enable organisms to initiate activity earlier in the spring and sustain activity later in the fall. Consistent with this expectation, the timing of biological events in spring has advanced considerably during recent decades. Of 385 species of British plants, 16% have advanced their date of flowering, whereas only 3% have delayed this date (Fitter and Fitter 2002). A global meta-analysis of 143 studies of amphibians, birds, invertebrates, trees, and other plants documented a significant advance of vernal activity (Root et al. 2003). For the affected species, the mean rate of advance was 5.1 days per decade. In a similar analysis of 677 species, 62% of species showed earlier activity in spring, such as breeding, nesting, flowering, migrating, or budding (Parmesan and Yohe 2003); on average, spring activity shifted 2.3 days per decade. These two estimates of phenological advance (5.1 days vs. 2.3 days) reflect both taxonomic and methodological differences between analyses. By combining the data, Parmesan (2007) estimated a mean advance of 2.8 days per decade, with amphibians advancing the most and plants advancing the least (Fig. 9.2). As we might expect from regional differences in warming, vernal activities have advanced more in some locations than in others. As expected, species at high latitudes have advanced more than species at low latitudes. Similarly, phenology has changed more in urban environments than it has in rural environments (Roetzer et al. 2000; White et al. 2002), revealing the additive effects of urban and global warming. Still, much of the variation in phenological change remains unexplained by these factors; for example, latitude accounted for less than 5% of the variation in the advance of vernal activity (Parmesan 2007).

These meta-analyses did not attempt to document temporal variation in the rate of phenological advance, but we should expect such variation for two reasons. First, the rate of warming experienced by particular stages of the life cycle has accelerated in recent decades. Case studies indicate that species respond to local rates of warming rather than the global rate. For example, some birds lay their eggs at times of the year that have been less affected by global warming; this difference in life history might explain why the laying dates of some North American species have advanced while the laying dates of others have not (Torti and Dunn 2005). Second, even a constant rate of warming can cause a variable rate of phenological advance. Because a linear increase in temperature causes a nonlinear increase in performance, the physiological consequences of constant warming will vary over time.

9.2.2 Shifts in geographic ranges

In addition to shifting their activity in time, organisms have also shifted their activity in space. The geographic ranges of many species have shifted northward or upward in recent decades. Of 35 species of butterflies, 63% have moved their entire range poleward while only 3% have moved southward (Parmesan et al. 1999). Of 434 species whose ranges have shifted historically (17–1000 years), 80% have shifted in accord with predictions based on climate change (Parmesan and Yohe 2003). For 99 species of birds, butterflies, and alpine herbs, range boundaries have shifted an average of 6 km northward or 6 m upward per decade (Parmesan and Yohe 2003). By no means have shifts in geographic ranges been confined to terrestrial environments. Over the last 40 years, warm-adapted copepods have moved 10°N and have begun to replace cold-adapted copepods (Beaugrand et al. 2002). Similarly, warm-adapted species of algae have shifted north, but cold-adapted species have shifted either north or south (Lima et al. 2007a).

Either range contraction or global extinction is a likely outcome for many species, particularly when body size, life history, or an ecological interaction constrains dispersal. Some species have already lost the southern or lower portion of their range (see, for example, Franco et al. 2006). Sadly, species at extreme latitudes or altitudes have nowhere to go; hence, many of these species will face extinction (Parmesan 2006). Thomas and colleagues (2004) used several methods to estimate the risk of extinction for more than 1000 species of animals and plants. They assumed a species–area relationship to link the overlap among species to the risk of extinction. First, they predicted future ranges from climate-envelope models (see Section 9.3). Then, they projected extinction risk for different scenarios of climate change. Because the potential for dispersal clearly affects the predicted shift in range, two extreme situations were compared: perfect dispersal and no dispersal. For perfect dispersal, a species' future range equaled its future climate envelope. For no dispersal, a species' future range was the intersection of its current and future climate envelopes. Given minimal warming (0.8–1.7°C), 9–13% of species were predicted to face extinction under the assumption of perfect dispersal; without dispersal, as many as 31% of species might become extinct. With more extreme warming (>2°C), these predictions increased to 21–32% with dispersal and 38–52% without dispersal. Extinctions caused by global climate change might even exceed those caused by local habitat destruction (Thomas et al. 2004). More realistically, global and local insults to the environment will interact to heighten the risk of extinction. In particular, the fragmentation of landscapes by humans will impair the dispersal of organisms during climate change (Parmesan and Galbraith 2004), generating an outcome closer to the scenario of no dispersal than the scenario of perfect dispersal.

9.2.3 Disruption of ecological interactions

Because shifts in phenologies and ranges have differed among species, some ecological interactions have become uncoupled (Harrington et al. 1999). In a review of 11 interspecific relationships, Visser and Both (2005) concluded that five species advanced too little and two species advanced too much compared with their prey. Uncoupling has occurred over space and time. For example, the northern spread of the brown argus butterfly has resulted in some release from parasitism (Menéndez et al. 2008). Similarly, the vernal activities of other insects have advanced more rapidly than their botanical

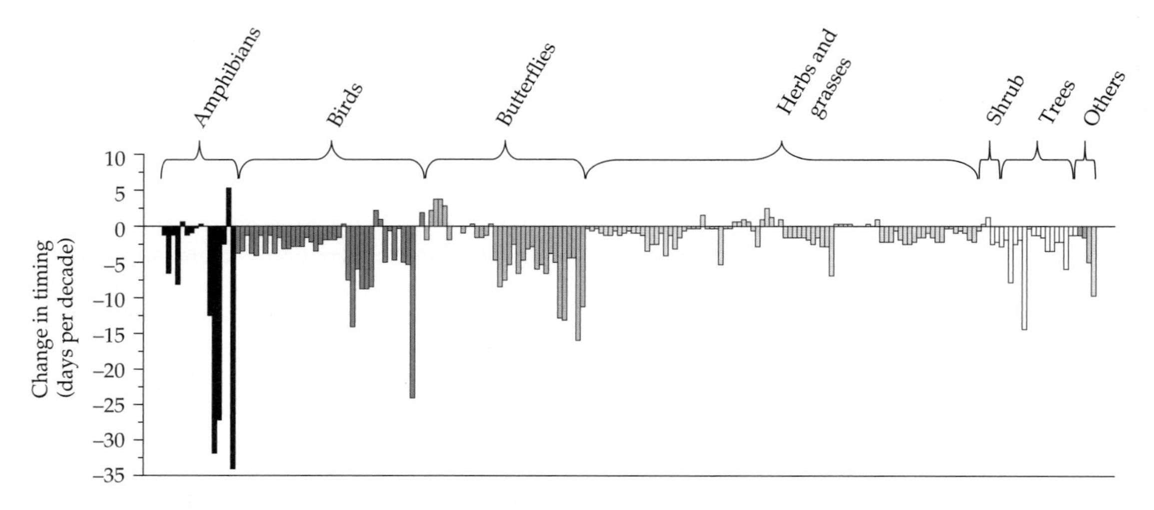

Figure 9.2 Changes in the timing of vernal activity have varied greatly among taxa. Bars represent the rate of advance (−) or delay (+) of vernal activity for individual species. The group labeled "Others" includes two species of fish, a species of flies, and a species of mammals. Adapted from Parmesan (2007) with permission from Blackwell Publishing.

hosts (Gordo and Sanz 2005) or their avian predators (Visser et al. 2006). Uncoupling of ecological interactions promises to become more severe as the compositions of local communities change (Peterson et al. 2002). Compositional changes seem inevitable because species differ in their ability to migrate and their susceptibility to extinction. The consequences for community dynamics will depend on whether invading species disrupt the interactions between native species or fill the vacuum created by extinction.

9.2.4 Changes in primary productivity

The organismal and populational impacts of climate change have altered the structure and function of ecosystems. Biologists care much about the consequences for primary productivity because of its potential feedback on global warming. Greater productivity of plants creates a sink for carbon dioxide, while less productivity creates a source. Both the emission of carbon dioxide and the increase in environmental temperature should enhance the productivity of plants in certain ecosystems (Kirschbaum 2004). High levels of carbon dioxide enable plants to increase photosynthesis without increasing transpiration. Warm soils promote the release of nutrients to sustain photosynthesis. Among 32 experiments in which ecosystems were warmed by 0.3–6°C for 2–9 years, warming increased plant productivity by 19% and soil respiration by 20% (Rustad et al. 2001); the productivity of plants in cold climates increased the most. Consistent with these experiments, recent warming has enhanced the growth of terrestrial forests, leading to greater sequestration of carbon. Nevertheless, this process will not create enough negative feedback to stabilize the global temperature. The growth of terrestrial forests should cease and may even reverse by the middle of this century (IPCC 2007). Furthermore, the increasing frequency of droughts has led to a drop in the primary productivity of other ecosystems (see, for example, Ciais et al. 2005). All things considered, global warming will continue despite the feedback caused by primary productivity.

9.3 Predicting ecological responses to global warming

Although we now recognize multiple impacts of global warming, what impacts should we expect to observe in the future? On one hand, we can certainly expect continued shifts in the phenologies and ranges of species. On the other hand, the magnitudes of these shifts have varied greatly among species in recent decades. Thus, we must move beyond simple inductive reasoning to accurately predict the impacts of global warming. In this section, we shall focus on mathematical models that provide insights about the ecological responses to climate change. By contrasting correlative and mechanistic models, I hope to convince you that mechanistic models offer the greater potential to incorporate evolutionary processes.

9.3.1 Correlative versus mechanistic models

Efforts to predict the biological impacts of climate change in general, and global warming in particular, have been based on the concept of the fundamental niche (Hutchinson 1957). The fundamental niche depicts the abiotic factors that limit the success of a genotype or species, having one dimension for every environmental variable we can imagine (see review by Wiens and Graham 2005). In practice, biologists use a limited number of variables to characterize these niches. Thermal variables, such as the mean and variance of environmental temperature, have played a prominent role in defining fundamental niches (Lima et al. 2007b; Walther et al. 2007).

To see how climate change might affect a species, researchers model the fundamental niche of a species and then identify the location of this niche at some time in the future. In recent years, biologists have used both correlative and mechanistic approaches to define fundamental niches (Pearson and Dawson 2003). Correlative models define the niche by relating current geographic distributions to local environmental conditions (see Elith et al. 2006). Assuming a species will track the geographic position of its niche, we can use a correlative model to predict a shift in a species' distribution during climate change (Hijmans and Graham 2006). Because correlative models require only environmental data associated with known localities (Graham et al. 2004; Guisan and Thuiller 2005), these models provide a convenient approach to predicting global distributions. Correlative models can also deal with species interactions (Araujo and Luoto 2007; Sutherst et al. 2007) and geographic variation (Murphy and Lovett-Doust 2007) to some degree. Nevertheless, these models conflate abiotic and biotic factors, fail to account for dispersal, and ignore potential evolutionary responses (Pearson and Dawson 2003; Soberón and Peterson 2005). In sharp contrast, mechanistic models explicitly link the environmental tolerances of individuals to the dynamics of populations. Their mathematical functions not only make the underlying assumptions transparent, they also provide opportunities to relax these assumptions and incorporate further biological detail. For example, we could modify a mechanistic model to account for biotic interactions, dispersal limitations, and evolutionary adaptation. Unfortunately, implementing a mechanistic model requires extensive knowledge of the behavior, physiology, and life history of a species.

Do we really need complicated mechanistic models, or can we rely on simpler correlative models to predict the impacts of climate change? No one can definitively answer this question at the moment, but recent attempts to hindcast responses to climate change have me leaning in one direction. When hindcasting, one predicts the current distribution of a species with a model based on environmental and biological data from the past (see Lima et al. 2007b; Walther et al. 2005). Araujo and colleagues (2005) conducted the most extensive hindcasting effort thus far, predicting the distributional changes of 116 species of birds between 1967 and 1991. For each species, these researchers applied four correlative models and several sets of assumptions. For 90% of the species, the models disagreed as to whether the geographic range should have expanded or contracted during the two decades. To make matters worse, when all models made similar predictions, these predictions were qualitatively wrong for half of the species.

Why do correlative models fail to accurately predict responses to climate change? Many factors—including biotic interactions, geographic dispersal, and evolutionary adaptation—interact with climate change to determine the success of a species (Jackson 1974; Pearson and Dawson 2003; Soberón and Peterson 2005). Current distributions appear limited by dispersal in some species and biotic interactions in others. For example, the limpet *Collisella scabra* grew well when transplanted either 100 or 200 km north of its current range boundary, suggesting barriers to dispersal limit a northern expansion (Gilman

2006). In contrast, the distribution of the butterfly *Aporia crataegi* appears constrained by the absence of host plants at high elevations. (Merrill et al. 2008). Recent experiments challenge the notion that we can cleanly separate biotic and abiotic determinants of the niche; for example, the outcomes of competitive interactions (Davis et al. 1998; Jiang and Kulczycki 2004; Jiang and Morin 2004) and host–parasite interactions (Harvell et al. 2002; Poulin 2006) depend on environmental temperature. Even if correlative models were to accurately capture the factors that limit the current distribution, adaptation to environmental change affects the future distribution.

These problems associated with correlative models can lead to overprediction or underprediction of a geographic distribution (Pulliam 2000; Soberón and Peterson 2005). A model would overpredict the distribution when individuals cannot disperse to all favorable environments and underpredict the distribution when individuals disperse routinely to unfavorable environments. The same model could overpredict or underpredict the distribution of a species when interactions among species occur. Finally, the model would underpredict the distribution when adaptation occurs. Therefore, the advancement of mechanistic models that account for these processes seems essential for making accurate predictions.

9.3.2 Mechanistic models of responses to environmental warming

Mechanistic models differ greatly in their complexity and generality. Some mechanistic models rely on an environmental threshold to define the fundamental niche, such as thermal constraints on survival, growth, development, or feeding (Cortemeglia and Beitinger 2006; Kearney and Porter 2004; Parker and Andrews 2007; Robinet et al. 2007; Sykes et al. 1996). Other models link these thermal constraints directly to the population dynamics of plants and animals (Dullinger et al. 2004; Edmunds 2005; Pulliam 2000). Here, we shall focus on two models that deal explicitly with population dynamics, because these models could eventually accommodate the potential for thermal adaptation. In the first model, environmental temperature determines the development, fecundity, and survivorship of a thermoconformer (Crozier and Dwyer 2006). In the second model, environmental temperature constrains the duration of activity by a thermoregulator (Buckley 2008). These models offer complementary perspectives of the impacts of environmental warming on the population dynamics and geographic distributions of ectotherms.

Crozier and Dwyer (2006) used performance curves to model the population growth and range limits of a thermoconformer in a seasonal environment. They split the life cycle into winter and summer stages. Winter temperatures determine the probability of surviving until summer (S). Summer temperatures determine the net reproduction (R), defined as the number of individuals at the end of the summer that descended from each female alive at the beginning of the summer. The growth rate of a local population (λ) equals the product of winter survivorship and net reproduction:

$$\lambda = S(T_{win}) \cdot R(T_{sum}), \tag{9.1}$$

where T_{win} and T_{sum} equal the mean temperatures of winter and summer, respectively, which are functions of latitude (L) and time (t). Because a population can persist only if $\lambda \geq 1$, Crozier and Dwyer could solve for the northerly limit of the range by calculating the latitude at which $\lambda = 1$. To predict the range shift in response to warming, they assumed a simple linear change in winter and summer temperatures (T_{win} and T_{sum}) over latitude and time:

$$T_{win} = a_w - b_w L + c_w t \tag{9.2}$$

and

$$T_{sum} = a_s - b_s L + c_s t, \tag{9.3}$$

where b_w and b_s determine the rates at which temperatures decrease with increasing latitude, and c_w and c_s determine the rates at which temperatures increase over time. Under these conditions, the change in the latitudinal limit with time was found:

$$\frac{dL}{dt} = \frac{c_w R(T_{sum}) \frac{dS}{dT_{win}} + c_s S(T_{win}) \frac{dR}{dT_{sum}}}{b_w R(T_{sum}) \frac{dS}{dT_{win}} + b_s S(T_{win}) \frac{dR}{dT_{sum}}} \tag{9.4}$$

If we assume net reproduction and winter survivorship decline exponentially with decreasing temperatures (i.e., $R \propto e^{\delta \cdot T_{sum}}$ and $S \propto e^{\rho \cdot T_{win}}$), eqn 9.4 becomes

$$\frac{dL}{dt} = \frac{\delta c_s + \rho c_w}{\delta b_s + \rho b_w}, \qquad (9.5)$$

where δ and ρ define the thermal sensitivities of R and S, respectively. Thus, the shift in the range limit depends not only on the rates of summer and winter warming, but also on the latitudinal clines in seasonal temperatures.

Crozier and Dwyer used their model to predict shifts in the northern limit of the sachem butterfly (*Atalopedes campestris*) under several warming scenarios. Values of $R(T_{sum})$ and $S(T_{win})$ were estimated from the thermal sensitivities of survivorship and development in butterflies from eastern Washington. The model predicted the current limit remarkably well in the western and eastern portions of the USA, but underpredicted this limit in the central portion (Fig. 9.3a). The predicted responses to warming scenarios were complex because winter and summer temperatures interact to alter the fundamental niche over time (Fig. 9.3b). For example, imagine the current range ends at the latitude where the winter temperature falls below 5°C. As temperatures rise steadily, the latitudinal boundary for this thermal threshold will shift northward. But the species will not track this boundary faithfully because summer temperatures also affect the range limit (see eqn 9.4). Because many environmental variables limit a species' distribution, such interactions should become an important focus of mechanistic models.

Taking a more complex mechanistic approach, Buckley (2008) modeled the constraints on foraging activity in a thermoregulating ectotherm. She envisioned a territorial, sit-and-wait forager that maintained a preferred body temperature during foraging through energetically inexpensive behaviors. Because the animal could forage only when the environment enabled thermoregulation, warming scenarios would alter the duration of foraging (t_f), and hence the assimilation of energy and the production of offspring. To model thermal constraints in each environment, Buckley calculated the operative temperatures in full sun and full shade; an individual could forage whenever the temperatures bounded by these extremes overlapped with the preferred range of temperatures. When foraging, the animal gained energy at a rate E, determined by the diameter of the foraging range (d) and the density of prey (a):

$$E = \frac{e_i - e_w t_w(d,a) - e_p t_p(d,a)}{t_w(d,a) + t_p(d,a)}, \qquad (9.6)$$

where e_i equals the energetic content of prey, e_w equals the energetic cost of waiting, e_p equals the energetic cost of pursuing, t_w equals the time spent waiting, and t_p equals the time spent pursuing. Buckley assumed animals maximize their rate of energy gain by foraging at an intermediate diameter (d^*). As the density of foragers (N) increases, the potential diameter of foraging becomes smaller than the energetically optimal diameter ($d < d^*$). This mechanism generates density-dependent growth of the population:

$$\Delta N = [m \cdot t_f \cdot E(d) - \mu]N, \qquad (9.7)$$

where μ equals the rate of mortality, t_f equals the time spent foraging, and m converts energy into offspring. Using these relationships, Buckley solved for the equilibrium density of the population.

Buckley used her model to predict current and future distributions of two species of lizards (*Sceloporus graciosus* and *Sceloporus undulatus*). *Sceloporus graciosus* occurs in western regions of the USA, whereas *S. undulatus* occurs throughout southeastern and central regions of this country. The extensive knowledge of the ecology and physiology of these species enabled Buckley to parameterize her mechanistic model. For example, energetic costs of waiting (e_w) and pursuing (e_p) were estimated as factors of the resting metabolic rate of each species. Prey densities were based on samples of insects from relevant locations. Operative environmental temperatures were calculated for the current climate and a hypothetical climate that warmed uniformly by 3°C. For most parameters, Buckley used mean values for each species or estimated values from data for a closely related species. For a few parameters of *S. undulatus*, she used a set of values representing lizards from different regions. By comparing the predictions derived from this

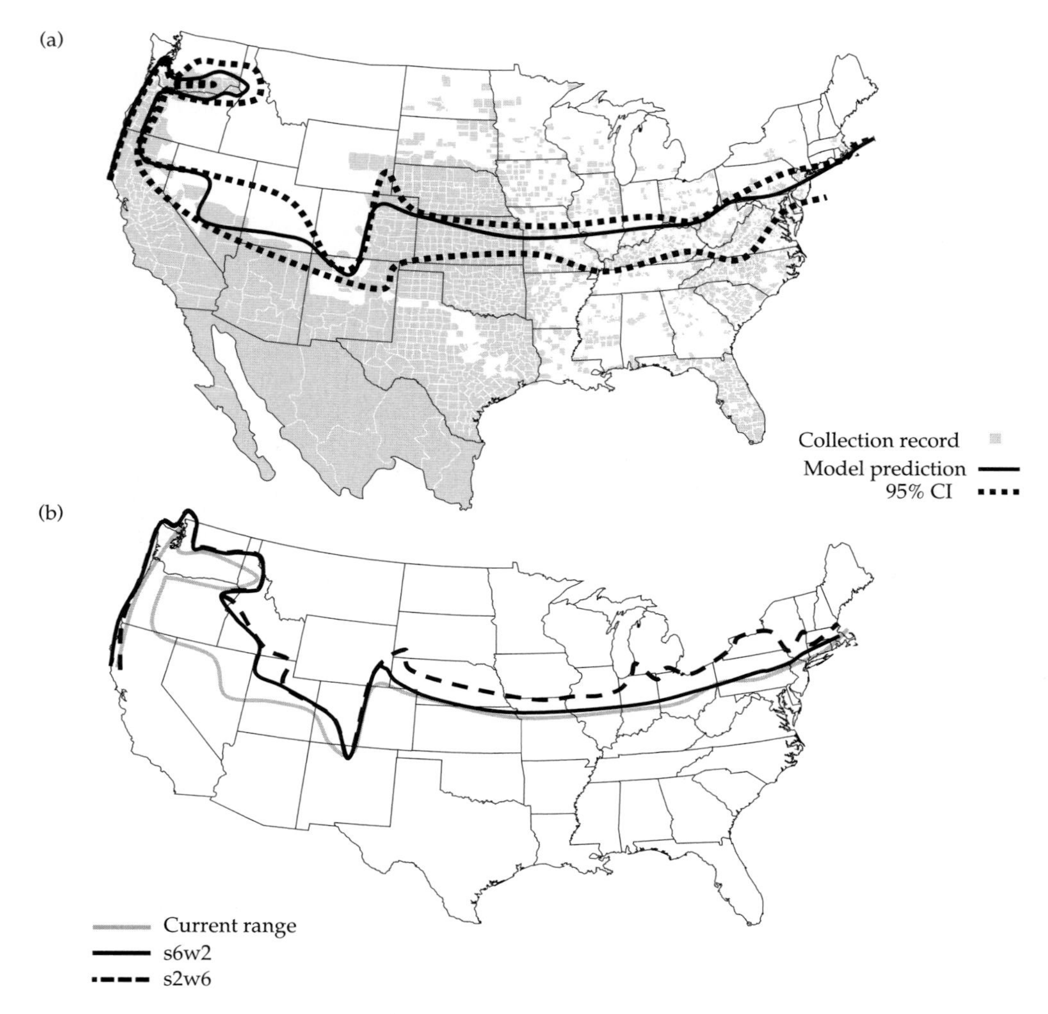

Figure 9.3 A mechanistic model was used to predict the northern limits of the sachem skipper (*Atalopedes campestris*) in current and future climates. (a) The predicted northern limit of the current range in relation to collection records. Dotted lines show the 95% confidence interval (CI) around the predicted range, derived from bootstrapping empirical data to generate different parameter values. (b) The predicted northern limit after 50 years for each of two warming scenarios: (1) 6°C per century in summer and 2°C per century in winter (s6w2), and (2) 2°C per century in summer and 6°C per century in winter (s2w6). For these predictions, the modelers assumed that (1) the winter survivorship was a logistic function of temperature and (2) the net reproductive rate depended on the thermal sensitivities of survivorship, fecundity, and voltinism. Reproduced from Crozier and Dwyer (2006, *The American Naturalist*, University of Chicago Press). © 2006 by The University of Chicago.

set, we can see how geographic variation in phenotypes might influence the response to climate change. For example, phenotypes in New Jersey were predicted to shift farther during warming than were phenotypes in Nebraska, largely because of a difference in body size (Fig. 9.4). Nevertheless, both parameterizations of the mechanistic model predicted more extensive shifts in the northern limit than did a correlative model based on thermal variables (Fig. 9.4). Because the mechanistic model generally predicts the current range more accurately than the correlative model does, we should heed the discrepancy between their predictions about the future.

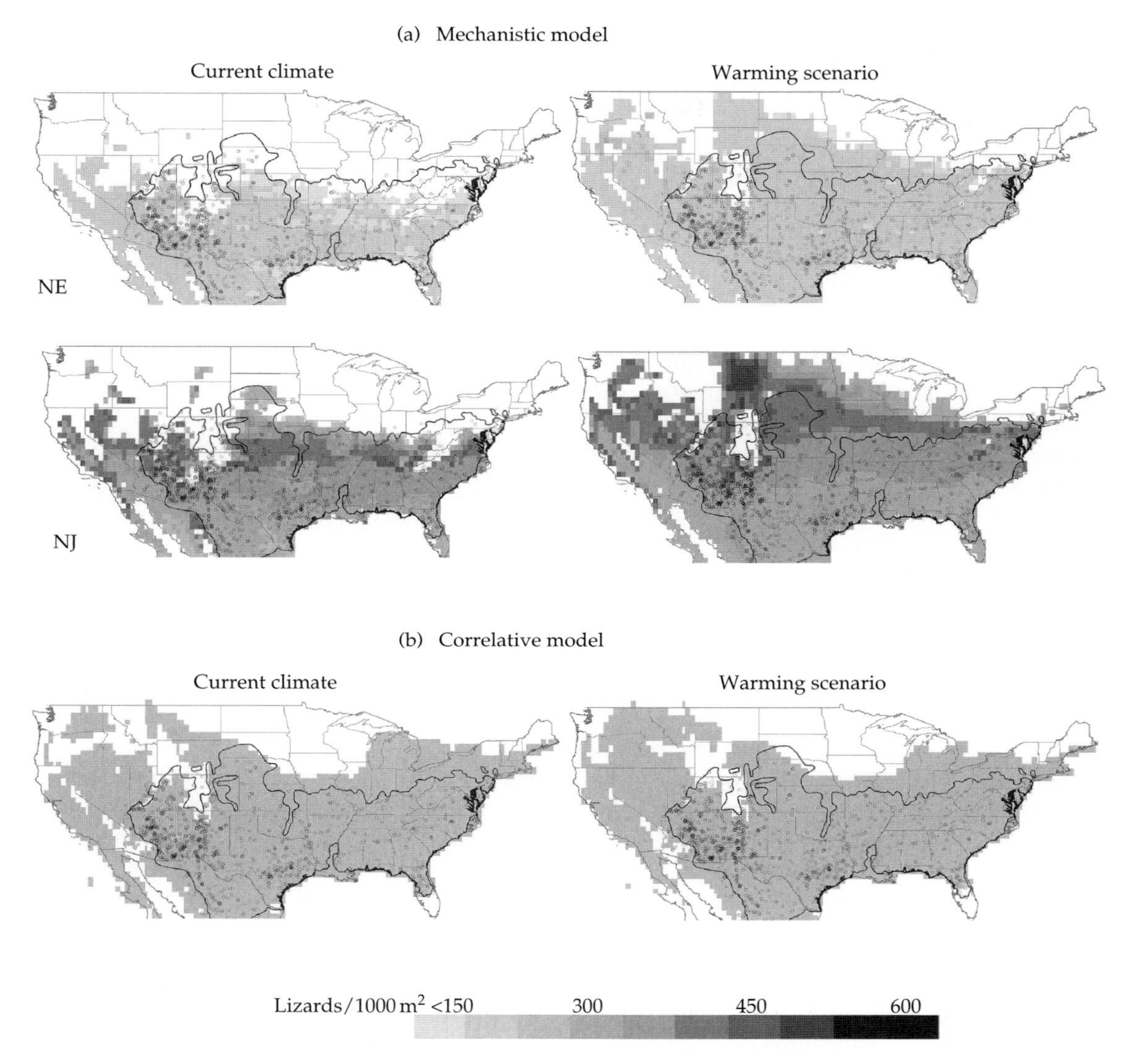

Figure 9.4 (a) A mechanistic model was used to predict the geographic distributions of the eastern fence lizard (*Sceloporus undulatus*) in current and future climates. A uniform warming of 3°C was applied to current temperatures to approximate a future climate. The predictions shown here were based on phenotypes in Nebraska (NE) and New Jersey (NJ). (b) For comparison, geographic distributions were also predicted by a correlative model (Genetic Algorithm for Rule-set Production). Adapted from Buckley (2008, *The American Naturalist*, University of Chicago Press). © 2008 by The University of Chicago.

9.3.3 Predicting differential responses of populations and species

Mechanistic models not only help us to infer distributional impacts of climate change, they can also reveal how adaptation to current environments determines the response to future environments. In the models described above, all individuals of a species shared the same thermal physiology; this assumption seems naïve given the focus of this book. Contrary to what the poet Gertrude Stein once penned, a rose is not a rose is not a rose. Rather, all species exhibit some genetic variation in thermal physiology, which should cause populations to respond differently over space and time, even if their environments warm similarly. In the example provided by Buckley (2008), we saw how geographic variation in body size affected the predicted response to climate change. Let us extend this focus to include geographic variation in thermal sensitivity.

To illustrate the problem, imagine we wish to parameterize a mechanistic model for a species that occurs throughout a broad latitudinal range. We could sample populations throughout this range, and use the mean values of each trait. However, we would probably find that certain parameters vary with latitude. Based on simple optimality models, the thermal optimum of performance should decrease as we sample populations from higher latitudes (see Fig. 3.19). We might also observe a corresponding increase in performance breadth. Which values of these parameters should we use to predict the response to climate change? Actually, choosing any one value would be a bad idea because local responses depend on this variation in thermal physiology.

Deutsch and colleagues (2008) illustrated the significance of heeding variation in thermal physiology when predicting the impacts of climate change. Using published data, they estimated the thermal sensitivities of fitness (r) in 46 species of insects. These fitness curves were used to calculate two indices: the warming tolerance and the safety margin. The warming tolerance was defined as the difference between the critical thermal maximum and the mean environmental temperature. The safety margin was defined as the difference between the thermal optimum for performance and the mean environmental temperature. Because both indices depend strongly on the seasonality of the environment, they both increase with increasing latitude. A similar relationship between seasonality and warming tolerance was found to exist among species of lizards, frogs, and turtles. Consequently, identical warming of tropical and temperate environments will have different biological consequences. Even when these researchers considered spatial heterogeneity of warming (scenario A2; IPCC 2007), tropical insects were predicted to suffer a greater reduction in fitness than were temperate insects

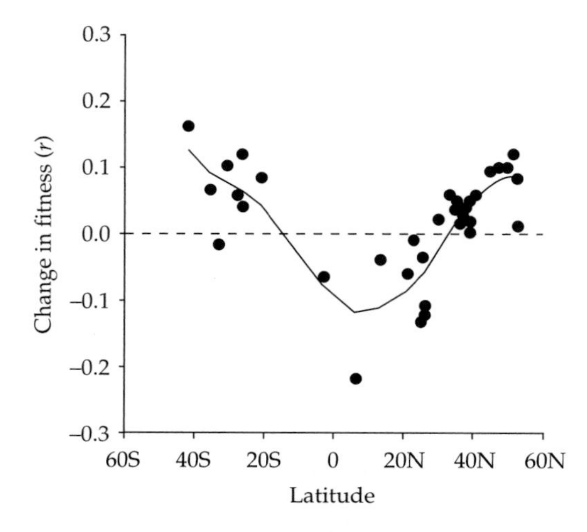

Figure 9.5 Environmental warming during the twenty-first century was predicted to increase the fitness (r) of temperate insects but decrease the fitness of tropical insects. Change in fitness was calculated using temperatures observed for 1950–1990 and those predicted for 2070–2100. Fitness at each time was calculated by integrating fitness curves over diel and seasonal cycles of temperature. The solid line shows a spline fit to the data. Adapted from Deutsch et al. (2008) with permission from The National Academy of Sciences. © 2008 by The National Academy of Sciences, USA.

(Fig. 9.5). Relaxing assumptions about seasonal activity and thermoregulatory behavior did not alter this prediction. Despite the fact that northern environments will continue to warm more rapidly than the tropics, tropical species may face the greatest risk of extinction. In the same vein, we should expect the most severe phenotypic selection to occur in tropical environments. Thus, the actual risk of extinction will depend on the demographic benefit of adaptation versus the demographic cost of environmental change. Given this reality, let us now consider the possibility of adaptation to global warming.

9.4 Adaptation to directional thermal change

Up to this point, we have not considered how adaptation might mitigate the impacts of global warming. Although the primary response during glacial retreats seems to have been shifts in geographic ranges (Parmesan 2006; Parmesan and Galbraith 2004), current barriers to dispersal may force adaptation or extinction. Given the range of adaptations discussed in this book, we should question efforts to predict biological responses in the absence of evolution. Indeed, based on the wealth of evidence for thermal adaptation within and among species, we should expect adaptation to current warming (Holt 1990; Jump and Penuelas 2005; Parmesan 2006; Skelly et al. 2007). Two kinds of adaptations to recent warming have been documented. First, natural selection alters phenological mechanisms, such as photoperiodic thresholds for activity and dormancy (reviewed by Bradshaw and Holzapfel 2006). For example, the photoperiodic threshold for diapause has evolved in the fall webworm (*Hyphantria cunea*) within the last 15 years (Gomi et al. 2007); worms collected in 2002 entered diapause at a shorter photoperiod than did worms collected in 1988 and 1995. Second, natural selection alters thermal physiology. Such physiological adaptations are more difficult to detect than phenological adaptations because historical records of thermal physiology pale in comparison to records of phenological events. Still, indirect evidence exists for rapid evolution of thermal physiology. Urbanization has warmed local environments as fast as or faster than greenhouse gases have warmed our planet (Oke 1997). Not surprisingly, the performance curves of several species of fungi differed between urban and rural genotypes (Fig. 9.6). Genetic markers associated with the heat tolerance of *Drosophila subobscura* have tracked climate warming on multiple continents (Balanya et al. 2006). Experimental studies of physiological adaptation have complemented these comparative studies. After translocating plants (*Umbilicus rupestris*) north of their range limit, Woodward

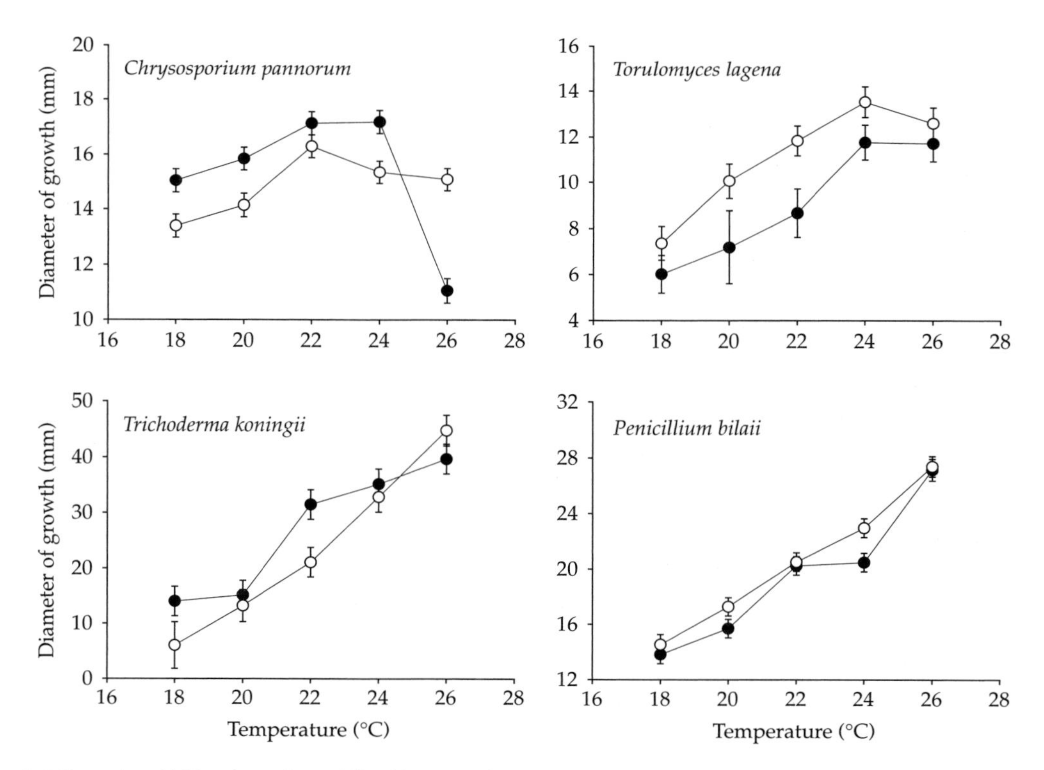

Figure 9.6 Thermal sensitivities of growth rate differed between urban and rural isolates of chitinolytic fungi. Urban isolates of *Chrysosporium pannorum* and *Trichoderma koningii* grew faster than rural isolates at 26°C, but grew slower than rural isolates at 18°C. In *Penicillium bilaii* and *Torulomyces lagena*, urban isolates grew as fast as or faster than rural isolates at each temperature. Urban and rural isolates are denoted by open and filled symbols, respectively. Error bars depict 95% confidence intervals. Adapted from McLean et al. (2005) with permission from Elsevier. © 2005 Elsevier Ltd.

(1990) observed the evolution of greater cold tolerance in just 9 years. Experimental warming of aquatic mesocosms also caused the rapid evolution of thermal physiology (Box 9.1). Clearly, we can expect both phenological and physiological adaptations to result from anthropogenic changes in temperature.

As we have discussed throughout this book, the rate of adaptation depends on many factors, including phenotypic plasticity, genetic variation, and evolutionary tradeoffs. Adaptive phenotypic plasticity, such as thermoregulatory behavior or physiological acclimation, will limit the strength of phenotypic selection (Huey et al. 2003; Jump and Penuelas 2005). However, these forms of plasticity impose costs (see Chapters 4 and 5). Furthermore, acclimation capacities should be most limited in tropical organisms (see Stillman 2003), which might face the greatest threat during climate change (Deutsch et al. 2008). The evolutionary response to phenotypic selection requires genetic variation. Such variation will deteriorate under strong selection, slowing the rate of adaptation (Bell and Collins 2008). Moreover, the death of individuals caused by selection will enhance genetic drift, leading to further loss of genetic variance (Holt 1990; Parmesan 2006). Immigration could enhance or inhibit the response to selection, depending on the fitness of the incoming genotypes (see Chapter 3). Finally, tradeoffs between traits can constrain the course of thermal adaptation; we can characterize

Box 9.1 Experimental warming in mesocosms

Although we generally think of adaptation as a process that requires considerable time, some organisms will adapt rapidly to changes in temperature. Van Doorslaer and colleagues (2007) demonstrated this fact by experimentally warming aquatic communities. They established 24 mesocosms containing sediment, plankton, fish, and nutrients. These mesocosms were exposed to either ambient temperatures or elevated temperatures. The two elevated thermal treatments corresponded to a likely scenario of warming during the next century (A2; IPCC 2007) and an additional increase of 50% (A2 + 50%). After 1 year, a species of water fleas (*Simocephalus vetulus*) was sampled from each of the mesocosms for a study of thermal reaction norms (Fig. 9.7). By raising two clones from each mesocosm

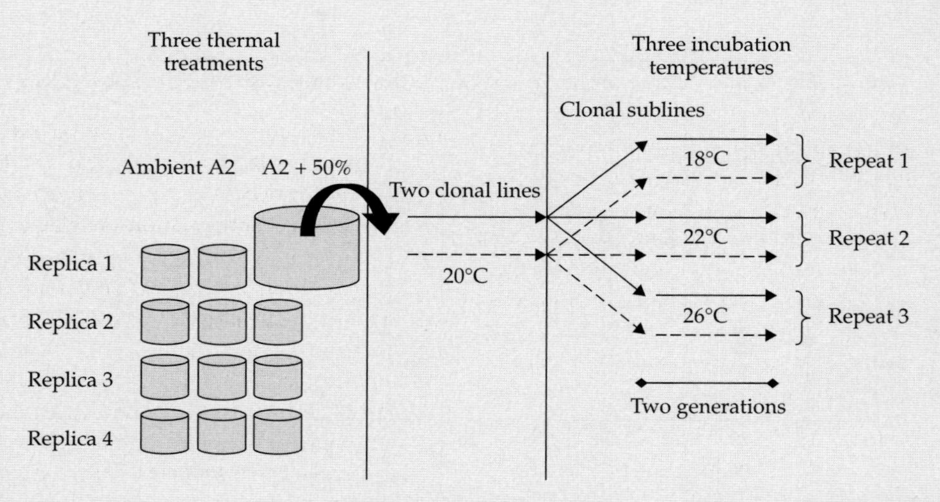

Figure 9.7 The design of a warming experiment conducted with aquatic mesocosms. From left to right, the diagram depicts the initial exposure of communities to several selective regimes and the subsequent sampling of water fleas (*Simocephalus vetulus*) for laboratory studies of thermal reaction norms. Reproduced from Van Doorslaer et al. (2007) with permission from Blackwell Publishing.

continues

Box 9.1 Continued

at three temperatures, the researchers documented genetic changes in thermal sensitivities. The most striking finding was an evolutionary increase in the thermal tolerance of *S. vetulus* in heated environments. Specifically, clones from the heated mesocosms survived better at 26°C than did clones from the unheated mesocosms (Fig. 9.8). No apparent tradeoffs with survivorship at low temperature occurred. Still, clones from all three mesocosms attained similar rates of population growth (r) at all three temperatures, suggesting some tradeoff was associated with the evolution of thermal tolerance. Although heat tolerance clearly evolved during warming, we must wonder whether adaptation would enable this population to persist given that the change in survivorship failed to affect fitness.

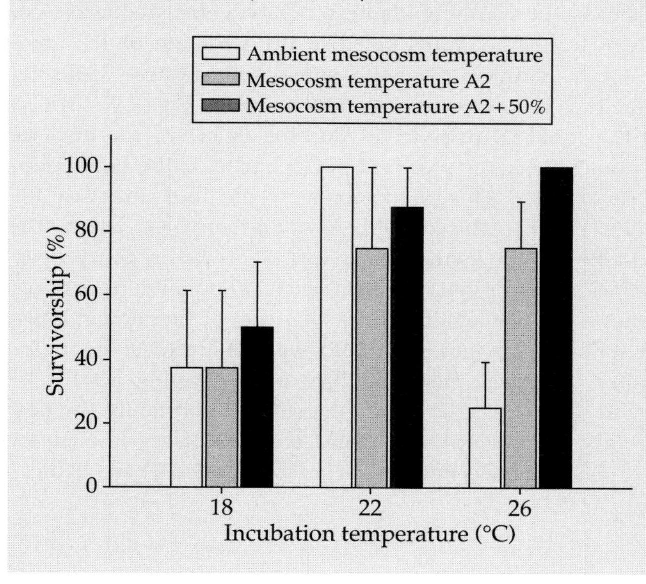

Figure 9.8 Experimental warming caused the evolution of thermal reaction norms for survivorship in *Simocephalus vetulus*. At 26°C, clones from heated mesocosms were more likely to survive from birth until the date of producing a second clutch of offspring. Error bars depict standard errors of the means. Adapted from Van Doorslaer et al. (2007) with permission from Blackwell Publishing.

these tradeoffs through studies of genetic covariances (see Chapter 1).

The failure to address thermal adaptation limits the generality of current models, whether they are correlative or mechanistic. Given the evidence on hand, it seems silly to ask whether species will adapt to anthropogenic warming. Rather, we should ask different questions. How quickly and in what ways will certain species adapt? Which species will adapt too slowly to avoid extinction? To answer these questions, we need to consider slightly different models than those presented in other parts of this book. In most models of thermal adaptation, the environmental temperature changes cyclically or stochastically. We need models in which the environmental temperature changes directionally. For convenience, I shall cast these models in terms of environmental warming, but the models apply equally well to environmental cooling.

9.4.1 Adaptation of thermoregulation

Many organisms rely primarily on thermoregulation to deal with thermal change (see Chapter 4). The flexibility of behavioral and physiological mechanisms of thermoregulation enables organisms to respond rapidly to warming. In tropical environments, both ectotherms and endotherms can ameliorate the negative physiological consequences of warming (Hertz 1981). At higher latitudes, ectotherms could thermoregulate to increase physiological performance during a warmer spring or avoid energetic stress during a milder winter. For endotherms, the warming at high latitudes should influence strategies of endothermy and torpor (Humphries et al. 2002). As in the current climate, thermoregulation should confer obvious benefits in future climates. But how will warming affect the costs of thermoregulation?

During global warming, opportunities to achieve preferred body temperatures will shift both dielly and seasonally. Quantifying these shifts requires detailed models of the physical environment (see Chapter 2). Recently, Michael Sears (unpublished observations) modeled the availability of preferred microclimates for a small lizard in the Mescalero Sand Dunes (New Mexico, USA) during a typical day in June. To simulate the effect of global warming, he then added 3°C to all operative temperatures and recalculated the available habitat. In the contemporary climate, habitat was most accessible in the early morning; only a few small patches remained usable by early afternoon (Fig. 9.9a). Under the global warming scenario, the abundance of preferred microclimates peaked about 40 minutes earlier in the day. Usable habitat became more limited and more fragmented at 0900 h, and disappeared completely by 1300 h (Fig. 9.9b). Because operative and preferred temperatures vary among species, temporal shifts in available habitat will differ for predators and prey. Any degree of warming can alter the potential overlap between predators and prey, as was predicted for fish and invertebrates in several lakes (Jansen and Hesslein 2004).

These modeling exercises tell us about the physical constraints on thermoregulation, but we still need to infer the costs of potential behaviors. With these costs, we can link temporal changes in thermal microclimates to adaptive shifts in thermoregulatory strategies. Remember, the cost of behavioral thermoregulation depends on the spatial distribution of operative temperatures, as well as their statistical distribution (see Box 4.4). Any scenario of warming changes not only the frequencies of temperatures but also the structure of the environment from the animal's perspective. Consider the example provided by Sears (Fig. 9.9). Applying a reasonable warming increased the distance between preferred microclimates in the early morning. This phenomenon would likely increase the energetic cost of behavioral thermoregulation. The exact increment in cost depends on the properties of the organism, such as mobility, perception, and memory. If the new cost of thermoregulation outweighs the benefit, we should expect the organism to thermoregulate poorly during activity or to cease activity altogether. Importantly, modest warming on a global scale could cause complex changes in behavior on local scales. These changes

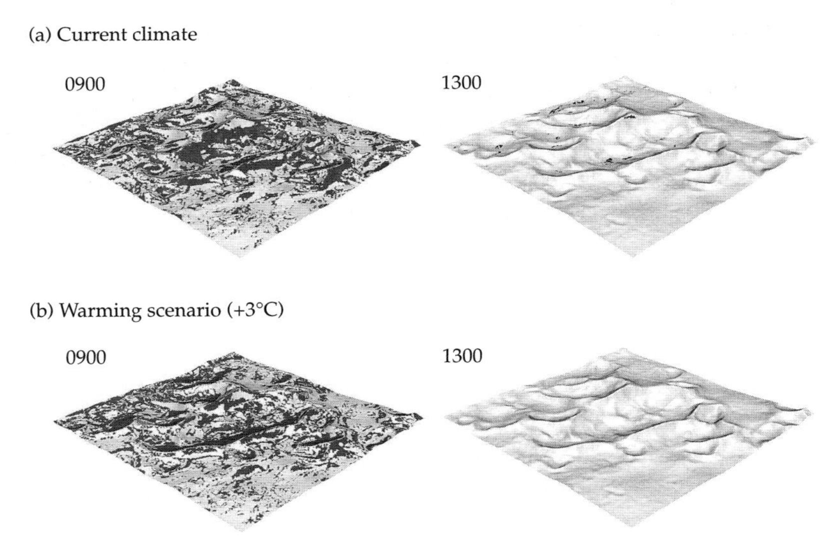

Figure 9.9 Global warming will alter the spatial structure of thermal environments. These maps depict the availability of preferred microclimates of *Sceloporus arenicolus* (shaded pixels) in the Mescalero sand dunes of New Mexico, USA (M. W. Sears, unpublished observations). Preferred microclimates include operative temperatures in the range of 29–37°C. In the current climate, the distribution of preferred microclimates becomes extremely rare by 1300 h. Awarming of 3°C would reduce preferred microclimates at 0900 h and eliminate them completely at 1300 h, potentially shifting activity to an earlier time of day.

will surely exert selective pressures on the thermal sensitivities of physiological performances.

9.4.2 Adaptation of the thermal optimum

How would gradual warming influence the evolution of performance curves? We can address this question with a model developed by Lynch and Lande (1993), who studied adaptation to directional environmental change. This model strongly resembles the one developed by Lynch and Gabriel (1987), which was covered in Chapter 3; however, Lynch and Lande directly modeled the environmental sensitivity of fitness. For a growing population, the instantaneous rate of increase of a genotype (r) provides an appropriate estimate of fitness (see Box 3.3). Lynch and Lande modeled fitness using a symmetrical function, which we shall interpret as a thermal reaction norm:

$$r(z, T_e) = r_m - \frac{(z - T_e)^2}{2\sigma^2}, \tag{9.8}$$

where r_m equals the maximal intrinsic rate of increase, z equals the thermal optimum, σ equals the thermal breadth, and T_e equals the operative environmental temperature. Clearly, this equation describes a perfect thermoconformer; to apply the model to a thermoregulator, one would simply replace the operative temperature (T_e) with the body temperature (T_b).

The mean and variance of the thermal optimum (z and V_z) depend on the additive genetic effect (g, with variance V_g), as well as environmental and nonadditive effects (e). Assuming e equals zero and has a variance of V_e, the growth rate of the population equals

$$r(z, T_e) = r_m - \frac{(\bar{g} - T_e)^2 + V_z}{2\sigma^2}, \tag{9.9}$$

where $V_z = V_g + V_e$. As in many models of evolution, a constant additive genetic variance was assumed. Note that the mean phenotypic value ($\bar{z}$) equals the mean genotypic value ($\bar{g}$).

Lynch and Lande assumed the environment changes at a constant rate (k), with some stochastic variation (ε) around this trend:

$$T_e(t) = kt + \varepsilon. \tag{9.10}$$

The stochastic component has a mean of zero and a variance equal to V_T. During environmental warming, the fitness of a previously adapted genotype declines and selection drives the thermal optimum toward the new operative temperature. Because the environment continues to warm, the thermal optimum will always lag behind the mean operative temperature. After some time, a balance between environmental change and natural selection enables the population to persist. If the environment warms too rapidly, however, this balance never occurs and the population declines to extinction.

To what extent can adaptation of the thermal optimum enable a population to persist during environmental warming? To answer this question, Lynch and Lande calculated the expected value of r for the population at an equilibrium maintained by selection, drift, and mutation. Obviously, this rate of growth must exceed zero for the population to persist.

Ideally, we should like to know the exact outcome of evolution in a warming environment. Unfortunately, environmental stochasticity and genetic drift prevent us from knowing this outcome in advance. Therefore, the predictions of interest become the probability distributions of the thermal optimum (z) and the rate of increase (r) at the evolutionary equilibrium. We can calculate the expected thermal optimum, $E(\bar{g})$, and its variance, $V(\bar{g})$, from the following equations:

$$E(\bar{g}) = kt - k\frac{\sigma^2}{V_g} \tag{9.11}$$

and

$$V(\bar{g}) = \frac{\sigma^2}{2N_e} + \frac{V_g V_T}{2\sigma^2}. \tag{9.12}$$

Equation 9.11 tells us that either a large performance breadth or a low genetic variance would cause the thermal optimum to lag far behind the operative temperature. Equation 9.12 tells us the outcome of evolution would be less predictable in smaller populations and more variable environments. Most of these insights should accord with our intuition, but let us briefly focus on one of them: the influence of the performance breadth on the expected thermal optimum. All other things being equal, a warming environment affects a thermal specialist more than it affects a thermal generalist. Thus, the thermal optimum of a generalist undergoes weaker selective pressure and lags farther behind the operative temperature.

Once we know the expected thermal optimum, we can calculate the expected growth rate of the population, $E(r)$, at the evolutionary equilibrium:

$$E(r) = r_m - \frac{V_z}{2\sigma^2} - \frac{k^2\sigma^2}{2(V_g)^2} - \frac{1}{4N_e} - \frac{V_T}{2\sigma^2}\left(\frac{V_g}{2\sigma^2} + 1\right). \tag{9.13}$$

From this equation, we see that the expected growth rate can never reach the maximal growth rate (r_m) because of directional and stochastic changes in temperature (k and V_T, respectively). The degree of maladaptation to directional change depends on the performance breadth (σ) and genetic drift ($\propto 1/N_e$). The degree of maladaptation to stochastic change depends on the performance breadth and the additive genetic variance (V_g). Equation 9.13 enables us to calculate the maximal rate of warming that permits the persistence of the population (Lynch and Lande 1993). Not surprisingly, this critical rate (k_c) depends on the same key factors that define the expected thermal optimum:

$$k_c = \frac{V_g}{\sigma}\left[2\bar{r}_m - \frac{1}{2N_e} - \frac{V_T}{\sigma^2}\left(\frac{V_g}{2\sigma^2} + 1\right)\right]^{1/2}, \tag{9.14}$$

where $\bar{r}_m$ equals the rate of population increase when the mean thermal optimum matches the mean operative temperature. Therefore, the likelihood of persistence under directional thermal change increases as (1) the genetic variance of the thermal optimum increases, (2) the performance breadth decreases, (3) the population size increases, and (4) the stochasticity of temperature decreases.

In a finite population, the risk of extinction during constant warming always exceeds that predicted by a deterministic model. For this reason, Burger and Lynch (1995) extended the model of Lynch and Lande (1993) to explore the effects of stochastic processes on the risk of extinction. This exercise required computer simulations under a set of assumptions that corresponded closely to those of Lynch and Lande's model.

Burger and Lynch considered a finite population of asexual, semelparous organisms with nonoverlapping generations. From our perspective, the phenotype under selection was the thermal optimum for survivorship; all individuals produced the same number of offspring if they survived to maturity. If the number of surviving offspring exceeded a hypothetical carrying capacity, excess individuals were culled randomly before mating. The thermal optimum was determined by additive genetic and environmental effects, whose variances possessed the same properties as they did in Lynch and Lande's deterministic model.

By coupling environmental and demographic stochasticities, Burger and Lynch quantified the way that adaptation affected the probability of persistence during directional environmental change. The distinction between this exercise and the one undertaken by Lynch and Lande (1993) should not be overlooked. Because Lynch and Lande derived the critical rate of environmental change for the expected phenotype, $E(r)$, their predictions hold only in a deterministic world. Specifically, their model predicts a population would persist unless the rate of environmental change exceeds the critical rate (k_c). But stochasticity dictates that an environment warming slowly over a long period can warm rapidly over a short period. A chance stretch of very warm years could lead to extinction even when the mean rate of change falls short of k_c. Furthermore, demographic stochasticity caused by density dependence can lead to maladaptation via genetic drift. The unpredictability of the real world makes the persistence of a population in a warming environment far more questionable. In this case, extinction always occurs if we consider a sufficient period of time. But we can ask the following question: what is the probability of extinction during a certain period of warming?

Computer simulations confirmed that environmental and demographic stochasticities drive a population to extinction even when the environment warms slower than k_c (Fig. 9.10). In each simulation, the mean phenotype eventually lagged behind the optimal phenotype and the population shrank. As each population dwindled through natural selection, genetic drift began to erode additive genetic variance, which in turn caused future maladaptation and more severe selection. Depending on the rate of environmental change, each population would either rebound or become extinct. The take-home message is clear. A real population, which experiences selection and drift in a stochastic environment, faces a greater risk of extinction than does an idealized population in a deterministic environment.

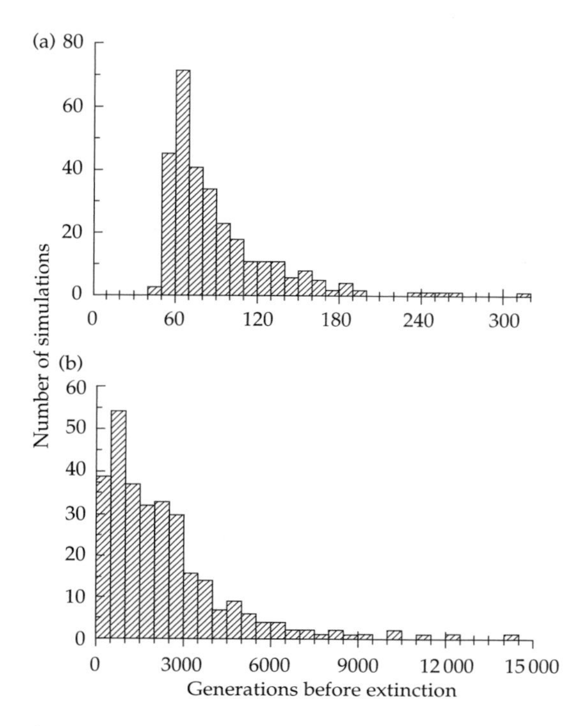

Figure 9.10 In simulations of evolution during directional environmental change, the likely time until extinction depended on the rate of change. (a) During rapid change ($k = 0.12$), the majority of populations were extinct within 100 generations. (b) During slow change ($k = 0.06$), populations persisted for much longer periods. Adapted from Burger and Lynch (1995) with permission from Blackwell Publishing.

9.4.3 Adaptation of the performance breadth

Huey and Kingsolver (1993) extended Lynch and Lande's analysis to consider two additional questions. First, how does directional change in temperature affect the optimal performance breadth? And second, how do specialist–generalist tradeoffs influence adaptation to directional change? To answer these questions, they permitted the performance breadth (σ) to evolve in addition to the maximal rate of increase (r_m). In one scenario, the evolution of σ imposed no effect on the evolution of r_m. In a second scenario, an increase in σ caused a decrease in r_m. With these minor modifications to the deterministic model, Huey and Kingsolver calculated the performance breadth that maximized the critical rate of warming (k_c).

Interestingly, genotypes with intermediate performance breadths would tolerate the fastest change in temperature (Huey and Kingsolver 1993). This result makes sense if we consider the details of the model. From eqn 9.11, we concluded that the thermal optimum of a generalist would lag farther behind the operative temperature than the thermal optimum of a specialist. Hence, a population of generalists has difficulty persisting during rapid warming (see eqn 9.14). Nevertheless, specialists face a different problem. If their performance breadth is too narrow, even a slow rise in temperature would lead to extinction. Consequently, an intermediate performance breath balances the needs to conserve fitness and facilitate adaptation during environmental change. A specialist–generalist tradeoff exacerbates the problem of a wide performance breadth because performance at the thermal optimum decreases as the performance breadth increases.

Adding environmental or demographic stochasticity to the model does not alter the basic result: genotypes with an intermediate performance breadth fare best in a changing environment. Huey and Kingsolver (1993) added stochasticity to the directional change in temperature. As thermal stochasticity increased, so did the performance breadth needed to maximize the critical rate of warming. Similarly, Burger and Lynch (1995) modeled the probability of extinction under environmental and demographic stochasticity. An intermediate performance breadth generally maximized the expected time until extinction (Fig. 9.11). The causes of these patterns differ slightly because of the genetic assumptions of each model. Huey and Kingsolver assumed a constant additive genetic variance, whereas Burger and Lynch modeled changes in genetic variance with a multilocus system of alleles. Therefore, the simulations of Burger and Lynch also revealed the potential for genetic constraints to limit adaptation. Under the strong selection resulting from a narrow performance breadth, a decrease in the additive genetic variance slowed the rate of adaptation and increased the probability of extinction. In fact, wider performance breadths always enhance persistence if they increase the additive genetic variance of the thermal optimum (Huey and Kingsolver 1993).

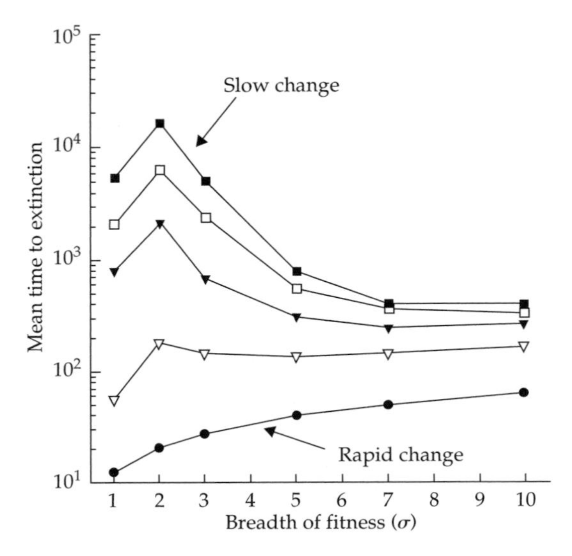

Figure 9.11 For most rates of warming, an intermediate thermal breadth would maximize the persistence of a population. Mean times to extinction were derived from simulated adaptation to various rates of directional environmental change ($k = 0.055 - 0.250$). If an environment warms rapidly, only an extremely wide thermal breadth would enable a population to persist for more than a few generations. Adapted from Burger and Lynch (1995) with permission from Blackwell Publishing.

All things considered, both the thermal optimum and the performance breadth influence the viability of a population during environmental warming.

This conclusion bolsters an earlier one based on a model described in Chapter 3. Specifically, Gilchrist (1995) found that generalists were more fit than specialists when temperature varied little within generations and greatly among generations. The seasonal variation among generations of Gilchrist's model functionally equals the steady increase in temperature of Huey and Kingsolver's model. The only difference being that the temperature rises and falls cyclically instead of rising continuously. Both patterns of thermal change favor generalists. Given this robust prediction, regions that have heated more rapidly in recent decades should contain genotypes that have wider performance breadths. In fact, global warming might explain why the performance breadths of real organisms often exceed those predicted by models of adaptation to stochastic change.

9.5 Thermal games in a warming world

In Chapter 8, we saw how biotic interactions can influence the course of thermal adaptation. What happens if we apply the concept of a thermal game to evolution in a warming environment? Consider a simple game between two or more competitors that was recently analyzed by Johansson (2008). With a few assumptions, this game can be used to infer the outcome of coevolution during warming. First, let us assume that the thermal sensitivity of fitness[16] follows a Gaussian function, and the environment conforms to a Gaussian distribution of operative temperatures. For simplicity, let us also assume that the competing species resemble one another physically, such that they experience identical operative temperatures in the same microhabitat. The presence of one species in a microhabitat prevents the other species from effectively using this site, thus altering the frequency of available temperatures (see Fig. 8.9). Thus, competition between species increases as their thermal optima become more similar; if they possess the same thermal optimum, the model effectively becomes an analysis of a single species.

Given these assumptions, the evolutionarily stable strategy in a constant environment resembles the one described in Chapter 8 (see Fig. 8.10). As long as the range of operative temperatures exceeds the thermal breadth of fitness, the two species can coexist. At an evolutionary equilibrium, their thermal optima will deviate symmetrically from the mean operative temperature. Species with lower or higher thermal optima would not perform well enough to invade the community, and those with a thermal optimum equal to the mean would face too much competition. As the thermal breadth of each species decreases, the number of coexisting species in the

[16] More precisely, Johansson (2008) modeled the thermal sensitivity of the carrying capacity of a species, but I consider the carrying capacity to be an estimate of fitness. This difference in terminology serves to simplify the explanation of the model without loss of generality.

evolutionarily stable strategy increases. For a community of three species, the thermal optimum of the third species would equal the mean operative temperature, while the thermal optima of the other two would deviate symmetrically from this temperature (as in the evolutionarily stable strategy for two species).

How might species coevolve during global warming? To answer this question, we can draw on Johansson's simulations of adaptation during directional environmental change. Johansson examined the coevolution of competing species using an individual-based model, which I interpret here in the language of thermal biology. First, Johansson let genetic variation build up during 20 000 generations of evolution in a constant environment. Then, he initiated environmental change by increasing the mean operative temperature at a constant rate. During warming, the thermal optima of the species were permitted to coevolve, but the thermal breadths remained constant. In simulations of a single species, the outcomes matched the prediction of Lynch and Lande's model (see Section 9.4.2); specifically, the thermal optimum lagged behind the mean operative temperature, but the species persisted as long as warming proceeded slowly. In simulations of two competing species, one species was very susceptible to extinction during warming (Fig. 9.12). This result stemmed from the different phenotypes of the species at the onset of warming. Initially, one species had a thermal optimum that exceeded the operative temperature, while the other species had a thermal optimum that did not. Therefore, warming immediately increased the fitness of one species while decreasing the fitness of the other species (Fig. 9.12a). For the advantaged species, the thermal optimum evolved ahead of but close to the mean operative temperature. For the disadvantaged species, the thermal optimum lagged far behind. The fate of the lagging species depended on the rate of warming, but extinction was much more likely than it was without competition. In fact, the lagging species became extinct at a rate of warming that was only one-fifth of the critical rate for a single species! When the community initially consisted of three species, the species with the lowest thermal optimum became extinct unless the environment warmed extremely slowly ($< 10\%$ of the critical rate for a single species; Fig. 9.12c and d). Upon extinction of the third species, the remaining two evolved thermal optima that tracked the mean operative temperature (Fig. 9.12d). Taken together, these simulations tell us that competition will exacerbate risks of extinction during global warming. Intuition suggests predation or parasitism would do the same.

9.6 Evolutionary consequences of gene flow in a warming world

Adaptation to constant warming depends not only on natural selection at the local scale but also on gene flow at the regional scale. Most species constitute a metapopulation, within which individuals migrate between locations with different thermal conditions. In Chapter 3, we briefly considered the role of gene flow during the evolution of performance curves (Section 3.9). Gene flow initially enhances adaptation to a novel environment, but ultimately prevents perfect adaptation. However, these models focus on the migration of genotypes from a source population to a sink population, as occurs during a range expansion. During climate change, local genotypes become maladapted, but preadapted genotypes might exist in some part of the geographic range (see, for example, Rehfeldt et al. 2001). The balance between the immigration of preadapted and maladapted genotypes determines the impact of gene flow on the adaptation and persistence of a population. In this final section, we shall examine the interaction between natural selection and gene flow in the context of global warming.

9.6.1 Spatially heterogeneous warming can reduce the flow of maladapted genotypes

A model developed by Kirkpatrick and Barton (1997) suggests that global warming will alter the effect of gene flow on local adaptation. Instead of focusing on environmental variation over time, Kirkpatrick and Barton focused on environmental variation over space. Let us assume that the operative environmental temperature varies linearly along a spatial gradient (e.g., latitude), such that the temperature at any location (x) can be

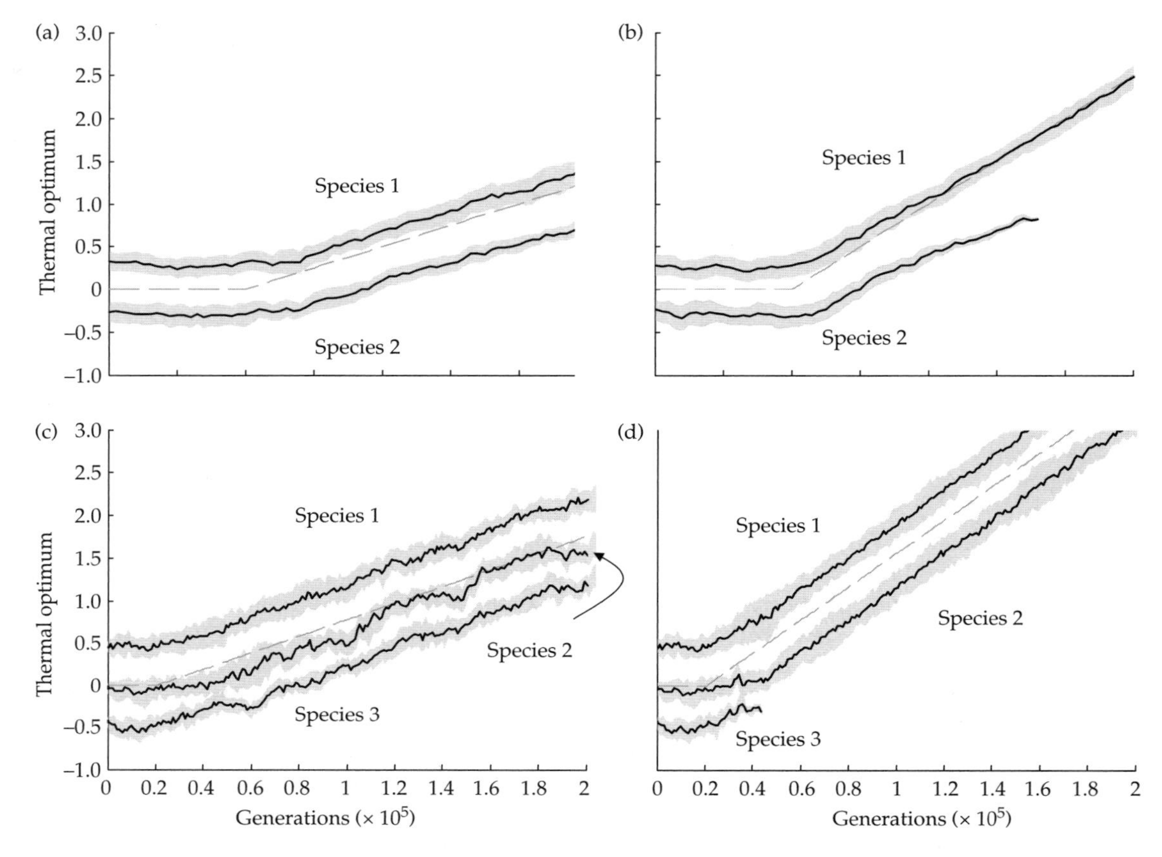

Figure 9.12 Competition influenced the simulated adaptation of a continuous trait, equivalent to the thermal optimum, during directional environmental change. (a) In a community of two species, the thermal optimum of one species (lower, solid line) lags behind the mean temperature (middle, dashed line), while the thermal optimum of the other species (upper, solid line) adapts to the changing environment. (b) If warming occurs more rapidly, the maladapted species may become extinct. The different responses of the two species occur because the fitness of one species initially increases during warming while that of the other species initially decreases. (c and d) In a community of three species, either the thermal optima of all three species parallel the mean temperature (c) or one species becomes extinct relatively soon after warming commences (d). Adapted from Johansson (2008) with permission from Blackwell Publishing.

calculated as

$$T_e(x) = ax + b, \tag{9.15}$$

where a and b represent constants. A metapopulation distributed along this cline would undergo thermal adaptation while exchanging genes by migration. To model local adaptation, Kirkpatrick and Barton used the same fitness function as did Lynch and Lande (eqn 9.8). Gene flow occurred via random dispersal prior to reproduction. A simple model of population dynamics linked adaptation to the density of individuals at each location, $N(x)$:

$$N(x) = ce^{\bar{r}d}, \tag{9.16}$$

where c and d represent constants. When $d = 0$, eqn 9.16 imposes a form of density dependence in which the population consists of c individuals at its equilibrium. Otherwise, the density at the equilibrium depends on the population's mean fitness ($\bar{r}$).

Depending on the magnitude of the thermal cline, the model predicts one of two outcomes (Fig. 9.13). A shallow cline enables a species to expand its range along the entire gradient and adapt perfectly to each local environment. In contrast, a steep cline limits the geographic range and prevents perfect

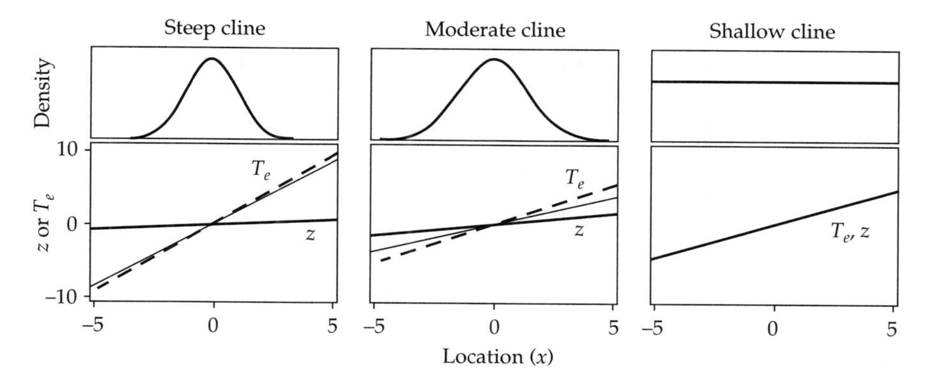

Figure 9.13 The steepness of a thermal cline greatly affects the predicted geographic distribution of a species. Bottom plots show the predicted clines in the thermal optimum, *z* (solid lines), for different clines in environmental temperature, T_e (dashed lines). Upper plots show the corresponding distributions of individuals. Along a shallow thermal cline, a species readily adapts to local environments and occupies the entire cline. As the thermal cline becomes steeper, the inability to adapt to local conditions constrains the range of the species. Adapted from Kirkpatrick and Barton (1997, *The American Naturalist*, University of Chicago Press). © 1997 by The University of Chicago.

adaptation at each location. The density of individuals follows a Gaussian distribution, with the greatest density occurring at the center of the range (Fig. 9.13). Gene flow from the center of the range limits adaptation at the edge. Kirkpatrick and Barton referred to the range boundary as soft because some number of individuals technically exists at any point along the linear gradient; yet reproductive constraints and demographic stochasticity would generate a hard range boundary. In a more complex model of population dynamics, gene flow can also cause extinction when the thermal cline becomes too steep relative to the thermal breadth of the species (Kirkpatrick and Barton 1997). Bear in mind that this model assumes the thermal breadth cannot evolve.

How might global warming affect the interaction between natural selection and gene flow? Kirkpatrick and Barton's model offers a tentative answer to this question. When gene flow limits adaptation, a change in the thermal cline would perturb the metapopulation from its equilibrial state. Imagine a homogeneous warming along the linear gradient (i.e., an increase in b of eqn 9.15). The range of the species would shift along the gradient, but the distribution of densities and the degree of adaptation should remain the same (Kirkpatrick and Barton 1997). But now imagine that warming occurs heterogeneously along the linear gradient. This phenomenon would alter the steepness of the thermal cline (i.e., change a as well as b). In the real world, latitudinal clines in temperature should become less steep because environments at high latitudes continue to warm faster than those at low latitudes (see Section 9.1). This dampening of thermal clines could facilitate adaptation to local environments. Of course, many other abiotic and biotic factors will change also, such that we cannot know whether the mechanism described by this model will actually enable species to persist (for a review of biotic limits to geographic ranges, see Case et al. 2005).

9.6.2 Spatially heterogeneous warming can increase the flow of preadapted genotypes

When environments warm at different rates, migration might enhance the rate of local adaptation. As the local environment warms, migration enables some individuals to escape local stress and find an environment to which they are preadapted. This shuffling of genotypes can maintain genetic diversity and speed local adaptation. This process was recently modeled by de Mazancourt and colleagues (2008). They created environments consisting of discrete patches, each of which underwent environmental change at a different rate. Individuals of one or more species inhabited these patches, where they grew for some period and then reproduced according to their biomass. The offspring were assigned a distribution of phenotypes with a mean equal to the mean parental phenotype in the patch. Offspring dispersed randomly among patches and the cycle continued. This process was simulated for communities of various sizes, ranging from one to 16 species.

The growth rate (and hence the fitness) of each individual depended on a Gaussian performance curve, similar to eqn 3.5. If we interpret this curve as the thermal sensitivity of growth rate, the model predicts the evolution of the thermal optimum in a warming environment. The thermal optimum of each species evolved by hard and soft selection. Hard selection occurred because the local environment determined the fecundity of each individual. Soft selection occurred because no patch could hold more than 10 000 individuals. If the number of residents plus immigrants exceeded this maximum, excess individuals were randomly culled from the population. Thus, recruitment of each phenotype in a patch was proportional to its abundance after dispersal. The performance breadth could not evolve, but a range of breadths were examined to see how this parameter influenced adaptation of the thermal optimum. At the beginning of each simulation, populations were allowed to adapt to a constant environment for 400 generations. Afterward, the researchers imposed 50 generations of spatially heterogeneous warming and 50 generations of recovery in a constant environment (warmer than the initial environment).

In the absence of competition, heterogeneous warming enhanced persistence and sometimes sped adaptation. Under certain conditions, patches in which individuals were initially adapted provided stepping stones for adaptation to other patches. Under other conditions, preadaptation coupled with migration slowed adaptation to novel environments. Overall, the effect of spatial heterogeneity depended on the dispersal rate and the performance breadth. When the performance breadth was wide and the dispersal rate was low, increasing spatial heterogeneity increased the rate of adaptation. Otherwise, adaptation decreased with increasing spatial heterogeneity; very high dispersal led to maladaptation in all patches.

In general, competition among species negated the benefits of migration during environmental change, increasing the chance of extinction and slowing the rate of adaptation. Individuals fared well in patches where they were preadapted, but were excluded from patches in which competitors were preadapted. The only exception occurred when performance breadths were narrow; in this case, extinction during warming was so severe that species rarely encountered one another. Both migration and competition combined to slow adaptation. High dispersal enabled species to sort themselves into patches where they were preadapted. In contrast, low dispersal forced species to adapt to changes in the local environment. Simulations of evolution in more complex communities could change our perception of the interaction between natural selection and gene flow during global warming.

9.7 Conclusions

Anthropogenic climate change poses a tremendous risk to biodiversity. Rates of warming during recent decades have exceeded those experienced during the previous millennia. These changes have led to advances in phenology, shifts in geographic ranges, and disruptions of ecological interactions. The current acceleration of global warming will exacerbate these effects in the coming decades. Mechanistic models provide a means to understand ecological impacts because they force researchers to make explicit assumptions about the biology of organisms. One can relax these assumptions to explore the relative contribution of different factors to changes in population dynamics and geographic distributions. Given this property, mechanistic models can ultimately be extended to deliver predictions about evolutionary impacts.

A growing body of evidence underscores the need to consider evolutionary responses to global warming. Both the physiological regulation of phenology and the thermal sensitivity of performance have evolved in warming environments. Nevertheless, quantitative genetic and allelic models suggest that rapid warming will lead to persistent maladaptation and certain extinction. The degree of maladaptation during warming depends on numerous factors, including the stochasticity of temperature, the size of the population, the additive genetic variance, the rate of gene flow, and the interactions between species. Although current models illustrate important constraints on adaptation, they are too simplistic to provide predictions about specific systems. By merging the ecological details of mechanistic models with the evolutionary processes of allelic models, we might arrive at an applied theory of thermal adaptation that more accurately predicts the impacts of anthropogenic warming.

Three endeavors will determine the success of an applied theory. First, we must accurately characterize the selective environments. This task requires attention to spatial and temporal variations in the rate of thermal change. Both the mean and variance of temperature will change heterogeneously, and some environments in the future may not resemble any environment on earth today (Williams et al. 2007). Second, we must link dynamic changes in thermal environments to natural selection. A recurring theme of this book was the mismatch between theoretical predictions and empirical observations, even in cases where selective environments were known. Clearly, we need a better understanding of the tradeoffs that constrain phenotypes; otherwise, we will be stuck with poor representations of fitness landscapes. Finally, we must accurately quantify the genetic constraints on thermal adaptation. In theory, directional selection, such as that caused by constant warming, should enhance the stability of genetic variances and covariances (Jones et al. 2004). In practice, we know far too little about the genetic architecture of thermal reaction norms to predict genetic constraints. Yet, these constraints ultimately determine whether adaptation will proceed rapidly enough to keep pace with environmental change.

In the opening chapter, I articulated a challenge for evolutionary thermal biology: to explain the diversity of strategies by which organisms cope with their thermal environments. How will thermal biologists respond to this challenge? I hope we respond by establishing a strong connection between theoretical and empirical research. This book was designed to facilitate that connection. The models that we have examined constitute a foundational theory of thermal adaptation. This theory should guide us toward critical experiments, which would enable us to evaluate and refine the models. The threat of rapid warming on a global scale reminds us of the challenges that remain and adds a sense of urgency to our mission. But we should not rush to simple conclusions or despair at slow progress. Given the thermal prognosis for this century, the need for an applied theory of thermal adaptation will certainly outlive our careers. Perhaps the next generation of researchers will achieve Levins' vision by dissolving the boundary between theoretical and empirical biology.

References

Ackermann, M., and M. Doebeli. 2004. Evolution of niche width and adaptive diversification. Evolution 58: 2599–2612.

Addo-Bediako, A., S. L. Chown, and K. J. Gaston. 2000. Thermal tolerance, climatic variability and latitude. Proceedings of the Royal Society B 267:739–745.

Adolph, S. C. 1990. Influence of behavioral thermoregulation on microhabitat use by two *Sceloporus* lizards. Ecology 71:315–327.

Albentosa, M., R. Beiras, and A. P. Camacho. 1994. Determination of optimal thermal conditions for growth of clam (*Venerupis pullastra*) seed. Aquaculture 126:315–328.

Almansa, E., J. J. Sanchez, S. Cozzi, C. Rodriguez, and M. Diaz. 2003. Temperature-activity relationship for the intestinal $Na^{+}-K^{+}$-ATPase of *Sparus aurata*. A role for the phospholipid microenvironment? Journal of Comparative Physiology B 173:231–237.

Alonzo, S. H., and M. Mangel. 2001. Survival strategies and growth of krill: avoiding predators in space and time. Marine Ecology Progress Series 209:203–217.

Alonzo, S. H., P. V. Switzer, and M. Mangel. 2003. Ecological games in space and time: the distribution and abundance of Antarctic krill and penguins. Ecology 84:1598–1607.

Ames, D. 1980. Thermal environment affects production efficiency of livestock. Bioscience 30:457–460.

Amo, L., P. Lopéz, and J. Martín. 2007a. Pregnant female lizards *Iberolacerta cyreni* adjust refuge use to decrease thermal costs for their body condition and cell-mediated immune response. Journal of Experimental Zoology 307A:106–112.

—. 2007b. Refuge use: a conflict between avoiding predation and losing mass in lizards. Physiology & Behavior 90:334–343.

Anderson, A. R., A. A. Hoffmann, and S. W. McKechnie. 2005. Response to selection for rapid chill-coma recovery in *Drosophila melanogaster*: physiology and life-history traits. Genetical Research 85:15–22.

Anderson, D. R., K. P. Burnham, and W. L. Thompson. 2000. Null hypothesis testing: problems, prevalence, and an alternative. Journal of Wildlife Management 64:912–923.

Anderson, J. L., L. Albergotti, S. Proulx, C. Peden, R. B. Huey, and P. C. Phillips. 2007. Thermal preference of *Caenorhabditis elegans*: a null model and empirical tests. Journal of Experimental Biology 210:3107–3116.

Andrews, R. M. 1998. Geographic variation in field body temperature of *Sceloporus* lizards. Journal of Thermal Biology 23:329–334.

Andrews, R. M., F. R. M. de la Cruz, and M. V. Santa Cruz. 1997. Body temperatures of female *Sceloporus grammicus*: thermal stress or impaired mobility? Copeia:108–115.

Andrews, R. M., F. R. Mendez-de la Cruz, M. Villagran-Santa Cruz, and F. Rodriguez-Romero. 1999. Field and selected body temperatures of the lizards *Sceloporus aeneus* and *Sceloporus bicanthalis*. Journal of Herpetology 33:93–100.

Angilletta, M. J. 2001. Thermal and physiological constraints on energy assimilation in a widespread lizard (*Sceloporus undulatus*). Ecology 82:3044–3056.

—. 2006. Estimating and comparing thermal performance curves. Journal of Thermal Biology 31:541–545.

Angilletta, M. J., A. F. Bennett, H. Guderley, C. A. Navas, F. Seebacher, and R. S. Wilson. 2006a. Coadaptation: a unifying principle in evolutionary thermal biology. Physiological and Biochemical Zoology 79:282–294.

Angilletta, M. J., and A. E. Dunham. 2003. The temperature-size rule in ectotherms: simple evolutionary explanations may not be general. American Naturalist 162: 332–342.

Angilletta, M. J., T. Hill, and M. A. Robson. 2002b. Is physiological performance optimized by thermoregulatory behavior?: a case study of the eastern fence lizard, *Sceloporus undulatus*. Journal of Thermal Biology 27:199–204.

Angilletta, M. J., P. H. Niewiarowski, A. E. Dunham, A. D. Leache, and W. P. Porter. 2004b. Bergmann's clines in ectotherms: illustrating a life-history perspective with sceloporine lizards. American Naturalist 164: E168–E183.

Angilletta, M. J., P. H. Niewiarowski, and C. A. Navas. 2002a. The evolution of thermal physiology in ectotherms. Journal of Thermal Biology 27:249–268.

Angilletta, M. J., C. E. Oufiero, and A. D. Leache. 2006b. Direct and indirect effects of environmental temperature on the evolution of reproductive strategies: An information-theoretic approach. American Naturalist 168:E123-E135.

Angilletta, M. J., and M. W. Sears. 2003. Is parental care the key to understanding endothermy in birds and mammals? American Naturalist 162:821–825.

Angilletta, M. J., T. D. Steury, and M. W. Sears. 2004a. Temperature, growth rate, and body size in ectotherms: fitting pieces of a life-history puzzle. Integrative and Comparative Biology 44:498–509.

Angilletta, M. J., and Y. L. Werner. 1998. Australian geckos do not display diel variation in thermoregulatory behavior. Copeia:736–742.

Angilletta, M. J., R. S. Wilson, C. A. Navas, and R. S. James. 2003. Tradeoffs and the evolution of thermal reaction norms. Trends in Ecology & Evolution 18:234–240.

Angilletta, M. J., R. S. Winters, and A. E. Dunham. 2000. Thermal effects on the energetics of lizard embryos: implications for hatchling phenotypes. Ecology 81: 2957–2968.

Araujo, M. B., and M. Luoto. 2007. The importance of biotic interactions for modelling species distributions under climate change. Global Ecology and Biogeography 16:743–753.

Araujo, M. B., R. J. Whittaker, R. J. Ladle, and M. Erhard. 2005. Reducing uncertainty in projections of extinction risk from climate change. Global Ecology and Biogeography 14:529–538.

Archer, M. A., J. P. Phelan, K. A. Beckman, and M. R. Rose. 2003. Breakdown in correlations during laboratory evolution. II. Selection on stress resistance in *Drosophila* populations. Evolution 57:536–543.

Armbruster, P., W. E. Bradshaw, K. Ruegg, and C. M. Holzapfel. 2001. Geographic variation and the evolution of reproductive allocation in the pitcher-plant mosquito, *Wyeomyia smithii*. Evolution 55:439–444.

Arnett, A. E., and N. J. Gotelli. 1999. Geographic variation in life-history traits of the ant lion, *Myrmeleon immaculatus*: evolutionary implications of Bergmann's rule. Evolution 53:1180–1188.

Arnfield, A. J. 2003. Two decades of urban climate research: a review of turbulence, exchanges of energy and water, and the urban heat island. International Journal of Climatology 23:1–26.

Arnold, S. J. 1992. Constraints on phenotypic evolution. American Naturalist 140.

—. 1994. Multivariate inheritance and evolution: a review of concepts *in* C. R. B. Boak, ed. Quantitative Genetic Studies of Behavioral Evolution. Chicago, University of Chicago Press.

Arnold, S. J., C. R. Peterson, and J. Gladstone. 1995. Behavioral variation in natural populations. VII. Maternal body temperature does not affect juvenile thermoregulation in a garter snake. Animal Behaviour 50:623–633.

Artacho, P., L. E. Castaneda, and R. F. Nespolo. 2005. The role of quantitative genetic studies in animal physiological ecology. Revista Chilena De Historia Natural 78:161–167.

Asbury, D. A., and S. C. Adolph. 2007. Behavioural plasticity in an ecological generalist: microhabitat use by western fence lizards. Evolutionary Ecology Research 9:801–815.

Ashton, K. G. 2002a. Do amphibians follow Bergmann's rule? Canadian Journal of Zoology 80:708–716.

—. 2002b. Patterns of within-species body size variation of birds: strong evidence for Bergmann's rule. Global Ecology and Biogeography 11:505–523.

—. 2004. Sensitivity of intraspecific latitudinal clines of body size for tetrapods to sampling, latitude and body size. Integrative and Comparative Biology 44:403–412.

Ashton, K. G., and C. R. Feldman. 2003. Bergmann's rule in nonavian reptiles: turtles follow it, lizards and snakes reverse it. Evolution 57:1151–1163.

Ashton, K. G., M. C. Tracy, and A. de Queiroz. 2000. Is Bergmann's rule valid for mammals? American Naturalist 156:390–415.

Atkin, O. K., B. R. Loveys, L. J. Atkinson, and T. L. Pons. 2006a. Phenotypic plasticity and growth temperature: understanding interspecific variability. Journal of Experimental Botany 57:267–281.

Atkin, O. K., I. Scheurwater, and T. L. Pons. 2006b. High thermal acclimation potential of both photosynthesis and respiration in two lowland *Plantago* species in contrast to an alpine congeneric. Global Change Biology 12:500–515.

Atkinson, D. 1994. Temperature and organism size: a biological law for ectotherms. Advances in Ecological Research 25:1–58.

—. 1996a. Ectotherm life history responses to developmental temperature, Pages 183–204 *in* I. A. Johnston, and A. F. Bennett, eds. Animals and temperature: phenotypic and evolutionary adaptation. Cambridge, Cambridge University Press.

—. 1996b. On the solutions to a major life-history puzzle. Oikos 77:359–365.

Atkinson, D., B. J. Ciotti, and D. J. S. Montagnes. 2003. Protists decrease in size linearly with temperature: ca. 2.5% degrees C-1. Proceedings of the Royal Society B 270:2605–2611.

Atkinson, D., S. A. Morley, D. Weetman, and R. N. Hughes. 2001. Offspring size responses to maternal temperature in ectotherms, Pages 269–285 *in* D. Atkinson, and

M. Thorndyke, eds. Environment and Animal Development: Genes, Life Histories and Plasticity. Oxford, Bios Scientific Publishers.

Avery, H. W., J. R. Spotila, J. D. Congdon, R. U. Fischer, E. A. Standora, and S. B. Avery. 1993. Roles of diet protein and temperature in the growth and nutritional energetics of juvenile slider turtles, *Trachemys scripta*. Physiological Zoology 66:902–925.

Ayers, D. Y., and R. Shine. 1997. Thermal influences on foraging ability: body size, posture and cooling rate of an ambush predator, the python *Morelia spilota*. Functional Ecology 11:342–347.

Ayres, M. P., and J. M. Scriber. 1994. Local adaptation to regional climates in *Papilio canadensis* (Lepidoptera, Papilionidae). Ecological Monographs 64:465–482.

Ayrinhac, A., V. Debat, P. Gibert, A. G. Kister, H. Legout, B. Moreteau, R. Vergilino et al. 2004. Cold adaptation in geographical populations of *Drosophila melanogaster*: phenotypic plasticity is more important than genetic variability. Functional Ecology 18:700–706.

Azevedo, R. B. R., V. French, and L. Partridge. 1996. Thermal evolution of egg size in *Drosophila melanogaster*. Evolution 50:2338–2345.

—. 1997. Life-history consequences of egg size in *Drosophila melanogaster*. American Naturalist 150:250–282.

Baer, C. F., and J. Travis. 2000. Direct and correlated responses to artificial selection on acute thermal stress tolerance in a livebearing fish. Evolution 54:238–244.

Baird, O. E., and C. C. Krueger. 2003. Behavioral thermoregulation of brook and rainbow trout: comparison of summer habitat use in an Adirondack River, New York. Transactions of the American Fisheries Society 132:1194–1206.

Bakken, G. S. 1976. A heat transfer analysis of animals: unifying concepts and the application of metabolism chamber data to field ecology. Journal of Theoretical Biology 60:337–384.

—. 1981. A two-dimensional operative-temperature model for thermal energy management by animals. Journal of Thermal Biology 6:23–30.

—. 1985. Operative and standard operative temperature: tools for thermal and energetic studies. American Zoologist 25:933–943.

—. 1989. Arboreal perch properties and the operative temperature experienced by small animals. Ecology 70:922–930.

—. 1992. Measurement and application of operative and standard operative temperatures in ecology. American Zoologist 32:194–216.

Bakken, G. S., and D. M. Gates. 1975. Heat-transfer analysis of animals: some implications for field ecology, physiology, and evolution, Pages 255–290 *in* D. M. Gates, and R. B. Schmerl, eds. Perspectives of Biophysical Ecology. New York, Springer-Verlag.

Balanya, J., J. M. Oller, R. B. Huey, G. W. Gilchrist, and L. Serra. 2006. Global genetic change tracks global climate warming in *Drosophila subobscura*. Science 313: 1773–1775.

Båmstedt, U., J. Lane, and M. B. Marinussen. 1999. Bioenergetics of ephyra larvae of the scyphozoan jellyfish *Aurelia aurita* in relation to temperature and salinity. Marine Biology 135.

Barber, B. J., and E. C. Crawford. 1977. Stochastic dual-limit hypothesis for behavioral thermoregulation in lizards. Physiological Zoology 50:53–60.

Bardoloi, S., and L. K. Hazarika. 1994. Body temperature and thermoregulation of *Antheraea assama* larvae. Entomologia Experimentalis et Applicata 72:207–217.

Barlow, C. A. 1962. The influence of temperature on the growth of experimental populations of *Myzus persicae* (Sulzer) and *Macrosiphum euphorbiae* (Thomas)(Aphididae). Canadian Journal of Zoology 40:145–156.

Bartelt, P. E., and C. R. Peterson. 2005. Physically modeling operative temperatures and evaporation rates in amphibians. Journal of Thermal Biology 30:93–102.

Bartholomew, G. A. 1981. A matter of size: an examination of endothermy in insects and terrestrial vertebrates, Pages 45–78 *in* B. Heinrich, ed. Insect Thermoregulation. New York, John Wiley & Sons.

—. 1982. Physiological control of body temperature, Pages 167–211 *in* C. Gans, and F. H. Pough, eds. Biology of the Reptilia. New York, Academic Press.

Bartlett, P. N., and D. M. Gates. 1967. Energy budget of a lizard on a tree trunk. Ecology 48:315–322.

Basheer, I. A., and M. Hajmeer. 2000. Artificial neural networks: fundamentals, computing, design, and application. Journal of Microbiological Methods 43:3–31.

Bashey, F., and A. E. Dunham. 1997. Elevational variation in the thermal constraints on and microhabitat preferences of the greater earless lizard *Cophosaurus texanus*. Copeia:725–737.

Baur, B., and C. Raboud. 1988. Life history of the land snail *Arianta arbustorum* along an altitudinal gradient. Journal of Animal Ecology 57:71–87.

Bauwens, D., A. M. Castilla, and P. L. N. Mouton. 1999. Field body temperatures, activity levels and opportunities for thermoregulation in an extreme microhabitat specialist, the girdled lizard (*Cordylus macropholis*). Journal of Zoology 249:11–18.

Bauwens, D., T. Garland, A. M. Castilla, and R. Van Damme. 1995. Evolution of sprint speed in lacertid lizards: morphological, physiological, and behavioral covariation. Evolution 49:848–863.

Bauwens, D., P. E. Hertz, and A. M. Castilla. 1996. Thermoregulation in a lacertid lizard: the relative contributions of distinct behavioral mechanisms. Ecology 77:1818–1830.

Bayne, B. L. 1999. Physiological components of growth differences between individual oysters (*Crassostrea gigas*) and a comparison with *Saccostrea commercialis*. Physiological and Biochemical Zoology 72:705–713.

—. 2000. Relations between variable rates of growth, metabolic costs and growth efficiencies in individual Sydney rock oysters (*Saccostrea commercialis*). Journal of Experimental Marine Biology and Ecology 251:185–203.

—. 2004. Phenotypic flexibility and physiological tradeoffs in the feeding and growth of marine bivalve molluscs. Integrative and Comparative Biology 44:425–432.

Beaugrand, G., P. C. Reid, F. Ibanez, J. A. Lindley, and M. Edwards. 2002. Reorganization of North Atlantic marine copepod biodiversity and climate. Science 296: 1692–1694.

Beaupre, S. J. 1995. Effects of geographically variable thermal environment on bioenergetics of mottled rock rattlesnakes. Ecology 76:1655–1665.

Beaupre, S. J., A. E. Dunham, and K. L. Overall. 1993. The effects of consumption rate and temperature on apparent digestibility coefficient, urate production, metabolizable energy coefficient and passage time in canyon lizards (*Sceloporus merriami*) from two populations. Functional Ecology 7:273–280.

Becker, U., G. Colling, P. Dostal, A. Jakobsson, and D. Matthies. 2006. Local adaptation in the monocarpic perennial *Carlina vulgaris* at different spatial scales across Europe. Oecologia 150:506–518.

Beitinger, T. L., W. A. Bennett, and R. W. McCauley. 2000. Temperature tolerances of North American freshwater fishes exposed to dynamic changes in temperature. Environmental Biology of Fishes 58:237–275.

Beitinger, T. L., and L. C. Fitzpatrick. 1979. Physiological and ecological correlates of preferred temperature in fish. American Zoologist 19:319–329.

Beitinger, T. L., and J. J. Magnuson. 1975. Influence of social rank and size on thermoselection behavior of bluegill (*Lepomis macrochirus*). Journal of the Fisheries Research Board of Canada 32:2133–2136.

Belk, M. C., and D. D. Houston. 2002. Bergmann's rule in ectotherms: a test using freshwater fishes. American Naturalist 160:803–808.

Belk, M. C., J. B. Johnson, K. W. Wilson, M. E. Smith, and D. D. Houston. 2005. Variation in intrinsic individual growth rate among populations of leatherside chub (*Snyderichthys copei* Jordan & Gilbert): adaptation to temperature or length of growing season? Ecology of Freshwater Fish 14:177–184.

Bell, G., and S. Collins. 2008. Adaptation, extinction and global change. Evolutionary Applications 1:3–16.

Belliure, J., and L. M. Carrascal. 2002. Influence of heat transmission mode on heating rates and on the selection of patches for heating in a Mediterranean lizard. Physiological and Biochemical Zoology 75:369–376.

Bennett, A. F. 1980. The thermal dependence of lizard behaviour. Animal Behaviour 28:752–762.

—. 1987a. The accomplishments of ecological physiology, Pages 1–8 *in* M. E. Feder, A. F. Bennett, R. B. Huey, and W. Burggren, eds. New Directions in Ecological Physiology. New York, Cambridge University Press.

—. 1987b. Evolution of the control of body temperature: Is warmer better?, Pages 421–431 *in* P. Dejours, L. Bolis, C. R. Taylor, and E. R. Weibel, eds. Comparative Physiology: Life in Water and on Land. Padova, Liviana Press.

—. 2003. Experimental evolution and the Krogh principle: generating biological novelty for functional and genetic analyses. Physiological and Biochemical Zoology 76: 1–11.

Bennett, A. F., J. W. Hicks, and A. J. Cullum. 2000. An experimental test of the thermoregulatory hypothesis for the evolution of endothermy. Evolution 54:1768–1773.

Bennett, A. F., and R. E. Lenski. 1993. Evolutionary adaptation to temperature. II. Thermal niches of experimental lines of *Escherichia coli*. Evolution 47:1–12.

—. 1996. Evolutionary adaptation to temperature. V. Adaptive mechanisms and correlated responses in experimental lines of *Escherichia coli*. Evolution 50: 493–503.

—. 1999. Experimental evolution and its role in evolutionary physiology. American Zoologist 39:346–362.

Bennett, A. F., R. E. Lenski, and J. E. Mittler. 1992. Evolutionary adaptation to temperature. I. Fitness responses of *Escherichia coli* to changes in its thermal environment. Evolution 46:16–30.

Bennett, A. F., and J. A. Ruben. 1979. Endothermy and activity in vertebrates. Science 206:649–654.

Berger, D., R. Walters, and K. Gotthard. 2008. What limits insect fecundity? Body size- and temperature-dependent egg maturation and oviposition in a butterfly. Functional Ecology 22:523–529.

Berrigan, D. 2000. Correlations between measures of thermal stress resistance within and between species. Oikos 89:301–304.

Berrigan, D., and E. L. Charnov. 1994. Reaction norms for age and size at maturity in response to temperature: a puzzle for life historians. Oikos 70:474–478.

Berrigan, D., and J. C. Koella. 1994. The evolution of reaction norms: simple models for age and size at maturity. Journal of Evolutionary Biology 7.

Berrigan, D., and S. M. Scheiner. 2004. Modeling the evolution of phenotypic plasticity, Pages 82–97 *in* T. J. DeWitt, and S. M. Scheiner, eds. Phenotypic Plasticity: Functional and Conceptual Approaches. Oxford, Oxford University Press.

Berry, J., and O. Bjorkman. 1980. Photosynthetic response and adaptation to temperature in higher plants. Annual Review of Plant Physiology 31:491–543.

Bertalanffy, von, L. 1960. Principles and theory of growth, Pages 137–259 *in* W. W. Nowinski, ed. Fundamental Aspects of Normal and Malignant Growth. New York, Elsevier.

Berven, K. A. 1982. The genetic basis of altitudinal variation in the wood frog *Rana sylvatica*. II. An experimental analysis of larval development. Oecologia 52: 360–369.

Berven, K. A., and D. E. Gill. 1983. Interpreting geographic variation in life-history traits. American Zoologist 23: 85–97.

Berven, K. A., D. E. Gill, and S. J. Smith-Gill. 1979. Countergradient selection in the green frog, *Rana clamitans*. Evolution 33:609–623.

Beuchat, C. A., and S. Ellner. 1987. A quantitative test of life-history theory: thermoregulation by a viviparous lizard. Ecological Monographs 57:45–60.

Bicego, K. C., R. C. H. Barros, and L. G. S. Branco. 2007. Physiology of temperature regulation: comparative aspects. Comparative Biochemistry and Physiology A 147:616–639.

Bilcke, J., S. Downes, and I. Buscher. 2006. Combined effect of incubation and ambient temperature on the feeding performance of a small ectotherm. Austral Ecology 31:937–947.

Billerbeck, J. M., T. E. Lankford, and D. O. Conover. 2001. Evolution of intrinsic growth and energy acquisition rates. I. Trade-offs with swimming performance in *Menidia menidia*. Evolution 55:1863–1872.

Billerbeck, J. M., E. T. Schultz, and D. O. Conover. 2000. Adaptive variation in energy acquisition and allocation among latitudinal populations of the Atlantic silverside. Oecologia 122:210–219.

Billings, W. D., P. J. Godfrey, B. F. Chabot, and D. P. Bourque. 1971. Metabolic acclimation to temperature in Arctic and Alpine ecotypes of *Oxyria digyna*. Arctic and Alpine Research 3:277–289.

Bishop, J. A., and W. S. Armbruster. 1999. Thermoregulatory abilities of Alaskan bees: effects of size, phylogeny and ecology. Functional Ecology 13:711–724.

Black-Samuelsson, S., and S. Andersson. 1997. Reaction norm variation between and within populations of two rare plant species, *Vicia pisiformis* and *V. dumetorum* (Fabaceae). Heredity 79:268–276.

Blackburn, T. M., K. J. Gaston, and N. Loder. 1999. Geographic gradients in body size: a clarification of Bergmann's rule. Diversity and Distributions 5:165–174.

Blanckenhorn, W. U. 2000. Temperature effects on egg size and their fitness consequences in the yellow dung fly *Scathophaga stercoraria*. Evolutionary Ecology 14: 627–643.

Blanckenhorn, W. U., and M. Demont. 2004. Bergmann and converse Bergmann latitudinal clines in arthropods: two ends of a continuum? Integrative and Comparative Biology 44:413–424.

Blanckenhorn, W. U., R. C. Stillwell, K. A. Young, C. W. Fox, and K. G. Ashton. 2006. When Rensch meets Bergmann: does sexual size dimorphism change systematically with latitude? Evolution 60:2004–2011.

Blázquez, M. C. 1995. Body temperature, activity patterns and movements by gravid and nongravid females of *Malpolon monspessulanus*. Journal of Herpetology 29: 264–266.

Block, B. A., and J. R. Finnerty. 1994. Endothermy in fishes: a phylogenetic analysis of constraints, predispositions, and selection pressures. Environmental Biology of Fishes 40:283–302.

Blouin-Demers, G., and P. Nadeau. 2005. The cost–benefit model of thermoregulation does not predict lizard thermoregulatory behavior. Ecology 86:560–566.

Blouin-Demers, G., and P. J. Weatherhead. 2001. An experimental test of the link between foraging, habitat selection and thermoregulation in black rat snakes *Elaphe obsoleta obsoleta*. Journal of Animal Ecology 70:1006–1013.

Blows, M. W., and A. A. Hoffman. 2005. A reassessment of genetic limits to evolutionary change. Ecology 86: 1371–1384.

Bochdanovits, Z., and G. de Jong. 2003. Experimental evolution in *Drosophila melanogaster*: interaction of temperature and food quality selection regimes. Evolution 57:1829–1836.

Boddy, L. 1983. Effect of temperature and water potential on growth rate of wood-rotting basidiomycetes. Transactions of the British Mycological Society 80: 141–149.

Boily, P. 2002. Individual variation in metabolic traits of wild nine-banded armadillos (*Dasypus novemcinctus*), and the aerobic capacity model for the evolution of endothermy. Journal of Experimental Biology 205: 3207–3214.

Bowker, R. G. 1984. Precision of thermoregulation of some African lizards. Physiological Zoology 57:401–412.

Bowler, K. 2005. Acclimation, heat shock and hardening. Journal of Thermal Biology 30:125–130.

Bradshaw, W. E., and C. M. Holzapfel. 2006. Evolutionary response to rapid climate change. Science 312:1477–1478.

Braña, F. 1993. Shifts in body temperature and escape behavior of female *Podarcis muralis* during pregnancy. Oikos 66:216–222.

Brattstrom, B. H. 1979. Amphibian temperature regulation studies in the field and laboratory. American Zoologist 19:345–356.

Brett, J. R., J. E. Shelbourn, and C. T. Shoop. 1969. Growth rate and body composition of fingerling sockeye salmon, *Oncorhynchus nerka*, in relation to temperature and ration size. Journal of the Fisheries Research Board of Canada 26:2363–2393.

Brock, T. D. 1994, Life at high temperatures. Yellowstone National Park, Yellowstone Association for Natural Science.

Brodie, E. D., A. J. Moore, and F. J. Janzen. 1995. Visualizing and quantifying natural selection. Trends in Ecology and Evolution 10:313–318.

Bronikowski, A. M., A. F. Bennett, and R. E. Lenski. 2001. Evolutionary adaptation to temperature. VII. Effects of temperature on growth rate in natural isolates of *Escherichia coli* and *Salmonella enterica* from different thermal environments. Evolution 55:33–40.

Brown, G. P., and P. J. Weatherhead. 2000. Thermal ecology and sexual size dimorphism in northern water snakes, *Nerodia sipedon*. Ecological Monographs 70:311–330.

Brown, J. H., J. F. Gillooly, A. P. Allen, V. M. Savage, and G. B. West. 2004. Toward a metabolic theory of ecology. Ecology 85:1771–1789.

Brown, R. P., and S. Griffin. 2005. Lower selected body temperatures after food deprivation in the lizard *Anolis carolinensis*. Journal of Thermal Biology 30: 79–83.

Bryant, S. R., and T. G. Shreeve. 2002. The use of artificial neural networks in ecological analysis: estimating microhabitat temperature. Ecological Entomology 27:424–432.

Bryant, S. R., C. D. Thomas, and J. S. Bale. 2000. Thermal ecology of gregarious and solitary nettle-feeding nymphalid butterfly larvae. Oecologia 122:1–10.

Bubli, O. A., A. G. Imasheva, and V. Loeschcke. 1998. Selection for knockdown resistance to heat in *Drosophila melanogaster* at high and low larval densities. Evolution 52:619–625.

Bubliy, O. A., A. Riihimaa, F. M. Norry, and V. Loeschcke. 2002. Variation in resistance and acclimation to low-temperature stress among three geographical strains of *Drosophila melanogaster*. Journal of Thermal Biology 27:337–344.

Buchner, O., and G. Neuner. 2003. Variability of heat tolerance in alpine plant species measured at different altitudes. Arctic Antarctic and Alpine Research 35: 411–420.

Buckley, L. B. 2008. Linking traits to energetics and population dynamics to predict lizard ranges in changing environments. American Naturalist 171:E1-E19.

Buckley, L. B., and J. Roughgarden. 2005. Lizard habitat partitioning on islands: the interaction of local and landscape scales. Journal of Biogeography 32:2113–2121.

Bull, J. J., M. R. Badgett, and H. A. Wichman. 2000. Big-benefit mutations in a bacteriophage inhibited with heat. Molecular Biology and Evolution 17:942–950.

Burger, R., and M. Lynch. 1995. Evolution and extinction in a changing environment: a quantitative-genetic analysis. Evolution 49:151–163.

Burggren, W. W., and W. E. Bemis. 1990. Studying physiological evolution: paradigms and pitfalls., Pages 191–228 *in* M. H. Nitecki, ed. Evolutionary Innovations. Chicago, University of Chicago Press.

Burnham, K. P., and D. R. Anderson. 2002, Model Selection and Multimodel Inference: A Practical Information-Theoretic Approach. New York, Springer.

Campbell, G. S. 1977, An Introduction to Environmental Biophysics. New York, Springer-Verlag.

Carlson, S. M., and T. R. Seamons. 2008. A review of quantitative genetic components of fitness in salmonids: implications for adaptation to future change. Evolutionary Applications 1:222–238.

Carrière, Y., and G. Boivin. 1997. Evolution of thermal sensitivity of parasitization capacity in egg parasitoids. Evolution 51:2028–2032.

—. 2001. Constraints on the evolution of thermal sensitivity of foraging in *Trichogramma*: genetic trade-offs and plasticity in maternal selection. American Naturalist 157:570–581.

Case, T. J., R. D. Holt, M. A. McPeek, and T. H. Keitt. 2005. The community context of species' borders: ecological and evolutionary perspectives. Oikos 108:28–46.

Casey, T. M. 1981. Behavioral mechanisms of thermoregulation, Pages 79–114 *in* B. Heinrich, ed. Insect Thermoregulation. New York, John Wiley & Sons.

—. 1992. Biophysical ecology and heat exchange in insects. American Zoologist 32:225–237.

Castañeda, L. E., M. A. Lardies, and F. Bozinovic. 2004. Adaptive latitudinal shifts in the thermal physiology of a terrestrial isopod. Evolutionary Ecology Research 6: 579–593.

—. 2005. Interpopulational variation in recovery time from chill coma along a geographic gradient: a study in the common woodlouse, *Porcellio laevis*. Journal of Insect Physiology 51:1346–1351.

Cerda, X., J. Retana, and A. Manzaneda. 1998. The role of competition by dominants and temperature in the foraging of subordinate species in Mediterranean ant communities. Oecologia 117:404–412.

Chamberlin, T. C. 1890. The method of multiple working hypotheses. Science 15:92–97.

Chapin, F. S., and M. C. Chapin. 1981. Ecotypic differentiation of growth processes in *Carex aquatilis* along latitudinal and local gradients. Ecology 62:1000–1009.

Chappell, M. A., and D. W. Whitman. 1990. Grasshopper thermoregulation, Pages 43–172 *in* R. F. Chapman, and A. Joern, eds. Biology of Grasshoppers. New York, Wiley and Sons.

Charland, M. B. 1995. Thermal consequences of reptilian viviparity: thermoregulation in gravid and nongravid garter snakes (*Thamnophis*). Journal of Herpetology 29:383–390.

Charland, M. B., and P. T. Gregory. 1990. The influence of female reproductive status on thermoregulation in a viviparous snake, *Crotalus viridis*. Copeia:1089–1098.

Charnov, E. L., and J. F. Gillooly. 2004. Size and temperature in the evolution of fish life histories. Integrative and Comparative Biology 44:494–497.

Chen, C. P., R. E. Lee, and D. L. Denlinger. 1990. A comparison of the responses of tropical and temperate flies (Diptera, Sarcophagidae) to cold and heat stress. Journal of Comparative Physiology B 160:543–547.

Chen, X. J., X. F. Xu, and X. Ji. 2003. Influence of body temperature on food assimilation and locomotor performance in white-striped grass lizards, *Takydromus wolteri* (Lacertidae). Journal of Thermal Biology 28:385–391.

Chintalapati, S., M. D. Kiran, and S. Shivaji. 2004. Role of membrane lipid fatty acids in cold adaptation. Cellular and Molecular Biology 50:631–642.

Chown, S. L., M. D. Le Lagadec, and C. H. Scholtz. 1999. Partitioning variance in a physiological trait: desiccation resistance in keratin beetles (Coleoptera, Trogidae). Functional Ecology 13:838–844.

Chown, S. L., and S. W. Nicolson. 2004, Insect Physiological Ecology: Mechanisms and Patterns. Oxford, Oxford University Press.

Christian, K., and G. Bedford. 1996. Thermoregulation by the spotted tree monitor, *Varanus scalaris*, in the seasonal tropics of Australia. Journal of Thermal Biology 21:67–73.

Christian, K. A., G. Bedford, B. Green, T. Schultz, and K. Newgrain. 1998. Energetics and water flux of the marbled velvet gecko (*Oedura marmorata*) in tropical and temperate habitats. Oecologia 116:336–342.

Christian, K. A., and G. S. Bedford. 1995. Seasonal changes in thermoregulation by the frillneck lizard, *Chlamydosaurus kingii*, in tropical Australia. Ecology 76:124–132.

Christian, K. A., and C. R. Tracy. 1981. The effect of the thermal environment on the ability of hatchling Galapagos land iguanas to avoid predation during dispersal. Oecologia 49:218–223.

Christian, K. A., C. R. Tracy, and W. P. Porter. 1986. The effect of cold exposure during incubation of*Sceloporus undulatus* eggs. Copeia 1986:1012–1014.

Christian, K. A., and B. W. Weavers. 1996. Thermoregulation of monitor lizards in Australia: an evaluation of methods in thermal biology. Ecological Monographs 66:139–157.

Ciais, P., M. Reichstein, N. Viovy, A. Granier, J. Ogee, V. Allard, M. Aubinet et al. 2005. Europe-wide reduction in primary productivity caused by the heat and drought in 2003. Nature 437:529–533.

Cichon, M., and J. Kozłowski. 2000. Ageing and typical survivorship curves result from optimal resource allocation. Evolutionary Ecology Research 2:857–870.

Clarke, A. 1991. What is cold adaptation and how should we measure it? American Zoologist 31:81–92.

Clarke, A., and K. J. Gaston. 2006. Climate, energy and diversity. Proceedings of the Royal Society B 273: 2257–2266.

Clarke, A., and P. Rothery. 2008. Scaling of body temperature in mammals and birds. Functional Ecology 22:58–67.

Cleavitt, N. 2004. Comparative ecology of a lowland and a subalpine species of *Mnium* in the northern Rocky Mountains. Plant Ecology 174:205–216.

Clusella-Trullas, S., J. S. Terblanche, T. M. Blackburn, and S. L. Chown. 2008. Testing the thermal melanism hypothesis: a macrophysiological approach. Functional Ecology 22:232–238.

Coelho, J. R. 2001. Behavioral and physiological thermoregulation in male cicada killers (*Sphecius speciosus*) during territorial behavior. Journal of Thermal Biology 26:109–116.

Cohen, M. P., and R. A. Alford. 1996. Factors affecting diurnal shelter use by the cane toad, *Bufo marinus*. Herpetologica 52:172–181.

Congdon, J. D., and J. W. Gibbons. 1987. Morphological constraint on egg size: a challenge to optimal egg size theory. Proceedings of the National Academy of Sciences 84:4145–4147.

Connell, J. H. 1961. The influence of interspecific competition and other factors on the distribution of the barnacle*Chthamalus stellatus*. Ecology 42:710–723.

Conner, J. K. 2003. Artificial selection: a powerful tool for ecologists. Ecology 84:1650–1660.

Conover, D. O. 1992. Seasonality and the scheduling of life history at different latitudes. Journal of Fish Biology 41:161–178.

—. 1998. Local adaptation in marine fishes: evidence and implications for stock enhancement. Bulletin of Marine Science 62:477–493.

Conover, D. O., and T. M. C. Present. 1990. Countergradient variation in growth rate: compensation for length of the growing season among Atlantic silversides from different latitudes. Oecologia 83:316–324.

Conover, D. O., and E. T. Schultz. 1995. Phenotypic similarity and the evolutionary significance of countergradient variation. Trends in Ecology & Evolution 10:248–252.

Cooper, B. S., B. H. Williams, and M. J. Angilletta. 2008. Unifying indices of heat tolerance in ectotherms. Journal of Thermal Biology 33:in press.

Cooper, V. S., A. F. Bennett, and R. E. Lenski. 2001. Evolution of thermal dependence of growth rate of *Escherichia coli* populations during 20 000 generations in a constant environment. Evolution 55:889–896.

Cooper, W. E., and N. Greenberg. 1992. Reptilian coloration and behavior, Pages 298–422 *in* C. Gans, and D. Crews, eds. Biology of the Reptilia, 18A: Hormones, Brain, and Behavior Chicago, University of Chicago.

Cortemeglia, C., and T. L. Beitinger. 2006. Projected US distributions of transgenic and wildtype zebra danios, *Danio rerio*, based on temperature tolerance data. Journal of Thermal Biology 31:422–428.

Cossins, A. R., and K. B. Bowler. 1987, Temperature Biology of Animals. New York, Chapman and Hall.

Crill, W. D., R. B. Huey, and G. W. Gilchrist. 1996. Within- and between-generation effects of temperature on the morphology and physiology of *Drosophila melanogaster*. Evolution 50:1205–1218.

Crockett, E. L. 1998. Cholesterol function in plasma membranes from ectotherms: membrane-specific roles in adaptation to temperature. American Zoologist 38: 291–304.

Crompton, A. W., C. R. Taylor, and J. A. Jagger. 1978. Evolution of homeothermy in mammals. Nature 272: 333–336.

Crowder, L. B., and J. J. Magnuson. 1983. Cost–benefit analysis of temperature and food resource use: a synthesis with examples from fishes, Pages 189–221 *in* W. P. Aspey, and S. I. Lustick, eds. Behavioral Energetics. Columbus, Ohio State University Press.

Crowley, S. R. 1987. The effect of desiccation upon the preferred body temperature and activity level of the lizard *Sceloporus undulatus*. Copeia:25–32.

Crozier, L., and G. Dwyer. 2006. Combining population-dynamic and ecophysiological models to predict climate-induced insect range shifts. American Naturalist 167:853–866.

Cunningham, S., and J. Read. 2003a. Comparison of temperate and tropical rainforest tree species: growth responses to temperature. Journal of Biogeography 30:143–153.

Cunningham, S. C., and J. Read. 2002. Comparison of temperate and tropical rainforest tree species: photosynthetic responses to growth temperature. Oecologia 133:112–119.

—. 2003b. Do temperate rainforest trees have a greater ability to acclimate to changing temperatures than tropical rainforest trees? New Phytologist 157:55–64.

Czarnołeski, M., and J. Kozłowski. 1998. Do Bertalanffy's growth curves result from optimal resource allocation? Ecology Letters 1:5–7.

Danks, H. V. 2004. Seasonal adaptations in arctic insects. Integrative and Comparative Biology 44:85–94.

Dark, J. 2005. Annual lipid cycles in hibernators: integration of physiology and behavior. Annual Review of Nutrition 25:469–497.

Dark, J., D. R. Miller, P. Licht, and I. Zucker. 1996. Glucoprivation counteracts effects of testosterone on daily torpor in Siberian hamsters. American Journal of Physiology 39:R398-R403.

Daut, E. F., and R. M. Andrews. 1993. The effect of pregnancy on thermoregulatory behavior of the viviparous lizard *Chalcides ocellatus*. Journal of Herpetology 27:6–13.

David, J. R., L. O. Araripe, M. Chakir, H. Legout, B. Lemos, G. Petavy, C. Rohmer et al. 2005. Male sterility at extreme temperatures: a significant but neglected phenomenon for understanding *Drosophila* climatic adaptations. Journal of Evolutionary Biology 18:838–846.

David, J. R., P. Gibert, E. Gravot, G. Petavy, J. P. Morin, D. Karan, and B. Moreteau. 1997. Phenotypic plasticity and developmental temperature in *Drosophila*: analysis and significance of reaction norms of morphometrical traits. Journal of Thermal Biology 22:441–451.

David, J. R., P. Gibert, B. Moreteau, G. W. Gilchrist, and R. B. Huey. 2003. The fly that came in from the cold: geographic variation of recovery time from low-temperature exposure in *Drosophila subobscura*. Functional Ecology 17:425–430.

David, J. R., H. Legout, and B. Moreteau. 2006. Phenotypic plasticity of body size in a temperate population of *Drosophila melanogaster*: when the temperature-size rule does not apply. Journal of Genetics 85:9–23.

Davis, A. J., L. S. Jenkinson, J. H. Lawton, B. Shorrocks, and S. Wood. 1998. Making mistakes when predicting shifts in species range in response to global warming. Nature 391:783–786.

Davison, I. R. 1991. Environmental effects on algal photosynthesis: temperature. Journal of Phycology 27:2–8.

Dawkins, R. 1980. Good strategy or evolutionarily stable strategy?, Pages 331–367 *in* G. W. Barlow, and J. Silverberg, eds. Sociobiology: Beyond Nature/Nurture? Boulder, Westview Press.

Dawson, W. R. 1975. On the physiological significance of the preferred body temperatures of reptiles, Pages 443–473 *in* D. M. Gates, and R. B. Schmerl, eds. Perspectives in Biophysical Ecology. Berlin, Springer-Verlag.

Day, T. 2000. Competition and the effect of spatial resource heterogeneity on evolutionary diversification. American Naturalist 155:790–803.

De Block, M., and R. Stoks. 2003. Adaptive sex-specific life history plasticity to temperature and photoperiod in a damselfly. Journal of Evolutionary Biology 16:986–995.

De Boer, M. K., E. M. Koolmees, E. G. Vrieling, A. M. Breeman, and M. Van Rijssel. 2005. Temperature responses of three *Fibrocapsa japonica* strains (Raphidophyceae) from different climate regions. Journal of Plankton Research 27:47–60.

de Jong, G. 1995. Phenotypic plasticity as a product of selection in a variable environment. American Naturalist 145:493–512.

—. 1999. Unpredictable selection in a structured population leads to local genetic differentiation in evolved reaction norms. Journal of Evolutionary Biology 12: 839–851.

de Mazancourt, C., E. Johnson, and T. G. Barraclough. 2008. Biodiversity inhibits species' evolutionary responses to changing environments. Ecology Letters 11:380–388.

de Queiroz, A., and K. G. Ashton. 2004. The phylogeny of a species-level tendency: species heritability and possible deep origins of Bergmann's rule in tetrapods. Evolution 58:1674–1684.

DeAngelis, D. L., and W. M. Mooij. 2005. Individual-based modeling of ecological and evolutionary processes. Annual Review of Ecology Evolution and Systematics 36:147–168.

Debat, V., and P. David. 2001. Mapping phenotypes: canalization, plasticity and developmental stability. Trends in Ecology & Evolution 16:555–561.

Deere, J. A., and S. L. Chown. 2006. Testing the beneficial acclimation hypothesis and its alternatives for locomotor performance. American Naturalist 168:630–644.

Deere, J. A., B. J. Sinclair, D. J. Marshall, and S. L. Chown. 2006. Phenotypic plasticity of thermal tolerances in five oribatid mite species from sub-Antarctic Marion Island. Journal of Insect Physiology 52:693–700.

Denny, M. W., and C. D. G. Harley. 2006. Hot limpets: predicting body temperature in a conductance-mediated thermal system. Journal of Experimental Biology 209:2409–2419.

Denoel, M., M. Mathieu, and P. Poncin. 2005. Effect of water temperature on the courtship behavior of the Alpine newt *Triturus alpestris*. Behavioral Ecology and Sociobiology 58:121–127.

Deutsch, C. A., J. J. Tewksbury, R. B. Huey, K. S. Sheldon, C. K. Ghalambor, D. C. Haak, and P. R. Martin. 2008. Impacts of climate warming on terrestrial ectotherms across latitude. Proceedings of the National Academy of Sciences 105:6669–6672.

DeWitt, T. J., and R. B. Langerhans. 2004. Integrated solutions to environmental: theory of multimoment reaction norms, Pages 272 *in* T. J. DeWitt, and S. M. Scheiner, eds. Phenotypic Plasticity: Functional and Conceptual Approaches Oxford, Oxford University Press.

DeWitt, T. J., and S. M. Scheiner. 2004. Phenotypic variation from single genotypes: a primer., Pages 1–9 *in* T. J. DeWitt, and S. M. Scheiner, eds. Phenotypic Plasticity: Functional and Conceptual Approaches. Oxford, Oxford University Press.

DeWitt, T. J., A. Sih, and D. S. Wilson. 1998. Costs and limits of phenotypic plasticity. Trends in Ecology & Evolution 13:77–81.

Diaz, F., E. Sierra, A. D. Re, and L. Rodriguez. 2002. Behavioural thermoregulation and critical thermal limits of *Macrobrachium acanthurus* (Wiegman). Journal of Thermal Biology 27:423–428.

Diaz, J. A. 1997. Ecological correlates of the thermal quality of an ectotherm's habitat: a comparison between two temperate lizard populations. Functional Ecology 11:79–89.

Diego-Rasilla, F. J. 2003. Influence of predation pressure on the escape behaviour of *Podarcis muralis* lizards. Behavioural Processes 63:1–7.

Diego-Rasilla, F. J., and V. Perez-Mellado. 2000. The effects of population density on time budgets of the Iberian wall lizard (*Podarcis hispanica*). Israel Journal of Zoology 46:215–229.

Dillon, M. E., M. R. Frazier, and R. Dudley. 2006. Into thin air: physiology and evolution of alpine insects. Integrative and Comparative Biology 46:49–61.

Dohm, M. R., W. J. Mautz, P. G. Looby, K. S. Gellert, and J. A. Andrade. 2001. Effects of ozone on evaporative water loss and thermoregulatory behavior of marine toads (*Bufo marinus*). Environmental Research 86:274–286.

Dorcas, M. E., and C. R. Peterson. 1998. Daily body temperature variation in free-ranging rubber boas. Herpetologica 54:88–103.

Dorcas, M. E., C. R. Peterson, and M. E. T. Flint. 1997. The thermal biology of digestion in rubber boas (*Charina bottae*): physiology, behavior, and environmental constraints. Physiological Zoology 70:292–300.

Downes, S. 2001. Trading heat and food for safety: costs of predator avoidance in a lizard. Ecology 82:2870–2881.

Downes, S., and D. Bauwens. 2002. An experimental demonstration of direct behavioural interference in two

Mediterranean lacertid lizard species. Animal Behaviour 63:1037–1046.

Downes, S., and A. M. Hoefer. 2004. Antipredatory behaviour in lizards: interactions between group size and predation risk. Animal Behaviour 67:485–492.

Downes, S., and R. Shine. 1998. Heat, safety or solitude? Using habitat selection experiments to identify a lizard's priorities. Animal Behaviour 55:1387–1396.

Drent, J. 2002. Temperature responses in larvae of *Macoma balthica* from a northerly and southerly population of the European distribution range. Journal of Experimental Marine Biology and Ecology 275:117–129.

Dreyer, B. S., P. Neuenschwander, B. Bouyjou, J. Baumgartner, and S. Dorn. 1997. The influence of temperature on the life table of *Hyperaspis notata*. Entomologia Experimentalis Et Applicata 84:85–92.

Du, W. G., Y. W. Lu, and J. Y. Shen. 2005. The influence of maternal thermal environments on reproductive traits and hatchling traits in a lacertid lizard, *Takydromus septentrionalis*. Journal of Thermal Biology 30:153–161.

Du, W. G., S. J. Yan, and X. Ji. 2000. Selected body temperature, thermal tolerance and thermal dependence of food assimilation and locomotor performance in adult blue-tailed skinks, *Eumeces elegans*. Journal of Thermal Biology 25:197–202.

Duellman, W. E., and L. Trueb. 1986, Biology of Amphibians. New York, McGraw Hill.

Dufresne, F., and P. D. N. Hebert. 1998. Temperature-related differences in life-history characteristics between diploid and polyploid clones of the *Daphnia pulex* complex. Ecoscience 5:433–437.

Dulai, S., I. Molnar, and E. Lehoczki. 1998. Effects of growth temperatures of 5 and 25 degrees C on long-term responses of photosystem II to heat stress in atrazine-resistant and susceptible biotypes of *Erigeron canadensis*. Australian Journal of Plant Physiology 25:145–153.

Dullinger, S., T. Dirnbock, and G. Grabherr. 2004. Modelling climate change-driven treeline shifts: relative effects of temperature increase, dispersal and invasibility. Journal of Ecology 92:241–252.

Dunham, A. E., B. W. Grant, and K. L. Overall. 1989. Interfaces between biophysical and physiological ecology and the population ecology of terrestrial vertebrate ectotherms. Physiological Zoology 62:335–355.

Dutton, R. H., L. C. Fitzpatrick, and J. L. Hughes. 1975. Energetics of the rusty lizard *Sceloporus olivaceus*. Ecology 56:1378–1387.

Dwyer, S. A., O. Ghannoum, A. Nicotra, and S. Von Caemmerer. 2007. High temperature acclimation of C-4 photosynthesis is linked to changes in photosynthetic biochemistry. Plant Cell and Environment 30: 53–66.

Dybdahl, M. F., and S. L. Kane. 2005. Adaptation vs. phenotypic plasticity in the success of a clonal invader. Ecology 86:1592–1601.

Dzialowski, E. M. 2005. Use of operative temperature and standard operative temperature models in thermal biology. Journal of Thermal Biology 30:317–334.

Dzialowski, E. M., and M. P. O'Connor. 2001. Physiological control of warming and cooling during simulated shuttling and basking in lizards. Physiological and Biochemical Zoology 74:679–693.

—. 2004. Importance of the limbs in the physiological control of heat exchange in *Iguana iguana* and *Sceloporus undulatus*. Journal of Thermal Biology 29:299–305.

Edmunds, P. J. 2005. The effect of sub-lethal increases in temperature on the growth and population trajectories of three scleractinian corals on the southern Great Barrier Reef. Oecologia 146:350–364.

Eggert, A., E. M. Burger, and A. M. Breeman. 2003a. Ecotypic differentiation in thermal traits in the tropical to warm-temperate green macrophyte *Valonia utricularis*. Botanica Marina 46:69–81.

Eggert, A., P. R. Van Hasselt, and A. M. Breeman. 2003b. Differences in thermal acclimation of chloroplast functioning in two ecotypes of *Valonia utricularis* (Chlorophyta). European Journal of Phycology 38:123–131.

Eggert, A., R. J. W. Visser, P. R. Van Hasselt, and A. M. Breeman. 2006. Differences in acclimation potential of photosynthesis in seven isolates of the tropical to warm temperate macrophyte*Valonia utricularis* (Chlorophyta). Phycologia 45:546–556.

Eggert, A., and C. Wiencke. 2000. Adaptation and acclimation of growth and photosynthesis of five Antarctic red algae to low temperatures. Polar Biology 23:609–618.

Ehleringer, J., and I. Forseth. 1980. Solar tracking by plants. Science 210:1094–1098.

Elith, J., C. H. Graham, R. P. Anderson, M. Dudik, S. Ferrier, A. Guisan, R. J. Hijmans et al. 2006. Novel methods improve prediction of species' distributions from occurrence data. Ecography 29:129–151.

Elliott, J. M. 1982. The effects of temperature and ration size on the growth and energetics of salmonids in captivity. Comparative Biochemistry and Physiology B 73:81–91.

Ellison, A. M. 2004. Bayesian inference in ecology. Ecology Letters 7:509–520.

Ellison, G. T. H., and J. D. Skinner. 1992. The influence of ambient temperature on spontaneous daily torpor in pouched mice (*Saccostomus campestris*: Rodentia—Cricetidae) from southern Africa. Journal of Thermal Biology 17:25–31.

Else, P. L., N. Turner, and A. J. Hulbert. 2004. The evolution of endothermy: role for membranes and molecular activity. Physiological and Biochemical Zoology 77:950–958.

Else, P. L., and B. J. Wu. 1999. What role for membranes in determining the higher sodium pump molecular activity of mammals compared to ectotherms? Journal of Comparative Physiology B 169:296–302.

Endler, J. A. 1986, Natural Selection in the Wild. Princeton, Princeton University Press.

Eppley, R. W. 1972. Temperature and phytoplankton growth in sea. Fishery Bulletin 70:1063–1085.

Falconer, D. S. 1952. The problem of environment and selection. American Naturalist 86:293–298.

—. 1989, Introduction to Quantitative Genetics. New York, John Wiley & Sons, Inc.

Fangue, N. A., M. Hofmeister, and P. M. Schulte. 2006. Intraspecific variation in thermal tolerance and heat shock protein gene expression in common killifish, *Fundulus heteroclitus*. Journal of Experimental Biology 209:2859–2872.

Farmer, C. G. 2000. Parental care: the key to understanding endothermy and other convergent features in birds and mammals. American Naturalist 155:326–334.

Feder, M. E., A. F. Bennett, and R. B. Huey. 2000. Evolutionary physiology. Annual Review of Ecology and Systematics 31:315–341.

Feder, M. E., and G. E. Hofmann. 1999. Heat-shock proteins, molecular chaperones, and the stress response: evolutionary and ecological physiology. Annual Review of Physiology 61:243–282.

Feder, M. E., and J. F. Lynch. 1982. Effects of latitude, season, elevation, and microhabitat on field body temperatures of neotropical and temperate zone salamanders. Ecology 63:1657–1664.

Felsenstein, J. 1985. Phylogenies and the comparative method. American Naturalist 125:1–15.

—. 2002. Contrasts for a within-species comparative method, Pages 118–129 *in* M. Slatkin, and M. Veuille, eds. Modern Developments in Theoretical Population Genetics. Oxford, Oxford University Press.

Fields, P. A. 2001. Protein function at thermal extremes: balancing stability and flexibility. Comparative Biochemistry and Physiology A 129:417–431.

Fischer, K., A. N. M. Bot, P. M. Brakefield, and B. J. Zwaan. 2003c. Fitness consequences of temperature-mediated egg size plasticity in a butterfly. Functional Ecology 17:803–810.

Fischer, K., P. M. Brakefield, and B. J. Zwaan. 2003b. Plasticity in butterfly egg size: why larger offspring at lower temperatures? Ecology 84:3138–3147.

Fischer, K., E. Eenhoorn, A. N. M. Bot, P. M. Brakefield, and B. J. Zwaan. 2003a. Cooler butterflies lay larger eggs: developmental plasticity versus acclimation. Proceedings of the Royal Society B 270:2051–2056.

Fischer, K., and K. Fiedler. 2000. Sex-related differences in reaction norms in the butterfly *Lycaena tityrus* (Lepidoptera: Lycaenidae). Oikos 90:372–380.

—. 2001a. Dimorphic growth patterns and sex-specific reaction norms in the butterfly *Lycaena hippothoe sumadiensis*. Journal of Evolutionary Biology 14:210–218.

—. 2001b. Egg weight variation in the butterfly Lycaena hippothoe: more small or fewer large eggs? Population Ecology 43:105–109.

—. 2002. Life-history plasticity in the butterfly *Lycaena hippothoe*: local adaptations and trade-offs. Biological Journal of the Linnean Society 75:173–185.

Fischer, K., B. J. Zwaan, and P. M. Brakefield. 2004. Genetic and environmental sources of egg size variation in the butterfly *Bicyclus anynana*. Heredity 92:163–169.

Fisher, R. A. 1930, The Genetical Theory of Natural Selection. Oxford, Oxford University Press.

Fitter, A. H., and R. S. R. Fitter. 2002. Rapid changes in flowering time in British plants. Science 296: 1689–1691.

Fleming, I. A., and M. R. Gross. 1990. Latitudinal clines: a trade-off between egg number and size in Pacific salmon. Ecology 71:1–11.

Folk, D. G., P. Zwollo, D. M. Rand, and G. W. Gilchrist. 2006. Selection on knockdown performance in *Drosophila melanogaster* impacts thermotolerance and heat-shock response differently in females and males. Journal of Experimental Biology 209:3964–3973.

Forchhammer, M. C., and E. Post. 2000. Climatic signatures in ecology. Trends in Ecology & Evolution 15: 286–286.

Forseth, I., and J. R. Ehleringer. 1980. Solar tracking response to drought in a desert annual. Oecologia 44:159–163.

Fox, C. W., and M. E. Czesak. 2000. Evolutionary ecology of progeny size in arthropods. Annual Review of Entomology 45:341–369.

Franco, A. M. A., J. K. Hill, C. Kitschke, Y. C. Collingham, D. B. Roy, R. Fox, B. Huntley et al. 2006. Impacts of climate warming and habitat loss on extinctions at species' low-latitude range boundaries. Global Change Biology 12:1545–1553.

Frazier, M., R. B. Huey, and D. Berrigan. 2006. Thermodynamics constrains the evolution of insect population growth rate: "warmer is better". American Naturalist 168:512–520.

Freidenburg, L. K., and D. K. Skelly. 2004. Microgeographical variation in thermal preference by an amphibian. Ecology Letters 7:369–373.

Fretwell, S. D., and H. L. Lucas. 1969. On theoretical behavior and other factors influencing habitat distribution in birds. Acta Biotheoretica 19.

Frid, L., and J. H. Myers. 2002. Thermal ecology of western tent caterpillars *Malacosoma californicum pluviale* and infection by nucleopolyhedrovirus. Ecological Entomology 27:665–673.

Fry, F. E. J., and J. S. Hart. 1948. Cruising speed of goldfish in relation to water temperature Journal of the Fisheries Research Board of Canada 7:169–175.

Full, R. J., and A. Tullis. 1990. Capacity for sustained terrestrial locomotion in an insect: energetics, thermal dependence, and kinematics. Journal of Comparative Physiology B 160:573–581.

Gabriel, W. 1988. Quantitative genetic models for parthenogenetic species, Pages 73–82 *in* G. de Jong, ed. Population Genetics and Evolution. Berlin, Springer-Verlag.

—. 1999. Evolution of reversible plastic responses: inducible defenses and environmental tolerance, Pages 286–305 *in* C. D. Harvell, and R. Tollrian, eds. The Ecology and Evolution of Inducible Defenses. Princeton, Princeton University Press.

—. 2005. How stress selects for reversible phenotypic plasticity. Journal of Evolutionary Biology 18:873–883.

—. 2006. Selective advantage of irreversible and reversible phenotypic plasticity. Archiv Fur Hydrobiologie 167: 1–20.

Gabriel, W., B. Luttbeg, A. Sih, and R. Tollrian. 2005. Environmental tolerance, heterogeneity, and the evolution of reversible plastic responses. American Naturalist 166:339–353.

Gabriel, W., and M. Lynch. 1992. The selective advantage of reaction norms for environmental tolerance. Journal of Evolutionary Biology 5:41–59.

Gadgil, M., and W. H. Bossert. 1970. Life historical consequences of natural selection. American Naturalist 104:1–24.

Gagliano, M., M. I. McCormick, and M. G. Meekan. 2007. Temperature-induced shifts in selective pressure at a critical developmental transition. Oecologia 152:219–225.

Galen, C. 2006. Solar furnaces or swamp coolers: costs and benefits of water use by solar-tracking flowers of the alpine snow buttercup, *Ranunculus adoneus*. Oecologia 148:195–201.

Garbuz, D., M. B. Evgenev, M. E. Feder, and O. G. Zatsepina. 2003. Evolution of thermotolerance and the heat-shock response: evidence from inter/intraspecific comparison and interspecific hybridization in the *virilis* species group of Drosophila. I. Thermal phenotype. Journal of Experimental Biology 206:2399–2408.

Garland, T., and S. C. Adolph. 1994. Why not to do two-species comparative studies: limitations on inferring adaptation. Physiological Zoology 67:797–828.

Garland, T., A. F. Bennett, and E. L. Rezende. 2005. Phylogenetic approaches in comparative physiology. Journal of Experimental Biology 208:3015–3035.

Garland, T., and P. A. Carter. 1994. Evolutionary physiology. Annual Review of Physiology 56:579–621.

Garland, T., P. H. Harvey, and A. R. Ives. 1992. Procedures for the analysis of comparative data using phylogenetically independent contrasts. Systematic Biology 41:18–32.

Garland, T., R. B. Huey, and A. F. Bennett. 1991. Phylogeny and coadaptation of thermal physiology in lizards: a reanalysis. Evolution 45:1969–1975.

Garland, T., P. E. Midford, and A. R. Ives. 1999. An introduction to phylogenetically based statistical methods, with a new method for confidence intervals on ancestral values. American Zoologist 39:374–388.

Gaston, K. J., and S. L. Chown. 1999. Elevation and climatic tolerance: a test using dung beetles. Oikos 86:584–590.

Gaston, K. J., and J. I. Spicer. 1998. Do upper thermal tolerances differ in geographically separated populations of the beachflea *Orchestia gammarellus* (Crustacea: Amphipoda)? Journal of Experimental Marine Biology and Ecology 229:265–276.

Gates, D. M. 1980, Biophysical Ecology. New York, Springer-Verlag.

Gavrilets, S. 1997. Coevolutionary chase in exploiter-victim systems with polygenic characters. Journal of Theoretical Biology 186:527–534.

Gavrilets, S., and S. M. Scheiner. 1993. The genetics of phenotypic plasticity. VI. Theoretical predictions for directional selection. Journal of Evolutionary Biology 6:49–68.

Gebhardt, M. D., and S. C. Stearns. 1993. Phenotypic plasticity for life-history traits in *Drosophila melanogaster*. I. Effect on phenotypic and environmental correlations. Journal of Evolutionary Biology 6:1–16.

Geiser, F. 1998. Evolution of daily torpor and hibernation in birds and mammals: importance of body size. Clinical and Experimental Pharmacology and Physiology 25:736–739.

Geiser, F., G. Kortner, and I. Schmidt. 1998. Leptin increases energy expenditure of a marsupial by inhibition of daily torpor. American Journal of Physiology 44:R1627–R1632.

Gerald, G. W., and L. C. Spezzano. 2005. The influence of chemical cues and conspecific density on the temperature selection of a freshwater snail (*Melanoides tuberculata*). Journal of Thermal Biology 30:237–245.

Ghalambor, C. K., R. B. Huey, P. R. Martin, J. J. Tewksbury, and G. Wang. 2006. Are mountain passes higher in the tropics? Janzen's hypothesis revisited. Integrative and Comparative Biology 46:5–17.

Gibert, P., and R. B. Huey. 2001. Chill-coma temperature in *Drosophila*: Effects of developmental temperature, latitude, and phylogeny. Physiological and Biochemical Zoology 74:429–434.

Gibert, P., B. Moreteau, J. R. David, and S. M. Scheiner. 1998. Describing the evolution of reaction norm shape: Body pigmentation in *Drosophila*. Evolution 52: 1501–1506.

Gibert, P., B. Moreteau, G. Petavy, D. Karan, and J. R. David. 2001. Chill-coma tolerance, a major climatic adaptation among *Drosophila* species. Evolution 55: 1063–1068.

Gilchrist, A. S., R. B. R. Azevedo, L. Partridge, and P. O'Higgins. 2000. Adaptation and constraint in the evolution of *Drosophila melanogaster* wing shape. Evolution & Development 2:114–124.

Gilchrist, G. W. 1995. Specialists and generalists in changing environments. I. Fitness landscapes of thermal sensitivity. American Naturalist 146:252–270.

—. 1996. A quantitative genetic analysis of thermal sensitivity in the locomotor performance curve of *Aphidius ervi*. Evolution 50:1560–1572.

—. 2000. The evolution of thermal sensitivity in changing environments *in* K. B. Storey, and J. M. Storey, eds. Cell and Molecular Responses to Stress. Volume 1. Environmental Stressors and Gene Responses. Amsterdam, Elsevier Science.

Gilchrist, G. W., and R. B. Huey. 1999. The direct response of *Drosophila melanogaster* to selection on knockdown temperature. Heredity 83:15–29.

—. 2001. Parental and developmental temperature effects on the thermal dependence of fitness in *Drosophila melanogaster*. Evolution 55:209–214.

—. 2004. Plastic and genetic variation in wing loading as a function of temperature within and among parallel clines in *Drosophila subobscura*. Integrative and Comparative Biology 44:461–470.

Gilchrist, G. W., R. B. Huey, J. Balanya, M. Pascual, and L. Serra. 2004. A time series of evolution in action: a latitudinal cline in wing size in South American *Drosophila subobscura*. Evolution 58:768–780.

Gilchrist, G. W., R. B. Huey, and L. Partridge. 1997. Thermal sensitivity of *Drosophila melanogaster*: evolutionary responses of adults and eggs to laboratory natural selection at different temperatures. Physiological Zoology 70:403–414.

Gilchrist, G. W., and J. G. Kingsolver. 2001. Is optimality over the hill?, Pages 219–241 *in* S. H. Orzack, and E. Sober, eds. Adaptationism and Optimality. Cambridge, Cambridge University Press.

Gillespie, J. 2004, Population Genetics: A Concise Guide. Baltimore, Johns Hopkins University Press.

Gilligan, M. R. 1991. Bergmann ecogeographic trends among the triplefin blennies (Teleostei: Tripterygiidae) in the Gulf of California, Mexico. Environmental Biology of Fishes 31:301–305.

Gillis, R. 1991. Thermal biology of two populations of red-chinned lizards (*Sceloporus undulatus erythrocheilus*) living in different habitats in southcentral Colorado. Journal of Herpetology 25:18–23.

Gillooly, J. F., J. H. Brown, G. B. West, V. M. Savage, and E. L. Charnov. 2001. Effects of size and temperature on metabolic rate. Science 293:2248–2251.

Gillooly, J. F., E. L. Charnov, G. B. West, V. M. Savage, and J. H. Brown. 2002. Effects of size and temperature on developmental time. Nature 417:70–73.

Gilman, S. E. 2006. Life at the edge: an experimental study of a poleward range boundary. Oecologia 148: 270–279.

Gilman, S. E., D. S. Wethey, and B. Helmuth. 2006. Variation in the sensitivity of organismal body temperature to climate change over local and geographic scales. Proceedings of the National Academy of Sciences 103:9560–9565.

Gittleman, J. L., and H. K. Luh. 1994. Phylogeny, evolutionary models and comparative methods: a simulation study, Pages 103–122 *in* E. P., and R. Vane-Wright, eds. Phylogenetics and Ecology. London, Academic Press.

Glanville, E. J., and F. Seebacher. 2006. Compensation for environmental change by complementary shifts of thermal sensitivity and thermoregulatory behaviour in an ectotherm. Journal of Experimental Biology 209: 4869–4877.

Gomes, F. R., J. G. Chaui-Berlinck, J. Bicudo, and C. A. Navas. 2004. Intraspecific relationships between resting and activity metabolism in anuran amphibians: influence of ecology and behavior. Physiological and Biochemical Zoology 77:197–208.

Gomi, T., M. Nagasaka, T. Fukuda, and H. Hagihara. 2007. Shifting of the life cycle and life-history traits of the fall webworm in relation to climate change. Entomologia Experimentalis et Applicata 125:179–184.

Gomulkiewicz, R., and M. Kirkpatrick. 1992. Quantitative genetics and the evolution of reaction norms. Evolution 46:390–411.

Good, D. S. 1993. Evolution of behaviours in *Drosophila melanogaster* in high temperatures: genetic and environmental effects. Journal of Insect Physiology 39:537–544.

Gordo, O., and J. J. Sanz. 2005. Phenology and climate change: a long-term study in a Mediterranean locality. Oecologia 146:484–495.

Gotthard, K. 2000. Increased risk of predation as a cost of high growth rate: an experimental test in a butterfly. Journal of Animal Ecology 69:896–902.

Gotthard, K., and S. Nylin. 1995. Adaptive plasticity and plasticity as an adaptation: a selective review of plasticity in animal morphology and life history. Oikos 74.

Graham, C. H., S. Ferrier, F. Huettman, C. Moritz, and A. T. Peterson. 2004. New developments in museum-based informatics and applications in biodiversity analysis. Trends in Ecology & Evolution 19:497–503.

Grant, B. W. 1990. Trade-offs in activity time and physiological performance for thermoregulating desert lizards, *Sceloporus merriami*. Ecology 71:2323–2333.

Grant, B. W., and A. E. Dunham. 1988. Thermally imposed time constraints on the activity of the desert lizard *Sceloporus merriami*. Ecology 69:167–176.

—. 1990. Elevational covariation in environmental constraints and life histories of the desert lizard *Sceloporus merriami*. Ecology 71:1765–1776.

Grant, B. W., and W. P. Porter. 1992. Modeling global macroclimatic constraints on ectotherm energy budgets. American Zoologist 32:154–178.

Gratani, L., and L. Varone. 2004. Adaptive photosynthetic strategies of the Mediterranean maquis species according to their origin. Photosynthetica 42:551–558.

Graves, B. M., and D. Duvall. 1993. Reproduction, rookery use, and thermoregulation in free-ranging, pregnant *Crotalus v. viridis*. Journal of Herpetology 27:33–41.

Greenwald, O. E. 1974. Thermal dependence of striking and prey capture by gopher snakes. Copeia 1974: 141–148.

Gregory, P. T., L. H. Crampton, and K. M. Skebo. 1999. Conflicts and interactions among reproduction, thermoregulation and feeding in viviparous reptiles: are gravid snakes anorexic? Journal of Zoology 248:231–241.

Grether, G. F. 2005. Environmental change, phenotypic plasticity, and genetic compensation. American Naturalist 166:E115-E123.

Grewal, P. S., S. Selvan, and R. Gaugler. 1994. Thermal adaptation of entomopathogenic nematodes: niche breadth for infection, establishment, and reproduction. Journal of Thermal Biology 19:245–253.

Griffiths, J. S., and D. F. Alderdice. 1972. Effects of acclimation and acute temperature experience on swimming speed of juvenile coho salmon. Journal of the Fisheries Research Board of Canada 29:251–264.

Grigg, G. C. 2004. An evolutionary framework for studies of hibernation and short-term torpor Pages 131–141 *in* B. M. Barnes, and H. V. Carey, eds. Life in the Cold: Evolution, Adaptation, Mechanisms and Applications. Twelfth International Hibernation Symposium, Biological Papers of the University of Alaska.

Grigg, G. C., and L. Beard. 2000. Hibernation by echidnas in mild climates: hints about the evolution of endothermy?, Pages 5–19 *in* G. Heldmaier, M. Klingenspor, and A. P. Beard, eds. Life In the Cold: Eleventh International Hibernation Symposium. Berlin, Springer.

Grigg, G. C., L. A. Beard, and M. L. Augee. 2004. The evolution of endothermy and its diversity in mammals and birds. Physiological and Biochemical Zoology 77:982–997.

Grigg, G. C., and F. Seebacher. 1999. Field test of a paradigm: hysteresis of heart rate in thermoregulation by a free-ranging lizard (*Pogona barbata*). Proceedings of the Royal Society B 266:1291–1297.

Grimm, N. B., S. H. Faeth, N. E. Golubiewski, C. L. Redman, J. G. Wu, X. M. Bai, and J. M. Briggs. 2008. Global change and the ecology of cities. Science 319: 756–760.

Guderley, H. 2004. Metabolic responses to low temperature in fish muscle. Biological Reviews 79:409–427.

Guderley, H., P. H. Leroy, and A. Gagne. 2001. Thermal acclimation, growth, and burst swimming of three-spine stickleback: enzymatic correlates and influence of photoperiod. Physiological and Biochemical Zoology 74:66–74.

Guisan, A., and W. Thuiller. 2005. Predicting species distribution: offering more than simple habitat models. Ecology Letters 8:993–1009.

Gunderson, C. A., R. J. Norby, and S. D. Wullschleger. 2000. Acclimation of photosynthesis and respiration to simulated climatic warming in northern and southern populations of *Acer saccharum*: laboratory and field evidence. Tree Physiology 20:87–96.

Gurevitch, J., and L. V. Hedges. 1993. Meta-analysis: combining the results of independent experiments, Pages 378–398 *in* S. Scheiner, and J. Gurevitch, eds. Design and analysis of experiments. New York, Chapman and Hall.

Gvoždík, L. 2002. To heat or to save time? Thermoregulation in the lizard *Zootoca vivipara* (Squamata: Lacertidae) in different thermal environments along an altitudinal gradient. Canadian Journal of Zoology 80:479–492.

—. 2003. Postprandial thermophily in the Danube crested newt, *Triturus dobrogicus*. Journal of Thermal Biology 28:545–550.

Gvoždík, L., and A. M. Castilla. 2001. A comparative study of preferred body temperatures and critical thermal tolerance limits among populations of *Zootoca vivipara* (Squamata: Lacertidae) along an altitudinal gradient. Journal of Herpetology 35:486–492.

Gvoždík, L., M. Puky, and M. Sugerkova. 2007. Acclimation is beneficial at extreme test temperatures in the Danube crested newt, *Triturus dobrogicus* (Caudata, Salamandridae). Biological Journal of the Linnean Society 90:627–636.

Hahn, M. W., and M. Pockl. 2005. Ecotypes of planktonic actinobacteria with identical 16S rRNA genes adapted to thermal niches in temperate, subtropical, and tropical freshwater habitats. Applied and Environmental Microbiology 71:766–773.

Haldimann, P., and U. Feller. 2005. Growth at moderately elevated temperature alters the physiological response of the photosynthetic apparatus to heat stress in pea (*Pisum sativum* L.) leaves. Plant Cell and Environment 28:302–317.

Hallas, R., M. Schiffer, and A. A. Hoffmann. 2002. Clinal variation in *Drosophila serrata* for stress resistance and body size. Genetical Research 79:141–148.

Hansen, T. F. 2006. The evolution of genetic architecture. Annual Review of Ecology, Evolution, and Systematics 37:123–157.

Harrington, R., I. Woiwod, and T. Sparks. 1999. Climate change and trophic interactions. Trends in Ecology & Evolution 14:146–150.

Harrison, J. F., J. H. Fewell, S. P. Roberts, and H. G. Hall. 1996. Achievement of thermal stability by varying metabolic heat production in flying honeybees. Science 274:88–90.

Harvell, C. D., C. E. Mitchell, J. R. Ward, S. Altizer, A. P. Dobson, R. S. Ostfeld, and M. D. Samuel. 2002. Climate warming and disease risks for terrestrial and marine biota. Science 296:2158–2162.

Harvey, P. H., and A. Purvis. 1991. Comparative methods for explaining adaptations. Nature 351:619–624.

Harwood, R. H. 1979. Effect of temperature on the digestive efficiency of 3 species of lizards, *Cnemidophorus tigris*, *Gerrhonotus multicarinatus* and *Sceloporus occidentalis*. Comparative Biochemistry and Physiology A 63:417–433.

Hawkins, A. J. S. 1991. Protein turnover: a functional appraisal. Functional Ecology 5:222–233.

—. 1995. Effects of temperature change on ectotherm metabolism and evolution: metabolic and physiological interrelations underlying the superiority of multilocus heterozygotes in heterogeneous environments. Journal of Thermal Biology 20:23–33.

Hayes, J. P., and T. Garland. 1995. The evolution of endothermy: testing the aerobic capacity model. Evolution 49:836–847.

Haynie, D. T. 2001, Biological Thermodynamics. Cambridge, Cambridge University Press.

Hazel, J. R. 1995. Thermal adaptation in biological membranes: is homeoviscous adaptation the explanation? Annual Review of Physiology 57:19–42.

Hazel, J. R., and E. E. Williams. 1990. The role of alterations in membrane lipid composition in enabling physiological adaptation of organisms to their physical environment. Progress in Lipid Research 29:167–227.

Hazel, W. N. 2002. The environmental and genetic control of seasonal polyphenism in larval color and its adaptive significance in a swallowtail butterfly. Evolution 56: 342–348.

Heath, J. E. 1964. Reptilian thermoregulation: evaluation of field studies. Science 146:784–785.

—. 1965. Temperature regulation and diurnal activity in horned lizards. University of California Publications in Zoology 64:97–136.

Heckrotte, C. 1983. The influence of temperature on the behavior of the Mexican jumping bean. Journal of Thermal Biology 8:333–335.

Heino, M., and V. Kaitala. 1999. Evolution of resource allocation between growth and reproduction in animals with indeterminate growth. Journal of Evolutionary Biology 12:423–429.

Heinrich, B. 1974. Thermoregulation in endothermic insects. Science 185:747–756.

—. 1981. Insect Thermoregulation. New York, John Wiley & Sons.

—. 1990. Is reflectance basking real? Journal of Experimental Biology 154:31–43.

Helmuth, B. S. T. 1998. Intertidal mussel microclimates: predicting the body temperature of a sessile invertebrate. Ecological Monographs 68:51–74.

Helmuth, B. S. T., and G. E. Hofmann. 2001. Microhabitats, thermal heterogeneity, and patterns of physiological stress in the rocky intertidal zone. Biological Bulletin 201:374–384.

Hendry, A. P., T. Day, and A. B. Cooper. 2001. Optimal size and number of propagules: allowance for discrete stages and effects of maternal size on reproductive output and offspring fitness. American Naturalist 157:387–407.

Henwood, K. 1975. A field-tested thermoregulation model for two diurnal Namib desert tenebrionid beetles. Ecology 56:1329–1342.

Hercus, M. J., V. Loeschcke, and S. I. S. Rattan. 2003. Lifespan extension of *Drosophila melanogaster* through hormesis by repeated mild heat stress. Biogerontology 4:149–156.

Herczeg, G., A. Gonda, J. Saarikivi, and J. Merila. 2006. Experimental support for the cost–benefit model of lizard thermoregulation. Behavioral Ecology and Sociobiology 60:405–414.

Herczeg, G., A. Herrero, J. Saarikivi, A. Gonda, M. Jantti, and J. Merila. 2008. Experimental support for the cost–benefit model of lizard thermoregulation: the effects of predation risk and food supply. Oecologia 155:1–10.

Hereford, J., T. F. Hansen, and D. Houle. 2004. Comparing strengths of directional selection: How strong is strong? Evolution 58:2133–2143.

Hernandez, R. M., R. F. Buckle, E. Palacios, and S. B. Baron. 2006. Preferential behavior of white shrimp *Litopenaeus vannamei* (Boone 1931) by progressive temperature-salinity simultaneous interaction. Journal of Thermal Biology 31:565–572.

Hertz, A., and D. Kobler. 2000. A framework for the description of evolutionary algorithms. European Journal of Operational Research 126:1–12.

Hertz, P. E. 1974. Thermal passivity of a tropical forest lizard, *Anolis polylepis*. Journal of Herpetology 8:323–327.

—. 1979. Sensitivity to high temperatures in three West Indian grass anoles (Sauria, Iguanidae), with a review of heat sensitivity in the genus *Anolis*. Comparative Biochemistry and Physiology A 63:217–222.

—. 1981. Adaptation to altitude in two West Indian anoles (Reptilia, Iguanidae): field thermal biology and physiological ecology. Journal of Zoology 195:25–37.

—. 1983. Eurythermy and niche breadth in West Indian *Anolis* lizards: a reappraisal, Pages 472–483 *in* G. J. Rhodin, and K. I. Miyata, eds. Advances in Herpetology and Evolutionary: Essays in Honor of Ernest E. Williams. Cambridge, Museum of Comparative Zoology, Harvard University.

—. 1992. Temperature regulation in Puerto Rican *Anolis* lizards: a field test using null hypotheses. Ecology 73:1405–1417.

Hertz, P. E., A. Arce-Hernandez, J. Ramirez-Vazquez, W. Tirado-Rivera, and L. Vazquez-Vives. 1979. Geographical variation of heat sensitivity and water loss rates in the tropical lizard, *Anolis gundlachi*. Comparative Biochemistry and Physiology A 62:947–953.

Hertz, P. E., and R. B. Huey. 1981. Compensation for altitudinal changes in the thermal environment by some *Anolis* lizards on Hispaniola. Ecology 62:515–521.

Hertz, P. E., R. B. Huey, and T. Garland. 1988. Time budgets, thermoregulation, and maximal locomotor performance: are reptiles Olympians or boy scouts? American Zoologist 28:927–938.

Hertz, P. E., R. B. Huey, and E. Nevo. 1983. Homage to Santa Anita: thermal sensitivity of sprint speed in agamid lizards. Evolution 37:1075–1084.

Hertz, P. E., R. B. Huey, and R. D. Stevenson. 1993. Evaluating temperature regulation by field-active ectotherms: the fallacy of the inappropriate question. American Naturalist 142:796–818.

Hijmans, R. J., and C. H. Graham. 2006. The ability of climate envelope models to predict the effect of climate change on species distributions. Global Change Biology 12:1–10.

Hilborn, R., and M. Mangel. 1997, The Ecological Detective. Princeton, Princeton University Press.

Hirshfield, M. F., and D. W. Tinkle. 1975. Natural selection and the evolution of reproductive effort. Proceedings of the National Academy of Sciences 72:2227–2231.

Hochochka, P. W., and G. N. Somero. 2002, Biochemical Adaptation. Oxford, Oxford University Press.

Hoffmann, A., A. Anderson, and R. Hallas. 2002. Opposing clines for high and low temperature resistance in *Drosophila melanogaster*. Ecology Letters 5:614–618.

Hoffmann, A. A. 1995. Acclimation: increasing survival at a cost. Trends in Ecology & Evolution 10:1–2.

Hoffmann, A. A., J. G. Sorensen, and V. Loeschcke. 2003. Adaptation of *Drosophila* to temperature extremes: bringing together quantitative and molecular approaches. Journal of Thermal Biology 28:175–216.

Hoffmann, A. A., and M. Watson. 1993. Geographical variation in the acclimation responses of *Drosophila* to temperature extremes. American Naturalist 142: S93-S113.

Hoffmann, M. H., J. Tomiuk, H. Schmuths, C. Koch, and K. Bachmann. 2005. Phenological and morphological responses to different temperature treatments differ among a world-wide sample of accessions of *Arabidopsis thaliana*. Acta Oecologica 28:181–187.

Holder, K. K., and J. J. Bull. 2001. Profiles of adaptation in two similar viruses. Genetics 159:1393–1404.

Holland, K. N., R. W. Brill, R. K. C. Chang, J. R. Sibert, and D. A. Fournier. 1992. Physiological and behavioral thermoregulation in bigeye tuna (*Thunnus obesus*). Nature 358:410–412.

Holt, R. D. 1990. The microevolutionary consequences of climate change. Trends in Ecology & Evolution 5: 311–315.

Holt, R. D., M. Barfield, and R. Gomulkiewicz. 2004. Temporal variation can facilitate niche evolution in harsh sink environments. American Naturalist 164:187–200.

Holt, R. D., R. Gomulkiewicz, and M. Barfield. 2003. The phenomenology of niche evolution via quantitative traits in a 'black-hole' sink. Proceedings of the Royal Society B 270:215–224.

Hopkin, R. S., S. Qari, K. Bowler, D. Hyde, and M. Cuculescu. 2006. Seasonal thermal tolerance in marine Crustacea. Journal of Experimental Marine Biology and Ecology 331: 74–81.

Howe, P. D., S. R. Bryant, and T. G. Shreeve. 2007. Predicting body temperature and activity of adult *Polyommatus icarus* using neural network models under current and projected climate scenarios. Oecologia 153:857–869.

Howe, R. W. 1962. The effects of temperature and humidity on the oviposition rate of *Tribolium castaneum* (Hbst.)(Coleoptera, Tenebrionidae). Bulletin of Entomological Research 53:301–310.

Howe, S. A., and A. T. Marshall. 2002. Temperature effects on calcification rate and skeletal deposition in the temperate coral, *Plesiastrea versipora* (Lamarck). Journal of Experimental Marine Biology and Ecology 275:63–81.

Huey, R. B. 1974. Behavioral thermoregulation in lizards: importance of associated costs. Science 184:1001–1003.

—. 1982. Temperature, physiology, and the ecology of reptiles, Pages 25–91 *in* C. Gans, and F. H. Pough, eds. Biology of the Reptilia. New York, Academic Press.

—. 1987. Phylogeny, history, and the comparative method, Pages 76–98 *in* M. E. Feder, A. F. Bennett, W. W. Burggren, and R. B. Huey, eds. New Directions in Ecological Physiology. Cambridge, Cambridge University Press.

—. 1991. Physiological consequences of habitat selection. American Naturalist 137:S91-S115.

Huey, R. B., and A. F. Bennett. 1987. Phylogenetic studies of coadaptation: preferred temperatures versus optimal performance temperatures of lizards. Evolution 41: 1098–1115.

Huey, R. B., and D. Berrigan. 1996. Testing evolutionary hypotheses of acclimation, Pages 205–237 *in* I. A. Johnston, and A. F. Bennett, eds. Animals and Temperature: Phenotypic and Evolutionary Adaptation. Cambridge, Cambridge University Press.

—. 2001. Temperature, demography, and ectotherm fitness. American Naturalist 158:204–210.

Huey, R. B., D. Berrigan, G. W. Gilchrist, and J. C. Herron. 1999. Testing the adaptive significance of acclimation: a strong inference approach. American Zoologist 39: 323–336.

Huey, R. B., W. D. Crill, J. G. Kingsolver, and K. E. Weber. 1992. A method for rapid measurement of heat or cold resistance of small insects. Functional Ecology 6:489–494.

Huey, R. B., G. W. Gilchrist, M. L. Carlson, D. Berrigan, and L. Serra. 2000. Rapid evolution of a geographic cline in size in an introduced fly. Science 287:308–309.

Huey, R. B., and P. E. Hertz. 1984. Is a jack-of-all-temperatures a master of none? Evolution 38:441–444.

Huey, R. B., P. E. Hertz, and B. Sinervo. 2003. Behavioral drive versus behavioral inertia in evolution: a null model approach. American Naturalist 161:357–366.

Huey, R. B., and J. G. Kingsolver. 1993. Evolution of resistance to high temperature in ectotherms. American Naturalist 142:S21-S46.

Huey, R. B., P. H. Niewiarowski, J. Kaufmann, and J. C. Herron. 1989. Thermal biology of nocturnal ectotherms: is sprint performance of geckos maximal at low body temperatures? Physiological Zoology 62:488–504.

Huey, R. B., L. Partridge, and K. Fowler. 1991. Thermal sensitivity of *Drosophila melanogaster* responds rapidly to laboratory natural selection. Evolution 45:751–756.

Huey, R. B., E. R. Pianka, and J. A. Hoffman. 1977. Seasonal variation in thermoregulatory behavior and body temperature of diurnal Kalahari lizards. Ecology 58:1066–1075.

Huey, R. B., and M. Slatkin. 1976. Cost and benefits of lizard thermoregulation. Quarterly Review of Biology 51:363–384.

Huey, R. B., and R. D. Stevenson. 1979. Integrating thermal physiology and ecology of ectotherms: discussion of approaches. American Zoologist 19:357–366.

Hughes, L. 2000. Biological consequences of global warming: is the signal already apparent? Trends in Ecology & Evolution 15:56–61.

Hughes, N. F. 1998. A model of habitat selection by drift-feeding stream salmonids at different scales. Ecology 79:281–294.

Hughes, N. F., and T. C. Grand. 2000. Physiological ecology meets the ideal-free distribution: predicting the distribution of size-structured fish populations across temperature gradients. Environmental Biology of Fishes 59:285–298.

Hugie, D. M., and L. M. Dill. 1994. Fish and game: a game-theoretic approach to habitat selection by predators and prey. Journal of Fish Biology 45:151–169.

Hulbert, A. J., and P. L. Else. 2000. Mechanisms underlying the cost of living in animals. Annual Review of Physiology 62:207–235.

Humphries, M. M., D. L. Kramer, and D. W. Thomas. 2003b. The role of energy availability in mammalian hibernation: an experimental test in free-ranging eastern chipmunks. Physiological and Biochemical Zoology 76:180–186.

Humphries, M. M., D. W. Thomas, and D. L. Kramer. 2003a. The role of energy availability in mammalian hibernation: a cost–benefit approach. Physiological and Biochemical Zoology 76:165–179.

Humphries, M. M., D. W. Thomas, and J. R. Speakman. 2002. Climate-mediated energetic constraints on the distribution of hibernating mammals. Nature 418:313–316.

Hung, H. W., C. F. Lo, C. C. Tseng, and G. H. Kou. 1997. Antibody production in Japanese eels, *Anguilla japonica* Temminck & Schlegel. Journal of Fish Diseases 20: 195–200.

Husak, J. F. 2006. Does speed help you survive? A test with collared lizards of different ages. Functional Ecology 20:174–179.

Hutchinson, G. E. 1957. Concluding remarks. Cold Spring Harbor Symposia on Quantitative Biology 22:415–427.

Hutchison, V. H., and J. D. Maness. 1979. Role of behavior in temperature acclimation and tolerance in ectotherms. American Zoologist 19:367–384.

Imsland, A. K., T. M. Jonassen, S. O. Stefansson, S. Kadowaki, and M. H. G. Berntssen. 2000. Intraspecific differences in physiological efficiency of juvenile Atlantic halibut *Hippoglossus hippoglossus* L. Journal of the World Aquaculture Society 31:285–296.

Intergovernmental Panel on Climate Change. 2000. Emissions Scenarios, A Special Report of the IPCC Working Group III.

Intergovernmental Panel on Climate Change. 2007. Climate Change 2007: Synthesis Report, Fourth Assessment Report of the Intergovernmental Panel on Climate Change.

Isaac, L. A., and P. T. Gregory. 2004. Thermoregulatory behaviour of gravid and non-gravid female grass snakes (*Natrix natrix*) in a thermally limiting high-latitude environment. Journal of Zoology 264:403–409.

Izem, R., and J. G. Kingsolver. 2005. Variation in continuous reaction norms: quantifying directions of biological interest. American Naturalist 166:277–289.

Jackson, J. B. C. 1974. Biogeographic consequences of eurytopy and stenotopy among marine bivalves and their evolutionary significance. American Naturalist 108: 541–560.

Jaenicke, R. 1991. Protein stability and molecular adaptation to extreme conditions. European Journal of Biochemistry 202:715–728.

—. 1993. Structure-function relationship of hyperthermophilic enzymes. ACS Symposium Series 516:53–67.

James, A. C., R. B. R. Azevedo, and L. Partridge. 1995. Cellular basis and developmental timing in a size cline of *Drosophila melanogaster*. Genetics 140:659–666.

—. 1997. Genetic and environmental responses to temperature of *Drosophila melanogaster* from a latitudinal cline. Genetics 146:881–890.

James, A. C., and L. Partridge. 1995. Thermal evolution of rate of larval development in *Drosophila melanogaster* in laboratory and field populations. Journal of Evolutionary Biology 8:315–330.

Jansen, W., and R. H. Hesslein. 2004. Potential effects of climate warming on fish habitats in temperate zone lakes with special reference to Lake 239 of the experimental lakes area (ELA), north-western Ontario. Environmental Biology of Fishes 70:1–22.

Jayne, B. C., and A. F. Bennett. 1990. Selection on locomotor performance capacity in a natural population of garter snakes. Evolution 44:1204–1229.

Jeffery, W. R. 2006. Regressive evolution of pigmentation in the cavefish *Astyanax*. Israel Journal of Ecology & Evolution 52:405–422.

Ji, X., W. G. Du, and P. Y. Sun. 1996. Body temperature, thermal tolerance and influence of temperature on sprint speed and food assimilation in adult grass lizards, *Takydromus septentrionalis*. Journal of Thermal Biology 21:155–161.

Ji, X., C.-X. Lin, L.-H. Lin, Q.-B. Qiu, and Y. Du. 2007. Evolution of vivparity in warm-climate lizards: an experimental test of the maternal manipulation hypothesis Journal of Evolutionary Biology 20: 1037–1045.

Ji, X., X. Zheng, Y. Xu, and R. Sun. 1995. Some aspects of thermal biology of the skink (*Eumeces chinensis*). Acta Zoologica Sinica 41:268–274.

Jiang, L., and A. Kulczycki. 2004. Competition, predation and species responses to environmental change. Oikos 106:217–224.

Jiang, L., and P. J. Morin. 2004. Temperature-dependent interactions explain unexpected responses to environmental warming in communities of competitors. Journal of Animal Ecology 73:569–576.

Jin, Y. M., and A. L. DeVries. 2006. Antifreeze glycoprotein levels in Antarctic notothenioid fishes inhabiting different thermal environments and the effect of warm acclimation. Comparative Biochemistry and Physiology B 144:290–300.

Joern, A., B. J. Danner, J. D. Logan, and W. Wolesensky. 2006. Natural history of mass-action in predator-prey models: a case study from wolf spiders and grasshoppers. American Midland Naturalist 156:52–64.

Johansson, J. 2008. Evolutionary responses to environmental changes: how does competition affect adaptation? Evolution 62:421–435.

John-Alder, H. B., and A. F. Bennett. 1987. Thermal adaptations in lizard muscle function. Journal of Comparative Physiology B 157:241–252.

John-Alder, H. B., P. J. Morin, and S. Lawler. 1988. Thermal physiology, phenology, and distribution of tree frogs. American Naturalist 132:506–520.

Johnsen, O., C. G. Fossdal, N. Nagy, J. Molmann, O. G. Daehlen, and T. Skroppa. 2005. Climatic adaptation in *Picea abies* progenies is affected by the temperature during zygotic embryogenesis and seed maturation. Plant, Cell and Environment 28:1090–1102.

Johnson, J. B., and K. S. Omland. 2004. Model selection in ecology and evolution. Trends in Ecology & Evolution 19:101–108.

Johnson, T. P., and A. F. Bennett. 1995. The thermal acclimation of burst escape performance in fish: an integrated study of molecular and cellular physiology and organismal performance. Journal of Experimental Biology 198:2165–2175.

Johnston, I. A., and R. S. Wilson. 2006. Temperature-induced developmental plasticity in ectotherms, Pages 124–138 *in* S. J. Warburton, W. W. Burggren, B. Pelster, C. L. Reiber, and J. Spicer, eds. Comparative Developmental Physiology.

Jonassen, T. M., A. K. Imsland, R. Fitzgerald, S. W. Bonga, E. V. Ham, G. Naevdal, M. O. Stefansson et al. 2000. Geographic variation in growth and food conversion efficiency of juvenile Atlantic halibut related to latitude. Journal of Fish Biology 56:279–294.

Jones, A. G., S. J. Arnold, and R. Borger. 2003. Stability of the G-matrix in a population experiencing pleiotropic mutation, stabilizing selection, and genetic drift. Evolution 57:1747–1760.

Jones, A. G., S. J. Arnold, and R. Burger. 2004. Evolution and stability of the G-matrix on a landscape with a moving optimum. Evolution 58:1639–1654.

Jonsson, B., T. Forseth, A. J. Jensen, and T. F. Naesje. 2001. Thermal performance of juvenile Atlantic salmon, *Salmo salar* L. Functional Ecology 15:701–711.

Jump, A. S., and J. Penuelas. 2005. Running to stand still: adaptation and the response of plants to rapid climate change. Ecology Letters 8:1010–1020.

Kalberer, S. R., M. Wisniewski, and R. Arora. 2006. Deacclimation and reacclimation of cold-hardy plants: current understanding and emerging concepts. Plant Science 171:3–16.

Kamil, A. C., J. R. Krebs, and H. R. Pulliam. 1987. Foraging Behavior, Page 676. New York, Plenum Press.

Kammer, A. E. 1981. Physiological mechanisms of thermoregulation, Pages 115–158 *in* B. Heinrich, ed. Insect Thermoregulation. New York, Wiley.

Karl, I., and K. Fischer. 2008. Why get big in the cold? Towards a solution to a life-history puzzle. Oecologia 155:215–225.

Karl, I., S. A. Janowitz, and K. Fischer. 2008. Altitudinal life-history variation and thermal adaptation in the copper butterfly *Lycaena tityrus*. Oikos 117:778–788.

Karlsson, B., and H. Van Dyck. 2005. Does habitat fragmentation affect temperature-related life-history traits? A laboratory test with a woodland butterfly. Proceedings of the Royal Society of London B 272:1257–1263.

Karlsson, B., and C. Wiklund. 2005. Butterfly life history and temperature adaptations: dry open habitats select for increased fecundity and longevity. Journal of Animal Ecology 74:99–104.

Kassen, R. 2002. The experimental evolution of specialists, generalists, and the maintenance of diversity. Journal of Evolutionary Biology 15:173–190.

Kauffman, A. S., M. J. Paul, and I. Zucker. 2004. Increased heat loss affects hibernation in golden-mantled ground squirrels. American Journal of Physiology 287: R167–R173.

Kaufmann, J. S., and A. F. Bennett. 1989. The effect of temperature and thermal acclimation on locomotor performance in *Xantusia vigilis*, the desert night lizard. Physiological Zoology 62:1047–1058.

Kaufmann, R., and F. H. Pough. 1982. The effect of temperature upon the efficiency of assimilation of preformed water by the desert iguana (*Dipsosaurus dorsalis*). Comparative Biochemistry and Physiology 72A: 221–224.

Kawecki, T. J., and D. Ebert. 2004. Conceptual issues in local adaptation. Ecology Letters 7:1225–1241.

Kearney, M., and W. P. Porter. 2004. Mapping the fundamental niche: physiology, climate, and the distribution of a nocturnal lizard. Ecology 85:3119–3131.

Kelty, J. D., and R. E. Lee. 2001. Rapid cold-hardening of *Drosophila melanogaster* (Diptera: Drosophilidae) during ecologically based thermoperiodic cycles. Journal of Experimental Biology 204:1659–1666.

Kemp, D. J., and A. K. Krockenberger. 2002. A novel method of behavioural thermoregulation in butterflies. Journal of Evolutionary Biology 15:922–929.

—. 2004. Behavioural thermoregulation in butterflies: the interacting effects of body size and basking posture in *Hypolimnas bolina* (L.) (Lepidoptera: Nymphalidae). Australian Journal of Zoology 52:229–236.

Kemp, T. S. 2006. The origin of mammalian endothermy: a paradigm for the evolution of complex biological structure. Zoological Journal of the Linnean Society 147:473–488.

Kerr, G. D., and C. M. Bull. 2006. Interactions between climate, host refuge use, and tick population dynamics. Parasitology Research 99:214–222.

Kerr, J. T., and M. Ostrovsky. 2003. From space to species: ecological applications for remote sensing. Trends in Ecology & Evolution 18:299–305.

Kessler, K., and W. Lampert. 2004. Fitness optimization of *Daphnia* in a trade-off between food and temperature. Oecologia 140:381–387.

Ketterson, E. D., and V. Nolan. 1992. Hormones and life histories: an integrative approach. American Naturalist 140:S33-S62.

Kiefer, M. C., M. Van Sluys, and C. F. D. Rocha. 2005. Body temperatures of *Tropidurus torquatus* (Squamata, Tropiduridae) from coastal populations: do body temperatures vary along their geographic range? Journal of Thermal Biology 30:449–456.

Kimura, M. T. 2004. Cold and heat tolerance of drosophilid flies with reference to their latitudinal distributions. Oecologia 140:442–449.

Kindlmann, P., A. F. G. Dixon, and I. Dostalkova. 2001. Role of ageing and temperature in shaping reaction norms and fecundity functions in insects. Journal of Evolutionary Biology 14:835–840.

Kingsolver, J. G. 1983. Thermoregulation and flight in *Colias* butterflies: elevational patterns and mechanistic limitations. Ecology 64:534–545.

—. 1987. Evolution and coadaptation of thermoregulatory behavior and wing pigmentation pattern in pierid butterflies. Evolution 41:472–490.

—. 2000. Feeding, growth, and the thermal environment of cabbage white caterpillars, *Pieris rapae* L. Physiological and Biochemical Zoology 73:621–628.

—. 2001. Mechanisms and patterns of selection on performance curves: thermal sensitivity of caterpillar growth *in* D. Atkinson, and M. Thorndyke, eds. Environment and Animal Development: Genes, Life Histories and Plasticity Oxford, BIOS Scientific Publishers, Ltd.

—. 2007. Relating environmental variation to selection on reaction norms: an experimental test American Naturalist 169:163–174.

Kingsolver, J. G., and R. Gomulkiewicz. 2003. Environmental variation and selection on performance curves. Integrative and Comparative Biology 43:470–477.

Kingsolver, J. G., R. Gomulkiewicz, and P. A. Carter. 2001a. Variation, selection and evolution of function-valued traits. Genetica 112:87–104.

Kingsolver, J. G., H. E. Hoekstra, J. M. Hoekstra, D. Berrigan, S. N. Vignieri, C. E. Hill, A. Hoang et al. 2001b. The strength of phenotypic selection in natural populations. American Naturalist 157:245–261.

Kingsolver, J. G., and R. B. Huey. 1998. Evolutionary analyses of morphological and physiological plasticity in thermally variable environments. American Zoologist 38:545–560.

Kingsolver, J. G., K. R. Massie, G. J. Ragland, and M. H. Smith. 2007. Rapid population divergence in thermal reaction norms for an invading species: breaking the temperature-size rule. Journal of Evolutionary Biology 20:892–900.

Kingsolver, J. G., G. J. Ragland, and J. G. Shlichta. 2004. Quantitative genetics of continuous reaction norms: thermal sensitivity of caterpillar growth rates. Evolution 58:1521–1529.

Kingsolver, J. G., J. G. Shlichta, G. J. Ragland, and K. R. Massie. 2006. Thermal reaction norms for caterpillar growth depend on diet. Evolutionary Ecology Research 8:703–715.

Kingsolver, J. G., and H. A. Woods. 1997. Thermal sensitivity of growth and feeding in *Manduca sexta* caterpillars. Physiological Zoology 70:631–638.

—. 1998. Interactions of temperature and dietary protein concentration in growth and feeding of *Manduca sexta* caterpillars. Physiological Entomology 23:354–359.

Kirkpatrick, M., and N. H. Barton. 1997. Evolution of a species' range. American Naturalist 150:1–23.

Kirschbaum, M. U. F. 2004. Direct and indirect climate change effects on photosynthesis and transpiration. Plant Biology 6:242–253.

Klok, C. J., and S. L. Chown. 2005. Inertia in physiolgical traits: *Embryonopsis halticella* caterpillars (Yponomeutidae) across the Antarctic polar frontal zone. Journal of Insect Physiology 51:87–97.

Klok, C. J., B. J. Sinclair, and S. L. Chown. 2004. Upper thermal tolerance and oxygen limitation in terrestrial arthropods. Journal of Experimental Biology 207: 2361–2370.

Kluger, M. J. 1979. Fever in ectotherms: evolutionary implications. American Zoologist 19:295–304.

Kluger, M. J., W. Kozak, C. A. Conn, L. R. Leon, and D. Soszynski. 1998. Role of fever in disease. Annals of the New York Academy of Sciences 856:224–233.

Knies, J. L., R. Izem, K. L. Supler, J. G. Kingsolver, and C. L. Burch. 2006. The genetic basis of thermal reaction norm evolution in lab and natural phage populations. PLoS Biology 4:1–8.

Kopp, M., and S. Gavrilets. 2006. Multilocus genetics and the coevolution of quantitative traits. Evolution 60: 1321–1336.

Kopp, M., and J. Hermisson. 2006. The evolution of genetic architecture under frequency-dependent disruptive selection. Evolution 60:1537–1550.

Koteja, P. 2000. Energy assimilation, parental care and the evolution of endothermy. Proceedings of the Royal Society B 267:479–484.

—. 2004. The evolution of concepts on the evolution of endothermy in birds and mammals. Physiological and Biochemical Zoology 77:1043–1050.

Kozłowski, J. 1992. Optimal allocation of resources to growth and reproduction: implications for age and size at maturity. Trends in Ecology & Evolution 7:15–19.

Kozłowski, J., M. Czarnołeski, and M. Dańko. 2004. Can optimal resource allocation models explain why ectotherms grow larger in cold? Integrative and Comparative Biology 44:480–493.

Krebs, R. A., and M. E. Feder. 1997. Natural variation in the expression of the heat-shock protein Hsp70 in a population of *Drosophila melanogaster* and its correlation with tolerance of ecologically relevant thermal stress. Evolution 51:173–179.

—. 1998. Experimental manipulation of the cost of thermal acclimation in *Drosophila melanogaster*. Biological Journal of the Linnean Society 63:593–601.

Krebs, R. A., and V. Loeschcke. 1994a. Costs and benefits of activation of the heat-shock response in *Drosophila melanogaster*. Functional Ecology 8:730–737.

—. 1994b. Effects of exposure to short-term heat stress on fitness components in *Drosophila melanogaster*. Journal of Evolutionary Biology 7:39–49.

—. 1995. Resistance to thermal stress in adult *Drosophila buzzatii*: acclimation and variation among populations.

Biological Journal of the Linnean Society 56: 505–515.

—. 1996. Acclimation and selection for increased resistance to thermal stress in *Drosophila buzzatii*. Genetics 142: 471–479.

—. 1999. A genetic analysis of the relationship between life-history variation and heat-shock tolerance in *Drosophila buzzatii*. Heredity 83:46–53.

Kreuger, B., and D. A. Potter. 2001. Diel feeding activity and thermoregulation by Japanese beetles (Coleoptera: Scarabaeidae) within host plant canopies. Environmental Entomology 30:172–180.

Krochmal, A. R., and G. S. Bakken. 2003. Thermoregulation is the pits: use of thermal radiation for retreat site selection by rattlesnakes. Journal of Experimental Biology 206:2539–2545.

Kronfeld-Schor, N., and T. Dayan. 2003. Partitioning of time as an ecological resource. Annual Review of Ecology Evolution and Systematics 34:153–181.

Ksiazek, A., M. Konarzewski, and I. B. Lapo. 2004. Anatomic and energetic correlates of divergent selection for basal metabolic rate in laboratory mice. Physiological and Biochemical Zoology 77:890–899.

Kudo, I., M. Miyamoto, Y. Noiri, and Y. Maita. 2000. Combined effects of temperature and iron on the growth and physiology of the marine diatom *Phaeodactylum tricornutum* (Bacillariophyceae). Journal of Phycology 36:1096–1102.

Lacey, E. P., and D. Herr. 2005. Phenotypic plasticity, parental effects and parental care in plants? I. An examination of spike reflectance in *Plantago lanceolata* (Plantaginaceae). American Journal of Botany 92:920–930.

Ladyman, M., and D. Bradshaw. 2003. The influence of dehydration on the thermal preferences of the western tiger snake, *Notechis scutatus*. Journal of Comparative Physiology B 173:239–246.

Lagerspetz, K. Y. H. 2006. What is thermal acclimation? Journal of Thermal Biology 31:332–336.

Lagerspetz, K. Y. H., and L. A. Vainio. 2006. Thermal behaviour of crustaceans. Biological Reviews 81: 237–258.

Lakoff, G., and R. E. Núñez. 2000, Where Mathematics Comes From: How the Embodied Mind Brings Mathematics into Being. New York, Basic Books.

Lamb, R. J., and G. H. Gerber. 1985. Effects of temperature on the development, growth, and survival of larvae and pupae of a north-temperate chrysomelid beetle. Oecologia 67:8–18.

Lamb, R. J., P. A. Mackay, and G. H. Gerber. 1987. Are development and growth of pea aphids, *Acyrthosiphon pisum*, in North America adapted to local temperatures? Oecologia 72:170–177.

Lampert, W. 1989. The adaptive significance of diel vertical migration of zooplankton. Functional Ecology 3: 21–27.

—. 2005. Vertical distribution of zooplankton: density dependence and evidence for an ideal free distribution with costs. BMC Biology 3.

Lampert, W., E. McCauley, and B. F. J. Manly. 2003. Trade-offs in the vertical distribution of zooplankton: ideal free distribution with costs? Proceedings of the Royal Society B 270:765–773.

Lande, R. 1979. Quantitative genetic analysis of multivariate evolution, applied to brain-body size allometry. Evolution 33:402–416.

—. 1984. The genetic correlation between characters maintained by selection, linkage and inbreeding. Genetical Research 44:309–320.

Lande, R., and S. J. Arnold. 1983. The measurement of selection on correlated characters. Evolution 37:1210–1226.

Lankford, T. E., J. M. Billerbeck, and D. O. Conover. 2001. Evolution of intrinsic growth and energy acquisition rates. II. Trade-offs with vulnerability to predation in *Menidia menidia*. Evolution 55:1873–1881.

Lankford, T. E., and T. E. Targett. 2001. Physiological performance of young-of-the-year Atlantic croakers from different Atlantic coast estuaries: implications for stock structure. Transactions of the American Fisheries Society 130:367–375.

Laptikhovsky, V. 2006. Latitudinal and bathymetric trends in egg size variation: a new look at Thorson's and Rass's rules. Marine Ecology 27:7–14.

Larimore, R. W., and M. J. Duever. 1968. Effects of temperature acclimation on swimming ability of smallmouth bass fry. Transactions of the American Fisheries Society 97:175–184.

Larsson, S., T. Forseth, I. Berglund, A. J. Jensen, I. Naslund, J. M. Elliott, and B. Jonsson. 2005. Thermal adaptation of Arctic charr: experimental studies of growth in eleven charr populations from Sweden, Norway and Britain. Freshwater Biology 50:353–368.

Laugen, A. T., A. Laurila, K. I. Jonsson, F. Soderman, and J. Merila. 2005. Do common frogs (*Rana temporaria*) follow Bergmann's rule? Evolutionary Ecology Research 7: 717–731.

Laugen, A. T., A. Laurila, and J. Merila. 2003b. Latitudinal and temperature-dependent variation in embryonic development and growth in *Rana temporaria*. Oecologia 135:548–554.

Laugen, A. T., A. Laurila, K. Rasanen, and J. Merila. 2003a. Latitudinal countergradient variation in the common frog (*Rana temporaria*) development rates - evidence for local adaptation. Journal of Evolutionary Biology 16:996–1005.

Law, R. 1979. Optimal life histories under age-specific predation. American Naturalist 114:399–417.

Le Galliard, J. F., M. Le Bris, and J. Clobert. 2003. Timing of locomotor impairment and shift in thermal preferences during gravidity in a viviparous lizard. Functional Ecology 17:877–885.

Lebedeva, L. I., and T. N. Gerasimova. 1985. Peculiarities of *Philodina roseola* (Ehrbg.) (Rotatoria, Bdelloida): growth and reproduction under various temperature conditions. Internationale Revue der gesamten Hydrobiologie 70:509–525.

Lee, R. E., and J. P. Costanzo. 1998. Biological ice nucleation and ice distribution in cold-hardy ectothermic animals. Annual Review of Physiology 60:55–72.

Lee, R. E., K. Damodaran, S. X. Yi, and G. A. Lorigan. 2006. Rapid cold-hardening increases membrane fluidity and cold tolerance of insect cells. Cryobiology 52:459–463.

Leffler, C. W. 1972. Some effects of temperature on the growth and metabolic rate of juvenile blue crabs, *Callinectes sapidus*, in the laboratory. Marine Biology 14: 104–110.

Lek, S., M. Delacoste, P. Baran, I. Dimopoulos, J. Lauga, and S. Aulagnier. 1996. Application of neural networks to modelling nonlinear relationships in ecology. Ecological Modelling 90:39–52.

Lek, S., and J. F. Guégan. 1999. Artificial neural networks as a tool in ecological modelling, an introduction. Ecological Modelling 120:65–73.

—. 2000, Artificial Neuronal Networks: Application to Ecology and Evolution. Berlin, Springer.

Lemos-Espinal, J. A., G. R. Smith, and R. E. Ballinger. 1997. Thermal ecology of the lizard, *Sceloporus gadoviae*, in an arid tropical scrub forest. Journal of Arid Environments 35:311–319.

Lenormand, T. 2002. Gene flow and the limits to natural selection. Trends in Ecology & Evolution 17:183–189.

Leroi, A. M., A. F. Bennett, and R. E. Lenski. 1994. Temperature-acclimation and competitive fitness: an experimental test of the beneficial acclimation assumption. Proceedings of the National Academy of Sciences 91:1917–1921.

Levins, R. 1968, Evolution in Changing Environments: Some Theoretical Explorations. Princeton, Princeton University Press.

Li, D. Q. 2002. The combined effects of temperature and diet on development and survival of a crab spider, *Misumenops tricuspidatus* (Fabricius) (Araneae: Thomisidae). Journal of Thermal Biology 27:83–93.

Li, X. C., and L. Z. Wang. 2005. Effect of temperature and thermal acclimation on locomotor performance of *Macrobiotus harmsworthi* Murray (Tardigrada, Macrobiotidae). Journal of Thermal Biology 30:588–594.

Licht, P. 1967. Thermal adaptation in the enzymes of lizards in relation to preferred body temperatures, Pages 131–146 *in* C. L. Prosser, ed. Molecular Mechanisms of Temperature Adaptation. Washington, D. C., American Association for the Advancement of Science.

—. 1968. Response of the thermal preferendum and heat resistance to thermal acclimation under different photoperiods in the lizard *Anolis carolinensis*. American Midland Naturalist 79:149–158.

Licht, P., W. R. Dawson, and V. H. Shoemaker. 1966a. Heat resistance of some Australian lizards. Copeia 1966: 162–169.

Licht, P., W. R. Dawson, V. H. Shoemaker, and A. R. Main. 1966b. Observations on the thermal relations of western Australian lizards. Copeia 1966:97–110.

Lillywhite, H. B., and C. A. Navas. 2006. Animals, energy, and water in extreme environments: perspectives from Ithala 2004. Physiological and Biochemical Zoology 79:265–273.

Lima, F. P., P. A. Ribeiro, N. Queiroz, S. J. Hawkins, and A. M. Santos. 2007a. Do distributional shifts in northern and southern species of algae match the warming pattern? Global Change Biology 13:2592–2604.

Lima, F. P., P. A. Ribeiro, N. Queiroz, R. Xavier, P. Tarroso, S. J. Hawkins, and A. M. Santos. 2007b. Modelling past and present geographical distribution of the marine gastropod *Patella rustica* as a tool for exploring responses to environmental change. Global Change Biology 13:2065–2077.

Lima, S. L. 2002. Putting predators back into behavioral predator-prey interactions. Trends in Ecology & Evolution 17:70–75.

Lima, S. L., and L. M. Dill. 1990. Behavioral decisions made under the risk of predation: a review and prospectus. Canadian Journal of Zoology 68:619–640.

Lindgren, B., and A. Laurila. 2005. Proximate causes of adaptive growth rates: growth efficiency variation among latitudinal populations of *Rana temporaria*. Journal of Evolutionary Biology 18:820–828.

Loeschcke, V., J. Bundgaard, and J. S. F. Barker. 1999. Reaction norms across and genetic parameters at different temperatures for thorax and wing size traits in *Drosophila aldrichi* and *D. buzzatii*. Journal of Evolutionary Biology 12:605–623.

Loeschcke, V., and R. A. Krebs. 1996. Selection for heat-shock resistance in larval and in adult *Drosophila buzzatii*: comparing direct and indirect responses. Evolution 50:2354–2359.

Loeschcke, V., R. A. Krebs, and J. S. F. Barker. 1994. Genetic variation for resistance and acclimation to high temperature stress in *Drosophila buzzatii*. Biological Journal of the Linnean Society 52:83–92.

Loeschcke, V., and J. G. Sorenson. 2005. Acclimation, heat shock and hardening—a response from evolutionary biology. Journal of Thermal Biology 30:255–257.

Lonsdale, D. J., and J. S. Levinton. 1985. Latitudinal differentiation in copepod growth: an adaptation to temperature. Ecology 66:1397–1407.

—. 1989. Energy budgets of latitudinally separated *Scottolana canadensis* (Copepoda, Harpacticoida). Limnology and Oceanography 34:324–331.

Loose, C. J., and P. Dawidowicz. 1994. Trade-offs in diel vertical migration by zooplankton: the costs of predator avoidance. Ecology 75:2255–2263.

Lorenzon, P., J. Clobert, A. Oppliger, and H. John-Alder. 1999. Effect of water constraint on growth rate, activity and body temperature of yearling common lizard (*Lacerta vivipara*). Oecologia 118:423–430.

Lourdais, O., B. Heulin, and D. F. Denardo. 2008. Thermoregulation during gravidity in the children's python (*Antaresia childreni*): a test of the preadaptation hypothesis for maternal thermophily in snakes. Biological Journal of the Linnean Society 93:499–508.

Luo, L. G., Y. F. Qu, and X. Ji. 2006. Thermal dependence of food assimilation and sprint speed in a lacertid lizard *Eremias argus* from northern China. Acta Zoologica 52:256–262.

Lustick, S. I. 1983. Cost–benefit of thermoregulation in birds: influences of posture, microhabitat selection, and color, Pages 265–294 *in* W. P. Aspey, and S. I. Lustick, eds. Behavioral Energetics. Columbus, Ohio State University Press.

Lutterschmidt, W. I., and V. H. Hutchison. 1997a. The critical thermal maximum: data to support the onset of spasms as the definitive end point. Canadian Journal of Zoology 75:1553–1560.

—. 1997b. The critical thermal maximum: history and critique. Canadian Journal of Zoology 75:1561–1574.

Lutterschmidt, W. I., and H. K. Reinert. 1990. The effect of ingested transmitters upon the temperature preference of the northern water snake, *Nerodia s. sipedon*. Herpetologica 46:39–42.

Lynch, M., and W. Gabriel. 1987. Environmental tolerance. American Naturalist 129:283–303.

Lynch, M., and R. Lande. 1993. Evolution and extinction in response to environmental change, Pages 234–250 *in* P. Kareiva, J. Kingsolver, and R. Huey, eds. Biotic Interactions and Global Change. Sunderland, Sinauer Assocs., Inc.

Lynch, M., and B. Walsh. 1998, Genetics and Analysis of Quantitative Traits. Sunderland, Sinauer Associates, Inc.

Lysenko, S., and J. E. Gillis. 1980. The effect of ingestive status on the thermoregulatory behavior of *Thamnophis sirtalis sirtalis* and *Thamnophis sirtalis parietalis*. Journal of Herpetology 14:155–159.

Mac, M. J. 1985. Effects of ration size on preferred temperature of lake charr *Salvelinus namaycush*. Environmental Biology of Fishes 14:227–231.

Macario, A. J. L., and E. C. de Macario. 2007. Molecular chaperones: multiple functions, pathologies, and potential applications. Frontiers in Bioscience 12:2588–2600.

Magnum, C. P., and P. W. Hochachaka. 1998. New directions in comparative physiology and biochemistry: mechanisms, adaptations, and evolution. Physiological Zoology 71:471–484.

Magnuson, J. J., L. B. Crowder, and P. A. Medvick. 1979. Temperature as an ecological resource. American Zoologist 19:331–343.

Mangel, M., and C. W. Clark. 1988, Dynamic Modeling in Behavioral Ecology. Princeton, Princeton University Press.

Martin, J., and P. Lopez. 2001. Repeated predatory attacks and multiple decisions to come out from a refuge in an alpine lizard. Behavioral Ecology 12:386–389.

Martin, T. L., and R. B. Huey. 2008. Why "suboptimal" is optimal: Jensen's inequality and ectotherm thermal preferences. American Naturalist 171:E102–E118.

Martins, E. P. 2000. Adaptation and the comparative method. Trends in Ecology & Evolution 15:296–299.

Martins, E. P., and T. Garland. 1991. Phylogenetic analyses of the correlated evolution of continuous characters: a simulation study. Evolution 45:534–557.

Martins, E. P., and T. F. Hansen. 1996. The statistical analysis of interspecific data: a review and evaluation of phylogenetic comparative methods, Pages 22–75 *in* E. P. Martins, ed. Phylogenies and the Comparative Method in Animal Behavior. Oxford, Oxford University Press.

—. 1997. Phylogenies and the comparative method: a general approach to incorporating phylogenetic information into the analysis of interspecific data. American Naturalist 149:646–667.

Marx, J. C., T. Collins, S. D'Amico, G. Feller, and C. Gerday. 2007. Cold-adapted enzymes from marine Antarctic microorganisms. Marine Biotechnology 9:293–304.

Masaki, S. 1967. Geographic variation and climatic adaptation in a field cricket (Orthoptera: Gryllidae). Evolution 21:725–741.

Mathies, T., and R. M. Andrews. 1997. Influence of pregnancy on the thermal biology of the lizard, *Sceloporus jarrovi*: why do pregnant females exhibit low body temperatures? Functional Ecology 11:498–507.

May, M. L. 1995. Simultaneous control of head and thoracic temperature by the green darner dragonfly *Anax junius*

(Odonata, Aeshnidae). Journal of Experimental Biology 198:2373–2384.

Maynard Smith, J. 1972. Game theory and the evolution of fighting, Pages 8–28 *in* J. Maynard Smith, ed. On Evolution. Edinburgh, Edinburgh University Press.

—. 1976. Evolution and the theory of games. American Scientist 64:41–45.

—. 1978. Optimization theory in evolution. Annual Review of Ecology and Systematics 9:31–56.

—. 1982, Evolution and the Theory of Games. Cambridge, Cambridge University Press.

Maynard Smith, J., and G. R. Price. 1973. The logic of animal conflict. Nature 246:15–18.

McCabe, J., and L. Partridge. 1997. An interaction between environmental temperature and genetic variation for body size for the fitness of adult female *Drosophila melanogaster*. Evolution 51:1164–1174.

McConnell, E., and A. G. Richards. 1955. How fast can a cockroach run? Bulletin of the Brooklyn Entomological Society 50:36–43.

McCullough, E. C., and W. P. Porter. 1971. Computing clear day solar radiation spectra for terrestrial ecological environment. Ecology 52:1008–1015.

McGinley, M. A., D. H. Temme, and M. A. Geber. 1987. Parental investment in offspring in variable environments: theoretical and empirical considerations. American Naturalist 130:370–398.

McKechnie, A. E., G. Kortner, and B. G. Lovegrove. 2006. Thermoregulation under semi-natural conditions in speckled mousebirds: the role of communal roosting. African Zoology 41:155–163.

McKechnie, A. E., and B. G. Lovegrove. 2002. Avian facultative hypothermic responses: a review. Condor 104:705–724.

McLean, M. A., M. J. Angilletta, and K. S. Williams. 2005. If you can't stand the heat, stay out of the city: thermal reaction norms of chitinolytic fungi in an urban heat island. Journal of Thermal Biology 30:384–391.

McNab, B. K. 2002, Physiological Ecology of Vertebrates. Ithaca, Cornell University Press.

McPeek, M. A. 2004. The growth/predation risk tradeoff: so what is the mechanism? American Naturalist 163: E88–E111.

Measey, G. J., and S. Van Dongen. 2006. Bergmann's rule and the terrestrial caecilian *Schistometopum thomense* (Amphibia: Gymnophiona: Caeciiidae). Evolutionary Ecology Research 8:1049–1059.

Medvick, P. A., J. J. Magnuson, and S. Sharr. 1981. Behavioral thermoregulation and social interactions of bluegills, *Lepomis macrochirus*. Copeia:9–13.

Meffe, G. K. 1990. Offspring size variation in eastern mosquitofish (*Gambusia holbrooki*: Poeciliidae) from contrasting thermal environments. Copeia 1990:10–18.

Meiri, S., and T. Dayan. 2003. On the validity of Bergmann's rule. Journal of Biogeography 30:331–351.

Mejia-Ortiz, L. M., and R. G. Hartnoll. 2005. Modifications of eye structure and integumental pigment in two cave crayfish. Journal of Crustacean Biology 25:480–487.

Menéndez, R., A. González-Megías, O. T. Lewis, M. R. Shaw, and C. D. Thomas. 2008. Escape from natural enemies during climate-driven range expansion: a case study. Ecological Entomology 33:413–421.

Merila, J., A. Laurila, A. T. Laugen, K. Rasanen, and M. Pahkala. 2000. Plasticity in age and size at metamorphosis in *Rana temporaria* – comparison of high and low latitude populations. Ecography 23:457–465.

Merrill, R. M., D. Gutiérrez, O. T. Lewis, J. Gutiérrez, S. B. Díez, and R. J. Wilson. 2008. Combined effects of climate and biotic interactions on the elevational range of a phytophagous insect. Journal of Animal Ecology 77:145–155.

Michod, R. E. 1979. Evolution of life histories in response to age-specific mortality factors. American Naturalist 113:531–550.

Miles, D. B., and A. E. Dunham. 1993. Historical perspectives in ecology and evolutionary biology: the use of phylogenetic comparative analyses. Annual Review of Ecology and Systematics 24:587–619.

Miller, K., and G. C. Packard. 1977. An altitudinal cline in critical thermal maxima of chorus frogs (*Pseudacris triseriata*). American Naturalist 111:267–277.

Miller, S. R., and R. W. Castenholz. 2000. Evolution of thermotolerance in hot spring cyanobacteria of the genus *Synechococcus*. Applied and Environmental Microbiology 66:4222–4229.

Minoli, S. A., and C. R. Lazzari. 2003. Chronobiological basis of thermopreference in the haematophagous bug *Triatoma infestans*. Journal of Insect Physiology 49: 927–932.

Misra, R. K., and E. C. R. Reeve. 1964. Clines in body dimensions in populations of *Drosophila subobscura*. Genetical Research 5:240–256.

Mitchell, J. W. 1976. Heat transfer from spheres and other animal forms. Biophysical Journal 16:561–569.

Mitchell, M., and C. E. Taylor. 1999. Evolutionary computation: an overview. Annual Review of Ecology and Systematics 30:593–616.

Mitchell, S. E., and W. Lampert. 2000. Temperature adaptation in a geographically widespread zooplankter, *Daphnia magna*. Journal of Evolutionary Biology 13:371–382.

Mitchell, W. A., and T. J. Valone. 1990. The optimization research program: studying adaptations by their function. Quarterly Review of Biology 65:43–52.

Mitten, J. B. 1997, Selection in Natural Populations. Oxford, Oxford University Press.

Møller, H., R. H. Smith, and R. M. Sibly. 1989. Evolutionary demography of a bruchid beetle. II. Physiological manipulations. Functional Ecology 3:683–691.

Mondal, S., and U. Rai. 2001. In vitro effect of temperature on phagocytic and cytotoxic activities of splenic phagocytes of the wall lizard, *Hemidactylus flaviviridis*. Comparative Biochemistry and Physiology A 129: 391–398.

Mongold, J. A., A. F. Bennett, and R. E. Lenski. 1999. Evolutionary adaptation to temperature. VII. Extension of the upper thermal limit of *Escherichia coli*. Evolution 53:386–394.

Monteith, J. J., and M. H. Unsworth. 1990, Principles of Environmental Physics. London, Edward Arnold.

Mooney, H. A. 1980. Photosynthetic plasticity of populations of *Heliotropium curassavicum* L. originating from differing thermal regimes. Oecologia 45:372–376.

Moore, C. M., and L. M. Sievert. 2001. Temperature-mediated characteristics of the dusky salamander (*Desmognathus fuscus*) of southern Appalachia. Journal of Thermal Biology 26:547–554.

Moore, J. A. 1940. Stenothermy and eurythermy of animals in relation to habitat. American Naturalist 74:188–192.

Moran, N. A. 1992. The evolutionary maintenance of alternative phenotypes. American Naturalist 139: 971–989.

Morgan, I. J., and N. B. Metcalfe. 2001. The influence of energetic requirements on the preferred temperature of overwintering juvenile Atlantic salmon (*Salmo salar*). Canadian Journal of Fisheries and Aquatic Sciences 58:762–768.

Morrison, C., and J.-M. Hero. 2003b. Geographic variation in life-history characteristics of amphibians: a review. Journal of Animal Ecology 72:270–279.

Morrison, C., and J. M. Hero. 2003a. Altitudinal variation in growth and development rates of tadpoles of *Litoria chloris* and *Litoria pearsoniana* in southeast Queensland, Australia. Journal of Herpetology 37:59–64.

Mousseau, T. A. 1997. Ectotherms follow the converse to Bergmann's Rule. Evolution 51:630–632.

Mullens, D. P., and V. H. Hutchison. 1992. Diel, seasonal, postprandial and food-deprived thermoregulatory behavior in tropical toads (*Bufo marinus*). Journal of Thermal Biology 17:63–67.

Munoz, J. L. P., G. R. Finke, P. A. Camus, and F. Bozinovic. 2005. Thermoregulatory behavior, heat gain and thermal tolerance in the periwinkle *Echinolittorina peruviana* in central Chile. Comparative Biochemistry and Physiology A 142:92–98.

Munro, D., D. W. Thomas, and M. M. Humphries. 2005. Torpor patterns of hibernating eastern chipmunks *Tamias striatus* vary in response to the size and fatty acid composition of food hoards. Journal of Animal Ecology 74:692–700.

Murphy, H. T., and J. Lovett-Doust. 2007. Accounting for regional niche variation in habitat suitability models. Oikos 116:99–110.

Navas, C. A. 1996b. Implications of microhabitat selection and patterns of activity on the thermal ecology of high elevation neotropical anurans. Oecologia 108:617–626.

—. 1996a. Metabolic physiology, locomotor performance, and thermal niche breadth in neotropical anurans. Physiological Zoology 69:1481–1501.

—. 2006. Patterns of distribution of anurans in high Andean tropical elevations: insights from integrating biogeography and evolutionary physiology. Integrative and Comparative Biology 46:82–91.

Navas, C. A., and C. Araujo. 2000. The use of agar models to study amphibian thermal ecology. Journal of Herpetology 34:330–334.

Navas, C. A., R. S. James, J. M. Wakeling, K. M. Kemp, and I. A. Johnston. 1999. An integrative study of the temperature dependence of whole animal and muscle performance during jumping and swimming in the frog *Rana temporaria*. Journal of Comparative Physiology B 169:588–596.

Neat, F., K. Fowler, V. French, and L. Partridge. 1995. Thermal evolution of growth efficiency in *Drosophila melanogaster*. Proceedings of the Royal Society B 260: 73–78.

Nelson, K. C., and M. A. Palmer. 2007. Stream temperature surges under urbanization and climate change: data, models, and responses. Journal of the American Water Resources Association 43:440–452.

New, M., M. Hulme, and P. Jones. 1999. Representing twentieth-century space-time climate variability. Part I: Development of a 1961–90 mean monthly terrestrial climatology. Journal of Climate 12:829–856.

—. 2000. Representing twentieth-century space-time climate variability. Part II: Development of 1901–96 monthly grids of terrestrial surface climate. Journal of Climate 13:2217–2238.

New, M., D. Lister, M. Hulme, and I. Makin. 2002. A high-resolution data set of surface climate over global land areas. Climate Research 21:1–25.

Newell, J. C., and T. P. Quinn. 2005. Behavioral thermoregulation by maturing adult sockeye salmon (*Oncorhynchus nerka*) in a stratified lake prior to spawning. Canadian Journal of Zoology 83:1232–1239.

Newell, R. C., L. G. Johnson, and L. H. Kofoed. 1977. Adjustment of the components of energy balance in response to temperature change in *Ostrea edulis*. Oecologia 30:97–110.

Newman, R. A. 1998. Ecological constraints on amphibian metamorphosis: interactions of temperature and larval density with responses to changing food level. Oecologia 115:9–16.

Nice, C. C., and J. A. Fordyce. 2006. How caterpillars avoid overheating: behavioral and phenotypic plasticity of pipevine swallowtail larvae. Oecologia 146:541–548.

Nicieza, A. G., L. Reiriz, and F. Brana. 1994. Variation in digestive performance between geographically disjunct populations of Atlantic salmon: countergradient in passage time and digestion rate. Oecologia 99: 243–251.

Norry, F. M., O. A. Bubliy, and V. Loeschcke. 2001. Developmental time, body size and wing loading in *Drosophila buzzatii* from lowland and highland populations in Argentina. Hereditas 135:35–40.

Norry, F. M., and V. Loeschcke. 2003. Heat-induced expression of a molecular chaperone decreases by selecting for long-lived individuals. Experimental Gerontology 38:673–681.

Nosil, P., B. J. Crespi, C. P. Sandoval, and M. Kirkpatrick. 2006. Migration and the genetic covariance between habitat preference and performance. American Naturalist 167:E66-E78.

Nowak, M. A., and K. Sigmund. 2004. Evolutionary dynamics of biological games. Science 303:793–799.

Nussear, K. E., R. E. Espinoza, C. M. Gubbins, K. J. Field, and J. P. Hayes. 1998. Diet quality does not affect resting metabolic rate or body temperatures selected by an herbivorous lizard. Journal of Comparative Physiology B 168:183–189.

O'Connor, M. P., and J. R. Spotila. 1992. Consider a spherical lizard: animals, models, and approximations. American Zoologist 32:179–193.

O'Steen, S., and A. F. Bennett. 2003. Thermal acclimation effects differ between voluntary, maximum, and critical swimming velocities in two cyprinid fishes. Physiological and Biochemical Zoology 76:484–496.

Ohman, M. D., B. W. Frost, and E. B. Cohen. 1983. Reverse diel vertical migration: an escape from invertebrate predators. Science 220:1404–1407.

Ohtsu, T., M. T. Kimura, and C. Katagiri. 1998. How *Drosophila* species acquire cold tolerance: qualitative changes of phospholipids. European Journal of Biochemistry 252:608–611.

Ojanguren, A. F., and F. Braña. 2000. Thermal dependence of swimming endurance in juvenile brown trout. Journal of Fish Biology 56:1342–1347.

Oke, T. R. 1973. City size and the urban heat island. Atmospheric Environment 7:769–779.

—. 1997. Urban climates and global change, Pages 273–287 *in* A. Perry, and R. Thompson, eds. Applied Climatology: Principles and Practice. London, Routledge.

Oreskes, N., K. Shraderfrechette, and K. Belitz. 1994. Verification, validation, and confirmation of numerical models in the earth sciences. Science 263:641–646.

Otterlei, E., G. Nyhammer, A. Folkvord, and S. O. Stefansson. 1999. Temperature- and size-dependent growth of larval and early juvenile Atlantic cod (*Gadus morhua*): a comparative study of Norwegian coastal cod and northeast Arctic cod. Canadian Journal of Fisheries and Aquatic Sciences 56:2099–2111.

Ouedraogo, R. M., M. Cusson, M. S. Goettel, and J. Brodeur. 2003. Inhibition of fungal growth in thermoregulating locusts, *Locusta migratoria*, infected by the fungus *Metarhizium anisopliae* var *acridum*. Journal of Invertebrate Pathology 82:103–109.

Ouedraogo, R. M., M. S. Goettel, and J. Brodeur. 2004. Behavioral thermoregulation in the migratory locust: a therapy to overcome fungal infection. Oecologia 138:312–319.

Oufiero, C. E., and M. J. Angilletta. 2006. Convergent evolution of embryonic growth and development in the eastern fence lizard (*Sceloporus undulatus*). Evolution 60:1066–1075.

Oufiero, C. E., A. J. Smith, and M. J. Angilletta. 2007. The importance of energetic versus pelvic constraints on reproductive allocation in the eastern fence lizard (*Sceloporus undulatus*). Biological Journal of the Linnean Society 91:513–521.

Overgaard, J., J. G. Sorensen, S. O. Petersen, V. Loeschcke, and M. Holmstrup. 2005. Changes in membrane lipid composition following rapid cold hardening in *Drosophila melanogaster*. Journal of Insect Physiology 51:1173–1182.

Packard, G. C., M. J. Packard, and L. L. McDaniel. 2001. Seasonal change in the capacity for supercooling by neonatal painted turtles. Journal of Experimental Biology 204:1667–1672.

Padilla, D. K., and S. C. Adolph. 1996. Plastic inducible morphologies are not always adaptive: the importance of time delays in a stochastic environment. Evolutionary Ecology 10:105–117.

Palaima, A. 2002. Tolerance to temperature: an experimental study of *Daphnia* (Crustacea: Cladocera). Ph.D. Dissertation, University of Miami, Coral Gables.

—. 2007. The fitness cost of generalization: present limitations and future possible solutions. Biological Journal of the Linnean Society 90:583–590.

Palaima, A., and K. Spitze. 2004. Is a jack-of-all-temperatures a master of none? An experimental test with *Daphnia pulicaria* (Crustacea: Cladocera). Evolutionary Ecology Research 6:215–225.

Parker, D. E. 2004. Large-scale warming is not urban. Nature 432:290–290.

Parker, G. A., and M. Begon. 1986. Optimal egg size and clutch size: effects of environment and maternal phenotype. American Naturalist 128:573–592.

Parker, S. L., and R. M. Andrews. 2007. Incubation temperature and phenotypic traits of *Sceloporus undulatus*: implications for the northern limits of distribution. Oecologia 151:218–231.

Parmesan, C. 2006. Ecological and evolutionary responses to recent climate change. Annual Review of Ecology Evolution and Systematics 37:637–669.

—. 2007. Influences of species, latitudes and methodologies on estimates of phenological response to global warming. Global Change Biology 13:1860–1872.

Parmesan, C., S. Gaines, L. Gonzalez, D. M. Kaufman, J. Kingsolver, A. Townsend Peterson, and R. Sagarin. 2005. Empirical perspectives on species borders: from traditional biogeography to global change. Oikos 108:58–75.

Parmesan, C., and H. Galbraith. 2004. Observed impacts of global climate change in the U.S., Pages 1–55. Arlington, Pew Center on Global Climate Change.

Parmesan, C., N. Ryrholm, C. Stefanescu, J. K. Hill, C. D. Thomas, H. Descimon, B. Huntley et al. 1999. Poleward shifts in geographical ranges of butterfly species associated with regional warming. Nature 399:579–583.

Parmesan, C., and G. Yohe. 2003. A globally coherent fingerprint of climate change impacts across natural systems. Nature 421:37–42.

Parry, D., R. A. Goyer, and G. J. Lenhard. 2001. Macrogeographic clines in fecundity, reproductive allocation, and offspring size of the forest tent caterpillar *Malacosoma disstria*. Ecological Entomology 26:281–291.

Partridge, L., B. Barrie, K. Fowler, and V. French. 1994. Evolution and development of body size and cell size in *Drosophila melanogaster* in response to temperature. Evolution 48:1269–1276.

Partridge, L., and J. A. Coyne. 1997. Bergmann's rule in ectotherms: is it adaptive? Evolution 51:632–635.

Parzefall, J. 2001. A review of morphological and behavioural changes in the cave molly, *Poecilia mexicana*, from Tabasco, Mexico. Environmental Biology of Fishes 62:263–275.

Patino, S., J. Grace, and H. Banziger. 2000. Endothermy by flowers of *Rhizanthes lowii* (Rafflesiaceae). Oecologia 124:149–155.

Pearcy, R. W. 1977. Acclimation of photosynthetic and respiratory carbon dioxide exchange to growth temperature in *Atriplex lentiformis* (Torr.) Wats. Plant Physiology 59:795–799.

Pearson, R. G., and T. P. Dawson. 2003. Predicting the impacts of climate change on the distribution of species: are bioclimate envelope models useful? Global Ecology and Biogeography 12:361–371.

Pérez, E., F. Díaz, and S. Espina. 2003. Thermoregulatory behavior and critical thermal limits of the angelfish *Pterophyllum scalare* (Lichtenstein) (Pisces: Cichlidae). Journal of Thermal Biology 28:531–537.

Perrin, N. 1988. Why are offspring born larger when it is colder? Phenotypic plasticity for offspring size in the cladoceran *Simocephalus vetulus* (Müller). Functional Ecology 2:283–288.

—. 1995. About Berrigan and Charnov's life history puzzle. Oikos 73:137–139.

Perrin, N., and J. F. Rubin. 1990. On dome-shaped reaction norms for age-to-size at maturity in fishes. Functional Ecology 4:53–57.

Perrin, N., and R. M. Sibly. 1993. Dynamic models of energy allocation and investment. Annual Review of Ecology and Systematics 24:379–410.

Petavy, G., J. P. Morin, B. Moreteau, and J. R. David. 1997. Growth temperature and phenotypic plasticity in two *Drosophila* sibling species: probable adaptive changes in flight capacities. Journal of Evolutionary Biology 10: 875–887.

Petersen, A. M., T. T. Gleeson, and D. A. Scholnick. 2003. The effect of oxygen and adenosine on lizard thermoregulation. Physiological and Biochemical Zoology 76:339–347.

Peterson, A. T., M. A. Ortega-Huerta, J. Bartley, V. Sanchez-Cordero, J. Soberon, R. H. Buddemeier, and D. R. B. Stockwell. 2002. Future projections for Mexican faunas under global climate change scenarios. Nature 416: 626–629.

Pfrender, M. E., and M. Lynch. 2000. Quantitative genetic variation in *Daphnia*: temporal changes in genetic architecture. Evolution 54:1502–1509.

Phillips, P. K., and J. E. Heath. 2001. Heat loss in Dumbo: a theoretical approach. Journal of Thermal Biology 26: 117–120.

Pianka, E. R. 1970. Comparative autecology of lizard *Cnemidophorus tigris* in different parts of its geographic range. Ecology 51:703–720.

Piccione, G., G. Caola, and R. Refinetti. 2002. Circadian modulation of starvation-induced hypothermia in sheep and goats. Chronobiology International 19: 531–541.

Pigliucci, M. 1996. How organisms respond to environmental changes: from phenotypes to molecules (and vice versa). Trends in Ecology & Evolution 11:168–173.

—. 2003. Phenotypic integration: studying the ecology and evolution of complex phenotypes. Ecology Letters 6:265–272.

Pigliucci, M., and C. D. Schlichting. 1997. On the limits of quantitative genetics for the study of phenotypic evolution. Acta Biotheoretica 45:143–160.

Pilon, J., and L. Santamaría. 2002. Clonal variation in the thermal response of the submerged aquatic macrophyte *Potamogeton pectinatus*. Journal of Ecology 90:141–152.

Piper, S. C., and E. F. Stewart. 1996. A gridded global data set of daily temperature and precipitation for terrestrial biospheric modeling. Global Biogeochemical Cycles 10:757–782.

Pires, H. H. R., C. R. Lazzari, P. E. Schilman, L. Diotaiuti, and M. G. Lorenzo. 2002. Dynamics of thermopreference in the Chagas disease vector *Panstrongylus megistus* (Hemiptera: Reduviidae). Journal of Medical Entomology 39:716–719.

Platt, J. R. 1964. Strong inference. Science:347–353.

Podrabsky, J. E., and G. N. Somero. 2006. Inducible heat tolerance in Antarctic notothenioid fishes. Polar Biology 30:39–43.

Polo, V., P. López, and J. Martín. 2005. Balancing the thermal costs and benefits of refuge use to cope with persistent attacks from predators: a model and an experiment with an alpine lizard. Evolutionary Ecology Research 7:23–35.

Porter, W. P., and D. M. Gates. 1969. Thermodynamic equilibria of animals with environment. Ecological Monographs 39:227–244.

Porter, W. P., J. W. Mitchell, W. A. Beckman, and C. B. Dewitt. 1973. Behavioral implications of mechanistic ecology: thermal and behavioral modeling of desert ectotherms and their microenvironment. Oecologia 13:1–54.

Pörtner, H. O. 2001. Climate change and temperature-dependent biogeography: oxygen limitation of thermal tolerance in animals. Naturwissenschaften 88:137–146.

—. 2002. Climate variations and the physiological basis of temperature dependent biogeography: systemic to molecular hierarchy of thermal tolerance in animals. Comparative Biochemistry and Physiology A 132: 739–761.

Pörtner, H. O., A. F. Bennett, F. Bozinovic, A. Clarke, M. A. Lardies, M. Lucassen, B. Pelster et al. 2006. Trade-offs in thermal adaptation: the need for a molecular to ecological integration. Physiological and Biochemical Zoology 79:295–313.

Pörtner, H. O., I. Hardewig, F. J. Sartoris, and P. L. M. Van Dijk. 1998. Energetic aspects of cold adaptation critical temperatures in metabolic, ionic and acid-base regulation?, Pages 88–120 *in* H. O. Pörtner, and R. Playle, eds. Cold Ocean Physiology Cambridge, Cambridge University Press.

Pörtner, H. O., P. L. M. Van Dijk, I. Hardewig, and A. Sommer. 2000. Levels of metabolic cold adaptation: tradeoffs in eurythermal and stenothermal ectotherms, Pages 109–122 *in* W. Davison, C. Howard-Williams, and P. Broady, eds. Antarctic Ecosystems: Models for Wider Ecological Understanding. Christchurch, Caxton Press.

Poulin, R. 2006. Global warming and temperature-mediated increases in cercarial emergence in trematode parasites. Parasitology 132:143–151.

Powers, D. A., and P. M. Schulte. 1998. Evolutionary adaptations of gene structure and expression in natural populations in relation to a changing environment: a multidisciplinary approach to address the million-year saga of a small fish. Journal of Experimental Zoology 282:71–94.

Prange, H. D., and G. B. Hamilton. 1992. Humidity selection by thermoregulating grasshoppers. Journal of Thermal Biology 17:353–355.

Prasad, N. G., and M. N. Shakarad. 2004. Genetic correlations: transient truths of adaptive evolution. Journal of Genetics 83:3–6.

Pravosudov, V. V., and J. R. Lucas. 2000. The costs of being cool: a dynamic model of nocturnal hypothermia by small food-caching birds in winter. Journal of Avian Biology 31:463–472.

Present, T. M. C., and D. O. Conover. 1992. Physiological basis of latitudinal growth differences in *Menidia menidia*: variation in consumption or efficiency. Functional Ecology 6:23–31.

Prinzinger, R., A. Pressmar, and E. Schleucher. 1991. Body temperature in birds. Comparative Biochemistry and Physiology A 99:499–506.

Proulx, S. R., and T. Day. 2001. What can invasion analyses tell us about evolution under stochasticity in finite populations? Selection 2:1–15.

Pulgar, J., F. Bozinovic, and F. P. Ojeda. 1999. Behavioral thermoregulation in the intertidal fish *Girella laevifrons* (Kyphosidae): the effect of starvation. Marine and Freshwater Behaviour and Physiology 32:27–38.

Pulgar, J. M., M. Aldana, F. Bozinovic, and F. P. Ojeda. 2003. Does food quality influence thermoregulatory behavior in the intertidal fish *Girella laevifrons*? Journal of Thermal Biology 28:539–544.

Pulliam, H. R. 2000. On the relationship between niche and distribution. Ecology Letters 3:349–361.

Qualls, F. J., and R. Shine. 1998. Geographic variation in lizard phenotypes: importance of the incubation environment. Biological Journal of the Linnean Society 64:477–491.

Quinn, J. F., and A. E. Dunham. 1983. On hypothesis testing in ecology and evolution. American Naturalist 122: 602–617.

Rausch, R. N., L. I. Crawshaw, and H. L. Wallace. 2000. Effects of hypoxia, anoxia, and endogenous ethanol on thermoregulation in goldfish, *Carassius auratus*. American Journal of Physiology 278:R545-R555.

Ray, C. 1960. The application of Bergmann's and Allen's rules to the poikilotherms. Journal of Morphology 106:85–108.

Reeve, M. W., K. Fowler, and L. Partridge. 2000. Increased body size confers greater fitness at lower experimental temperature in male *Drosophila melanogaster*. Journal of Evolutionary Biology 13:836–844.

Regal, P. J. 1966. Thermophilic response following feeding in certain reptiles. Copeia:588–590.

—. 1971. Long term studies with operant conditioning techniques of temperature regulation patterns in reptiles. Journal de Physiologie 63:403–406.

Rehfeldt, G. E., W. R. Wykoff, and C. C. Ying. 2001. Physiologic plasticity, evolution, and impacts of a changing climate on *Pinus contorta*. Climatic Change 50: 355–376.

Reisen, W. K. 1995. Effect of temperature on *Culex trasalis* (Diptera: Culicidae) from the Coachella and San Joaquin Valleys of California. Journal of Medical Entomology 32:636–645.

Renaud, J. M., and E. D. Stevens. 1983. The extent of long-term temperature compensation for jumping distance in the frog, *Rana pipiens*, and the toad, *Bufo americanus*. Canadian Journal of Zoology 61:1284–1287.

Reyes, M. A., L. J. Corcuera, and L. Cardemil. 2003. Accumulation of HSP70 in *Deschampsia antarctica* Desv. leaves under thermal stress. Antarctic Science 15:345–352.

Reynolds, W. W., and M. E. Casterlin. 1979. Behavioral thermoregulation and the final preferendum paradigm. American Zoologist 19:211–224.

Reznick, D., L. Nunney, and A. Tessier. 2000. Big houses, big cars, superfleas and the costs of reproduction. Trends in Ecology & Evolution 15:421–425.

Rhen, T., and J. W. Lang. 1999. Temperature during embryonic and juvenile development influences growth in hatchling snapping turtles, *Chelydra serpentina*. Journal of Thermal Biology 24:33–41.

Ricci, C. 1991. Comparison of five strains of a parthenogenetic species, *Macrotrachela quadricornifera* (Rotifera, Bdelloidea). Hydrobiologia 211:147–155.

Rice, S. 2004, Evolutionary Theory: Mathematical and Conceptual Foundations. Sunderland, Sinauer Associates, Inc.

Riechert, S. E., and P. Hammerstein. 1983. Game theory in the ecological context. Annual Review of Ecology and Systematics 14:377–409.

Riechert, S. E., and C. R. Tracy. 1975. Thermal balance and prey availability: bases for a model relating web-site characteristics to spider reproductive success. Ecology 56:265–284.

Riley, P. A. 1997. Melanin. International Journal of Biochemistry & Cell Biology 29:1235–1239.

Roberts, S. P. 2005. Effects of flight behaviour on body temperature and kinematics during inter-male mate competition in the solitary desert bee *Centris pallida*. Physiological Entomology 30:151–157.

Roberts, S. P., and M. E. Feder. 2000. Changing fitness consequences of *hsp70* copy number in transgenic *Drosophila* larvae undergoing natural thermal stress. Functional Ecology 14:353–357.

Roberts, S. P., and J. F. Harrison. 1998. Mechanisms of thermoregulation in flying bees. American Zoologist 38:492–502.

Robinet, C., P. Baier, J. Pennerstorfer, A. Schopf, and A. Roques. 2007. Modelling the effects of climate change on the potential feeding activity of *Thaumetopoea pityocampa* (Den. & Schiff.) (Lep., Notodontidae) in France. Global Ecology and Biogeography 16:460–471.

Robinson, S. J. W., and L. Partridge. 2001. Temperature and clinal variation in larval growth efficiency in *Drosophila melanogaster*. Journal of Evolutionary Biology 14:14–21.

Robson, M. A., and D. B. Miles. 2000. Locomotor performance and dominance in male tree lizards, *Urosaurus ornatus*. Functional Ecology 14:338–344.

Rock, J., R. M. Andrews, and A. Cree. 2000. Effects of reproductive condition, season, and site on selected temperatures of a viviparous gecko. Physiological and Biochemical Zoology 73:344–355.

Roetzer, T., M. Wittenzeller, H. Haeckel, and J. Nekovar. 2000. Phenology in central Europe – differences and trends of spring phenophases in urban and rural areas. International Journal of Biometeorology 44:60–66.

Roff, D. A. 1980. Optimizing development time in a seasonal environment: the 'ups and downs' of clinal variation. Oecologia 45:202–208.

—. 1994. Optimality modeling and quantitative genetics: a comparison of the two approaches, Pages 49–66 *in* C. R. B. Boake, ed. Quantitative Genetic Studies of Behavioral Evolution. Chicago, University of Chicago Press.

—. 2002, Life History Evolution. Sunderland, Sinauer Associates, Inc.

Roff, D. A., and T. A. Mousseau. 1999. Does natural selection alter genetic architecture? An evaluation of quantitative genetic variation among populations of *Allonemobius socius* and *A. fasciatus*. Journal of Evolutionary Biology 12:361–369.

Rogers, K. D., F. Seebacher, and M. B. Thompson. 2004. Biochemical acclimation of metabolic enzymes in response to lowered temperature in tadpoles of *Limnodynastes peronii*. Comparative Biochemistry and Physiology A 137:731–738.

Rohmer, C., J. R. David, B. Moreteau, and D. Joly. 2004. Heat induced male sterility in *Drosophila melanogaster*: adaptive genetic variations among geographic populations and role of the Y chromosome. Journal of Experimental Biology 207:2735–2743.

Romare, P., and L. A. Hansson. 2003. A behavioral cascade: top-predator induced behavioral shifts in planktivorous fish and zooplankton. Limnology and Oceanography 48:1956–1964.

Rome, L. C., P. T. Loughna, and G. Goldspink. 1985. Temperature acclimation: improved sustained swimming performance in carp at low temperatures. Science 228: 194–196.

Rome, L. C., E. D. Stevens, and H. B. John-Alder. 1992. The influence of temperature and thermal acclimation on physiological function, Pages 183–205 *in* M. E. Feder, and W. W. Burggren, eds. Environmental physiology of the Amphibians. Chicago, The University of Chicago Press.

Ronce, O., and M. Kirkpatrick. 2001. When sources become sinks: migrational meltdown in heterogeneous habitats. Evolution 55:1520–1531.

Root, T. L., J. T. Price, K. R. Hall, S. H. Schneider, C. Rosenzweig, and J. A. Pounds. 2003. Fingerprints of global warming on wild animals and plants. Nature 421:57–60.

Rose, M. R., T. J. Nusbaum, and A. K. Chippindale. 1996. Laboratory evolution: the experimental Wonderland and the Cheshire Cat *in* M. R. Rose, and G. V. Lauder, eds. Adaptation. New York, Academic Press.

Rosenheim, J. A. 2004. Top predators constrain the habitat selection games played by intermediate predators and their prey. Israel Journal of Zoology 50:129–138.

Rossetti, Y., and M. Cabanac. 2006. Light versus temperature: an intersensitivity conflict in a gastropod (*Lymnaea auricularia*). Journal of Thermal Biology 31:514–520.

Roth, M. 2002, Effects of cities on local climates Workshop for the IGES/APN Mega-City Project:1–13.

Row, J. R., and G. Blouin-Demers. 2006. Thermal quality influences effectiveness of thermoregulation, habitat use, and behaviour in milk snakes. Oecologia 148: 1–11.

Rungruangsak-Torrissen, K., G. M. Pringle, R. Moss, and D. F. Houlihan. 1998. Effects of varying rearing temperatures on expression of different trypsin isozymes, feed conversion efficiency and growth in Atlantic salmon (*Salmo salar* L.). Fish Physiology and Biochemistry 19:247–255.

Rustad, L. E., J. L. Campbell, G. M. Marion, R. J. Norby, M. J. Mitchell, A. E. Hartley, J. H. C. Cornelissen et al. 2001. A meta-analysis of the response of soil respiration, net nitrogen mineralization, and aboveground plant growth to experimental ecosystem warming. Oecologia 126: 543–562.

Sack, R., A. Gochberg-Sarver, U. Rozovsky, M. Kedmi, S. Rosner, and A. Orr-Urtreger. 2005. Lower core body temperature and attenuated nicotine-induced hypothermic response in mice lacking the beta 4 neuronal nicotinic acetylcholine receptor subunit. Brain Research Bulletin 66:30–36.

Sadowska, E. T., M. K. Labocha, K. Baliga, A. Stanisz, A. K. Wroblewska, W. Jagusiak, and P. Koteja. 2005. Genetic correlations between basal and maximum metabolic rates in a wild rodent: consequences for evolution of endothermy. Evolution 59:672–681.

Saelim, S., and J. J. Zwiazek. 2000. Preservation of thermal stability of cell membranes and gas exchange in high temperature-acclimated *Xylia xylocarpa* seedlings. Journal of Plant Physiology 156:380–385.

Samietz, J., S. Kroder, D. Schneider, and S. Dorn. 2006. Ambient temperature affects mechanosensory host location in a parasitic wasp. Journal of Comparative Physiology A 192:151–157.

Samietz, J., M. A. Salser, and H. Dingle. 2005. Altitudinal variation in behavioural thermoregulation: local adaptation vs. plasticity in California grasshoppers. Journal of Evolutionary Biology 18:1087–1096.

Sanborn, A. F., M. H. Villet, and P. K. Phillips. 2003. Hot-blooded singers: endothermy facilitates crepuscular signaling in African platypleurine cicadas (Hemiptera: Cicadidae: *Platypleura* spp.). Naturwissenschaften 90:305–308.

—. 2004. Endothermy in African platypleurine cicadas: the influence of body size and habitat (Hemiptera: Cicadidae). Physiological and Biochemical Zoology 77: 816–823.

Santamaría, L., and M. J. M. Hootsmans. 1998. The effect of temperature on the photosynthesis, growth and reproduction of a Mediterranean submerged macrophyte, *Ruppia drepanensis*. Aquatic Botany 60:169–188.

Santamaría, L., and W. van Vierssen. 1997. Photosynthetic temperature responses of fresh- and brackish-water macrophytes: a review. Aquatic Botany 58: 135–150.

Santos, M. 2007. Evolution of total net fitness in thermal lines: *Drosophila subobscura* likes it 'warm'. Journal of Evolutionary Biology 20:2361–2370.

Santos, M., D. Brites, and H. Laayouni. 2006. Thermal evolution of pre-adult life history traits, geometric size and shape, and developmental stability in *Drosophila subobscura*. Journal of Evolutionary Biology 19:2006–2021.

Sargent, R. C., P. D. Taylor, and M. R. Gross. 1987. Parental care and the evolution of egg size in fishes. American Naturalist 129:32–46.

Sarkar, S. 2004. From the *Reaktionsnorm* to the evolution of adaptive plasticity: a historical sketch, 1909–1999, Pages 10–30 *in* T. J. DeWitt, and S. M. Scheiner, eds. Phenotypic Plasticity: Functional and Conceptual Approaches. Oxford, Oxford University Press.

Sarkar, S., and T. Fuller. 2003. Generalized norms of reaction for ecological developmental biology. Evolution & Development 5:106–115.

Sartorius, S. S., J. P. S. do Amaral, R. D. Durtsche, C. M. Deen, and W. I. Lutterschmidt. 2002. Thermoregulatory accuracy, precision, and effectiveness in two sand-dwelling lizards under mild environmental conditions. Canadian Journal of Zoology 80:1966–1976.

Sarup, P., J. G. Sorensen, K. Dimitrov, J. S. F. Barker, and V. Loeschcke. 2006. Climatic adaptation of *Drosophila buzzatii* populations in southeast Australia. Heredity 96:479–486.

Savage, V. M., J. F. Gillooly, J. H. Brown, G. B. West, and E. L. Charnov. 2004. Effects of body size and temperature on population growth. American Naturalist 163: E429–E441.

Schaefer, J., and A. Ryan. 2006. Developmental plasticity in the thermal tolerance of zebrafish *Danio rerio*. Journal of Fish Biology 69:722–734.

Schäuble, C. S., and G. C. Grigg. 1998. Thermal ecology of the Australian agamid *Pogona barbata*. Oecologia 114:461–470.

Scheiner, S. M. 1993. Genetics and evolution of phenotypic plasticity. Annual Review of Ecology and Systematics 24:35–68.

—. 1998. The genetics of phenotypic plasticity. VII. Evolution in a spatially-structured environment. Journal of Evolutionary Biology 11:303–320.

—. 2002. Selection experiments and the study of phenotypic plasticity. Journal of Evolutionary Biology 15: 889–898.

Scheiner, S. M., and R. F. Lyman. 1991. The genetics of phenotypic plasticity. I. Heritability. Journal of Evolutionary Biology 2:95–107.

Schlichting, C. D., and M. Pigliucci. 1998, Phenotypic Plasticity: A Reaction Norm Perspective. Sunderland, Sinauer Associates, Inc.

Schluter, D. 2000, The Ecology of Adaptive Radiation. Oxford, Oxford University Press.

Schoenauer, M., and Z. Michalewicz. 1997. Evolutionary computation. Control and Cybernetics 26:307–338.

Scholander, P. F., R. Hock, V. Walters, F. Johnson, and L. Irving. 1950a. Heat regulation in some arctic and tropical mammals and birds. Biological Bulletin 99: 237–258.

Scholander, P. F., V. Walters, R. Hock, and L. Irving. 1950b. Body insulation of some Arctic and tropical mammals and birds. Biological Bulletin 99:225–236.

Schoolfield, R. M., P. J. H. Sharpe, and C. E. Magnuson. 1981. Non-linear regression of biological temperature-dependent rate models based on absolute reaction-rate theory. Journal of Theoretical Biology 88:719–731.

Schulte, P. M. 2004. Changes in gene expression as biochemical adaptations to environmental change: a tribute to Peter Hochachka. Comparative Biochemistry and Physiology B 139:519–529.

Schultz, E. T., K. E. Reynolds, and D. O. Conover. 1996. Countergradient variation in growth among newly hatched *Fundulus heteroclitus*: Geographic differences revealed by common-environment experiments. Functional Ecology 10:366–374.

Schurmann, H., J. F. Steffensen, and J. P. Lomholt. 1991. The influence of hypoxia on the preferred temperature of rainbow trout *Oncorhynchus mykiss*. Journal of Experimental Biology 157:75–86.

Schutze, M. K., and A. R. Clarke. 2008. Converse Bergmann cline in a *Eucalyptus* herbivore, *Paropsis atomaria* Olivier (Coleoptera: Chrysomelidae): phenotypic plasticity or local adaptation? Global Ecology and Biogeography 17:424–431.

Scott, J. R., C. R. Tracy, and D. Pettus. 1982. A biophysical analysis of daily and seasonal utilization of climate space by a montane snake. Ecology 63:482–493.

Scott, M., D. Berrigan, and A. A. Hoffmann. 1997. Costs and benefits of acclimation to elevated temperature in *Trichogramma carverae*. Entomologia Experimentalis et Applicata 85:211–219.

Scribner, S. J., and P. J. Weatherhead. 1995. Locomotion and antipredator behavior in three species of semiaquatic snakes. Canadian Journal of Zoology 73:321–329.

Sears, M. W. 2006. Proceed with caution: invalidating tests of the cost–benefit model of thermoregulation with spatially-explicit movement simulations. Integrative and Comparative Biology 46:E127.

Sears, M. W., and M. J. Angilletta. 2007. Evaluating the costs of thermoregulation: simulating animal movements through spatially-structured environments define cost curves for small lizards. Integrative and Comparative Biology 47:E228.

Sears, M. W., G. S. Bakken, M. J. Angilletta, and L. A. Fitzgerald. 2004. Using artificial neural networks to model the operative temperatures of small animals in a spatially-explicit context. Integrative and Comparative Biology 44:745.

Seebacher, F., W. Davison, C. J. Lowe, and C. E. Franklin. 2005. A falsification of the thermal specialization paradigm: compensation for elevated temperatures in Antarctic fishes. Biology Letters 1:151–154.

Seebacher, F., and G. C. Grigg. 1997. Patterns of body temperature in wild freshwater crocodiles, *Crocodylus johnstoni*: thermoregulation versus thermoconformity, seasonal acclimatization, and the effect of social interactions. Copeia:549–557.

—. 2000. Social interactions compromise thermoregulation in crocodiles *Crocodylus johnstoni* and *Crocodylus porosus* Pages 310–316 *in* G. C. Grigg, F. Seebacher, and C. E. Franklin, eds. Crocodilian Biology and Evolution. Chipping Norton, Surrey Beatty and Sons Pty Ltd.

—. 2001. Changes in heart rate are important for thermoregulation in the varanid lizard *Varanus varius*. Journal of Comparative Physiology B 171:395–400.

Seebacher, F., G. C. Grigg, and L. A. Beard. 1999. Crocodiles as dinosaurs: behavioural thermoregulation in very large ectotherms leads to high and stable body temperatures. Journal of Experimental Biology 202:77–86.

Seebacher, F., and R. Shine. 2004. Evaluating thermoregulation in reptiles: the fallacy of the inappropriately applied method. Physiological and Biochemical Zoology 77:688–695.

Seebacher, F., and R. S. Wilson. 2006. Fighting fit: thermal plasticity of metabolic function and fighting success in the crayfish *Cherax destructor*. Functional Ecology 20:1045–1053.

Selong, J. H., T. E. McMahon, A. V. Zale, and F. T. Barrows. 2001. Effect of temperature on growth and survival of bull trout, with application of an improved method for determining thermal tolerance in fishes. Transactions of the American Fisheries Society 130:1026–1037.

Seymour, R. S., and A. J. Blaylock. 1999. Switching off the heater: influence of ambient temperature on thermoregulation by eastern skunk cabbage*Symplocarpus foetidus*. Journal of Experimental Botany 50:1525–1532.

Sharpe, P. J. H., and D. W. DeMichele. 1977. Reaction kinetics of poikilotherm development. Journal of Theoretical Biology 64:649–670.

Sherman, E., L. Baldwin, G. Fernandez, and E. Deurell. 1991. Fever and thermal tolerance in the toad *Bufo marinus*. Journal of Thermal Biology 16:297–301.

Shine, R. 1980. Costs of reproduction in reptiles. Oecologia 46:92–100.

Shine, R., P. S. Harlow, M. J. Elphick, M. M. Olsson, and R. T. Mason. 2000. Conflicts between courtship and thermoregulation: the thermal ecology of amorous male garter snakes (*Thamnophis sirtalis parietalis*, Colubridae). Physiological and Biochemical Zoology 73:508–516.

Shoemaker, V. H., L. L. McClanahan, P. C. Withers, S. S. Hillman, and R. C. Drewes. 1987. Thermoregulatory response to heat in the waterproof frogs *Phyllomedusa* and *Chiromantis*. Physiological Zoology 60:365–372.

Shudo, E., P. Haccou, and Y. Iwasa. 2003. Optimal choice between feedforward and feedback control in gene expression to cope with unpredictable danger. Journal of Theoretical Biology 223:149–160.

Sibly, R., and P. Calow. 1983. An integrated approach to life-cycle evolution using selective landscapes. Journal of Theoretical Biology 102:527–547.

Sibly, R. M., and D. Atkinson. 1994. How rearing temperature affects optimal adult size in ectotherms. Functional Ecology 8:486–493.

Sibly, R. M., D. Barker, J. Hone, and M. Pagel. 2007. On the stability of populations of mammals, birds, fish and insects. Ecology Letters 10:970–976.

Sievert, L. M., and P. Andreadis. 1999. Specific dynamic action and postprandial thermophily in juvenile northern water snakes, *Nerodia sipedon*. Journal of Thermal Biology 24:51–55.

Sievert, L. M., and V. H. Hutchison. 1988. Light versus heat: thermoregulatory behavior in a nocturnal lizard (*Gekko gecko*). Herpetologica 44:266–273.

Sievert, L. M., D. M. Jones, and M. W. Puckett. 2005. Postprandial thermophily, transit rate, and digestive efficiency of juvenile cornsnakes, *Pantherophis guttatus*. Journal of Thermal Biology 30:354–359.

Sigmund, K. 2005. John Maynard Smith and evolutionary game theory. Theoretical Population Biology 68:7–10.

Sih, A. 1998. Game theory and predator-prey response races, Pages 221–238 *in* L. A. Dugatkin, and H. K. Reeve, eds. Game Theory and Animal Behavior. Oxford, Oxford University Press.

Silbermann, R., and M. Tatar. 2000. Reproductive costs of heat shock protein in transgenic *Drosophila melanogaster*. Evolution 54:2038–2045.

Sims, D. W., V. J. Wearmouth, E. J. Southall, J. M. Hill, P. Moore, K. Rawlinson, N. Hutchinson et al. 2006. Hunt warm, rest cool: bioenergetic strategy underlying diel vertical migration of a benthic shark. Journal of Animal Ecology 75:176–190.

Sinclair, B. J. 1997. Seasonal variation in freezing tolerance of the New Zealand alpine cockroach *Celatoblatta quinquemaculata*. Ecological Entomology 22: 462–467.

Sinclair, B. J., and S. P. Roberts. 2005. Acclimation, shock and hardening in the cold. Journal of Thermal Biology 30:557–562.

Sinclair, B. J., P. Vernon, C. J. Klok, and S. L. Chown. 2003. Insects at low temperatures: an ecological perspective. Trends in Ecology & Evolution 18:257–262.

Sinervo, B. 1990. The evolution of maternal investment in lizards: an experimental and comparative analysis of egg size and its effects on offspring performance. Evolution 44:279–294.

Sinervo, B., P. Doughty, R. B. Huey, and K. Zamudio. 1992. Allometric engineering: a causal analysis of natural selection on offspring size. Science 258:1927–1930.

Skelly, D. K. 2004. Microgeographic countergradient variation in the wood frog, *Rana sylvetica*. Evolution 58: 160–165.

Skelly, D. K., L. N. Joseph, H. P. Possingham, L. K. Freidenburg, T. J. Farrugia, M. T. Kinnison, and A. P. Hendry. 2007. Evolutionary responses to climate change. Conservation Biology 21:1353–1355.

Slatyer, R. O. 1977. Altitudinal variation in photosynthetic characteristics of snow gum, *Eucalyptus pauciflorac* Sieb. ex Spreng. IV. Temperature response of four populations grown at different temperatures. Australian Journal of Plant Physiology 4:583–594.

Slip, D. J., and R. Shine. 1988. Thermophilic response to feeding of the diamond python, *Morelia s. spilota* (Serpentes, Boidae). Comparative Biochemistry and Physiology A 89:645–650.

Smith, C. C., and S. D. Fretwell. 1974. Optimal balance between size and number of offspring. American Naturalist 108:499–506.

Smith, E. M., and E. B. Hadley. 1974. Photosynthetic and respiratory acclimation to temperature in*Ledum groenlandicum* populations. Arctic and Alpine Research 6:13–27.

Smith, E. N. 1979. Behavioral and physiological thermoregulation of crocodilians. American Zoologist 19:239–247.

Smith, G. R., and R. E. Ballinger. 1994. Thermal ecology of *Sceloporus virgatus* from southeastern Arizona, with comparison to *Urosaurus ornatus*. Journal of Herpetology 28:65–69.

—. 1995. Temperature relationships of the tree lizard, *Urosaurus ornatus*, from desert and low-elevation montane populations in the southwestern USA. Journal of Herpetology 29:126–129.

Smith, G. R., R. E. Ballinger, and J. D. Congdon. 1993. Thermal ecology of the high-altitude bunch grass lizard, *Sceloporus scalaris*. Canadian Journal of Zoology 71: 2152–2155.

Smith, L. M., A. G. Appel, T. P. Mack, and G. J. Keever. 1999. Preferred temperature and relative humidity of males of two sympatric *Periplaneta* cockroaches (Blattodea: Blattidae) denied access to water. Environmental Entomology 28:935–942.

Snucins, E. J., and J. M. Gunn. 1995. Coping with a warm environment: behavioral thermoregulation by lake trout. Transactions of the American Fisheries Society 124:118–123.

Snyder, G. K., and W. W. Weathers. 1975. Temperature adaptations in amphibians. American Naturalist 109:93–101.

Soberón, J., and A. T. Peterson. 2005. Interpretation of models of fundamental ecological niches and species' distributional areas. Biodiversity Informatics 2:1–10.

Somero, G. N. 1995. Proteins and temperature. Annual Review of Physiology 57:43–68.

—. 2000. University in diversity: a perspective on the methods, contributions, and future of comparative physiology. Annual Review of Physiology 62:927–937.

—. 2002. Thermal physiology and vertical zonation of intertidal animals: optima, limits, and costs of living. Integrative and Comparative Biology 42:780–789.

—. 2003. Protein adaptations to temperature and pressure: complementary roles of adaptive changes in amino acid sequence and internal milieu. Comparative Biochemistry and Physiology B 136:577–591.

—. 2004. Adaptation of enzymes to temperature: searching for basic "strategies". Comparative Biochemistry and Physiology B 139:321–333.

Somero, G. N., E. Dahlhoff, and J. J. Lin. 1996. Stenotherms and eurytherms: mechanisms establishing thermal optima and tolerance ranges, Pages 53–78 *in* I. A. Johnston, and A. F. Bennett, eds. Animals and Temperature. Cambridge, Cambridge University Press.

Somero, G. N., and P. W. Hochachka. 1971. Biochemical adaptation to environment. American Zoologist 11: 159–167.

Sorci, G., J. Clobert, and S. Belichon. 1996. Phenotypic plasticity of growth and survival in the common lizard *Lacerta vivipara*. Journal of Animal Ecology 65:781–790.

Sorensen, J. G., J. Dahlgaard, and V. Loeschcke. 2001. Genetic variation in thermal tolerance among natural populations of *Drosophila buzzatii*: down regulation of Hsp70 expression and variation in heat stress resistance traits. Functional Ecology 15:289–296.

Sorensen, J. G., T. N. Kristensen, and V. Loeschcke. 2003. The evolutionary and ecological role of heat shock proteins. Ecology Letters 6:1025–1037.

Sorensen, J. G., F. M. Norry, A. C. Scannapieco, and V. Loeschcke. 2005. Altitudinal variation for stress resistance traits and thermal adaptation in adult *Drosophila buzzatii* from the New World. Journal of Evolutionary Biology 18:829–837.

Sorenson, J. G., and V. Loeschcke. 2002. Natural adaptation to environmental stress via physiological clock-regulation of stress resistance in *Drosophila*. Ecology Letters 5:16–19.

Soriano, P. J., A. Ruiz, and A. Arends. 2002. Physiological responses to ambient temperature manipulation by three species of bats from Andean cloud forests. Journal of Mammalogy 83:445–457.

Spellerberg, I. F. 1972. Temperature tolerances of southeast Australian reptiles examined in relation to reptile thermoregulatory behaviour and distribution. Oecologia 9:23–46.

Spotila, J. R., O. C. M. P., and G. S. Bakken. 1992. Biophysics of heat and mass transfer, Pages 59–80 *in* M. E. Feder, and W. W. Burggren, eds. Environmental Physiology of the Amphibians. Chicago, University of Chicago Press.

Spotila, J. R., O. H. Soule, and D. M. Gates. 1972. The biophysical ecology of the alligator: heat energy budgets and climate spaces. Ecology 53:1094–1102.

Stahlberg, F., M. Olsson, and T. Uller. 2001. Population divergence of developmental thermal optima in Swedish common frogs, *Rana temporaria*. Journal of Evolutionary Biology 14:755–762.

Stamper, J. L., I. Zucker, D. A. Lewis, and J. Dark. 1998. Torpor in lactating Siberian hamsters subjected to glucoprivation. American Journal of Physiology 43:R46-R51.

Stanwell-Smith, D., and L. S. Peck. 1998. Temperature and embryonic development in relation to spawning and field occurrence of larvae of three Antarctic echinoderms. Biological Bulletin 194:44–52.

Stapley, J. 2006. Individual variation in preferred body temperature covaries with social behaviours and colour in male lizards. Journal of Thermal Biology 31:362–369.

Stearns, S. C. 1989. The evolutionary significance of phenotypic plasticity. Bioscience 39:436–445.

—. 1992, The Evolution of Life Histories. Oxford, Oxford University Press.

—. 2000. Life history evolution: successes, limitations, and prospects Naturwissenschaften 87:476 486.

Steigenga, M. J., B. J. Zwaan, P. M. Brakefield, and K. Fischer. 2005. The evolutionary genetics of egg size plasticity in a butterfly. Journal of Evolutionary Biology 18:281–289.

Stelzer, C. P. 2002. Phenotypic plasticity of body size at different temperatures in a planktonic rotifer: mechanisms and adaptive significance. Functional Ecology 16:835–841.

Stenseng, E., C. E. Braby, and G. N. Somero. 2005. Evolutionary and acclimation-induced variation in the thermal limits of heart function in congeneric marine snails (genus *Tegula*): implications for vertical zonation. Biological Bulletin 208:138–144.

Stephens, D. W., J. S. Brown, and R. C. Ydenberg. 2007. Foraging: Behavior and Ecology, Pages 576. Chicago, University Of Chicago Press.

Stephens, D. W., and J. R. Krebs. 1986, Foraging Theory. Princeton, Princeton University Press.

Steppan, S. J., P. C. Phillips, and D. Houle. 2002. Comparative quantitative genetics: evolution of the G matrix. Trends in Ecology & Evolution 17:320–327.

Stevenson, R. D. 1985. Body size and limits to the daily range of body temperature in terrestrial ectotherms. American Naturalist 125:102–117.

Stevenson, R. D., C. R. Peterson, and J. S. Tsuji. 1985. The thermal dependence of locomotion, tongue flicking, digestion, and oxygen consumption in the wandering garter snake. Physiological Zoology 58:46–57.

Stillman, J. H. 2003. Acclimation capacity underlies susceptibility to climate change. Science 301:65.

Stillman, J. H., and G. N. Somero. 2000. A comparative analysis of the upper thermal tolerance limits of eastern Pacific porcelain crabs, genus *Petrolisthes*: Influences of latitude, vertical zonation, acclimation, and phylogeny. Physiological and Biochemical Zoology 73:200–208.

Stillwell, R. C., and C. W. Fox. 2005. Complex patterns of phenotypic plasticity: interactive effects of temperature during rearing and oviposition. Ecology 86:924–934.

Stoks, R., M. De Block, F. Van de Meutter, and F. Johansson. 2005. Predation cost of rapid growth: behavioural coupling and physiological decoupling. Journal of Animal Ecology 74:708–715.

Strong, K. W., and G. R. Daborn. 1980. The influence of temperature on energy budget variables, body size, and seasonal occurrence of the isopod *Idotea baltica* (Pallas). Canadian Journal of Zoology 58:1992–1996.

Sun, H. J., and E. I. Friedmann. 2005. Communities adjust their temperature optima by shifting producer-to-consumer ratio, shown in lichens as models: II. Experimental verification. Microbial Ecology 49:528–535.

Sutherst, R. W., G. F. Maywald, and A. S. Bourne. 2007. Including species interactions in risk assessments for global change. Global Change Biology 13:1843–1859.

Sykes, M. T., I. C. Prentice, and W. Cramer. 1996. A bioclimatic model for the potential distributions of north European tree species under present and future climates. Journal of Biogeography 23:203–233.

Talloen, W., H. Van Dyck, and L. Lens. 2004. The cost of melanization: butterfly wing coloration under environmental stress. Evolution 58:360–366.

Tansey, M. R., and T. D. Brock. 1972. The upper temperature limit for eukaryotic organisms. Proceedings of the National Academy of Sciences 69:2426–2428.

Tatar, M. 1999. Transgenes in the analysis of life span and fitness. American Naturalist 154:S67-S81.

—. 2000. Transgenic organisms in evolutionary ecology. Trends in Ecology & Evolution 15:207–211.

Taylor, P. D., and G. C. Williams. 1984. Demographic parameters at evolutionary equilibrium. Canadian Journal of Zoology 62:2264–2271.

Temple, G. K., and I. A. Johnston. 1998. Testing hypotheses concerning the phenotypic plasticity of escape performance in fish of the family Cottidae. Journal of Experimental Biology 201:317–331.

Templeton, A. R. 1986. Coadaptation and outbreeding depression, Pages 105–121 *in* M. E. Soulé, ed. Conservation Biology: The Science of Scarcity and Diversity. Sunderland, Sinauer Associates, Inc.

Ten Hwang, Y., S. Lariviere, and F. Messier. 2007. Energetic consequences and ecological significance of heterothermy and social thermoregulation in striped skunks (*Mephitis mephitis*). Physiological and Biochemical Zoology 80:138–145.

Terblanche, J. S., and S. L. Chown. 2006. The relative contributions of developmental plasticity and adult acclimation to physiological variation in the tsetse fly, *Glossina pallidipes* (Diptera, Glossinidae). Journal of Experimental Biology 209:1064–1073.

Thomas, C. D., A. Cameron, R. E. Green, M. Bakkenes, L. J. Beaumont, Y. C. Collingham, B. F. N. Erasmus et al. 2004. Extinction risk from climate change. Nature 427: 145–148.

Thomson, L. J., M. Robinson, and A. A. Hoffmann. 2001. Field and laboratory evidence for acclimation without costs in an egg parasitoid. Functional Ecology 15: 217–221.

Torti, V. M., and P. O. Dunn. 2005. Variable effects of climate change on six species of North American birds. Oecologia 145:486–495.

Tracy, C. R. 1976. A model of the dynamic exchanges of water and energy between a terrestrial amphibian and its environment. Ecological Monographs 46:293–326.

Tracy, C. R., and K. A. Christian. 1986. Ecological relations among space, time, and thermal niche axes. Ecology 67:609–615.

—. 2005. Preferred temperature correlates with evaporative water loss in hylid frogs from northern Australia. Physiological and Biochemical Zoology 78:839–846.

Tracy, C. R., F. H. Vanberkum, J. S. Tsuji, R. D. Stevenson, J. A. Nelson, B. M. Barnes, and R. B. Huey. 1984. Errors resulting from linear approximations in energy balance equations. Journal of Thermal Biology 9:261–264.

Travis, J., and D. N. Reznick. 1998. Experimental approaches to the study of evolution, Pages 437–459 *in* W. J. Resetarits Jr., and J. Bernardo, eds. Experimental Ecology: Issues and Perspectives. Oxford, Oxford University Press.

Tsai, M. L., C. F. Dai, and H. H. Chen. 1998. Desiccation resistance of two semiterrestrial isopods, *Ligia exotica* and *Ligia taiwanensis* (Crustacea) in Taiwan. Comparative Biochemistry and Physiology A 119: 361–367.

Tsai, T. S., and M. C. Tu. 2005. Postprandial thermophily of Chinese green tree vipers, *Trimeresurus s. stejnegeri*: interfering factors on snake temperature selection in a thigmothermal gradient. Journal of Thermal Biology 30:423–430.

Tsuji, J. S. 1988. Thermal acclimation of metabolism in *Sceloporus* lizards from different latitudes. Physiological Zoology 61:241–253.

Tu, M. C., and V. H. Hutchison. 1994. Influence of pregnancy on thermoregulation of water snakes (*Nerodia rhombifera*). Journal of Thermal Biology 19:255–259.

Tveiten, H., S. E. Solevåg, and H. K. Johnsen. 2001. Holding temperature during the breeding season influences final maturation and egg quality in common wolffish. Journal of Fish Biology 58:374–385.

Underwood, B. A. 1991. Thermoregulation and energetic decision-making by the honeybees *Apis cerana*, *Apis dorsata* and *Apis laboriosa*. Journal of Experimental Biology 157:19–34.

van't Land, J., P. van Putten, B. Zwaan, A. Kamping, and W. van Delden. 1999. Latitudinal variation in wild populations of *Drosophila melanogaster*: heritabilities and reaction norms. Journal of Evolutionary Biology 12:222–232.

van Berkum, F. H. 1986. Evolutionary patterns of the thermal sensitivity of sprint speed in *Anolis* lizards. Evolution 40:594–604.

—. 1988. Latitudinal patterns of the thermal sensitivity of sprint speed in lizards. American Naturalist: 327–343.

van Berkum, F. H., R. B. Huey, and B. A. Adams. 1986. Physiological consequences of thermoregulation in a tropical lizard (*Ameiva festiva*). Physiological Zoology 59:464–472.

Van Damme, R., D. Bauwens, A. M. Castilla, and R. F. Verheyen. 1989. Altitudinal variation of the thermal biology and running performance in the lizard *Podarcis tiliguerta*. Oecologia 80:516–524.

—. 1990. Comparative thermal ecology of the sympatric lizards *Podarcis tiliguerta* and *Podarcis sicula*. Acta Oecologica 11:503–512.

Van Damme, R., D. Bauwens, and R. F. Verheyen. 1991. The thermal dependence of feeding behavior, food consumption and gut passage time in the lizard *Lacerta vivipara* Jacquin. Functional Ecology 5:507–517.

van der Have, T. M. 2002. A proximate model for thermal tolerance in ectotherms. Oikos 98:141–155.

van Dijk, P. L. M., G. Staaks, and I. Hardewig. 2002. The effect of fasting and refeeding on temperature preference, activity and growth of roach, *Rutilus rutilus*. Oecologia 130:496–504.

Van Doorslaer, W., and R. Stoks. 2005a. Growth rate plasticity to temperature in two damselfly species differing in latitude: contributions of behaviour and physiology. Oikos:599–605.

van Doorslaer, W., and R. Stoks. 2005b. Thermal reaction norms in two *Coenagrion* damselfly species: contrasting embryonic and larval life-history traits. Freshwater Biology 50:1982–1990.

Van Doorslaer, W., R. Stoks, E. Jeppesen, and L. De Meester. 2007. Adaptive microevolutionary responses to simulated global warming in *Simocephalus vetulus*: a mesocosm study. Global Change Biology 13: 878–886.

van Huis, A., P. W. Arendse, M. Schilthuizen, P. P. Wiegers, H. Heering, M. Hulshof, and N. K. Kaashoek. 1994. *Uscana lariophaga*, egg parasitoid of bruchid beetle storage pests of cowpea in West Africa: the effect of temperature and humidity. Entomologia experimentalis et applicata 70:41–53.

van Tienderen, P. H. 1991. Evolution of generalists and specialists in spatially heterogeneous environments. Evolution 45:1317–1331.

Vannote, R. L., and B. W. Sweeney. 1980. Geographic analysis of thermal equilibria: a conceptual model for evaluating the effect of natural and modified thermal regimes on aquatic insect communities. American Naturalist 115:667–695.

Vargas, R. I., W. A. Walsh, E. B. Jang, J. W. Armstrong, and D. T. Kanehisa. 1996. Survival and development of immature stages of four Hawaiian fruit flies (Diptera: Tephritidae) reared at five constant temperatures. Annals of the Entomological Society of America 89:64–69.

Vázquez, D. P., and R. D. Stevens. 2004. The latitudinal gradient in niche breadth: concepts and evidence. American Naturalist 164:E1-E19.

Verdu, J. R., A. Diaz, and E. Galante. 2004. Thermoregulatory strategies in two closely related sympatric*Scarabaeus* species (Coleoptera: Scarabaeinae). Physiological Entomology 29:32–38.

Vernberg, F. J. 1962. Comparative physiology: latitudinal effects on physiological properties of animal populations. Annual Review of Physiology 24:517–544

Via, S. 1987. Genetic constraints on the evolution of phenotypic plasticity, Pages 47–71 *in* V. Loeschcke, ed. Genetic Constraints on Adaptive Evolution. Berlin, Springer-Verlag.

—. 1994. The evolution of phenotypic plasticity: what do we really know?, Pages 35–57 *in* L. A. Real, ed. Ecological Genetics. Princeton, Princeton University Press.

Via, S., R. Gomulkiewicz, G. de Jong, S. M. Scheiner, C. D. Schlichting, and P. H. van Tienderen. 1995. Adaptive phenotypic plasticity: consensus and controversy. Trends in Ecology & Evolution 10:212–217.

Via, S., and R. Lande. 1985. Genotype-environment interaction and the evolution of phenotypic plasticity. Evolution 39:505–522.

Visser, M. E., and C. Both. 2005. Shifts in phenology due to global climate change: the need for a yardstick. Proceedings of the Royal Society B 272:2561–2569.

Visser, M. E., L. J. M. Holleman, and P. Gienapp. 2006. Shifts in caterpillar biomass phenology due to climate change and its impact on the breeding biology of an insectivorous bird. Oecologia 147:164–172.

Vitt, L. J., and S. S. Sartorius. 1999. HOBOs, Tidbits and lizard models: the utility of electronic devices in field studies of ectotherm thermoregulation. Functional Ecology 13:670–674.

Vona, V., V. D. Rigano, O. Lobosco, S. Carfagna, S. Esposito, and C. Rigano. 2004. Temperature responses of growth, photosynthesis, respiration and NADH: nitrate reductase in crophilic and mesophilic algae. New Phytologist 163:325–331.

Waldschmidt, S. R., S. M. Jones, and W. P. Porter. 1986. The effect of body temperature and feeding regime on activity, passage time, and digestive coefficient in the lizard *Uta stansburiana*. Physiological Zoology 59: 376–383.

Wallman, H. L., and W. A. Bennett. 2006. Effects of parturition and feeding on thermal preference of Atlantic stingray,*Dasyatis sabina* (Lesueur). Environmental Biology of Fishes 75:259–267.

Walsberg, G. E., and B. O. Wolf. 1996. A test of the accuracy of operative temperature thermometers for studies of small ectotherms. Journal of Thermal Biology 21:275–281.

Walther, G. R., S. Berger, and M. T. Sykes. 2005. An ecological 'footprint' of climate change. Proceedings of the Royal Society B 272:1427–1432.

Walther, G. R., E. S. Gritti, S. Berger, T. Hickler, Z. Y. Tang, and M. T. Sykes. 2007. Palms tracking climate change. Global Ecology and Biogeography 16:801–809.

Walther, G. R., E. Post, P. Convey, A. Menzel, C. Parmesan, T. J. C. Beebee, J. M. Fromentin et al. 2002. Ecological responses to recent climate change. Nature 416:389–395.

Walton, B. M., and A. F. Bennett. 1993. Temperature dependent color change in Kenyan chameleons. Physiological Zoology 66:270–287.

Walvoord, M. E. 2003. Cricket frogs maintain body hydration and temperature near levels allowing maximum jump performance. Physiological and Biochemical Zoology 76:825–835.

Warner, R. R. 1980. The coevolution of behavioral and life history characteristics, Pages 151–188 *in* G. W. Barlow, and J. Silverberg, eds. Sociobiology: Beyond Nature/Nurture? Boulder, Westview Press.

Warren, C. E., and G. E. Davis. 1967. Laboratory studies on the feeding, bioenergetics, and growth of fish, Pages 175–214 *in* S. D. Gerking, ed. The Biological Basis of Freshwater Fish Production. Oxford, Blackwell Scientific Publications.

Warren, M., M. A. McGeoch, S. W. Nicolson, and S. L. Chown. 2006. Body size patterns in *Drosophila* inhabiting a mesocosm: interactive effects of spatial variation in temperature and abundance. Oecologia 149: 245–255.

Warwick, C. 1991. Observations on disease-associated preferred body temperatures in reptiles. Applied Animal Behaviour Science 28:375–380.

Webb, J. K., and M. J. Whiting. 2005. Why don't small snakes bask? Juvenile broad-headed snakes trade thermal benefits for safety. Oikos 110:515–522.

Weetman, D., and D. Atkinson. 2002. Antipredator reaction norms for life history traits in *Daphnia pulex*: dependence on temperature and food. Oikos 98:299–307.

Weng, Q. H., D. S. Lu, and J. Schubring. 2004. Estimation of land surface temperature-vegetation abundance relationship for urban heat island studies. Remote Sensing of Environment 89:467–483.

Werner, Y. L. 1973. Optimal temperatures for inner-ear performance in gekkonoid lizards. Journal of Experimental Zoology 195:319–352.

—. 1990. Do gravid females of oviparous gekkonid. lizards maintain elevated body temperatures? *Hemidactylus frenatus* and *Lepidodactylus lugubris* on Oahu. Amphibia-Reptilia 11:200–204.

West-Eberhard, M. J. 2003, Developmental Plasticity and Evolution. Oxford, Oxford University Press.

Westman, W., and F. Geiser. 2004. The effect of metabolic fuel availability on thermoregulation and torpor in a marsupial hibernator. Journal of Comparative Physiology B 174:49–57.

Whitaker, B. D., and K. L. Poff. 1980. Thermal adaptation of thermosensing and negative thermotaxis in *Dictyostelium*. Experimental Cell Research 128:87–93.

White, M. A., R. R. Nemani, P. E. Thornton, and S. W. Running. 2002. Satellite evidence of phenological differences between urbanized and rural areas of the eastern United States deciduous broadleaf forest. Ecosystems 5: 260–273.

Whitlock, M. C., P. C. Phillips, F. B. G. Moore, and S. J. Tonsor. 1995. Multiple fitness peaks and epistasis. Annual Review of Ecology and Systematics 26:601–629.

Widdows, J., and B. L. Bayne. 1971. Temperature acclimation of *Mytilus edulis* with reference to its energy budget. Journal of the Marine Biological Association of the United Kingdom 51:827–843.

Wiens, J. J., and C. H. Graham. 2005. Niche conservatism: integrating evolution, ecology, and conservation biology. Annual Review of Ecology and Systematics 36:519–539.

Wiens, J. J., C. H. Graham, D. S. Moen, S. A. Smith, and T. W. Reeder. 2006. Evolutionary and ecological causes of the latitudinal diversity gradient in hylid frogs: treefrog trees unearth the roots of high tropical diversity. American Naturalist 168:579–596.

Wieser, W. 1994. Cost of growth in cells and organisms: general rules and comparative aspects. Biological Reviews 69:1–33.

Wiggins, P. R., and P. B. Frappell. 2000. The influence of haemoglobin on behavioural thermoregulation and oxygen consumption in *Daphnia carinata*. Physiological and Biochemical Zoology 73:153–160.

Wildhaber, M. L. 2001. The trade-off between food and temperature in the habitat choice of bluegill sunfish. Journal of Fish Biology 58:1476–1478.

Wildhaber, M. L., and P. J. Lamberson. 2004. Importance of the habitat choice behavior assumed when modeling the effects of food and temperature on fish populations. Ecological Modelling 175:395–409.

Wilhelm, F. M., and D. W. Schindler. 2000. Reproductive strategies of *Gammarus lacustris* (Crustacea: Amphipoda) along an elevational gradient. Functional Ecology 14.

Williams, J. W., S. T. Jackson, and J. E. Kutzbacht. 2007. Projected distributions of novel and disappearing climates by 2100 AD. Proceedings of the National Academy of Sciences 104:5738–5742.

Wills, C. A., and S. J. Beaupre. 2000. An application of randomization for detecting evidence of thermoregulation in timber rattlesnakes (*Crotalus horridus*) from northwest Arkansas. Physiological and Biochemical Zoology 73:325–334.

Wilson, R. S. 2001. Geographic variation in thermal sensitivity of jumping performance in the frog *Limnodynastes peronii*. Journal of Experimental Biology 204:4227–4236.

—. 2005. Temperature influences the coercive mating and swimming performance of male eastern mosquitofish. Animal Behaviour 70:1387–1394.

Wilson, R. S., and C. E. Franklin. 1999. Thermal acclimation of locomotor performance in tadpoles of the frog *Limnodynastes peronii*. Journal of Comparative Physiology B 169:445–451.

—. 2000a. Effect of ontogenetic increases in body size on burst swimming performance in tadpoles of the striped marsh frog, *Limnodynastes peronii*. Physiological and Biochemical Zoology 73:142–152.

—. 2000b. Inability of adult *Limnodynastes peronii* (Amphibia: Anura) to thermally acclimate locomotor performance. Comparative Biochemistry and Physiology A 127:21–28.

—. 2002. Testing the beneficial acclimation hypothesis. Trends in Ecology & Evolution 17:66–70.

Wilson, R. S., C. E. Franklin, W. Davison, and P. Kraft. 2001. Stenotherms at sub-zero temperatures: thermal dependence of swimming performance in Antarctic fish. Journal of Comparative Physiology B 171:263–269.

Wilson, R. S., E. Hammill, and I. A. Johnston. 2007. Competition moderates the benefits of thermal acclimation to reproductive performance in male eastern mosquitofish. Proceedings of the Royal Society B 274:1199–1204.

Wilson, R. S., R. S. James, and I. A. Johnston. 2000. Thermal acclimation of locomotor performance in tadpoles and adults of the aquatic frog *Xenopus laevis*. Journal of Comparative Physiology B 170:117–124.

Windig, J. J., C. G. F. de Kovel, and G. de Jong. 2004. Genetics and mechanics of plasticity, Pages 31–49 *in* T. J. DeWitt, and S. M. Scheiner, eds. Phenotypic Plasticity: Functional and Conceptual Approaches. Oxford, Oxford University Press.

Winne, C. T., and M. B. Keck. 2005. Intraspecific differences in thermal tolerance of the diamondback watersnake (*Nerodia rhombifer*): effects of ontogeny, latitude, and sex. Comparative Biochemistry and Physiology A 140:141–149.

Withers, P. C., and J. D. Campbell. 1985. Effects of environmental cost on thermoregulation in the desert iguana. Physiological Zoology 58:329–339.

Witters, L. R., and L. Sievert. 2001. Feeding causes thermophily in the woodhouse's toad (*Bufo woodhousii*). Journal of Thermal Biology 26:205–208.

Wolf, J. B., and E. D. Brodie. 1998. The coadaptation of parental and offspring characters. Evolution 52: 299–308.

Woods, H. A. 1999. Egg-mass size and cell size: effects of temperature on oxygen distribution. American Zoologist 39:244–252.

Woodward, F. I. 1990. The impact of low temperatures in controlling the geographical distribution of plants. Philosophical Transactions of the Royal Society B 326:585–593.

Wu, B. J., P. L. Else, L. H. Storlien, and A. J. Hulbert. 2001. Molecular activity of Na+/K+-ATPase from different sources is related to the packing of membrane lipids. Journal of Experimental Biology 204:4271–4280.

Wu, B. J., A. J. Hulbert, L. H. Storlien, and P. L. Else. 2004. Membrane lipids and sodium pumps of cattle and crocodiles: an experimental test of the membrane pacemaker theory of metabolism. American Journal of Physiology 287:R633–R641.

Xiong, F. S. S., E. C. Mueller, and T. A. Day. 2000. Photosynthetic and respiratory acclimation and growth response of Antarctic vascular plants to contrasting temperature regimes. American Journal of Botany 87:700–710.

Yadav, J. P., and B. N. Singh. 2005. Evolutionary genetics of *Drosophila ananassae* III. Effect of temperature on certain fitness trait's in two natural populations. Journal of Thermal Biology 30:457–466.

Yamahira, K., and D. O. Conover. 2002. Intra- vs. interspecific latitudinal variation in growth: adaptation to temperature or seasonality? Ecology 83:1252–1262.

Yamahira, K., M. Kawajiri, K. Takeshi, and T. Irie. 2007. Inter- and intrapopulation variation in thermal reaction norms for growth rate: evolution of latitudinal compensation in ectotherms with a genetic constraint. Evolution 61:1577–1589.

Yamasaki, T., T. Yamakawa, Y. Yamane, H. Koike, K. Satoh, and S. Katoh. 2002. Temperature acclimation of photosynthesis and related changes in photosystem II electron transport in winter wheat. Plant Physiology 128: 1087–1097.

Yamori, W., K. Suzuki, K. Noguchi, M. Nakai, and I. Terashima. 2006. Effects of Rubisco kinetics and Rubisco activation state on the temperature dependence of the photosynthetic rate in spinach leaves from contrasting growth temperatures. Plant Cell and Environment 29:1659–1670.

Yampolsky, L. Y., and S. M. Scheiner. 1996. Why larger offspring at lower temperatures? A demographic approach. American Naturalist 147:86–100.

Yang, J., Y. Y. Sun, H. An, and X. Ji. 2008. Northern grass lizards (*Takydromus septentrionalis*) from different populations do not differ in thermal preference and thermal tolerance when acclimated under identical thermal conditions. Journal of Comparative Physiology B 178:343–349.

Zamudio, K. R., R. B. Huey, and W. D. Crill. 1995. Bigger isn't always better: body size, developmental and parental temperature and male territorial success in *Drosophila melanogaster*. Animal Behaviour 49:671–677.

Zari, T. A. 1998. Effects of sexual condition on food consumption and temperature selection in the herbivorous desert lizard, *Uromastyx philbyi*. Journal of Arid Environments 38:371–377.

Zbikowska, E. 2004. Does behavioural fever occur in snails parasitised with trematode larvae? Journal of Thermal Biology 29:675–679.

Zeilstra, I., and K. Fischer. 2005. Cold tolerance in relation to developmental and adult temperature in a butterfly. Physiological Entomology 30:92–95.

Zera, A. J., and L. G. Harshman. 2001. The physiology of life history trade-offs in animals. Annual Review of Ecology and Systematics 32:95–126.

Zhang, Y. P., and X. A. Ji. 2004. The thermal dependence of food assimilation and locomotor performance in southern grass lizards, *Takydromus sexlineatus* (Lacertidae). Journal of Thermal Biology 29:45–53.

Zhao, C. Y., Z. R. Nan, and G. D. Cheng. 2005. Methods for modelling of temporal and spatial distribution of air temperature at landscape scale in the southern Qilian mountains, China. Ecological Modelling 189: 209–220.

Zhou, X. H., X. Z. Liu, L. L. Wallace, and Y. Q. Luo. 2007. Photosynthetic and respiratory acclimation to experimental warming for four species in a tallgrass prairie ecosystem. Journal of Integrative Plant Biology 49:270–281.

Author Index

Species Index

Subject Index